AF807069

Springer Geography

Advisory Editors

Mitja Brilly, Faculty of Civil and Geodetic Engineering, University of Ljubljana, Ljubljana, Slovenia

Nancy Hoalst-Pullen, Department of Geography and Anthropology, Kennesaw State University, Kennesaw, USA

Michael Leitner, Department of Geography and Anthropology, Louisiana State University, Baton Rouge, USA

Mark W. Patterson, Department of Geography and Anthropology, Kennesaw State University, Kennesaw, USA

Márton Veress, Department of Physical Geography, University of West Hungary, Szombathely, Hungary

The Springer Geography series seeks to publish a broad portfolio of scientific books, aiming at researchers, students, and everyone interested in geographical research.

The series includes peer-reviewed monographs, edited volumes, textbooks, and conference proceedings. It covers the major topics in geography and geographical sciences including, but not limited to; Economic Geography, Landscape and Urban Planning, Urban Geography, Physical Geography and Environmental Geography.

Springer Geography — now indexed in Scopus

Xiao Huang • Siqin Wang • John Wilson •
Peter Kedron
Editors

GeoAI and Human Geography

The Dawn of a New Spatial Intelligence Era

Springer

Editors
Xiao Huang
Department of Environmental Sciences
Emory University
Atlanta, GA, USA

Siqin Wang
Spatial Sciences Institute
University of Southern California
Los Angeles, CA, USA

John Wilson
Spatial Sciences Institute
University of Southern California
Los Angeles, CA, USA

Peter Kedron
Department of Geography
University of California, Santa Barbara
Santa Barbara, CA, USA

ISSN 2194-315X ISSN 2194-3168 (electronic)
Springer Geography
ISBN 978-3-031-87420-8 ISBN 978-3-031-87421-5 (eBook)
https://doi.org/10.1007/978-3-031-87421-5

© The Editor(s) (if applicable) and The Author(s), under exclusive license to Springer Nature Switzerland AG 2025

This work is subject to copyright. All rights are solely and exclusively licensed by the Publisher, whether the whole or part of the material is concerned, specifically the rights of translation, reprinting, reuse of illustrations, recitation, broadcasting, reproduction on microfilms or in any other physical way, and transmission or information storage and retrieval, electronic adaptation, computer software, or by similar or dissimilar methodology now known or hereafter developed.
The use of general descriptive names, registered names, trademarks, service marks, etc. in this publication does not imply, even in the absence of a specific statement, that such names are exempt from the relevant protective laws and regulations and therefore free for general use.
The publisher, the authors and the editors are safe to assume that the advice and information in this book are believed to be true and accurate at the date of publication. Neither the publisher nor the authors or the editors give a warranty, expressed or implied, with respect to the material contained herein or for any errors or omissions that may have been made. The publisher remains neutral with regard to jurisdictional claims in published maps and institutional affiliations.

This Springer imprint is published by the registered company Springer Nature Switzerland AG
The registered company address is: Gewerbestrasse 11, 6330 Cham, Switzerland

If disposing of this product, please recycle the paper.

Preface

Geospatial artificial intelligence (GeoAI), the rapidly evolving fusion of geospatial technologies and artificial intelligence, is reshaping the way researchers and practitioners grapple with questions central to human geography. The potential of this interdisciplinary field is both transformative and expansive, offering novel ways to understand the social, economic, political, cultural, and spatial dimensions of our everyday environments. By integrating advanced computing with place-based knowledge, GeoAI allows us to identify patterns and relationships that traditional methods often miss, and it prompts us to rethink core geographic concepts of scale, distance, and spatial interaction.

This volume, *GeoAI and Human Geography*, encompasses 28 chapters that illustrate how these innovations unfold across a broad spectrum of themes. Early chapters introduce the essential frameworks of human geography in the context of the GeoAI era, from the evolution of geographic thought to the theoretical pillars that link artificial intelligence with geographical inquiry. Subsequent chapters describe major methodological breakthroughs—from explainable AI in spatial analysis to the use of natural language processing for interpreting textual data, to human-centered computer vision for urban sensing, to advances in social network analysis and high-performance computing for large-scale geospatial problems. These contributions showcase the technological strides that are redefining how data is gathered, processed, and ultimately interpreted.

Another large portion of this book focuses on traditional branches of human geography—cultural, economic, political, health, tourism, cartography, transport, and urban studies—illustrating how each area is adopting GeoAI tools to enrich understanding and generate new insights. Whether exploring the intersection of economic activity and digital geographies, contextualizing food environments with advanced analytics, or weaving spatial intelligence into transport planning and urban development, these chapters demonstrate the versatility of GeoAI in addressing both longstanding and emerging research questions. They also reinforce the importance of integrating local context, critical perspectives, and diverse data sources, ensuring that the technological dimension remains rooted in real-world social and environmental concerns.

As the book moves toward its conclusion, readers encounter chapters that envision future trajectories. These contributions delve into topics such as identifying gentrification by synthesizing physical and socioeconomic data, probing the spatial biases of generative AI models, fostering ethical and sustainable approaches to geospatial decision-making, and examining the political dimensions that shape the production and application of spatial knowledge. Discussions on the digital divide, climate change adaptation, and cross-disciplinary research illustrate the expansive possibilities for applying GeoAI to pressing global challenges. The final chapters articulate a forward-looking stance on how GeoAI may continue to influence human geography, urging scholars, practitioners, and policymakers to harness its power responsibly and equitably.

Collectively, this book offers a comprehensive map of how GeoAI-based approaches are reshaping the knowledge horizons of human geography. Each chapter underscores the importance of maintaining a balanced perspective that respects both the rigor of computational methods and the complexities of social and cultural phenomena. We hope this work will inspire researchers, students, and professionals to collaborate more deeply across disciplinary boundaries, spark thoughtful conversations on the ethical and methodological intricacies of using AI for spatial inquiries, and foster innovative applications that enrich our understanding of people and places.

Finally, the editors would like to extend heartfelt thanks to all contributors who poured their expertise and creativity into these pages. In presenting this volume, we invite readers to explore the new frontiers of GeoAI and human geography, reflect on the novel insights emerging from data-driven analysis, and envision the collective responsibility we share for shaping an inclusive, sustainable, and thriving future.

Atlanta, GA, USA — Xiao Huang
Los Angeles, CA, USA — Siqin Wang
Los Angeles, CA, USA — John Wilson
Santa Barbara, CA, USA — Peter Kedron

Contents

About the Editors

Xiao Huang is an Assistant Professor in the Department of Environmental Sciences at Emory University. His research expertise encompasses human–environment interaction, computational social sciences, urban informatics, disaster mapping and mitigation, GeoAI, and disaster remote sensing. Dr. Huang has contributed extensively to his field, authoring over 200 peer-reviewed journal articles and over 20 book chapters and playing a pivotal role in editing five books. He is among the world's top 2% scientists by Stanford/Elsevier's rankings. In his professional capacity, he serves as an Associate Editor for Computational Urban Science and is a member of the editorial boards of several prestigious journals. His research has garnered significant attention and received coverage in renowned media outlets, such as Nature News, NASA, NBC, and Fox. His work has attracted substantial funding from NSF, NASA, the Bill & Melinda Gates Foundation, the National Academies, and the Centers for Medicare and Medicaid Services.

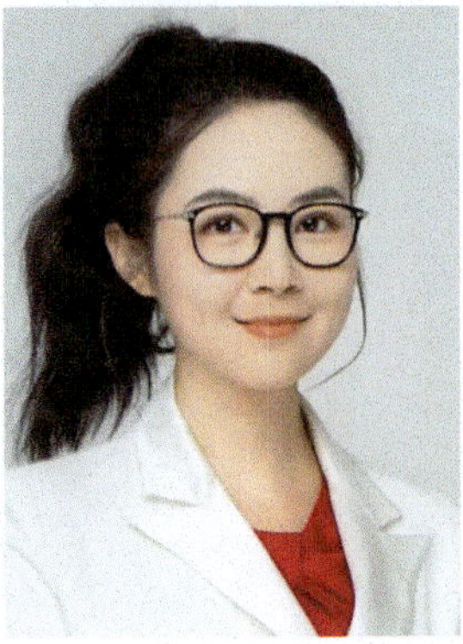

Siqin Wang is an Associate Professor (Teaching) of Spatial Sciences at the Spatial Sciences Institute within the Dornsife College of Letters, Arts and Sciences at the University of Southern California. Her research interests are in GIScience, spatiotemporal big data analytics, computational social science, digital health geography, human-centered GeoAI, human mobility and migration, smart cities, and human–climate interactions and she publishes extensively in these areas. She was named one of the 50 Rising Stars by Geospatial World in 2024. Her 2021 co-first-authored paper on human mobility and COVID-19 transmission received the 2022 Best Paper Award and the most Downloaded Paper from the Annals of GIS in 2023. Her professional roles include serving as the Associated Chair for the Spatial Data Lab affiliated with Harvard University and as Vice Chair of the Young Scientist Innovation Network for Digital Earth of the International Society for Digital Earth. Prior to joining the USC Spatial Sciences Institute, Dr. Wang was an Adjunct Senior Lecturer at the Royal Melbourne Institute of Technology, Australia, and as an Honorary Lecturer and Research Fellow at the University of Queensland, Australia. From April 2022 to July 2023, she also was a JSPS Research Fellow at the University of Tokyo, Japan, where she researched healthcare access and social vulnerability in post-COVID Japan.

John Wilson is Professor of Spatial Sciences and Sociology at the University of Southern California (USC), where he founded the Spatial Sciences Institute. He also holds appointments in USC's School of Architecture, the Keck School of Medicine, and the Viterbi School of Engineering. His research leverages GIS tools, fieldwork, and computer modeling to investigate the interplay among environmental systems, society, and human health. Dr. Wilson has published extensively, including edited volumes such as *Terrain Analysis: Principles and Applications*, the *Handbook of Geographic Information Science*, and *Environmental Applications of Digital Terrain Modeling*. His numerous honors include the UCGIS Lifetime Achievement in GIScience Education Award, his election as a Fellow of UCGIS, and a Lifetime Achievement Award from the International Society for Geomorphometry. He founded the journal *Transactions in GIS* and continues as its Editor-in-Chief. He has served on the editorial boards of several leading journals, was President of UCGIS, and leads the GIS&T Body of Knowledge 2.0 project. Formerly the Chair of USC's Department of Geography, he has also held visiting appointments at institutions worldwide.

Peter Kedron is a human geographer and GIScientist with expertise in economic geography, spatial analysis, and the design and evaluation of research. Dr. Kedron joined UCSB as an Associate Professor and member of the Center for Spatial Data Science after previously holding faculty positions at Arizona State University (2018–2023), Oklahoma State University (2016–2018), and Ryerson University (2012–2016). He earned his Ph.D. in Geography from the State University of New York at Buffalo, an MA in Economics from the University of Michigan, and B.A.s in Economics and Psychology from the State University of New York at Buffalo. His research develops and uses spatial analytical methods to explain geographic variation in social and ecological processes. Dr. Kedron's recent work focuses on the use of replication as a means of evaluating geographic research and on developing statistical approaches to improve the accumulation of evidence collected from different locations. To date, he has authored over 55 peer-reviewed articles, delivered over 100 conference presentations and public talks, and supervised over 20 graduate students and postdoctoral scholars.

Part I
Foundations of Human Geography and GeoAI

Chapter 1
Human Geography in the Era of Big Data and AI

Jonathan Corcoran, Siqin Wang, and Xiao Huang

Abstract This chapter explores the transformative impact of geospatial artificial intelligence (GeoAI) on human geography. GeoAI integrates artificial intelligence, geospatial big data, and high-performance computing to address complex spatial challenges. Tracing its evolution from early artificial intelligence (AI) applications in the 1980s to its formal introduction in 2018, the chapter highlights how advancements in AI, big data, and digital infrastructures have revolutionized geospatial analysis. It discusses the democratization of GeoAI through open-source tools, making advanced geospatial analysis accessible and fostering collaboration across diverse research communities. By interweaving tools and techniques, GeoAI enables researchers to unravel intricate socio-spatial complexities, leading to innovative applications in urban planning, cultural geography, and population studies. The chapter also examines future opportunities, such as enhanced decision-making, improved predictive capacities, and real-time monitoring. It discusses challenges including data quality, skill gaps, implementation costs, reproducibility concerns, and ethical issues like bias, privacy, and the rise of "deep fake geography." Emphasizing the need for responsible practices, the chapter underscores the importance of balancing innovation with ethical considerations to fully harness GeoAI's potential.

Keywords GeoAI · Human Geography · Big Data

J. Corcoran (✉)
School of the Environment, The University of Queensland, St. Lucia, Brisbane, QLD, Australia
e-mail: jj.corcoran@uq.edu.au

S. Wang
Spatial Sciences Institute, University of Southern California, Los Angeles, CA, USA
e-mail: siqinwan@usc.edu

X. Huang
Department of Environmental Sciences, Emory University, Atlanta, GA, USA
e-mail: xiao.huang2@emory.edu

© The Author(s), under exclusive license to Springer Nature Switzerland AG 2025

X. Huang et al. (eds.), *GeoAI and Human Geography*, Springer Geography,
https://doi.org/10.1007/978-3-031-87421-5_1

1.1 Introduction

Human geography has embraced the big data era with a growing diversity of studies spanning understanding city liveliness in developing countries (Liu et al., 2020) to quantifying conflict intensity (Levin et al., 2018) and understanding cultural ties through human mobility (Wu et al., 2016) to automating the extraction of features from historic maps (Duan et al., 2020) to name just a select few. Fuelled by the rapidly expanding and increasingly diverse landscape of spatially explicit data that is collected with greater frequency and resolution (Perez & Sengupta, 2024) underpinned by advancements in high-performance computing, the field is experiencing unprecedented growth. This momentum is further propelled by the efforts of dedicated research groups, the availability of open-source toolkits, and the support of various journal outlets, together shaping and sustaining its continued evolution.

The term *Geospatial artificial intelligence* or *GeoAI* refers to a fast-emerging field that "*integrates artificial intelligence, geospatial big data, and high-performance computing for geospatial problem solving*" (Li, 2022). Despite being more than 30 years in the making (Scheider and Richter, 2023) with studies back in the 1980s first detailing how spatial problems might be approached using artificial intelligence approaches (Smith, 1984; Couclelis, 1986; Openshaw, 1992), it has only been in the last five or so years that we have seen the birth then the growth of GeoAI across many subdomains that comprise human geography (Wang et al., 2024).

The origins of *Artificial Intelligence* can be traced back to the mid-twentieth century, when in the summer of 1956, a project titled "Artificial Intelligence" for a workshop was proposed to be held at Dartmouth, marking the birthplace of artificial intelligence (AI) (Negnevitsky, 2024). In their proposal, they defined artificial intelligence as "*the science and engineering of making intelligent machines.*" They also claimed that every aspect of learning or any other feature of intelligence can in principle be so precisely described that a machine can be made to simulate it. It was the first time the term "Artificial Intelligence" was used, marking a new era in technological advancement (Cordeschi, 2007). Early AI research focused on symbolic reasoning and problem-solving, with significant contributions from Alan Turing, who proposed the concept of a universal machine capable of simulating any human intelligence (Brooks, 1999). The initial AI systems were designed to solve mathematical problems, play games like chess, and perform simple logical tasks. Despite early optimism, progress was slow due to limited computational power and the complexity of replicating human cognition.

The term "GeoAI" was formally introduced in 2018 during a workshop on "*Artificial Intelligence for Geospatial Data Analysis*" held at the International Conference on Geographic Information Science (GIScience), emerging from the convergence of advancements in AI and geospatial science (Gao, 2021). The concept of GeoAI, however, evolved over several years through the integration of AI with geospatial sciences and its relevant disciplines (e.g., physical geography, remote sensing, and environmental science). The development and application of AI techniques in geospatial analysis began earlier, but the explicit recognition and formalization of GeoAI as a distinct field gained prominence with the growing use of AI in processing and interpreting geospatial data (Couclelis, 2023). The foundation of GeoAI can be traced to the early use of AI in spatial analysis. In the late twentieth century, the development of geographic information systems (GIS) revolutionized the way geospatial data was collected, analyzed, and visualized.

As AI technologies progressed, they were increasingly and profoundly integrated with GIS to enhance spatial data analysis and predictive modeling. This integration gained momentum with the proliferation of big data, which provided extensive geospatial datasets and further fuelled the capabilities of AI models (Li & Hsu, 2022). The rise of big data, characterized by vast volumes, high velocity, and diverse types of spatial information, has provided GeoAI with unprecedented opportunities for enhanced analysis and decision-making. This integration began as geospatial technologies (e.g., GIS and remote sensing) started to handle increasingly large datasets, including satellite imagery, sensor data, social media feeds, mobile phone signal data, smart-card transportation, point of interests and crowdsourcing data more broadly (Huang et al., 2022, 2024). AI and machine learning algorithms, which thrive on large datasets, have been employed to extract patterns, trends, and insights from this expansive geospatial data. By leveraging big data, GeoAI can perform more intelligent spatial analyses to reveal the complex and sophisticated relationship among human (i.e., community and society) and environment (i.e., built and natural environment), such as real-time monitoring of people's response to environmental changes, predictive modelling of urban and population growth, and precise capturing human activities (Wang et al., 2024). This synergy enables more accurate and actionable insights, driving innovations in the subdomain of human geography and social science more broadly, such as urban geography, social geography, cultural geography, urban planning and health intervention. The continuous growth of big data and advancements in AI techniques are further propelling the capabilities and applications of GeoAI, making it a crucial tool for understanding and managing the dynamic aspects of our cities and societies.

1.2 The Expansion of GeoAI: Transforming Human Geography

1.2.1 Riding the Technological Wave: AI, Big Data, and Digital Infrastructures

GeoAI is founded on several key pillars that are essential for its development and success. The convergence of Artificial Intelligence (AI), big data, and robust technological infrastructures has ushered in a new era of geospatial analysis and application, profoundly transforming the field of geography. The unprecedented growth of data, fuelled by advancements in sensor technologies, satellite imagery, social media, and mobile devices, has provided a rich repository of spatial information (Huang et al., 2024). This wealth of data, often referred to as "big data," presents both opportunities and challenges for geographers seeking to understand complex socio-spatial phenomena.

AI has become an essential tool in harnessing the potential of big data. Techniques such as deep learning, natural language processing, and computer vision enable the extraction of meaningful patterns and insights from vast and heterogeneous datasets. These advancements have made it possible to analyze and interpret spatial data at unprecedented scales and resolutions, offering new perspectives on traditional geographical questions (Janowicz et al., 2022). The integration of AI and big data is supported by advancements in computational infrastructures, such as cloud computing and high-performance computing. These infrastructures provide the necessary computational power and storage capacity to process and analyze large volumes of spatial data efficiently. Cloud platforms, like Google Earth Engine and Amazon Web Services, offer scalable solutions that democratize access to powerful geospatial analysis tools, enabling researchers from diverse backgrounds and institutions to collaborate and innovate. Furthermore, the development of sophisticated geospatial databases and data management systems has facilitated the organization, retrieval, and dissemination of spatial information. Spatial data infrastructures and geospatial data repositories, such as the ArcGIS Living Atlas of the World, Global Earth Observation System of Systems, and the Copernicus Open Access Hub, provide centralized access to diverse geospatial datasets, promoting data sharing and interoperability. These infrastructures are crucial for enabling the seamless integration of geospatial data from various sources, supporting comprehensive and multi-scale analyses.

The synergy between AI, big data, and technological infrastructures is expected to drive and continue driving significant advancements in human geography, particularly as the data and problems human geographers face become increasingly large, complex, and spatially intricate. This integration facilitates the ability to process and analyze vast amounts of spatial data efficiently, enabling geographers to gain deeper insights into complex socio-spatial phenomena.

1.2.2 Democratising GeoAI: The Open-Source Revolution

The open-source movement has fundamentally democratized the field of GeoAI, making advanced geospatial analysis accessible to a broader and more diverse audience. Historically, geospatial technologies were proprietary and expensive, limiting their use to well-funded institutions and organizations (Yang et al., 2010). The advent of open-source tools and platforms has significantly lowered these barriers, fostering a more inclusive and collaborative environment for geospatial research and applications. Platforms like GitHub have been instrumental in this transformation by providing a space for developers to share and collaborate on code, making it a hub for innovation in GeoAI. Open-source libraries, such as QGIS (Moyroud & Portet, 2018), GeoPandas (Jordahl et al., 2021), PySAL (Rey & Anselin, 2009), Rasterio (Gillies, 2019), and geemap (Wu, 2020), among many, have empowered researchers and practitioners to perform sophisticated geospatial analyses without costly software licenses. These tools have enabled tasks ranging from basic mapping and spatial data management to advanced spatial statistics and machine learning applications.

The impact of this democratization is profound. Researchers from institutions with limited resources can now access and utilize the same advanced tools as those from well-funded universities and corporations, leading to a more equitable distribution of knowledge and capability across the geospatial community. Moreover, the open-source movement has fostered a culture of collaboration and knowledge sharing. Online forums, workshops, and conferences dedicated to open-source geospatial tools have proliferated, creating spaces where users can exchange ideas, troubleshoot problems, and develop new applications. This collaborative ethos has accelerated the pace of innovation in GeoAI, as demonstrated during the COVID-19 pandemic, where the rapid development and deployment of open-source tools were crucial in tracking and analyzing the spread of the virus (Li et al., 2021; Brovelli & Coetzee, 2021; Malarvizhi et al., 2022).

Furthermore, the open-source movement has played a critical role in capacity building and education. Students and early-career professionals can gain hands-on experience with state-of-the-art geospatial technologies without financial constraints. This exposure enhances their technical skills and prepares them to contribute to the field of GeoAI with innovative solutions and diverse perspectives. The flexibility and adaptability of open-source GeoAI tools support interdisciplinary research, allowing for integration with various data sources and analytical frameworks to address complex socio-spatial issues.

1.2.3 Interweaving Tools and Techniques: Unravelling Socio-Spatial Complexities

GeoAI is inherently collaborative; its strength does not lie solely in individual tools and technologies but in the ability to integrate these elements, providing unique capabilities to address complex challenges. In an era characterized by intricate socio-spatial phenomena, the integrative nature of GeoAI enables researchers to achieve a more comprehensive and nuanced understanding of geographical issues (Wang et al., 2024). This approach is crucial for tackling the multifaceted nature of human geography, a domain previously untouched by the AI revolution. By leveraging the synergistic potential of AI and geospatial technologies, GeoAI transcends traditional boundaries, enabling more sophisticated analyses and solutions.

Geographers have capitalized on the opportunity to integrate the latest advancements in AI with geospatial data and spatial laws, developing innovative data-handling techniques that significantly benefit the field. Human geographers, in particular, have embraced these advancements. For instance, cultural geographers utilize natural language processing techniques to analyze social media data, uncovering patterns in cultural expressions, behaviours, and trends across different regions (Kitchin et al., 2013). Political geographers employ AI-driven spatial analysis to study electoral geography, border conflicts, and geopolitical dynamics (Chauhan et al., 2021; Zhang, 2021). Urban geographers leverage GeoAI to optimize urban planning and management by analysing traffic patterns, land use, and infrastructure development (Liu & Biljecki, 2022). Population geographers take advantage of GeoAI to predict demographic changes, migration patterns, and population distribution (Zoraghein & O'Neill, 2020; Huang et al., 2021).

In addition, GeoAI serves as a bridge connecting various fields, fostering interdisciplinary collaboration and innovation. Its integrative capabilities enable the seamless combination of geospatial data, AI techniques, and domain-specific knowledge, making it a powerful tool for addressing diverse and complex socio-spatial issues. The cutting-edge data and algorithmic solutions brought by GeoAI and the big data era enhance the precision and depth of geographical analysis while facilitating the cross-pollination of ideas and methods from different disciplines. This interdisciplinary synergy leads to groundbreaking insights and solutions. By breaking down silos and encouraging collaboration, GeoAI is poised to drive significant advancements in understanding and managing socio-spatial challenges.

1.3 Futures

We continue to experience growth and diversity in methods, data, and application areas. As GeoAI continues to develop across the spatial sciences, its application within human geography is set to broaden and deepen. There exists a number of opportunities and challenges as this new field evolves. We discuss each of these in turn.

1.3.1 Opportunities

There exists an exciting number of opportunities that can leverage ongoing development in GeoAI. The first includes enhanced decision-making, through advances in spatial analytics, more efficient management of finite resources, and progress in urban planning and environmental protection. The study by Zhai and colleagues (2019) developed a new model to delineate functional regions of urban areas using geo-text mining on POI data holding important implications for urban planning. The second concerns improvements in predictive capacities, and the capacity to unveil previously hidden relationships to underpin accurate forecasting. Application areas include climate and environmental modeling, spatial epidemiology, and economic development. The study by Blanford (2024) showcases the predictive power of GeoAI in its capacity to reduce the burden of vector-borne diseases (Blanford, 2024). The third is the possibility of real-time monitoring. One example of this is the Google-based flood forecasting initiative that employs real-time GeoAI to predict and communicate flood risks to vulnerable communities (Nearing et al., 2024).

1.3.2 Challenges

Several challenges for GeoAI lay ahead. The first of these involves data quality and ensuring the accuracy and consistency of the diverse datasets is both acknowledged and managed appropriately. Achieving this will help ensure outputs are accurate, trustworthy, and equitable. The second involves the skill hurdle (Rowe, 2021), wherein the traditional analytic and computational frameworks that human geographers receive training in are typically indifferent to those required for work in the GeoAI space, calling for new training to be embedded within current teaching programs. A third challenge relates to costs, where implementing GeoAI-based solutions can often be expensive and might act to restrict their wider application to individuals, organizations, and developing regions. Finally, there is the issue of reproducibility, as GeoAI models and workflows often rely on opaque algorithms and custom datasets that are difficult to replicate, leading to concerns about the reliability and generalizability of findings. Addressing these reproducibility challenges is critical to building trust and fostering broader adoption of GeoAI methodologies.

1.3.3 Concerns

Several concerns have emerged and been the subject of discussion. The first involves bias and discrimination, where algorithms may inadvertently act to reinforce existing biases in data that can lead to flawed predictions and decision-making. One example of this involved issues in the quality of mapping that led to a misidentification of slum areas (Lobo, 2023). This underscores the critical importance of robust data governance and ethical considerations. The second concern is the privacy issue, given that GeoAI-based approaches often rely on sensitive locational data. There is an urgent need to have in place careful management protocols to protect the rights of individuals. Another pressing concern is the rise of "deep fake geography," where synthetic geospatial data, such as manipulated satellite imagery or falsified maps, can be generated to mislead and misinform, potentially undermining trust in GeoAI applications (Zhao et al., 2021). A final concern is that of ethics that involves privacy concerns around consent and the ownership of data especially for marginalized groups and indigenous communities, wherein algorithmic biases might act to exacerbate social inequalities. An essential dogma to harness the full potential of GeoAI will be balancing innovation with responsible practices.

1.4 Conclusion

In conclusion, the rapidly emerging field of GeoAI holds immense potential to drive innovation and improve efficiency across a wide range of domains within human geography. The next 5 years will be exciting ones as we will no doubt see transformative applications emerging across urban planning, environmental monitoring, disaster management, and socio-economic research to note a select few. Set alongside this bright future, the adoption of GeoAI (both within human geography and beyond) needs a thoughtful approach that directly addresses its inherent challenges and concerns—ensuring we progress towards responsible GeoAI (Marasinghe et al., 2024). As this exciting field continues to evolve, fostering collaboration among technologists, policymakers, and communities will be an important feature to help ensure its benefits are maximized whilst mitigating its risks.

References

Blanford, J. I. (2024). Managing vector-borne diseases in a geoAI-enabled society. Malaria as an example. *Acta Tropica, 260*, 107406.

Brooks, R. A. (1999). *Cambrian intelligence: The early history of the new AI*. MIT Press.

Brovelli, M. A., & Coetzee, S. (2021). Open geospatial data for responding to the COVID-19 challenge. In *COVID-19 pandemic, geospatial information, and community resilience* (pp. 31–54). CRC Press.

Chauhan, P., Sharma, N., & Sikka, G. (2021). The emergence of social media data and sentiment analysis in election prediction. *Journal of Ambient Intelligence and Humanized Computing, 12*, 2601–2627.

Cordeschi, R. (2007). AI turns fifty: Revisiting its origins. *Applied Artificial Intelligence, 21*(4–5), 259–279.

Couclelis, H. (1986). Artificial intelligence in geography: Conjectures on the shape of things to come. *The Professional Geographer, 38*(1), 1–11.

Couclelis, H. (2023). GeoAI's thousand-year history. In *Handbook of geospatial artificial intelligence* (pp. 17–25). CRC Press.

Duan, W., Chiang, Y. Y., Leyk, S., Uhl, J. H., & Knoblock, C. A. (2020). Automatic alignment of contemporary vector data and georeferenced historical maps using reinforcement learning. *International Journal of Geographical Information Science, 34*(4), 824–849.

Gao, S. (2021). *Geospatial artificial intelligence (GeoAI)* (Vol. 10). Oxford University Press.

Gillies, S. (2019). *Rasterio documentation* (p. 23). San Francisco, CA, USA.

Huang, X., Zhu, D., Zhang, F., Liu, T., Li, X., & Zou, L. (2021). Sensing population distribution from satellite imagery via deep learning: Model selection, neighboring effects, and systematic biases. *IEEE Journal of Selected Topics in Applied Earth Observations and Remote Sensing, 14*, 5137–5151.

Huang, X., Wang, S., Zhang, M., Hu, T., Hohl, A., She, B., & Li, Z. (2022). Social media mining under the COVID-19 context: Progress, challenges, and opportunities. *International Journal of Applied Earth Observation and Geoinformation, 113*, 102967.

Huang, X., Wang, S., Yang, D., Hu, T., Chen, M., Zhang, M., & Hohl, A. (2024). Crowdsourcing geospatial data for earth and human observations: A review. *Journal of Remote Sensing, 4*, 0105.

Janowicz, K., Sieber, R., & Crampton, J. (2022). GeoAI, counter-AI, and human geography: A conversation. *Dialogues in Human Geography, 12*(3), 446–458.

Jordahl, K., Van den Bossche, J., Wasserman, J., McBride, J., Gerard, J., Tratner, J., & Farmer, C. (2021). *geopandas/geopandas: v0. 5.0*. Zenodo.

Kitchin, R., Linehan, D., O'Callaghan, C., & Lawton, P. (2013). Public geographies through social media. *Dialogues in Human Geography, 3*(1), 56–72.

Levin, N., Ali, S., & Crandall, D. (2018). Utilizing remote sensing and big data to quantify conflict intensity: The Arab spring as a case study. *Applied Geography, 94*, 1–17.

Li, W. (2022). GeoAI in social science. In *Handbook of spatial analysis in the social sciences* (pp. 291–304). Edward Elgar.

Li, W., & Hsu, C. Y. (2022). GeoAI for large-scale image analysis and machine vision: Recent progress of artificial intelligence in geography. *ISPRS International Journal of Geo-Information, 11*(7), 385.

Li, Z., Huang, X., Hu, T., Ning, H., Ye, X., Huang, B., & Li, X. (2021). ODT FLOW: Extracting, analyzing, and sharing multi-source multi-scale human mobility. *PLoS One, 16*(8), e0255259.

Liu, P., & Biljecki, F. (2022). A review of spatially-explicit GeoAI applications in urban geography. *International Journal of Applied Earth Observation and Geoinformation, 112*, 102936.

Liu, W., Wu, W., Thakuriah, P., & Wang, J. (2020). The geography of human activity and land use: A big data approach. *Cities, 97*, 102523.

Lobo, T. (2023). The ethics of mapping slums—And how AI complicates the picture. *Philosophy of the City Journal, 1*(1), 127–134.

Malarvizhi, A. S., Liu, Q., Sha, D., Lan, H., & Yang, C. (2022). An open-source workflow for spatiotemporal studies with COVID-19 as an example. *ISPRS International Journal of Geo-Information, 11*(1), 13.

Marasinghe, R., Yigitcanlar, T., Mayere, S., et al. (2024). Towards responsible urban geospatial AI: Insights from the white and Grey literatures. *Journal of Geovisualization and Spatial Analysis, 8*, 24. https://doi.org/10.1007/s41651-024-00184-2

Moyroud, N., & Portet, F. (2018). Introduction to QGIS. In *QGIS and generic tools* (Vol. 1, pp. 1–17).

Nearing, G., Cohen, D., Dube, V., et al. (2024). Global prediction of extreme floods in ungauged watersheds. *Nature, 627*, 559–563. https://doi.org/10.1038/s41586-024-07145-1

Negnevitsky, M. (2024). The history of artificial intelligence or from the WIT transactions on information and communication technologies, WIT Press 19.

Openshaw, S. (1992). Some suggestions concerning the development of artificial intelligence tools for spatial modelling and analysis in GIS. *The Annals of Regional Science, 26*, 35–51.

Perez, L., & Sengupta, R. (2024). Big data (R) evolution in geography: Complexity modelling in the last two decades. *Geography Compass, 18*(11), e70009.

Rey, S. J., & Anselin, L. (2009). PySAL: A python library of spatial analytical methods. In *Handbook of applied spatial analysis: Software tools, methods and applications* (pp. 175–193). Springer.

Rowe, F. (2021). *Big data and human geography*. Concise Encyclopedia of Human Geography.

Scheider, S., & Richter, K. F. (2023). Pragmatic GeoAI: Geographic information as externalized practice. *KI-Künstliche Intelligenz, 37*(1), 17–31.

Smith, T. R. (1984). Artificial intelligence and its applicability to geographical problem solving. *The Professional Geographer, 36*(2), 147–158.

Wang, S., Huang, X., Liu, P., Zhang, M., Biljecki, F., Hu, T., & Bao, S. (2024). Mapping the landscape and roadmap of geospatial artificial intelligence (GeoAI) in quantitative human geography: An extensive systematic review. *International Journal of Applied Earth Observation and Geoinformation, 128*, 103734.

Wu, Q. (2020). Geemap: A python package for interactive mapping with Google earth engine. *Journal of Open Source Software, 5*(51), 2305.

Wu, W., Wang, J., & Dai, T. (2016). The geography of cultural ties and human mobility: Big data in urban contexts. *Annals of the American Association of Geographers, 106*(3), 612–630.

Yang, C., Raskin, R., Goodchild, M., & Gahegan, M. (2010). Geospatial cyberinfrastructure: Past, present and future. *Computers, Environment and Urban Systems, 34*(4), 264–277.

Zhang, S. I. (2021). Mediatization of conflict in the social media era: A case study of Sino-Indian border crisis in 2017. *Journalism, 22*(10), 2618–2636.

Zhao, B., Zhang, S., Xu, C., Sun, Y., & Deng, C. (2021). Deep fake geography? When geospatial data encounter artificial intelligence. *Cartography and Geographic Information Science, 48*(4), 338–352.

Zoraghein, H., & O'Neill, B. C. (2020). A spatial population downscaling model for integrated human-environment analysis in the United States. *Demographic Research, 43*, 1563–1606.

Chapter 2
The Evolution of Geospatial Artificial Intelligence

Gengchen Mai, Yiqun Xie, Xiaowei Jia, Ni Lao, Jinmeng Rao, Qing Zhu, and Zeping Liu

Abstract Although studies applying artificial intelligence technologies to geographic problems date back to the 1980s, Geospatial Artificial Intelligence (GeoAI) continues to emerge as one of the fastest-developing research directions in spatial data science and geography today. In this chapter, we first review the recent GeoAI research advancements in various geospatial subdomains. We further discuss several important future research directions of GeoAI. We hope this chapter can provide a meaningful overview of the research landscape of GeoAI.

Keywords GeoAI · Spatial Data Science · Geospatial Challenges · Spatial Representation Learning · Privacy-Aware GeoAI

G. Mai (✉) · Z. Liu
SEAI Lab, Department of Geography and the Environment, The University of Texas at Austin, Austin, TX, USA
e-mail: gengchen.mai@austin.utexas.edu; zeping.liu@utexas.edu

Y. Xie
Department of Geographical Sciences, University of Maryland, College Park, College Park, MD, USA
e-mail: xie@umd.edu

X. Jia
Department of Computer Science, University of Pittsburgh, Pittsburgh, PA, USA
e-mail: xiaowei@pitt.edu

N. Lao · J. Rao
Google LLC, Mountain View, CA, USA
e-mail: nlao@google.com; jinmengrao@google.com

Q. Zhu
Climate and Ecosystem Sciences, Lawrence Berkeley National Laboratory, Berkeley, CA, USA
e-mail: QZhu@lbl.gov

© The Author(s), under exclusive license to Springer Nature Switzerland AG 2025
X. Huang et al. (eds.), *GeoAI and Human Geography*, Springer Geography,
https://doi.org/10.1007/978-3-031-87421-5_2

2.1 Introduction

Geospatial Artificial Intelligence (GeoAI) has been widely recognized as a rapidly evolving, cutting-edge research domain within Spatial Data Science, as well as the broader fields of geography and geoscience in recent years (Reichstein et al., 2019; Janowicz et al., 2020; Gao et al., 2024). The rapid development of GeoAI can be largely attributed to the increasing availability of high-quality geospatial data (e.g., location-based social media data, remote sensing imagery, geo-tagged ground-level images), the significant advancements in machine learning hardware (such as NVIDIA's GPU and Google's TPU), and the advancement of artificial intelligence (AI) and GeoAI models.

Given the rapidly evolving nature of GeoAI, we have reached a critical point of time to understand the nature of recent developments and speculate where the field is going in the near future. Therefore, we first summarize the fundamental problems GeoAI is trying to solve and analyze GeoAI research nowadays, which covers several application domains including geospatial image processing (Li et al., 2021; Mai et al., 2023c; Xie et al., 2023d; Li et al., 2023c), urban computing (Cai et al., 2020; Rao et al., 2020, 2023b; Hu et al., 2023a; Bao et al., 2020, 2022), earth system science (Jia et al., 2021b; Chen et al., 2023), and geospatial semantics (Yan et al., 2019; Mai et al., 2020a; Wang et al., 2020b; Mai et al., 2021; Scheider et al., 2021; Xu et al., 2023; Hu et al., 2023b). We finally highlight unique challenges in GeoAI research that remain to be solved in the future which distinguish GeoAI from general AI research such as spatial fairness, geoprivacy preservation, knowledge-guided GeoAI development, spatial representation learning, etc.

2.2 Major Research Directions of GeoAI

Given the vast amount of literature in GeoAI research, using an enumeration approach is impractical. A more meaningful method would be to explore its recent advancements by discussing several representative works across various geospatial subdomains, such as geospatial image processing, urban computing, earth system science, and geospatial semantics. So in the following, we will briefly discuss representative GeoAI research based on their main application area.

2.2.1 Geospatial Image Processing

GeoAI has been extensively applied to geospatial imagery such as satellite imagery, aerial remote sensing imagery, street view imagery, etc., for tasks such as scene classification (Cheng et al., 2020), semantic (Liu & Tang, 2023) and instance segmentation (Liu et al., 2022b), geospatial object detection (Zhang et al., 2023b), remote sensing time-series analysis (Zhou et al., 2022), change detection (Zheng et al., 2024), image depth estimation (Zheng et al., 2022), and 3D reconstruction (Li et al., 2023b). Given the inherently geospatial nature of these tasks, incorporating geographic priors is crucial for achieving accurate results. For instance, building detection from aerial imagery requires introducing regularized building structures to ensure reliable outcomes (Liu et al., 2023b). Similarly, identifying crop boundaries from satellite imagery must account for seasonal variations (Ghosh et al., 2021). For cloud detection in remote sensing imagery, geographic events such as fires should be considered to differentiate between smoke and clouds (Zhu & Woodcock, 2012). The integration of geographic knowledge into AI algorithms to address such geospatial challenges remains a central focus of GeoAI research.

2.2.2 *Urban Computing*

Urban computing represents a vital research direction of GeoAI, focusing on understanding and addressing urban challenges through the integration of spatial data, AI, and city-scale computational models (Zheng et al., 2014; Zheng, 2019; Cugurullo, 2020; Zou et al., 2025).

In transportation research, GeoAI-enhanced urban computing significantly improves traffic management by reducing congestion and optimizing public transport. For example, GeoAI can help dynamically alter traffic signals on real-time traffic data to reduce traffic congestion and air pollution (Yang et al., 2021). GeoAI can also refine public transit scheduling and rule planning (Gangwani & Gangwani, 2021). In ride-sharing services, GeoAI can perform demand forecasting and driver-passenger matching to improve service quality and system efficiency (Wang et al., 2020a).

Environmental sustainability is another key topic of urban computing in which GeoAI can leverage geospatial observation data of the urban environment from various data sources to mitigate the impacts of urbanization on the environment. AI can facilitate real-time air quality mapping (Kaginalkar et al., 2021), urban heat island (UHI) identification (Almeida et al., 2021), environmental hazard prediction, and disaster management (Yang et al., 2022).

In addition, GeoAI can also be utilized to improve community well-being (Zhang et al., 2024). For example, GeoAI has been deployed to analyze social media data and geospatial data (e.g., human mobility data) for crowd behavior and mobility pattern prediction (Zhu et al., 2024), crime hotspot prediction (Luca et al., 2023; Zhou et al., 2023), disease outbreak tracking and spread understanding (Gao et al., 2020b,a).

2.2.3 *Earth System Science*

Earth System Science (ESS) is an interdisciplinary field focused on understanding the complex interactions among the atmosphere, land surface, oceans, sea ice, land ice, and human systems. Historically, the study of ESS has heavily relied on process-based Earth System Models (ESMs) (Calvin & Bond-Lamberty, 2018) such as the Coupled Model Intercomparison Project (CMIP) (Taylor et al., 2012).

Over the past decade, advances in GeoAI have started to transform ESS research by complementing and enhancing traditional process-based earth system models (Irrgang et al., 2021). This practice has significantly improved the accuracy of ESMs by capturing the key processes more accurately (Kwiatkowski et al., 2020). Moreover, GeoAI has also enabled the diversification of ESMs to meet the dedicated needs of stakeholders and scientific communities (Ruiz et al., 2023).

Another emerging application of GeoAI in ESS is the near-real-time monitoring of greenhouse gas (GHG) emissions. Most anthropogenic GHGs are released by burning fossil fuels which reached 9.9 ± 0.5 Pg C yr^{-1} (1 Pg = 10^{15} g) in 2022 by recent estimation. However, monitoring GHG emissions is challenging. With the help of GeoAI, high-resolution mapping of CO_2 and methane (CH_4) emissions has become possible (Wang et al., 2019; Bonire & Gbenga-Ilori, 2021).

In addition, GeoAI has proven to be effective in improving the resilience of Earth systems (Tripathy et al., 2023). AI-powered early warning systems (Sharma et al., 2020) provide critical information about climate extreme and natural hazards (Materia et al., 2024). Moreover, GeoAI plays an important role in climate mitigation (Ramya et al., 2024) and helps maximize climate benefits and enhance GHG sequestration efforts (Amiri et al., 2024).

2.2.4 Geospatial Semantics

Geospatial semantics is an interdisciplinary research area combining cognitive science, geographic information science, artificial intelligence, computational linguistics, information retrieval, and Semantics Web (Janowicz et al., 2012; Kuhn, 2005). Major research topics include geo-ontology engineering (Battle & Kolas, 2011; Mai et al., 2022), geo-text mining (Hu et al., 2017, 2019), geospatial knowledge graphs (Regalia et al., 2018; Qi et al., 2023), geographic information retrieval (Jiang et al., 2018; Purves et al., 2018; Mai et al., 2020b), geographic question answering (Chen, 2014; Scheider et al., 2021; Mai et al., 2018; Hamzei et al., 2020; Mai et al., 2020a, 2021; Xu et al., 2023), etc. We will use geographic knowledge graphs and geographic question answering as two exemplar tasks to discuss the recent advancements of GeoAI in the geospatial semantics domain.

Knowledge graphs (KGs) are a novel paradigm for the representation, retrieval, and integration of data from highly heterogeneous sources into an interlinked cross-domain data repository in a human- and machine-readable and reasonable way (Janowicz et al., 2022). Geospatial knowledge graphs (GeoKGs) (Zhu, 2024) emphasize the role of space and time in the knowledge graph paradigm (Janowicz, 2010). Until now, many geospatial knowledge graphs have been developed to support geospatial data organization, retrieval, and synthesis such as YAGO2 (Hoffart et al., 2013), LinkedGeoData (Auer et al., 2009), GNIS-LD (Regalia et al.,

2018), EVKG (Qi et al., 2023), KnowWhereGraph (Janowicz et al., 2022), etc. Despite all these successes, many challenges exist such as spatiotemporal scoping of GeoKGs, GeoKG decentralization, automatic GeoKG construction and update, etc.

Geographic question answering (GeoQA) (Mai et al., 2021) is a subfield of question answering aiming at answering geographic questions automatically without human intervention. Many pioneer works in GeoQA focus on answering simple factoid geographic questions (Zelle & Mooney, 1996; Punjani et al., 2018). Recent works started to work on more complex geographic questions such as scenario-based geographic question answering (Contractor et al., 2021; Huang et al., 2019) and geo-analytical question answering (Scheider et al., 2021; Xu et al., 2023). The newly developed large language models (LLM) have also been incorporated into the GeoQA research. Recent research utilized LLM agents (e.g., LangChain, AutoGPT) to answer rather complex geographic questions by using LLMs to automatically generate the geoprocessing workflow for question answering such as GeoGPT (Zhang et al., 2023a) and LLMGeo (Li & Ning, 2023). However, developing a "natural language interface for a geographic information system" requires not only a more rigorous definition of spatial operations and concepts but also more annotated data samples for a comprehensive evaluation of such system.

2.3 Future Challenges and Opportunities of GeoAI

Despite these recent successes discussed in Sect. 2.2, there exist many challenges that remain to be solved which we will point out in the following.

2.3.1 Heterogeneity-Aware GeoAI

While aiming to learn a general machine learning model that works effectively across different spatial regions is ideal, this goal may not be technically achievable due to the inherent spatial heterogeneity (Goodchild, 2004). From a machine learning perspective, this means that the relationships between input variables and outputs often differ by geographic areas (Xie et al., 2021b; Li et al., 2023a). Indeed, spatial heterogeneity violates the typical independence and identical distribution (i.i.d.) assumption underlying most learning models. Past studies have explored the possibility of developing heterogeneity-aware frameworks for traditional machine learning models, including the geographically weighted regression (GWR) (Brunsdon et al., 1999) as well as its variants for random forest (Georganos et al., 2021), and more. However, these formulations are not directly compatible with the more recent deep learning and foundation models in terms of model sizes and performance. These recent models are substantially greater in size, making it infeasible to learn individual models at each location as GWR does. To spatialize these newer generations of learning models, recent efforts have explored new model-agnostic formulations using adaptive space partitioning. For example, a spatial transformation framework was proposed for general deep learning models, which can automatically recognize spatial heterogeneity during the training process and eliminate it by identifying optimal space partitions (Xie et al., 2021b, 2023c).

Despite the progress, there are many challenges remaining to be addressed. First, new designs are needed to improve the explainability of space partitions in the context of various physical and social factors. Second, more advanced parameter-sharing and co-training methods (Xie et al., 2023a, 2021c; Li et al., 2023c) are highly desired to reduce the extra training and storage cost of the heterogeneity-aware models in the era of larger foundation models (Mai et al., 2024b; Xie et al., 2023b; Rao et al., 2023a). Third, mapping bias can be easily caused by heterogeneity, and fairness-aware learning methods are needed for geospatial problems (Xie et al., 2022; He et al., 2023). Finally, easy-to-use infrastructures are needed around the spatialized learning frameworks to support smoother adoption and deployment of the techniques in real applications and further advancement of the algorithms.

2.3.2 Privacy-Aware GeoAI

GeoAI often relies on potentially personal or sensitive spatial data such as individual mobility patterns and geo-tagged images, which may raise privacy concerns if not handled properly (McKenzie et al., 2023). One main challenge in privacy-aware GeoAI is balancing the trade-off between spatial data utility and privacy protection. Techniques such as geomasking (Houfaf-Khoufaf et al., 2024), K-anonymization (McKenzie & Zhang, 2023), and differential privacy (Miranda-Pascual et al., 2023) are effective in protecting privacy but may reduce the accuracy of GeoAI models. Another main challenge is handling the endogenous privacy risks in spatial data and model centralization (Rao et al., 2021, 2023a). The centralized training and serving of GeoAI models may cause reidentification and privacy leaks through spatial correlations, model leaks, cross-referencing with other datasets, reverse engineering, and malicious prompting (Shao et al., 2023; Rao et al., 2023a).

Alongside the challenges, new opportunities have emerged with the advancement of privacy-aware GeoAI. One emerging trend is the use of GeoAI models to generate privacy-preserved synthetic spatial data that better balances the utility-privacy trade-offs (Rao et al., 2023b; Zhu et al., 2023). Another trend is the integration of decentralization techniques into GeoAI models to mitigate privacy concerns and threats inherent in centralized structures (Rao et al., 2021). In addition to technological solutions, regulatory and ethical frameworks are also necessary to ensure that privacy protections in GeoAI align with societal expectations (Rao et al., 2023a).

2.3.3 *Knowledge-Guided GeoAI*

Although machine learning methods have found immense success in modeling complex massive datasets, they often exhibit limited generalizability to unseen scenarios such as new spatial regions with different characteristics. Other than the above-mentioned spatial heterogeneity, this is also usually caused by limited and less representative training samples due to the substantial cost required for field study or sensor deployment. Moreover, many traditional models are not designed to capture nonstationary and multi-scale processes, and thus they often produce results that are inconsistent with established theories and knowledge in the target domain. One promising direction to address this issue is to incorporate scientific knowledge in the target domain to guide the construction of machine learning models, also known as knowledge-guided machine learning (KGML) (Karpatne et al., 2024), which has shown encouraging results in a range of applications, such as climate science (Kashinath et al., 2021; Xie et al., 2021a), hydrology (Chen et al., 2023, 2022; Read et al., 2019; Jia et al., 2021c), agriculture (Liu et al., 2022a, 2024; He et al., 2023), and geology (Sun et al., 2021). By incorporating additional knowledge in the model architecture (Liu et al., 2024; Khandelwal et al., 2020) or training process (Jia et al., 2021c,a; He et al., 2024; Chen et al., 2023), many prior works have shown that KGML models can achieve better generalizability to different spatial regions or time periods (Read et al., 2019; Chen et al., 2023).

Despite these progresses, there remain many opportunities to further enhance KGML methods toward geospatial applications. A key challenge is that scientific knowledge often provides only approximations of reality or omits fine-scale processes due to excessive computational cost, leading to inconsistencies with true observations. Excessive reliance on such imperfect knowledge could introduce additional model bias. Hence, new mechanisms are needed to harness imperfect or partial knowledge. Finally, many KGML methods are designed toward a specific application, which requires extensive effort and domain expertise. Future research could explore the development of general paradigms or foundation model frameworks to systematically incorporate various forms of knowledge (e.g., equations, causal dependencies, and knowledge graphs) into machine learning.

2.3.4 *Spatial Representation Learning*

Traditionally, due to the lack of representation learning techniques for various types of spatial data such as geospatial vector data, triangulated irregular networks (TINs), 3D LiDAR point clouds, 3D building models, etc., many pioneering GeoAI research resorted to some "compromised" approaches. They either performed feature engineering on spatial data (e.g., building polygons) based on domain knowledge (Jochem et al., 2018; Yan et al., 2022) or converted spatial data from their original formats (e.g., geospatial vector data) into formats that are easier to process by artificial neural networks (e.g., voxels (Zhou & Tuzel, 2018), raster image tiles (Balsebre et al., 2023), and graphs (He et al., 2018)). The feature engineering approach is domain knowledge dependent and hard to generalize to new tasks. The data conversion approach will lead to reduced data precision and increased data storage.

To overcome these drawbacks, spatial representation learning (SRL) is proposed as a new general-purpose representation learning framework for spatial data (Mai et al., 2024a), which aims at directly learning neural spatial representations of various types of spatial data in their native data format without the need for any feature engineering or data conversion stage. Some initial successes have been achieved such as location encoding (Mac Aodha et al., 2019; Mai et al., 2023d,b; Rußwurm et al., 2023; Cole et al., 2023; Wu et al., 2024), polyline encoding (Rao et al., 2020, 2023b), polygon encoding (Mai et al., 2023a; Yu et al., 2024), polyline decoding (Yu & Chen, 2022; Rao et al., 2023b), polygon decoding (Jiang et al., 2019; Liu et al., 2023a), spatial graph encoding/decoding (Mai et al., 2020a; Klemmer et al., 2023), etc.

However, there are several challenges remaining to be solved. First, most existing spatial representation learning models treat polylines and polygons as a finite sequence of vertices, while they in fact represent continuous geographic features. Models that can learn continuous neural representations of these spatial data are in need to capture the intrinsic nature of these spatial data. Second, representation learning models for different spatial data modalities are currently developed separately as different models, while a unified representation learning model is needed to seamlessly handle various spatial data formats simultaneously. This can both ease the GeoAI model development process and facilitate knowledge sharing across geospatial datasets in different data formats.

2.4 Conclusion

In this chapter, we provide an overview and discussion of the present and future of GeoAI. We first discuss the recent GeoAI achievements in several representative domains such as geospatial image processing, urban computing, earth system science, and geospatial semantics. Then, we highlight the future challenges and opportunities of GeoAI—heterogeneity-aware GeoAI, knowledge-guided GeoAI, privacy-aware GeoAI, and spatial representation learning. We believe that, as the fastest-developing research direction in spatial data science, GeoAI will continue to be the major driver behind many innovative geospatial applications.

References

Almeida, C. R. d., Teodoro, A. C., & Gonçalves, A. (2021). Study of the urban heat island (UHI) using remote sensing data/techniques: A systematic review. *Environments, 8*(10), 105.

Amiri, Z., Heidari, A., & Navimipour, N. J. (2024). Comprehensive survey of artificial intelligence techniques and strategies for climate change mitigation. *Energy, 308*, 132827.

Auer, S., Lehmann, J., & Hellmann, S. (2009). LinkedGeoData: Adding a spatial dimension to the web of data. In *ISWC 2009* (pp. 731–746). Springer.

Balsebre, P., Huang, W., Cong, G., & Li, Y. (2023). CityFM: City foundation models to solve urban challenges. *arXiv preprint arXiv:2310.00583*.

Bao, H. et al. (2022). Storm-GAN: spatio-temporal meta-GAN for cross-city estimation of human mobility responses to covid-19. In *ICDM 2022* (pp. 1–10). IEEE.

Bao, H., Zhou, X., Zhang, Y., Li, Y., & Xie, Y. (2020). Covid-GAN: Estimating human mobility responses to covid-19 pandemic through spatio-temporal conditional generative adversarial networks. In *ACM SIGSPATIAL 2020* (pp. 273–282).

Battle, R., & Kolas, D. (2011). GeoSPARQL: enabling a geospatial semantic web. *Semantic Web Journal, 3*(4), 355–370.

Bonire, G., & Gbenga-Ilori, A. (2021). Towards artificial intelligence-based reduction of greenhouse gas emissions in the telecommunications industry. *Scientific African, 12*, e00823.

Brunsdon, C., Fotheringham, A. S., & Charlton, M. (1999). Some notes on parametric significance tests for geographically weighted regression. *Journal of Regional Science, 39*(3), 497–524.

Cai, L. et al. (2020). Traffic transformer: Capturing the continuity and periodicity of time series for traffic forecasting. *Transactions in GIS, 24*(3), 736–755.

Calvin, K., & Bond-Lamberty, B. (2018). Integrated human-earth system modeling—state of the science and future directions. *Environmental Research Letters, 13*(6), 063006.

Chen, S., Xie, Y., Li, X., Liang, X., & Jia, X. (2023). Physics-guided meta-learning method in baseflow prediction over large regions. In *SDM 2023* (pp. 217–225). SIAM.

Chen, S., Zwart, J. A., & Jia, X. (2022). Physics-guided graph meta learning for predicting water temperature and streamflow in stream networks. In *KDD 2022* (pp. 2752–2761).

Chen, W. (2014). Parameterized spatial SQL translation for geographic question answering. In *2014 IEEE international conference on semantic computing* (pp. 23–27). IEEE.

Cheng, G., Xie, X., Han, J., Guo, L., & Xia, G.-S. (2020). Remote sensing image scene classification meets deep learning: Challenges, methods, benchmarks, and opportunities. *IEEE Journal of Selected Topics in Applied Earth Observations and Remote Sensing, 13*, 3735–3756.

Cole, E. et al. (2023). Spatial implicit neural representations for global-scale species mapping. In *International Conference on Machine Learning* (pp. 6320–6342). PMLR.

Contractor, D., Goel, S., Mausam, & Singla, P. (2021). Joint spatio-textual reasoning for answering tourism questions. In *WWW 2021* (pp. 1978–1989).

Cugurullo, F. (2020). Urban artificial intelligence: From automation to autonomy in the smart city. *Frontiers in Sustainable Cities, 2*, 38.

Gangwani, D., & Gangwani, P. (2021). Applications of machine learning and artificial intelligence in intelligent transportation system: A review. *ICAAAIML 2020* (pp. 203–216).

Gao, S. et al. (2020a). Association of mobile phone location data indications of travel and stay-at-home mandates with covid-19 infection rates in the US. *JAMA Network Open, 3*(9), e2020485–e2020485.

Gao, S., Hu, Y., & Li, W. (2024). Introduction to geospatial artificial intelligence (GeoAI). In *Handbook of geospatial artificial intelligence* (pp. 3–16). CRC Press.

Gao, S., Rao, J., Kang, Y., Liang, Y., & Kruse, J. (2020b). Mapping county-level mobility pattern changes in the United States in response to covid-19. *SIGSpatial Special, 12*(1), 16–26.

Georganos, S., Grippa, T., Niang Gadiaga, A., Linard, C., Lennert, M., Vanhuysse, S., Mboga, N., Wolff, E., & Kalogirou, S. (2021). Geographical random forests: a spatial extension of the random forest algorithm to address spatial heterogeneity in remote sensing and population modelling. *Geocarto International, 36*(2), 121–136.

Ghosh, R., Ravirathinam, P., Jia, X., Khandelwal, A., Mulla, D., & Kumar, V. (2021). Calcrop21: A georeferenced multi-spectral dataset of satellite imagery and crop labels. In *2021 IEEE International Conference on Big Data (Big Data)* (pp. 1625–1632). IEEE.

Goodchild, M. F. (2004). The validity and usefulness of laws in geographic information science and geography. *Annals of the Association of American Geographers, 94*(2), 300–303.

Hamzei, E., Li, H., Vasardani, M., Baldwin, T., Winter, S., & Tomko, M. (2020). Place questions and human-generated answers: A data analysis approach. In *AGILE 2020* (pp. 3–19). Springer.
He, E., Xie, Y., Liu, L., Chen, W., Jin, Z., & Jia, X. (2023). Physics guided neural networks for time-aware fairness: An application in crop yield prediction. In *AAAI 2023*.
He, E., Xie, Y., Liu, L., Jin, Z., Zhang, D., & Jia, X. (2024). Knowledge guided machine learning for extracting, preserving, and adapting physics-aware features. In *SDM 2024* (pp. 715–723). SIAM.
He, X., Zhang, X., & Xin, Q. (2018). Recognition of building group patterns in topographic maps based on graph partitioning and random forest. *ISPRS Journal of Photogrammetry and Remote Sensing, 136*, 26–40.
Hoffart, J., Suchanek, F. M., Berberich, K., & Weikum, G. (2013). YAGO2: A spatially and temporally enhanced knowledge base from wikipedia. *Artificial Intelligence, 194*, 28–61.
Houfaf-Khoufaf, W., Touya, G., & Le Guilcher, A. (2024). Geographically masking addresses to study covid-19 clusters. *Cartography and Geographic Information Science, 51*(2), 242–256.
Hu, M. et al. (2023a). Self-supervised pre-training for robust and generic spatial-temporal representations. In *ICDM 2023* (pp. 150–159). IEEE.
Hu, Y. et al. (2023b). Geo-knowledge-guided GPT models improve the extraction of location descriptions from disaster-related social media messages. *International Journal of Geographical Information Science, 37*(11), 2289–2318.
Hu, Y., Mao, H., & McKenzie, G. (2019). A natural language processing and geospatial clustering framework for harvesting local place names from geotagged housing advertisements. *International Journal of Geographical Information Science, 33*(4), 714–738.
Hu, Y., Ye, X., & Shaw, S.-L. (2017). Extracting and analyzing semantic relatedness between cities using news articles. *International Journal of Geographical Information Science, 31*(12), 2427–2451.
Huang, Z. et al. (2019). Geosqa: A benchmark for scenario-based question answering in the geography domain at high school level. In *EMNLP-IJCNLP 2019* (pp. 5866–5871).
Irrgang, C. et al. (2021). Towards neural earth system modelling by integrating artificial intelligence in earth system science. *Nature Machine Intelligence, 3*(8), 667–674.
Janowicz, K. (2010). The role of space and time for knowledge organization on the semantic web. *Semantic Web, 1*(1-2), 25–32.
Janowicz, K., Gao, S., McKenzie, G., Hu, Y., & Bhaduri, B. (2020). GeoAI: Spatially explicit artificial intelligence techniques for geographic knowledge discovery and beyond. *International Journal of Geographical Information Science, 34*(4), 625–636.
Janowicz, K., Hitzler, P., Li, W., Rehberger, D., Schildhauer, M., Zhu, R., Shimizu, C., Fisher, C., Cai, L., Mai, G., et al. (2022). Know, know where, knowwheregraph: A densely connected, cross-domain knowledge graph and geo-enrichment service stack for applications in environmental intelligence. *AI Magazine, 43*(1), 30–39.
Janowicz, K., Scheider, S., Pehle, T., & Hart, G. (2012). Geospatial semantics and linked spatiotemporal data–past, present, and future. *Semantic Web, 3*(4), 321–332.
Jia, X. et al. (2021a). Physics-guided machine learning from simulation data: An application in modeling lake and river systems. In *ICDM 2021* (pp. 270–279). IEEE.
Jia, X. et al. (2021b). Physics-guided recurrent graph model for predicting flow and temperature in river networks. In *SDM 2021* (pp. 612–620). SIAM.
Jia, X., Willard, J., Karpatne, A., Read, J. S., Zwart, J. A., Steinbach, M., & Kumar, V. (2021c). Physics-guided machine learning for scientific discovery: An application in simulating lake temperature profiles. *ACM/IMS Transactions on Data Science, 2*(3), 1–26.
Jiang, C. M. et al. (2019). Convolutional neural networks on non-uniform geometrical signals using Euclidean spectral transformation. In *ICLR 2019*.
Jiang, Y., Li, Y., Yang, C., Hu, F., Armstrong, E. M., Huang, T., Moroni, D., McGibbney, L. J., & Finch, C. J. (2018). Towards intelligent geospatial data discovery: A machine learning framework for search ranking. *International Journal of Digital Earth, 11*(9), 956–971.

Jochem, W. C., Bird, T. J., & Tatem, A. J. (2018). Identifying residential neighbourhood types from settlement points in a machine learning approach. *Computers, Environment and Urban Systems, 69*, 104–113.

Kaginalkar, A., Kumar, S., Gargava, P., & Niyogi, D. (2021). Review of urban computing in air quality management as smart city service: An integrated IoT, AI, and cloud technology perspective. *Urban Climate, 39*, 100972.

Karpatne, A., Jia, X., & Kumar, V. (2024). Knowledge-guided machine learning: Current trends and future prospects. *arXiv preprint arXiv:2403.15989*.

Kashinath, K., Mustafa, M., Albert, A., Wu, J., Jiang, C., Esmaeilzadeh, S., Azizzadenesheli, K., Wang, R., Chattopadhyay, A., Singh, A., et al. (2021). Physics-informed machine learning: case studies for weather and climate modelling. *Philosophical Transactions of the Royal Society A, 379*(2194), 20200093.

Khandelwal, A., Xu, S., Li, X., Jia, X., Stienbach, M., Duffy, C., Nieber, J., & Kumar, V. (2020). Physics guided machine learning methods for hydrology. *arXiv preprint arXiv:2012.02854*.

Klemmer, K., Safir, N. S., & Neill, D. B. (2023). Positional encoder graph neural networks for geographic data. In *International Conference on Artificial Intelligence and Statistics* (pp. 1379–1389). PMLR.

Kuhn, W. (2005). Geospatial semantics: why, of what, and how? In *Journal on data semantics III* (pp. 1–24). Springer.

Kwiatkowski, L., Torres, O., Bopp, L., Aumont, O., Chamberlain, M., Christian, J. R., Dunne, J. P., Gehlen, M., Ilyina, T., John, J. G., et al. (2020). Twenty-first century ocean warming, acidification, deoxygenation, and upper-ocean nutrient and primary production decline from CMIP6 model projections. *Biogeosciences, 17*(13), 3439–3470.

Li, H., Wang, J., Zollner, J. M., Mai, G., Lao, N., & Werner, M. (2023a). Rethink geographical generalizability with unsupervised self-attention model ensemble: A case study of OpenStreetMap missing building detection in Africa. In *ACM SIGSPATIAL 2023* (pp. 1–9).

Li, Q., Mou, L., Hua, Y., Shi, Y., Chen, S., Sun, Y., & Zhu, X. X. (2023b). 3dcentripetalnet: Building height retrieval from monocular remote sensing imagery. *International Journal of Applied Earth Observation and Geoinformation, 120*, 103311.

Li, W., Hsu, C.-Y., & Hu, M. (2021). Tobler's first law in GeoAI: A spatially explicit deep learning model for terrain feature detection under weak supervision. *Annals of the American Association of Geographers, 111*(7), 1887–1905.

Li, Z., & Ning, H. (2023). Autonomous GIS: The next-generation AI-powered GIS. *International Journal of Digital Earth, 16*(2), 4668–4686.

Li, Z., Xie, Y., Jia, X., Stuart, K., Delaire, C., & Skakun, S. (2023c). Point-to-region co-learning for poverty mapping at high resolution using satellite imagery. In *AAAI 2023*.

Liu, J., Ding, H., Cai, Z., Zhang, Y., Satzoda, R. K., Mahadevan, V., & Manmatha, R. (2023a). PolyFormer: Referring image segmentation as sequential polygon generation. In *CVPR 2023* (pp. 18653–18663).

Liu, L., Xu, S., Tang, J., Guan, K., Griffis, T. J., Erickson, M. D., Frie, A. L., Jia, X., Kim, T., Miller, L. T., et al. (2022a). KGML-ag: A modeling framework of knowledge-guided machine learning to simulate agroecosystems: a case study of estimating N2O emission using data from mesocosm experiments. *Geoscientific Model Development, 15*(7), 2839–2858.

Liu, L., Zhou, W., Guan, K., Peng, B., Xu, S., Tang, J., Zhu, Q., Till, J., Jia, X., Jiang, C., et al. (2024). Knowledge-guided machine learning can improve carbon cycle quantification in agroecosystems. *Nature Communications, 15*(1), 357.

Liu, Z., & Tang, H. (2023). Learning sparse geometric features for building segmentation from low-resolution remote-sensing images. *Remote Sensing, 15*(7), 1741.

Liu, Z., Tang, H., Feng, L., & Lyu, S. (2023b). China building rooftop area: the first multi-annual (2016–2021) and high-resolution (2.5 m) building rooftop area dataset in China derived with super-resolution segmentation from sentinel-2 imagery. *Earth System Science Data, 15*(8), 3547–3572.

Liu, Z., Tang, H., & Huang, W. (2022b). Building outline delineation from VHR remote sensing images using the convolutional recurrent neural network embedded with line segment information. *IEEE Transactions on Geoscience and Remote Sensing, 60*, 1–13.

Luca, M., Campedelli, G. M., Centellegher, S., Tizzoni, M., & Lepri, B. (2023). Crime, inequality and public health: a survey of emerging trends in urban data science. *Frontiers in Big Data, 6*, 1124526.

Mac Aodha, O., Cole, E., & Perona, P. (2019). Presence-only geographical priors for fine-grained image classification. In *ICCV 2019* (pp. 9596–9606).

Mai, G. et al. (2024a). SRL: Towards a general-purpose framework for spatial representation learning. In *ACM SIGSPATIAL 2024*.

Mai, G., Huang, W., Cai, L., Zhu, R., & Lao, N. (2022). Narrative cartography with knowledge graphs. *Journal of Geovisualization and Spatial Analysis, 6*(1), 4.

Mai, G., Huang, W., Sun, J., Song, S., Mishra, D., Liu, N., Gao, S., Liu, T., Cong, G., Hu, Y., et al. (2024b). On the opportunities and challenges of foundation models for GeoAI (vision paper). *ACM Transactions on Spatial Algorithms and Systems, 10*(2), 1–46.

Mai, G., Janowicz, K., Cai, L., Zhu, R., Regalia, B., Yan, B., Shi, M., & Lao, N. (2020a). SE-KGE: A location-aware knowledge graph embedding model for geographic question answering and spatial semantic lifting. *Transactions in GIS, 24*(3), 623–655.

Mai, G., Janowicz, K., He, C., Liu, S., & Lao, N. (2018). POIReviewQA: A semantically enriched poi retrieval and question answering dataset. In *Proceedings of the 12th Workshop on Geographic Information Retrieval* (pp. 1–2).

Mai, G., Janowicz, K., Prasad, S., Shi, M., Cai, L., Zhu, R., Regalia, B., & Lao, N. (2020b). Semantically-enriched search engine for geoportals: A case study with ArcGIS online. *AGILE: GIScience Series, 1*, 13.

Mai, G., Janowicz, K., Zhu, R., Cai, L., & Lao, N. (2021). Geographic question answering: challenges, uniqueness, classification, and future directions. *AGILE: GIScience Series, 2*, 8.

Mai, G., Jiang, C., Sun, W., Zhu, R., Xuan, Y., Cai, L., Janowicz, K., Ermon, S., & Lao, N. (2023a). Towards general-purpose representation learning of polygonal geometries. *GeoInformatica, 27*(2), 289–340.

Mai, G., Lao, N., He, Y., Song, J., & Ermon, S. (2023b). CSP: Self-supervised contrastive spatial pre-training for geospatial-visual representations. In *ICML 2023*. PMLR.

Mai, G., Lao, N., Sun, W., Ma, Y., Song, J., Meng, C., Ma, H., Rao, J., Li, Z., & Ermon, S. (2023c). SSIF: Learning continuous image representation for spatial-spectral super-resolution. *arXiv preprint arXiv:2310.00413*.

Mai, G., Xuan, Y., Zuo, W., He, Y., Song, J., Ermon, S., Janowicz, K., & Lao, N. (2023d). Sphere2vec: A general-purpose location representation learning over a spherical surface for large-scale geospatial predictions. *ISPRS Journal of Photogrammetry and Remote Sensing, 202*, 439–462.

Materia, S., García, L. P., van Straaten, C., O, S., Mamalakis, A., Cavicchia, L., Coumou, D., de Luca, P., Kretschmer, M., & Donat, M. (2024). Artificial intelligence for climate prediction of extremes: State of the art, challenges, and future perspectives. In *Wiley Interdisciplinary Reviews: Climate Change* (p. e914).

McKenzie, G., & Zhang, H. (2023). Platial k-anonymity: Improving location anonymity through temporal popularity signatures. In *GIScience 2023*. Schloss-Dagstuhl-Leibniz Zentrum für Informatik.

McKenzie, G., Zhang, H., & Gambs, S. (2023). Privacy and ethics in GeoAI. In *Handbook of Geospatial Artificial Intelligence* (pp. 388–405). CRC Press.

Miranda-Pascual, À., Guerra-Balboa, P., Parra-Arnau, J., Forné, J., & Strufe, T. (2023). SoK: differentially private publication of trajectory data. In *Proceedings on Privacy Enhancing Technologies*.

Punjani, D., Singh, K., Both, A., Koubarakis, M., Angelidis, I., Bereta, K., Beris, T., Bilidas, D., Ioannidis, T., Karalis, N., et al. (2018). Template-based question answering over linked geospatial data. In *Proceedings of the 12th workshop on geographic information retrieval* (pp. 1–10).

Purves, R. S., Clough, P., Jones, C. B., Hall, M. H., Murdock, V., et al. (2018). Geographic information retrieval: Progress and challenges in spatial search of text. *Foundations and Trends® in Information Retrieval, 12*(2-3), 164–318.

Qi, Y., Mai, G., Zhu, R., & Zhang, M. (2023). EVKG: An interlinked and interoperable electric vehicle knowledge graph for smart transportation system. *Transactions in GIS, 27*(4), 949–974.

Ramya, T., Charan, S. S., Amith, S., Kumar, T. A., Dondy, S., & Pallavi, L. (2024). AI-powered climate change adaptation and mitigation. In *2024 7th International Conference on Circuit Power and Computing Technologies (ICCPCT)* (Vol. 1, pp. 1325–1330). IEEE.

Rao, J., Gao, S., Kang, Y., & Huang, Q. (2020). LSTM-TrajGAN: A deep learning approach to trajectory privacy protection. In *GIScience 2021*. Schloss-Dagstuhl-Leibniz Zentrum für Informatik.

Rao, J., Gao, S., Li, M., & Huang, Q. (2021). A privacy-preserving framework for location recommendation using decentralized collaborative machine learning. *Transactions in GIS, 25*(3), 1153–1175.

Rao, J., Gao, S., Mai, G., & Janowicz, K. (2023a). Building privacy-preserving and secure geospatial artificial intelligence foundation models (vision paper). In *Proceedings of the 31st ACM International Conference on Advances in Geographic Information Systems* (pp. 1–4).

Rao, J., Gao, S., & Zhu, S. (2023b). CATS: Conditional adversarial trajectory synthesis for privacy-preserving trajectory data publication using deep learning approaches. *International Journal of Geographical Information Science, 37*(12), 2538–2574.

Read, J. S., Jia, X., Willard, J., Appling, A. P., Zwart, J. A., Oliver, S. K., Karpatne, A., Hansen, G. J., Hanson, P. C., Watkins, W., et al. (2019). Process-guided deep learning predictions of lake water temperature. *Water Resources Research, 55*(11), 9173–9190.

Regalia, B., Janowicz, K., Mai, G., Varanka, D., & Usery, E. L. (2018). GNIS-LD: Serving and visualizing the geographic names information system gazetteer as linked data. In *ESWC 2018* (pp. 528–540). Springer.

Reichstein, M., Camps-Valls, G., Stevens, B., Jung, M., Denzler, J., Carvalhais, N., & Prabhat, F. (2019). Deep learning and process understanding for data-driven earth system science. *Nature, 566*(7743), 195–204.

Ruiz, I. et al. (2023). Combined artificial intelligence, sustainable land management, and stakeholder engagement for integrated landscape management in Mediterranean watersheds. *Environmental Science & Policy, 145*, 217–227.

Rußwurm, M., Klemmer, K., Rolf, E., Zbinden, R., & Tuia, D. (2023). Geographic location encoding with spherical harmonics and sinusoidal representation networks. In *The Twelfth International Conference on Learning Representations*.

Scheider, S., Nyamsuren, E., Kruiger, H., & Xu, H. (2021). Geo-analytical question-answering with GIS. *International Journal of Digital Earth, 14*(1), 1–14.

Shao, J., Wang, Y., Gao, H., Shi, B., Shen, H., & Cheng, X. (2023). AsyLink: User identity linkage from text to geo-location via sparse labeled data. *Neurocomputing, 515*, 174–184.

Sharma, E., Deo, R. C., Prasad, R., & Parisi, A. V. (2020). A hybrid air quality early-warning framework: An hourly forecasting model with online sequential extreme learning machines and empirical mode decomposition algorithms. *Science of the Total Environment, 709*, 135934.

Sun, J., Innanen, K. A., & Huang, C. (2021). Physics-guided deep learning for seismic inversion with hybrid training and uncertainty analysis. *Geophysics, 86*(3), R303–R317.

Taylor, K. E., Stouffer, R. J., & Meehl, G. A. (2012). An overview of CMIP5 and the experiment design. *Bulletin of the American Meteorological Society, 93*(4), 485–498.

Tripathy, K. P., Mukherjee, S., Mishra, A. K., Mann, M. E., & Williams, A. P. (2023). Climate change will accelerate the high-end risk of compound drought and heatwave events. *Proceedings of the National Academy of Sciences, 120*(28), e2219825120.

Wang, C., Hou, Y., & Barth, M. (2020a). Data-driven multi-step demand prediction for ride-hailing services using convolutional neural network. In *Advances in Computer Vision: Proceedings of the 2019 Computer Vision Conference (CVC)* (Vol. 21, pp. 11–22). Springer.

Wang, J., Hu, Y., & Joseph, K. (2020b). Neurotpr: A neuro-net toponym recognition model for extracting locations from social media messages. *Transactions in GIS, 24*(3), 719–735.

Wang, Y. et al. (2019). A global map of emission clumps for future monitoring of fossil fuel CO 2 emissions from space. *Earth System Science Data, 11*(2), 687–703.
Wu, N., Cao, Q., Wang, Z., Liu, Z., Qi, Y., Zhang, J., Ni, J., Yao, X., Ma, H., Mu, L., et al. (2024). TorchSpatial: A location encoding framework and benchmark for spatial representation learning. *arXiv preprint arXiv:2406.15658*.
Xie, K., Liu, P., Zhang, J., Han, D., Wang, G., & Shen, C. (2021a). Physics-guided deep learning for rainfall-runoff modeling by considering extreme events and monotonic relationships. *Journal of Hydrology, 603*, 127043.
Xie, Y., Chen, W., He, E., Jia, X., Bao, H., Zhou, X., Ghosh, R., & Ravirathinam, P. (2023a). Harnessing heterogeneity in space with statistically-guided meta-learning. *Knowledge and Information Systems, 65*, 2699–2729.
Xie, Y. et al. (2023b). Geo-foundation models: Reality, gaps and opportunities. In *ACM SIGSPATIAL 2023* (pp. 1–4).
Xie, Y., He, E., Jia, X., Bao, H., Zhou, X., Ghosh, R., & Ravirathinam, P. (2021b). A statistically-guided deep network transformation and moderation framework for data with spatial heterogeneity. In *ICDM 2021* (pp. 767–776). IEEE.
Xie, Y., He, E., Jia, X., Chen, W., Skakun, S., Bao, H., Jiang, Z., Ghosh, R., & Ravirathinam, P. (2022). Fairness by "where": A statistically-robust and model-agnostic bi-level learning framework. *AAAI 2022, 36*(11), 12208–12216.
Xie, Y., Jia, X., Bao, H., Zhou, X., Yu, J., Ghosh, R., & Ravirathinam, P. (2021c). Spatial-net: A self-adaptive and model-agnostic deep learning framework for spatially heterogeneous datasets. In *ACM SIGSPATIAL 2021* (pp. 313–323).
Xie, Y., Jia, X., Chen, W., & He, E. (2023c). Heterogeneity-aware deep learning in space: Performance and fairness. In *Handbook of Geospatial Artificial Intelligence* (pp. 151–176). CRC Press.
Xie, Y., Li, Z., Bao, H., Jia, X., Xu, D., Zhou, X., & Skakun, S. (2023d). Auto-CM: Unsupervised deep learning for satellite imagery composition and cloud masking using spatio-temporal dynamics. In *AAAI 2023*.
Xu, H., Nyamsuren, E., Scheider, S., & Top, E. (2023). A grammar for interpreting geo-analytical questions as concept transformations. *International Journal of Geographical Information Science, 37*(2), 276–306.
Yan, B., Janowicz, K., Mai, G., & Zhu, R. (2019). A spatially explicit reinforcement learning model for geographic knowledge graph summarization. *Transactions in GIS, 23*(3), 620–640.
Yan, X., Ai, T., Yang, M., Tong, X., & Liu, Q. (2022). A graph deep learning approach for urban building grouping. *Geocarto International, 37*(10), 2944–2966.
Yang, L., Liu, X., Zhu, W., Zhao, L., & Beroza, G. C. (2022). Toward improved urban earthquake monitoring through deep-learning-based noise suppression. *Science Advances, 8*(15), eabl3564.
Yang, X., Xu, Y., Kuang, L., Wang, Z., Gao, H., & Wang, X. (2021). An information fusion approach to intelligent traffic signal control using the joint methods of multiagent reinforcement learning and artificial intelligence of things. *IEEE Transactions on Intelligent Transportation Systems, 23*(7), 9335–9345.
Yu, D., Hu, Y., Li, Y., & Zhao, L. (2024). Polygongnn: Representation learning for polygonal geometries with heterogeneous visibility graph. In *KDD 2024* (pp. 4012–4022).
Yu, W., & Chen, Y. (2022). Filling gaps of cartographic polylines by using an encoder–decoder model. *International Journal of Geographical Information Science, 36*(11), 2296–2321.
Zelle, J. M., & Mooney, R. J. (1996). Learning to parse database queries using inductive logic programming. In *Proceedings of the National Conference on Artificial Intelligence* (pp. 1050–1055).
Zhang, Y. et al. (2023a). Geogpt: Understanding and processing geospatial tasks through an autonomous GPT. *arXiv preprint arXiv:2307.07930*.
Zhang, Y., Guo, W., Wu, C., Li, W., & Tao, R. (2023b). FANET: An arbitrary direction remote sensing object detection network based on feature fusion and angle classification. *IEEE Transactions on Geoscience and Remote Sensing, 61*, 1–11.

Zhang, Y., Song, S., Li, X., Gao, S., & Raubal, M. (2024). Leveraging context-adjusted nighttime light data for socioeconomic explanations of global urban resilience. *Sustainable Cities and Society, 114*, 105739.

Zheng, Y. (2019). *Urban computing*. MIT Press.

Zheng, Y., Capra, L., Wolfson, O., & Yang, H. (2014). Urban computing: Concepts, methodologies, and applications. *ACM Transactions on Intelligent Systems and Technology (TIST), 5*(3), 1–55.

Zheng, Y., Liu, P., Qian, L., Qin, S., Liu, X., Ma, Y., & Cheng, G. (2022). Recognition and depth estimation of ships based on binocular stereo vision. *Journal of Marine Science and Engineering, 10*(8), 1153.

Zheng, Z., Zhong, Y., Zhang, L., & Ermon, S. (2024). Segment any change. *arXiv preprint arXiv:2402.01188.*

Zhou, B., Chen, L., Zhao, S., Zhou, F., Li, S., & Pan, G. (2023). Spatio-temporal analysis of urban crime leveraging multisource crowdsensed data. *Personal and Ubiquitous Computing* (pp. 1–14).

Zhou, Q., Zhu, Z., Xian, G., & Li, C. (2022). A novel regression method for harmonic analysis of time series. *ISPRS Journal of Photogrammetry and Remote Sensing, 185*, 48–61.

Zhou, Y., & Tuzel, O. (2018). VoxelNet: End-to-end learning for point cloud based 3d object detection. In *CVPR 2018* (pp. 4490–4499).

Zhu, R. (2024). Geospatial knowledge graphs. In *International Encyclopedia of Geography: People, the Earth, Environment and Technology*.

Zhu, Y., Ni, K., Li, X., Zaman, A., Liu, X., & Bai, Y. (2024). Artificial intelligence aided crowd analytics in rail transit station. *Transportation Research Record, 2678*(2), 481–492.

Zhu, Y., Ye, Y., Zhang, S., Zhao, X., & Yu, J. (2023). DiffTraj: Generating GPS trajectory with diffusion probabilistic model. *NeurIPS 2023, 36*, 65168–65188.

Zhu, Z., & Woodcock, C. E. (2012). Object-based cloud and cloud shadow detection in Landsat imagery. *Remote Sensing of Environment, 118*, 83–94.

Zou, X., Yan, Y., Hao, X., Hu, Y., Wen, H., Liu, E., Zhang, J., Li, Y., Li, T., Zheng, Y., et al. (2025). Deep learning for cross-domain data fusion in urban computing: Taxonomy, advances, and outlook. *Information Fusion, 113*, 102606.

Chapter 3
Pillars of GeoAI in Human Geography

Yue Lin and Junghwan Kim

Abstract This chapter examines the three foundational pillars—data, algorithms, and computational infrastructure—that support the emergence and growth of geospatial artificial intelligence (GeoAI) in human geography. We trace their historical evolution, from the era of quantitative geography and early computational methods to the recent surge in high-performance computing and advanced machine-learning models. Drawing on examples such as street-view image analysis for urban built environment auditing, we illustrate how these pillars enable more efficient and nuanced spatial analyses and decision-making processes. While GeoAI tools offer powerful means to uncover previously hidden patterns and improve our understanding of socio-spatial phenomena, they also prompt critical reflection. As GeoAI increasingly engages with sensitive issues related to bias, privacy, representational fairness, and the environmental costs of training large models, human geographers are examining its theoretical and ethical implications through critical perspectives and social theories. Recognizing the potential synergies—and the tensions—between innovative computational methods and critical inquiry fosters a more inclusive, responsible, and socially engaged framework for integrating GeoAI into human geography research and practice.

Keywords GeoAI · Human Geography · Spatial Big Data · Critical Geographies

3.1 Introduction

Human geography has a history of engaging with new technological advancements within the broader discipline. Since the late twentieth century, despite debates,

Y. Lin (✉)
Center for Spatial Data Science, University of Chicago, Chicago, IL, USA
e-mail: liny2@uchicago.edu

J. Kim
Department of Geography, Virginia Tech, Blacksburg, VA, USA
e-mail: junghwankim@vt.edu

© The Author(s), under exclusive license to Springer Nature Switzerland AG 2025
X. Huang et al. (eds.), *GeoAI and Human Geography*, Springer Geography,
https://doi.org/10.1007/978-3-031-87421-5_3

the automation and new insights introduced by geographical information systems (GIS) have significantly transformed human geography research and ushered in new modes of scientific discovery (Dobson, 1983). Over the past 5 years, geospatial artificial intelligence (GeoAI) has emerged as a major disruptive technology in human geography, integrating spatial, temporal, and place-based components into artificial intelligence methods (Janowicz et al., 2020; Gao et al., 2023). Compared to previous technologies, such as GIS, which rely on human-coded rules for automation, GeoAI has the ability to learn and adapt from training data for autonomous decision-making, demonstrating the potential for human-like intelligence in human geography research.

Data, algorithms, and computational infrastructure are the three pillars that form the foundation of GeoAI (Hu et al., 2019). Data are used to train GeoAI models and for analysis. Algorithms include key methodologies such as transformers and graph neural networks, through which data are processed, analyzed, and interpreted. Computational infrastructure includes the hardware, software, and technological systems that support the functioning and execution of GeoAI applications. Driven by these three pillars, GeoAI is increasingly being integrated into human geography in two significant ways. The first one uses GeoAI as a tool to enhance spatial analysis, modeling, and visualization, covering topics across all subfields of human geography, such as health, culture, economic, and political geography (Wang et al., 2024). In this context, leveraging GeoAI enables researchers to uncover previously hidden patterns and trends and allows for more nuanced and comprehensive analyses of spatial phenomena, which may not be observed through classical methods such as regression analysis. The second approach studies the theoretical and philosophical foundations of GeoAI within human geography, informed by critical theories such as feminism, poststructuralism, and postcolonialism (Janowicz et al., 2022; Janowicz, 2023; Lin & Zhao, 2025). This line of research critically examines the implications of GeoAI, questioning how it might reinforce or challenge existing power structures and inequalities (Kim et al., 2024a).

These engagements of GeoAI have established the foundations for integrating new technologies with human geography. As GeoAI continues to reshape the contemporary landscape of geographic inquiry, it is important to reflect on its growing integration. This chapter aims to trace the historical development of GeoAI and its varied engagement within human geography. In the remainder of this chapter, we will begin by examining the three pillars of GeoAI—data, algorithms, and computational infrastructure—through a historical lens, as well as how they have shaped the current landscape of GeoAI. We will then examine how these pillars of GeoAI are reflected in contemporary human geography research: both as tools that help researchers gain insights and as subjects for being critically examined at a fundamental level.

3.2 The Pillars of GeoAI: A Historical Lens

In this section, we will review how the three pillars of GeoAI—data, algorithms, and computational infrastructure—have evolved within geography and shaped the current landscape of GeoAI.

3.2.1 Data

Human geography has long been rooted in qualitative data and methodologies, such as ethnography, interviews, and content analysis. However, in the mid-twentieth century, the field experienced a shift with the advent of quantitative geography, which, driven by the quantitative revolution in geography, aimed to respond to the growing demand for systematic, empirical, and data-driven approaches to studying geographic phenomena (Berry, 1993). As a result, the use of quantitative data in human geography witnessed a significant increase, with surveys such as censuses becoming key sources of information. In the United States, for example, a notable development occurred in 1960 when the Census Bureau first mailed a combined population and housing questionnaire to all urban residents, making census data a primary resource for studying large-scale, nationwide geographic phenomena.

Spatial big data have become prevalent since the early twenty-first century with the rise of satellite technology, GPS-enabled devices, social media platforms, and sensor networks (Miller, 2010). Compared to survey data, spatial big data provide fine-grained geographic and temporal information, significantly impacting various subfields of human geography such as health, transportation, and environmental geography. For example, mobile phone data have been used to track disease spread patterns and inform prevention strategies (Zhou et al., 2020b), GPS and traffic sensor data are widely used to understand traffic patterns and human movement flow (Torre-Bastida et al., 2018), and satellite imagery and sensor networks provide crucial information on changing climate patterns and natural resources (Huang & Wang, 2020). The growing volume and diversity of spatial big data require more sophisticated methodological and technical approaches to extract meaningful insights. GeoAI has emerged as one such approach, enabling effective processing, analysis, visualization, and decision-making support of spatial big data (Li, 2020).

Besides being the subject of analysis by GeoAI, spatial big data serve as a critical source for training GeoAI models, enabling them to learn and develop their analytical capabilities. For example, recent studies have shown how satellite imagery, combined with other data sources such as historical topographic maps and 3D elevation models, is prepared to train models for automatically identifying natural features in new, unseen images (Arundel et al., 2020; Li et al., 2023). In addition, large-scale datasets comprising Google street-view images and other camera system data are developed to train AI models for analyzing urban environments and street scenes (Cordts et al., 2016). These training datasets enable GeoAI models to effectively interpret and classify a wide range of geographic and urban features, improving their performance and reliability in real-world applications.

3.2.2 Algorithms

The use of computational algorithms in geography can be traced back to the pursuit of automated geography, which emerged from the quantitative revolution and aimed to integrate new analytical methods and computer technology to address geographic problems (Dobson, 1983). Early algorithms applied in geography focused on computationally feasible tasks due to technological limitations at the time. In 1959, for example, Waldo Tobler introduced the "Map In-Map Out" (MIMO) system, which used explicit rules coded as algorithms to automate the transformation of input data (map in) into maps (map out) (Tobler, 1959). Since the 1990s, an emerging field of geocomputation has explored more complex computational algorithms for spatial analysis. Numerous methods have been developed to automate spatial data collection and processing, as well as for advanced spatial analysis such as geostatistical modeling (Curran & Atkinson, 1998), geometric processing (Xiao, 2016), and spatial optimization (Tong & Murray, 2012). The emergence of geocomputation marks a trend toward increased automation and reduced human intervention in various aspects of geographic studies.

GeoAI, which emerged in the 1980s but surged in prominence since the late 2010s, represents a significant advancement over previous computational algorithms in geography because of its ability to simulate human decision-making and potentially surpass human intelligence without relying on explicitly programmed rules (Openshaw & Openshaw, 1997; Gao et al., 2023). Foundational GeoAI algorithms include convolutional neural networks (CNNs), recurrent neural networks (RNNs), transformers, graph neural networks (GNNs), and other advanced developments in artificial intelligence that can be applied to geographic problems. In addition, there is an increasing focus on spatially explicit GeoAI models that incorporate locational information and spatial effects such as spatial dependence and heterogeneity (Janowicz et al., 2020; Jin et al., 2023; Liu & Biljecki, 2022). More recently, the development of foundation models such as generative pre-trained transformers (GPT) and DALL·E has accelerated a paradigm shift in GeoAI, moving from task-specific models to general-purpose models that can be adapted to perform geographic question answering, generate maps, and more (Xie et al., 2023).

3.2.3 Computing Infrastructure

The initial use of computer programs in geography is closely related to the advent of personal computers (PC) in the mid-twentieth century. PCs provided geographers with affordable and accessible computing power, allowing them to conduct spatial analysis and mapping tasks without relying on mainframe computers. Tobler's MIMO system is one example of early automated geography programs on PC. Another important development was by Harvard's Laboratory for Computer Graphics and Spatial Analysis, which developed the Synagraphic Mapping Package (SYMAP) in 1963, among others, for computer-assisted mapping and spatial analysis (Chrisman, 2006). These early programs laid the groundwork for the development of more widely used PC-based GIS that emerged in the 1970s and beyond, including GRASS GIS (1984), MapInfo (1986), ArcView (1990s), and QGIS (2002).

The rapid development of GeoAI in recent years has been significantly driven by the increasing availability of high-performance computing (HPC). Compared to PC, which typically have one or a few processors and may have a single GPU, HPC is often built as clusters comprising multiple processors and GPUs connected through high-speed networks. These clusters are designed to perform complex calculations and process large datasets quickly, significantly reducing the often-lengthy training times for GeoAI models. HPC has already been integrated into GIS, forming what is known as CyberGIS when combined with data storage, networks, software, and collaboration tools (Wang, 2010). Many commercial platforms now offer HPC solutions for GeoAI. For example, in 2018, Microsoft and Esri launched the GeoAI Data Science Virtual Machine (DSVM) as part of the Data Science Virtual Machine/Deep Learning Virtual Machine family of products on Azure. In addition, Amazon's SageMaker supports GeoAI capabilities by providing access to geospatial data sources, specialized processing operations, pre-trained machine-learning models, and built-in visualization tools to run GeoAI applications faster and at scale.

3.3 Applying GeoAI in Human Geography: A Case Study

Urban geography is an important subfield of human geography that examines the various interactions between built environment and human behaviors in urban environments. In this section, we present a case study where GeoAI is applied to urban geography research to demonstrate the critical role of the three pillars of GeoAI—data, algorithms, and computing infrastructure. Notably, we illustrate how geographers have utilized deep learning-based computer vision models to efficiently audit urban built environment elements, such as sidewalks, roads, benches, trees, traffic signals, and bus shelters (Kim et al., 2024b). Information obtained from auditing is foundational for examining citizens' perception of the useability of urban built environments, such as walkability (i.e., how easily an individual can walk on a specific sidewalk segment), street safety, and so on (Biljecki & Ito, 2021).

Traditionally, built environment auditing is considered a resource-intensive process because it is done manually by human surveyors. Human surveyors are asked to visit the field and visually observe the built environment by completing a questionnaire that contains a list of guiding questions to evaluate built environments

(Hanibuchi et al., 2019). For instance, when assessing sidewalks' walkability, questions include whether climate-comfort elements (e.g., trees, grasses) are located, whether amenities (e.g., benches, trash bins) are presented, maintenance quality, and so on. Human surveyors must complete the questionnaire, which would be burdensome. However, recent advances in GeoAI provided researchers with an innovative means to overcome the limitations of conventional methods (Koo et al., 2022). In what follows, we describe how advances in each pillar of GeoAI—data, algorithms, and computing infrastructure—contribute to applying GeoAI to urban built environment auditing tasks.

Regarding data, street-view images have positively contributed to GeoAI applications in urban built environment auditing. Street-view images are photos that capture eye-level perspectives of street landscape. A famous example is Google Maps street-view images collected and maintained by Google (Anguelov et al., 2010). A special camera mounted on a vehicle on roads takes panoramic photographs of street landscapes. Researchers can readily obtain street-view images through an application programming interface (API) at reasonable costs. Although Google street-view images are valuable data for built environment auditing as they provide detailed visual information, recent studies have criticized several limitations of Google-based images (Curtis et al., 2013; Fry et al., 2020; Ki et al., 2023; Kim & Jang, 2023; Rui, 2023).

First, images are taken from vehicles on streets, which do not necessarily reflect street landscapes from pedestrians' perspectives (Ki et al., 2023; Rui, 2023). Built environment auditing based on vehicles' perspectives might distort our true understanding of how individuals interact with built environment elements. Second, since images are collected and maintained by Google Maps, images may be unavailable and outdated, particularly for disadvantaged neighborhoods (Fry et al., 2020; Kim & Jang, 2023), such as rural towns and small-and medium-sized cities, compared to big cities in large metropolitan areas. Researchers have explored alternative methods, such as Mapiliary data (Biljecki & Ito, 2021; Mahabir et al., 2020), which is crowdsourced data that users voluntarily contribute. Another alternative method is collecting original, up-to-date street-view images using a low-cost camera mounted on a mobility scooter that can be operated on sidewalks (Fig. 3.1), developed by the Smart Cities for Good at Virginia Tech. This new data collection method is affordable and overcomes severe limitations of commercial street-view images, which eventually provides more reliable and comprehensive data for GeoAI-based urban built environment auditing than commercial methods.

In addition to data, algorithm advances have contributed to GeoAI applications in built environment auditing. Specifically, deep learning-based computer vision models are compelling tools for researchers to quickly process large sets of street-view images for conducting built environment audits (Kang et al., 2020). Without advanced algorithms, auditing would take considerable time. For instance, an in-person survey of approximately 3000 bus stops in San Francisco took about 3 months (Moran, 2022). This labor-intensive process can be efficiently handled by deep learning-based computer vision models, such as detection and semantic segmentation models, which can identify specific objects from a given street-view

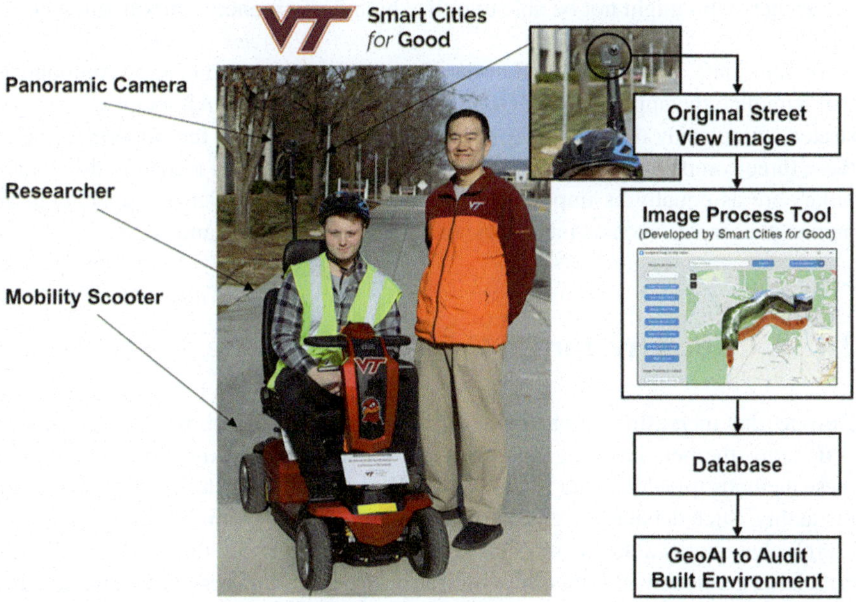

Fig. 3.1 A mobility scooter with GoPro camera to collect up-to-date, original, and pedestrian-centric street-view images. (Figure Source: Junghwan Kim)

image. Because of the efficiency and flexibility of those deep learning models, geographers have applied the models to various topics in built environment auditing, such as detecting traffic signals (Li et al., 2022), bus shelters (Kim et al., 2024b), walkability (Koo et al., 2022), and green space exposures (Gyanwali et al., 2024; Jang & Kim, 2025a). Recently, researchers have explored an innovative approach: integrating multimodal Large Language Models (LLMs) with built environment auditing (Akinboyewa et al., 2024; Jang & Kim, 2025b). Compared to deep learning-based methods that ask researchers to know advanced programming skills, using commercial LLMs (e.g., OpenAI's ChatGPT, Google's Gemini) would broaden the application of GeoAI in built environment auditing research because novice users can readily utilize conversation-based LLMs (Jang & Kim, 2025b).

The last pillar is computing infrastructure. With proper computing infrastructure, researchers can execute advanced algorithms by utilizing emerging data. It is not surprising to say that advances in computing infrastructure play a critical role in GeoAI applications in built environment auditing. Specifically, the development of relatively affordable GPUs has accelerated the application of deep learning methods in large-scale auditing tasks that require processing millions of street-view images. Ample memory, storage, and high-speed wireless networks are equally essential in promoting GeoAI applications in human geography as they enable researchers to swiftly access, transfer, and store big spatial data. The recent development of cloud computing services (e.g., Amazon Web Services) also provides a flexible means for researchers who might not be able to utilize high-performance computing resources locally.

In sum, data, algorithms, and computing infrastructure are crucial components that stimulate the application of GeoAI in human geography. Although we demonstrated a case study of built environment auditing to highlight the important role of these three components for GeoAI-powered human geography research, these three pillars are as equally as important when it comes to other GeoAI applications in human geography, such as inequality, mobility, public health, and so on.

3.4 A Critical (re)Turn of GeoAI?

Despite the increasing application of new quantitative technologies in human geography, the field has long examined and sometimes questioned the adequacy of these methods for addressing issues in critical geographies, such as social justice and inequality. Such debates have been ongoing in the development of GIS. Schuurman (2000) characterized the critiques of GIS since the 1990s into three waves: initially criticizing its positivism, then moving toward cooperation with GIS scholars, and eventually turning to more nuanced analyses of the effects of GIS. As GeoAI emerges as a new technological advancement in geography, it raises the question of how a critical lens evolves in its context. Does GeoAI face similar scrutiny, and how should it adapt to address the same concerns of social justice and inequality?

The ethical considerations surrounding the three pillars of GeoAI—data, algorithms, and computational infrastructure—have been a focus of attention since its inception. Unlike the early development of critical GIS, where there was antagonism between GIS and critical geographies, GeoAI scholars are actively reflecting on and contributing to the integration of GeoAI with critical sensibilities and advocating for its use for fostering social and political change (Lin & Zhao, 2025). For example, Janowicz et al. (2022) discuss issues related to data representation and bias, as well as the black-boxing nature of algorithms. In addition, Janowicz (2023) discusses the philosophical foundations and research ethics of GeoAI, highlighting key issues such as sustainability, which relates to the environmental impact of the computing infrastructure used to train AI models, including resource consumption and carbon dioxide emissions.

In a similar vein, other research also explores ethical considerations about the data, algorithms, and computing infrastructure involved in GeoAI. This includes the privacy implications of data and algorithms (McKenzie et al., 2023; Lin & Xiao, 2023; Lin, 2024; Wang et al., 2022), the potential misuse of algorithms for deepfake geography and misinformation (Zhao et al., 2021), and biases and inaccuracies in AI-driven geographic question answering (Atkins et al., 2024; Kim et al., 2024a; Lin & Grekousis, 2025; Renshaw et al., 2025). For instance, Kim et al. (2024a) demonstrated that ChatGPT (GPT-4) is more limited in providing local-specific information about environmental justice for poor and rural counties than rich and urban counties in the United States, implying potential disparities in training data between urban and rural regions. Considering recently developed large language models (LLMs) primarily rely on open-source data (e.g., Wikipedia pages) that are already not free from geographic biases and misrepresentations (Graham et al., 2014), advances in GeoAI algorithms would be prone to errors and inefficiencies for specific regions of the world if those issues in data are not mitigated. To sum up, existing studies demonstrate how GeoAI, as a new quantitative technology in geography, is increasingly aligned with the premises of critical geographies, which prompts reflection on its positionality and effects on society.

3.5 Concluding Remarks

In this chapter, we reviewed the evolution of the three pillars of GeoAI—data, algorithms, and computing infrastructure—and how they are integrated into human geography. We examined their role of not only enhancing the methodologies for different subfields of human geography, but also serving as subjects for critical examination by critical geographers. Within the broader discipline of geography, there has long been debate about whether new quantitative technologies can effectively address social justice and inequality, while this persistent critical-quantitative divide is also argued to be ultimately detrimental to the field (Kwan & Schwanen, 2009). In the current development of GeoAI, we observe its use not only to complement or even enhance traditional methods in human geography through enriched quantitative analysis but also as a focal point for ethical discussions. These discussions infuse GeoAI's pillars with critical sensibility and emphasize the importance of social responsibility in its application. This dual approach of recognizing the potential and positive impact of GeoAI while advocating for its cautious and reflective use is essentially helpful to embracing new technological disruptions in the field.

References

Akinboyewa, T., Ning, H., Lessani, M. N., & Li, Z. (2024). Automated floodwater depth estimation using large multimodal model for rapid flood mapping. *Computational Urban Science, 4*(1), 12.

Anguelov, D., Dulong, C., Filip, D., Frueh, C., Lafon, S., Lyon, R., & Weaver, J. (2010). Google street view: Capturing the world at street level. *Computer, 43*(6), 32–38.

Arundel, S. T., Li, W., & Wang, S. (2020). GeoNat v1.0: A dataset for natural feature mapping with artificial intelligence and supervised learning. *Transactions in GIS, 24*(3), 556–572. https://doi.org/10.1111/tgis.12633

Atkins, C., Girgente, G., Shirzaei, M., & Kim, J. (2024). Generative AI tools can enhance climate literacy but must be checked for biases and inaccuracies. *Communications Earth & Environment, 5*(1), 226.

Berry, B. J. L. (1993). Geography's quantitative revolution: Initial conditions, 1954-1960. A personal memoir. *Urban Geography, 14*(5), 434–441. https://doi.org/10.2747/0272-3638.14.5.434

Biljecki, F., & Ito, K. (2021). Street view imagery in urban analytics and GIS: A review. *Landscape and Urban Planning, 215*, 104217.

Chrisman, N. (2006). *Charting the unknown: How computer mapping at Harvard became GIS*. Esri Press.

Cordts, M., Omran, M., Ramos, S., Rehfeld, T., Enzweiler, M., Benenson, R., Franke, U., Roth, S., & Schiele, B. (2016). The cityscapes dataset for semantic urban scene understanding. In *Proceedings of the IEEE Conference on Computer Vision and Pattern Recognition (CVPR)* (pp. 3213–3223). IEEE. https://doi.org/10.1109/CVPR.2016.350

Curran, P. J., & Atkinson, P. M. (1998). Geostatistics and remote sensing. *Progress in Physical Geography, 22*(1), 61–78.

Curtis, J. W., Curtis, A., Mapes, J., Szell, A. B., & Cinderich, A. (2013). Using google street view for systematic observation of the built environment: Analysis of spatio-temporal instability of imagery dates. *International Journal of Health Geographics, 12*, 1–10.

Dobson, J. E. (1983). Automated geography. *Professional Geographer, 35*(2), 135–143. https://doi.org/10.1111/j.0033-0124.1983.00135.x

Fry, D., Mooney, S. J., Rodríguez, D. A., Caiaffa, W. T., & Lovasi, G. S. (2020). Assessing Google street view image availability in Latin American cities. *Journal of Urban Health, 97*, 552–560.

Gao, S., Hu, Y., & Li, W. (2023). *Handbook of geospatial artificial intelligence*. Routledge.

Graham, M., Hogan, B., Straumann, R. K., & Medhat, A. (2014). Uneven geographies of user-generated information: Patterns of increasing informational poverty. *Annals of the Association of American Geographers, 104*(4), 746–764.

Gyanwali, S., Karki, S., Jang, K. M., Crawford, T., Zhang, M., & Kim, J. (2024). Implications for spatial non-stationarity and the neighborhood effect averaging problem (NEAP) in green inequality research: evidence from three states in the USA. *Journal of Geographical Systems*, 1–24. https://doi.org/10.1007/s10109-024-00448-x

Hanibuchi, T., Nakaya, T., & Inoue, S. (2019). Virtual audits of streetscapes by crowdworkers. *Health & Place, 59*, 102203.

Hu, Y., Gao, S., Lunga, D., Li, W., Newsam, S., & Bhaduri, B. (2019). GeoAI at ACM SIGSPATIAL. *SIGSPATIAL Special, 11*(2), 5–15. https://doi.org/10.1145/3377000.3377002

Huang, B., & Wang, J. (2020). Big spatial data for urban and environmental sustainability. *Geo-spatial Information Science, 23*(2), 125–140. https://doi.org/10.1080/10095020.2020.1754138

Jang, K. M., & Kim, J. (2025a). Social inequalities in green exposure in small-and medium-sized US cities: A mobility-based approach. *Social Science Research, 127*, 103142. https://doi.org/10.1016/j.ssresearch.2025.103142

Jang, K. M., & Kim, J. (2025b). Multimodal large language models as built environment auditing tools. *The Professional Geographer, 77*(1), 84–90. https://doi.org/10.1080/00330124.2024.2404894

Janowicz, K. (2023). Philosophical foundations of GeoAI: exploring sustainability, diversity, and bias in GeoAI and spatial data science. In *Handbook of geospatial artificial intelligence* (pp. 26–42). https://doi.org/10.1201/9781003308423-3

Janowicz, K., Gao, S., McKenzie, G., Hu, Y., & Bhaduri, B. (2020). GeoAI: Spatially explicit artificial intelligence techniques for geographic knowledge discovery and beyond. *International Journal of Geographical Information Science, 34*(4), 625–636. https://doi.org/10.1080/13658816.2019.1684500

Janowicz, K., Sieber, R., & Crampton, J. (2022). GeoAI, counter-AI, and human geography: A conversation. *Dialogues in Human Geography, 12*(3), 446–458. https://doi.org/10.1177/20438206221132510

Jin, C., Park, S., Ha, H. J., Lee, J., Kim, J., Hutchenreuther, J., & Nara, A. (2023). Predicting households' residential mobility trajectories with geographically localized interpretable model-agnostic explanation (GLIME). *International Journal of Geographical Information Science, 37*(12), 2597–2619.

Kang, Y., Zhang, F., Gao, S., Lin, H., & Liu, Y. (2020). A review of urban physical environment sensing using street view imagery in public health studies. *Annals of GIS, 26*(3), 261–275.

Ki, D., Park, K., & Chen, Z. (2023). Bridging the gap between pedestrian and street views for human-centric environment measurement: A GIS-based 3D virtual environment. *Landscape and Urban Planning, 240*, 104873.

Kim, J., & Jang, K. M. (2023). An examination of the spatial coverage and temporal variability of Google street view (GSV) images in small-and medium-sized cities: A people-based approach. *Computers, Environment and Urban Systems, 102*, 101956.

Kim, J., Lee, J., Jang, K. M., & Lourentzou, I. (2024a). Exploring the limitations in how ChatGPT introduces environmental justice issues in the United States: A case study of 3,108 counties. *Telematics and Informatics, 86*, 102085. https://doi.org/10.1016/j.tele.2023.102085

Kim, J., Park, J., Lee, J., & Jang, K. M. (2024b). Examining the socio-spatial patterns of bus shelters with deep learning analysis of street-view images: A case study of 20 cities in the US. *Cities, 148*, 104852.

Koo, B. W., Guhathakurta, S., & Botchwey, N. (2022). Development and validation of automated microscale walkability audit method. *Health & Place, 73*, 102733.

Kwan, M. P., & Schwanen, T. (2009). Quantitative revolution 2: The critical (re) turn. *The Professional Geographer, 61*(3), 283–291.

Li, W. (2020). GeoAI: Where machine learning and big data converge in GIScience. *Journal of Spatial Information Science, 20*(20), 71–77. https://doi.org/10.5311/JOSIS.2020.20.658

Li, X., Ning, H., Huang, X., Dadashova, B., Kang, Y., & Ma, A. (2022). Urban infrastructure audit: An effective protocol to digitize signalized intersections by mining street view images. *Cartography and Geographic Information Science, 49*(1), 32–49. https://doi.org/10.1080/15230406.2021.1992299

Li, W., Wang, S., Arundel, S. T., & Hsu, C. Y. (2023). GeoImageNet: A multi-source natural feature benchmark dataset for GeoAI and supervised machine learning. *GeoInformatica, 27*(3), 619–640. https://doi.org/10.1007/s10707-022-00476-z

Lin, Y. (2024). Moving beyond anonymity: Embracing a collective approach to location privacy in data-intensive geospatial analytics. *Environment and Planning F, 3*, 45. https://doi.org/10.1177/26349825231224029

Lin, Y., & Xiao, N. (2023). A computational framework for preserving privacy and maintaining utility of geographically aggregated data: A stochastic spatial optimization approach. *Annals of the American Association of Geographers, 113*(5), 1035–1056.

Lin, Y., & Grekousis, G. (2025). Can geodemographic clustering be fair? Incorporating social fairness in crisp and fuzzy approaches through a unified framework. *International Journal of Geographical Information Science*, 1–26. https://doi.org/10.1080/13658816.2024.2444525

Lin, Y., & Zhao, B. (2025). Posthuman Cartography? Rethinking Artificial Intelligence, Cartographic Practices, and Reflexivity. *Annals of the American Association of Geographers, 115*(3), 499–512. https://doi.org/10.1080/24694452.2024.2435920

Liu, P., & Biljecki, F. (2022). A review of spatially-explicit GeoAI applications in urban geography. *International Journal of Applied Earth Observation and Geoinformation, 112*, 102936. https://doi.org/10.1016/j.jag.2022.102936

Mahabir, R., Schuchard, R., Crooks, A., Croitoru, A., & Stefanidis, A. (2020). Crowdsourcing street view imagery: A comparison of mapillary and OpenStreetCam. *ISPRS International Journal of Geo-Information, 9*(6), 341.

McKenzie, G., Zhang, H., & Gambs, S. (2023). Privacy and ethics in GeoAI. In *Handbook of geospatial Artificial intelligence* (pp. 388–405). https://doi.org/10.1201/9781003308423-19

Miller, H. J. (2010). The data avalanche is here. Shouldn't we be digging? *Journal of Regional Science, 50*(1), 181–201. https://doi.org/10.1111/j.1467-9787.2009.00641.x

Moran, M. E. (2022). Are shelters in place? Mapping the distribution of transit amenities via a bus-stop census of San Francisco. *Journal of Public Transportation, 24*, 100023.

Openshaw, S., & Openshaw, C. (1997). *Artificial intelligence in geography*. Wiley.

Renshaw, A., Lourentzou, I., Lee, J., Crawford, T., & Kim, J. (2025). Comparing the Spatial Querying Capacity of Large Language Models: OpenAI's ChatGPT and Google's Gemini Pro. *The Professional Geographer*, 1–13. https://doi.org/10.1080/00330124.2024.2434455

Rui, J. (2023). Measuring streetscape perceptions from driveways and sidewalks to inform pedestrian-oriented street renewal in Düsseldorf. *Cities, 141*, 104472.

Tobler, W. R. (1959). Automation and cartography. *Geographical Review, 49*(4), 526–534. https://doi.org/10.1002/9780470979587.ch19

Tong, D., & Murray, A. T. (2012). Spatial optimization in geography. *Annals of the Association of American Geographers, 102*(6), 1290–1309. https://doi.org/10.1080/00045608.2012.685044

Torre-Bastida, A. I., Del Ser, J., Laña, I., Ilardia, M., Bilbao, M. N., & Campos-Cordobés, S. (2018). Big data for transportation and mobility: Recent advances, trends and challenges. *IET Intelligent Transport Systems, 12*(8), 742–755. https://doi.org/10.1049/iet-its.2018.5188

Wang, S. (2010). A CyberGIS framework for the synthesis of cyberinfrastructure, GIS, and spatial analysis. *Annals of the Association of American Geographers, 100*(3), 535–557. https://doi.org/10.1080/00045601003791243

Wang, J., Kim, J., & Kwan, M. P. (2022). An exploratory assessment of the effectiveness of geomasking methods on privacy protection and analytical accuracy for individual-level geospatial data. *Cartography and Geographic Information Science, 49*(5), 385–406.

Zhou, C., Su, F., Pei, T., Zhang, A., Du, Y., Luo, B., Cao, Z., Wang, J., Yuan, W., Zhu, Y., Song, C., Chen, J., Xu, J., Li, F., Ma, T., Jiang, L., Yan, F., Yi, J., Hu, Y., ... Xiao, H. (2020a). COVID-19: Challenges to GIS with big data. Geography and Sustainability, 1(1), 77–87. https://doi.org/10.1016/j.jag.2024.103734.

Xiao, N. (2016). *GIS algorithms*. SAGE Publications.

Xie, Y., Wang, Z., Mai, G., Li, Y., Jia, X., Gao, S., & Wang, S. (2023). Geo-foundation models: Reality, gaps and opportunities. In *The 31st ACM international conference on advances in geographic information systems (SIGSPATIAL '23)* (pp. 11–14). https://doi.org/10.1145/3589132.3625616

Zhao, B., Zhang, S., Xu, C., Sun, Y., & Deng, C. (2021). Deep fake geography? When geospatial data encounter artificial intelligence. *Cartography and Geographic Information Science, 48*(4), 338–352. https://doi.org/10.1080/15230406.2021.1910075

Zhou, C., Su, F., Pei, T., Zhang, A., Du, Y., Luo, B., Cao, Z., Wang, J., Yuan, W., Zhu, Y., Song, C., Chen, J., Xu, J., Li, F., Ma, T., Jiang, L., Yan, F., Yi, J., Hu, Y., ... Xiao, H. (2020b). COVID-19: Challenges to GIS with big data. Geography and Sustainability, 1(1), 77–87. https://doi.org/10.1016/j.geosus.2020.03.005.

Part II
Major GeoAI Advances That Reshape Human Geography

Chapter 4
Integrating Natural Language Processing in Human Geography

Bing Zhou, Binbin Lin, Lei Zou, Mingzheng Yang, Hao Tian, and Heng Cai

Abstract Human geography places attention on human activities and interactions with the physical world, which covers a wide range of topics such as cultural, economic, political, social, health, and transportation. The application of natural language processing (NLP) enables fine-grained information retrieval from an array of textual big data available nowadays. It offers a novel avenue to understand human perspectives, feedback, and behaviors at a larger scale and higher efficiency. This chapter illustrates how NLP can be integrated into human geography by providing two major ways of interaction: geoparsing from texts and knowledge discovery from geographic narratives. This book chapter demonstrates the use of NLP techniques in human geography through two case studies relevant to a global pandemic, COVID-19. The first one locates the vulnerable communities with geoparsing tools and identifies the urgent needs of them during strict lockdown by investigating the content of the online help requests using semantic analysis. The second case study explores the public emotions of different demographic groups by sentiment analysis and social media data and proposes solutions to alleviate the estimation bias due to the discrepancy between real-world populations and social media users. The challenges and future directions of leveraging NLP in human geography are discussed at the end of the chapter. This chapter forms an overarching understanding of the fusion of NLP and human geography and underscores promising future research and practice in this direction.

Keywords Natural language processing (NLP) · Geoparsing · Sentiment analysis · Textual big data

---

B. Zhou (✉)
Department of Geography, Texas A&M University, College Station, TX, USA

Department of Geography, Pennsylvania State University, University Park, PA, USA
e-mail: bbz5159@psu.edu

B. Lin · L. Zou · M. Yang · H. Tian · H. Cai
Department of Geography, Texas A&M University, College Station, TX, USA

© The Author(s), under exclusive license to Springer Nature Switzerland AG 2025
X. Huang et al. (eds.), *GeoAI and Human Geography*, Springer Geography,
https://doi.org/10.1007/978-3-031-87421-5_4

4.1 Introduction

Human geography is a branch of geography that focuses on the study of human and their interactions with the environment, as well as how these factors are distributed across space and place, which encompasses a wide range of topics and can be divided into several subfields, which include cultural, economic, political, historical, urban, population, social, health, rural, regional, tourism, behavioral, environmental, and transport geography (Hoggart, 2002; Wang et al., 2024). Human geography bridges the gap between the social sciences and the earth sciences, offering a holistic understanding of the world by considering both the physical and human aspects of geography.

In the past several decades, the emergence and development of big data, especially geospatial big data, has significantly influenced the field of human geography, providing new opportunities for research and insights into spatial patterns, human behavior, and societal dynamics. It transformed human geography from using statistical and mathematical methods which were introduced during the quantitative revolution in the 20th century to contemporary human geography which is characterized by interdisciplinarity and the use of advanced technologies such as geographic information systems (GIS), deep learning, and high-performance computing (Rubenstein, 2022). Generally, the sources of big data can be categorized into directed, automated, and volunteered (Kitchin, 2013) and many of them come in textual form. Social media data, news media data, online review data, and traffic reports are examples of commonly used big textual data. These textual data provide an avenue for researchers and practitioners to understand the human perspective and behaviors during a certain event and its context information at a large scale.

Natural language processing (NLP) is a branch of artificial intelligence that centers on the interaction between computers and humans through natural language. It encompasses the development of algorithms and models that enable machines to understand, interpret, and generate human language in a way that is both meaningful and useful (Chowdhary, 2020). Meanwhile, NLP has experienced rapid advancements, driven by improvements in computational power, the availability of large datasets, and the development of sophisticated machine learning techniques. The introduction of neural network architectures, such as recurrent neural networks (RNNs) and convolutional neural networks (CNNs), revolutionized the field. More recently, transformer models like bidirectional encoder representations from transformers (BERT) and generative pre-trained transformer (GPT) have set new benchmarks in NLP tasks. These models leverage vast amounts of data and powerful computational resources to achieve unprecedented levels of accuracy and fluency in understanding and generating human language (Vaswani et al., 2017; Devlin et al., 2019). The applicability of NLP has been pushed to new heights by the deployment of numerous types of large language models (LLMs), which are built upon billions or even trillions of parameters. NLP can smoothly fit into the scope of human geography, especially for studying events that require a prompt understanding of human feedback through language understanding and textual data

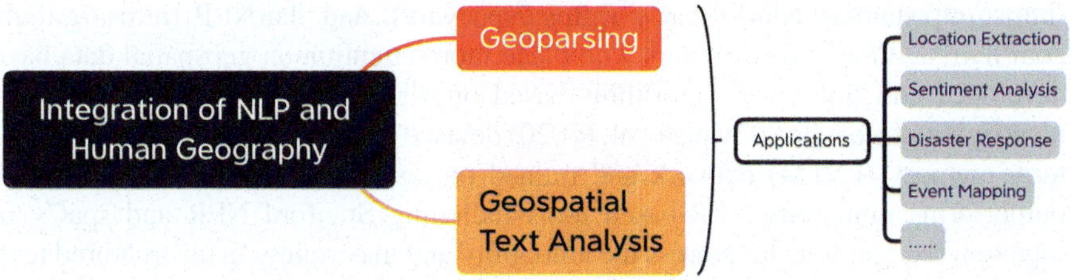

Fig. 4.1 Conceptual diagram of the integration of NLP and human geography

mining, such as disaster management and pandemics. It can be leveraged to generate geospatial knowledge/information by identifying location information from textual data or assessing human behaviors and perspectives by analyzing human-generated textual content associated with locations.

Geoparsing and geospatial text analysis serve as the methodological foundation for integrating NLP in human geography (Fig. 4.1). These two techniques facilitate a wide range of applications in human geography, e.g., location extraction, sentiment analysis, real-time disaster response, and event mapping.

Section 4.2 provides insight into how NLP can interact with human geography through the lens of geoparsing and analyzing textual data, which underscores promising future research and practice in this direction. Two case studies are provided under the context of the COVID-19 pandemic in Sects. 4.3 and 4.4 to exemplify the interaction. Challenges and future directions are discussed at the end of Sect. 4.3.

4.2 Interaction Between NLP and Human Geography

4.2.1 Geoparsing

Geoparsing is a general geographic task to extract geographic information from unstructured text. This is a two-step approach that involves identifying place names (toponym recognition) and other location-specific references in the text and converting them into structured data (toponym resolution), often in the form of geographic coordinates (Gelernter & Balaji, 2013).

The first step of geoparsing, toponym recognition, has witnessed transformative enhancements after combining NLP techniques. After the introduction of transformer-based deep learning models, the capability of NLP to understand and capture hidden information from unstructured text leaped, which concurrently brought opportunities to the improvement of toponym recognition. Named entity recognition (NER) is a general NLP task that can be turned into a toponym recognition tool by applying a filter to exclude named entities other than place names. It can be implemented by leveraging open-source Python packages such as Stanford NLP (https://nlp.stanford.edu/), spaCy (https://spacy.io/), and flairNLP (https://github.com/flairNLP/flair). Researchers who place more attention on geospatial data have developed and fine-tuned algorithms based on NLP which target specifically on identifying place names. Wang et al. (2020) devised NeuroTPR based on long short-term memory (LSTM) network and trained on social media datasets. NeuroTPR outperforms numerous NER-based tools including Stanford NER and spaCy in toponym recognition. To address the ambiguity and uncertainty in unstructured text, a hybrid approach named GazPNE was developed by combining rule-based filters, gazetteer search, and LSTM neural network (Hu et al., 2022a). The authors later optimized the model performance of GazPNE by swapping LSTM with transformer-based models which led to a higher F1-score of 0.84 (Hu et al., 2022b). Also built upon transformer-based models, Zhou et al. (2023) developed a ready-to-use Python tool, TopoBERT, and achieved state-of-the-art performance. TopoBERT incorporates Tobler's first law when designing the convolutional neural network classifier and stacks it upon the transformer-based model to recognize toponyms from texts. The most recent advancement in the integration of LLMs in an array of NLP downstream tasks has also brought shifts and changes to the conventional way of toponym recognition. It turns a token-level classification problem into a question-and-answer problem. Hu et al. (2023) experimented with the performance of toponym recognition on GPT models by designing prompts in a few-shot learning manner and achieved an over 40% performance gain over typically used NER approaches.

Geoparsing plays a crucial role in contemporary human geography, and its advancement has strengthened the significance from multiple aspects. First, reliable geoparsing tools facilitate geospatial data enrichment and spatial analysis. By converting textual data into spatial data, geoparsing enables the integration of diverse data sources with GIS. This allows geographers to perform spatial analysis on large volumes of text-based human-related data, diversifying traditional geographic datasets (Gelernter & Balaji, 2013). Second, it enables near real-time information retrieval and dynamic mapping. For example, in disaster management, geoparsing helps map current events, identify affected areas, and assess the impact of events by extracting information from news reports, social media posts, and other real-time communications (MacEachren et al., 2011; Sui et al., 2012). Third, it fosters an understanding of human dynamics. In health geography, geoparsing helps track the spread of diseases by extracting location-specific health information from reports, news, and social media. This aids in identifying disease hotspots and planning public health interventions (Kamel Boulos & Yang, 2013). Simultaneously, analyzing geolocated review data on healthcare facilities can help better depicting the character and quality of healthcare providers. This information provides insights into spatial disparities in proper healthcare access and informs efforts to improve healthcare equity (Garimella et al., 2024).

4.2.2 Geographical Analysis of Textual Data

Human geography proceeds on the base of understanding human perceptions, behaviors, and social dynamics influenced by the virtual and physical world. Unstructured textual data provide content to the researchers and solidify this foundation. Textual data come in various forms, including social media posts, news articles, review information, policy papers, traffic accident reports, and health records. Each type offers unique perspectives and information (Sui et al., 2012; Hu et al., 2019). The integration of NLP techniques with geographic methods supports the analysis of large-scale, unstructured text data and enables geographers to uncover insights about public sentiment, mobility patterns, place-based perceptions, and socio-cultural dynamics across different regions, which results in deeper insights, more comprehensive understandings, and supporting more informed decision-making (Gregory & Ell, 2007; Sui et al., 2012).

For instance, social media data are one of the most commonly used data sources under this scenario (Arthur & Williams, 2018). Sit et al. (2019) leveraged NLP models to harvest infrastructure damage-related social media data for Hurricane Irma to investigate areas with potentially highly affected individuals. An automated tool, VictimFinder, was developed based on pretrained language models and CNN classifiers to accurately identify help requests from Twitter data during Hurricane Harvey, which enables first responders and volunteers to react and execute rescue missions promptly during disasters (Zhou et al., 2022). Li et al. (2014) developed a novel algorithm to quantify public depressive emotions toward events related to the COVID-19 pandemic by analyzing the content of social media data. Results indicate that people's risk perception is sensitive to the release of COVID-19-related public news and media messages. With the help of NLP, new media can become another source of spatiotemporal data with rich context information. Geolocated news media was used to visualize the regional distribution of risk and help with news themes associated with reporting on marine aquaculture, which is a growing domestic industry with complex social and ecological impacts (Duffy et al., 2020). Sodoge et al. (2023) propose a method to evaluate drought impact by analyzing newspaper articles with NLP algorithms. The algorithm can precisely identify different socio-economic impacts (e.g., agriculture, forestry, livestock, fires) so that the spatiotemporal scope of the impacts can be scrutinized. Concurrently, longitudinal multisectoral datasets can be generated with this method. Traffic reports serve as another ideal material for NLP to extract valuable information for geographic analysis because they often comprise spatial, temporal, and rich descriptive texts about traffic accidents. Scholars have taken advantage of NLP in mining traffic reports to evaluate the severity of road accidents (Valcamonico et al., 2022), predict injury severity for cyclists in crashes (Janstrup et al., 2023), and detect fallen objects on the road (Lee et al., 2023).

4.3 Case Study 1: NLP for Mapping Humanitarian Needs

This section demonstrates the use of semantic analysis with NLP techniques on the data collected from an online mutual help-requesting platform during the 2022 Shanghai COVID-19 lockdown. The self-reported data act as a real-time agent to reflect public response to the prompt lockdown policies. Semantic analysis was performed on the textual content of the help request to identify the emergent needs of the underprepared, vulnerable communities. Based on the insights derived from these analyses, policymakers can better anticipate and address the urgent residential needs that may arise in future lockdowns. This knowledge can contribute to more effective planning and preparedness efforts to mitigate the impact on vulnerable populations and ensure their well-being during similar crises.

4.3.1 Data Collection

The core dataset for this study comprises humanitarian assistance requests made during the Shanghai lockdown. This data was gathered from "Let Us Help You," an open-source platform created and managed by the nonprofit organization Daohouer. As the largest and most recognized help request website during the lockdown, it received widespread promotion from several prominent Chinese portals. The platform has since been discontinued following the end of the lockdown. Figure 4.2 demonstrates the user interface of the website and several sample data.

During the data collection process, a web scraper was developed to extract data from the mutual help website by breaking down its structured HTML content. Each raw data entry includes the district where the request was made, a summary and

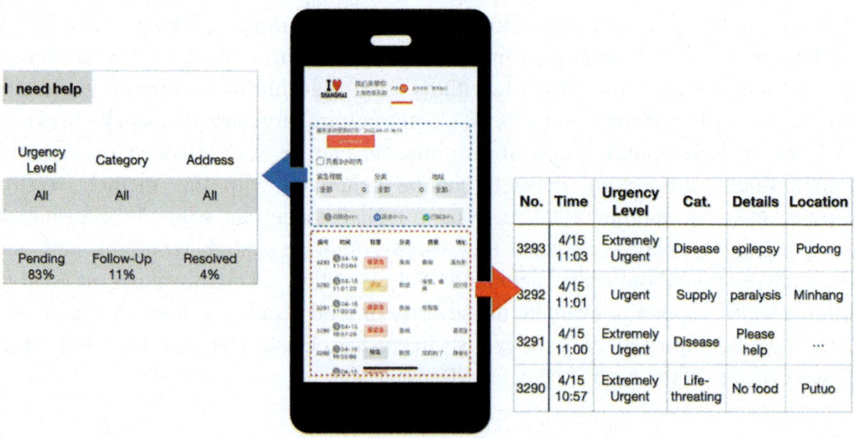

Fig. 4.2 User interface of the help request website and sample data

details of the request, the urgency level (extremely urgent, urgent, normal), and the category (serious disease, disease, supply, others). Some help requests contained precise addresses and were processed using the Google geocoding service. For records that could not be geocoded, manual correction was performed to fix typos and format addresses so the Google service could recognize them. This process assigned precise coordinates to the original data. After automated geocoding and manual adjustments, 4,396 help request records were obtained. Of these, 1,805 (41.1%) had precise geographical coordinates, like points of interest (POI) or street-level centers, while 2,591 (58.9%) included only district information. Of the 4,396 requests, 2,653 (60.4%) were self-categorized as extremely urgent, 1,191 (27.1%) were urgent, and 552 (12.5%) were normal. Additionally, 1,220 (27.8%) posts were self-identified as related to serious disease, 1,307 (29.7%) to disease, 1,124 (25.6%) to supply shortages, and 744 (16.9%) to other issues.

4.3.2 Methodology

To examine the semantic patterns of the online help requests, word cloud visualization and latent Dirichlet allocation (LDA) were leveraged. Word cloud visualization is a popular method for extracting information and highlighting keywords in texts. It displays words in various sizes based on their frequency, with larger and bolder words appearing more often and indicating higher importance (Yum, 2020). However, Chinese, being an ideographic language without word delimiters, poses a unique challenge (Li and Yuan, 1998). To address this, we used Jieba, a widely used Chinese word segmentation Python package, which employs a dynamic programming search algorithm based on Chinese grammar and vocabulary for efficient and reliable performance (Zhu et al., 2020). Stop words were removed before creating the word cloud. To delve deeper into the information from the help requests, Latent Dirichlet allocation (LDA) was employed to uncover hidden topics. LDA is a Bayesian-based probabilistic unsupervised topic modeling algorithm that identifies latent topics as a multinomial distribution over words (Blei, 2003). LDA calculates the probability of a word belonging to a topic with Eq. 4.1.

$$P\left(Z|W,D\right)=\frac{W_{z+\beta w}}{\text{number of tokens in } Z+\beta}\times D_{z} \tag{4.1}$$

The topics are denoted as Z. The probability that a word W belongs to topic Z is calculated by multiplying the normalized word frequency of W in topic Z by the number of other words in document D that belong to topic Z (Blei, 2003). B and βw are hyperparameters that adjust the probability of word W belonging exclusively to topic Z. LDA processes documents iteratively, reassigning each word to a topic until the model stabilizes and reaches equilibrium, meaning topics with high-frequency words are more prevalent in the documents (Underwood, 2012). This method relies on accurate word segmentation, following the same approach as used in word cloud generation.

4.3.3 Semantic Analysis Results

The generated word clouds are shown in Fig. 4.3, where Chinese words have been translated into English for better visualization and understanding. The word sizes in the clouds reflect their frequencies in help requests under different categories. Figure 4.3a displays the word cloud for help requests categorized as serious disease, where the most frequent words include hospital, tablet, medicine, patient, nucleic acid, need, and medicine. Figure 4.3b presents the word cloud for the disease category, with frequent terms like need, tablet, medicine, boxes, capsule, hospital, drug, and elderly. Figure 4.3c illustrates the supply category, prominently featuring supplies,

Fig. 4.3 Word cloud visualization under each category: (**a**) serious disease; (**b**) disease; (**c**) supply; and (**d**) others

vegetables, food, and community. Lastly, Figure 4.3d shows the others category, where keywords like quarantine, Shanghai, nucleic acid, hotel, and building are more evenly distributed.

LDA was employed to discover hidden topics. The number of topics, a crucial hyperparameter in LDA, needed to be defined before running the algorithm. Initial tests were conducted with 3, 5, 8, and 10 topics. Based on data visualization and qualitative interpretability, three topics were selected to avoid overly segmented topics and ensure meaningful, coherent results. The results of the LDA analysis are summarized in Table 4.1, providing a clear view of the dominant terms in each topic. These results offer valuable insights into the nature of the help requests. The elderly population was significantly impacted by the lockdown, with many requests focusing on the need for long-term periodic treatment, chemotherapy, dialysis, and urgent hospital transfers for COVID-positive residents. Additionally, there was a high demand for specific medications, including insulin, hydrochloric acid, citalopram, oxalic acid, and fumaric acid.

Less urgent topics were primarily concerned with grocery shortages, with numerous requests for items such as milk powder, vegetables, rice, bread, and instant noodles. There were also frequent mentions of immediate quarantine needs and requests to return home. By scrutinizing the topics extracted from the LDA, it is clear that the most pressing issues during the 2022 Shanghai lockdown were related to the needs of patients with chronic diseases requiring ongoing treatment and those who tested positive for COVID-19 and needed medical care.

The findings underscore the importance of preparing for future health crises by ensuring adequate medical supplies and resources when planning and executing lockdown policies. For example, having extra medical supplies on hand for each apartment complex and designating specific hospitals for patients requiring periodic treatments could alleviate some of the challenges faced during such lockdowns. NLP plays a crucial role in this study by geolocating the help request for spatial analysis and extracting latent topics from the help request that summarize the urgent needs of the underrepresented communities during the lockdown.

4.4 Case Study 2: NLP for Demographic and Sentiment Analysis

This section presents a case study demonstrating the application of NLP in analyzing social media data to capture user demographics and track spatial–temporal trends in sentiment toward COVID-19. We collected geotagged textual data from Twitter/X in the USA from January 2020 to December 2021. NLP-based models were employed to infer demographic information and emotional states of users discussing the COVID-19 pandemic. The spatial–temporal patterns of public sentiment related to COVID-19 were visualized by age, gender, and state. Additionally, we developed an

Table 4.1 LDA-detected topics under each category

Category	Topic Number	Topics in Chinese	Topics in English
Extremely urgent	3	老人需要化疗	Aged people need chemotherapy
		核酸阳性需隔离转运	COVID test positive; need transfer to hospital
		家里需要配药	Need medicine at home
Urgent	3	家中母亲需要胰岛素	Mother at home need insulin
		需要奶粉配送	Require milk powder delivery
		家中需要物资及药物	Need supplies and medication
Normal	3	蔬菜泡面鸡蛋	Vegetables, instant noodles, eggs
		居委会核酸检测	Community COVID test
		隔离物资及药物	Quarantine supplies and medicine
Disease	3	药品 精神	Psychotropic drugs
		胰岛素、盐酸、高血压、注射液	Insulin, hydrochloric acid, high blood pressure, injection
		普兰，西酞，富马酸，草酸，胰岛素	Citalopram, oxalic acid, fumaric acid, insulin
Serious disease	3	老人需要化疗	Aged people need chemotherapy
		老人病重透析	Aged people need dialysis
		老人核酸阳性	Aged people COVID test positive
Supply	3	奶粉泡面卫生巾牛奶	Milk powder, instant noodles, milk, sanitary pad
		蔬菜面包大米	Vegetables, bread, rice
		团购蔬菜水果	Group purchase vegetables and fruit
Others	3	希望隔离酒店	Wish quarantine hotel
		社区接种疫苗	Vaccination by community service
		隔离返乡火车站	Isolation, train station back home

adjusted sentiment measure, the sentiment adjusted by demographics (SAD) index, using a post-stratification approach, which was then used to estimate US state-level public sentiment over the two-year period.

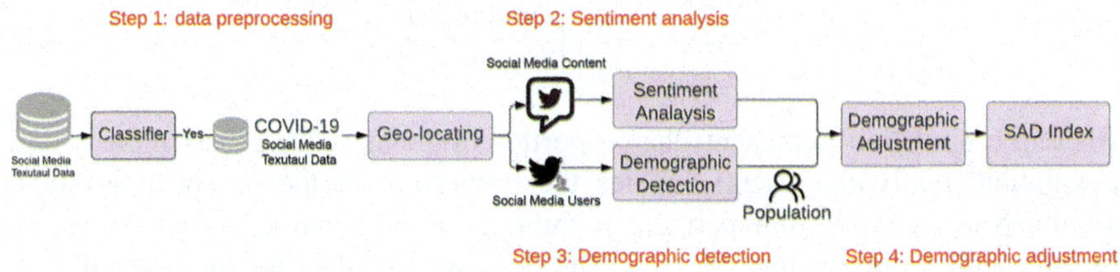

Fig. 4.4 Framework of assessing demographic-adjusted public sentiment using NLP and social media

4.4.1 Data and Methods

The workflow, shown in Fig. 4.4, involves four primary steps: (1) social media data preprocessing, (2) NLP for sentiment analysis, (3) NLP for demographic detection, and (4) assessing and adjusting sentiment by demographics.

The first step is gathering and preprocessing Twitter/X data. A total of 4,822,802 geotagged tweets posted from the USA from January 2020 to December 2021 were collected. The second step is sentiment analysis. Each tweet's sentiment can be assessed based on tweet text through NLP models. Conventional NLP methods for sentiment analysis can be categorized into lexicon-based, machine learning-based, and hybrid methods. This case study leverages the VADER package developed by Hutto and Gilbert (2014), a lexicon-based tool that provides efficient and consistent sentiment analysis. It generates an index ranging from -1 to 1 based on texts, with -1 representing the most negative sentiment and 1 representing the most positive sentiment. In the third step, Twitter users' age and gender were detected by the M3 (multimodal, multilingual, and multi-attribute) model proposed by Wang et al. (2019) based on users' screen names, usernames, profile images, and biographies. This model is a fusion of NLP and deep learning. M3 categorized users into eight distinct groups: males aged 18 and under, males aged 19–29, males aged 30–39, males aged 40 and above, females aged 18 and under, females aged 19–29, females aged 30–39, and females aged 40 and above. The final step estimates and adjusts public sentiment by demographics. We calculated each user's sentiment as the average sentiment of all COVID-19-related tweets posted by the user over a certain time period (e.g., 1 month or 1 year). The adjusted negative sentiment index is the weighted proportion of Twitter/X users within a specific community and timeframe who display predominantly negative emotions. If the index sits between -0.05 and 0.05, it is regarded as neutral. If it goes beyond 0.05, it is considered positive, while the index below -0.05 is regarded as negative. We used the post-stratification method, a weighting adjustment technique commonly employed in survey data processing, to correct deviations caused by demographic bias (Dargatz & Hill, 1996; Kalton & Flores-Cervantes, 2003; Vaske et al., 2011). The adjustment weighting was computed based on the demographic differences between Twitter/X users and the general population using Eq. 4.2:

$$w_{ij} = \frac{p_i\,(\text{population})}{p_{ij}\,(\text{Twitter users})}, \tag{4.2}$$

where p_i(population) represents the proportion of the i_{th} social group in the overall population, p_{ij}(Twitter users) denotes the proportion of the i_{th} social group in Twitter/X users during time period j (monthly or yearly), and w_{ij} means the weight for i_{th} social group in time period j. The demographic data for the population in 2020 was sourced from the US Census Bureau and utilized to mitigate biases in social media data (Bureau, 2021).

Subsequently, we quantified the counts of Twitter/X users across various sentiment levels within each time period j and applied the weights calculated from Equation 4.2 to adjust the distributions of Twitter/X users across these sentiment categories. In Equation 4.3, the variable l designates sentiment levels categorized as negative (1), neutral (2), and positive (3). N_{lij} represents the count of Twitter/X users within social group i exhibiting sentiment level l during timeframe j. The resulting metric Adjusted percentage$_{lj}$ signifies the overall percentage of Twitter users displaying sentiment level l during timeframe j after demographic bias adjustments. Specifically, the sentiment adjusted by demographics (SAD) index for a given timeframe j is defined as the percentage of Twitter/X users expressing an overall negative sentiment toward COVID-19 after demographic adjustment (Eq. 4.4).

$$\text{Adjusted percentage}_{lj} = \frac{\sum_{i=1}^{8} N_{lij} w_{ij}}{\sum_{i=1}^{8} \left(\left(\sum_{l=1}^{3} \text{N}_{lij} \right) w_{ij} \right)} \tag{4.3}$$

$$SAD_{lj} = \frac{\sum_{i=1}^{8} N_{lij} w_{ij}}{\sum_{i=1}^{8} \left(\left(\sum_{l=1}^{3} N_{lij} \right) w_{ij} \right)} \quad (l = \text{negative}) \tag{4.4}$$

4.4.2 Demographic Bias of Twitter/X Users

Figure 4.5 illustrates demographic disparities between Twitter/X users discussing COVID-19 and the actual US population by gender and age. Twitter/X users under 18 represent 14.03%, notably lower than the corresponding US population segment (23.36%). The age group of 19 to 29 constitutes 35.35% of Twitter/X users, which is more than twice the percentage of the same age group in the US population (14.90%). Twitter/X users aged 30 to 39 make up 18.76%, exceeding the US population percentage in this range (13.56%). It is worth noting that the females over 40 are significantly underrepresented on Twitter (9.43%) compared to their presence in the US population (25.31%). These findings quantitatively unravel the significant demographic biases in Twitter/X data, highlighting the need to rectify

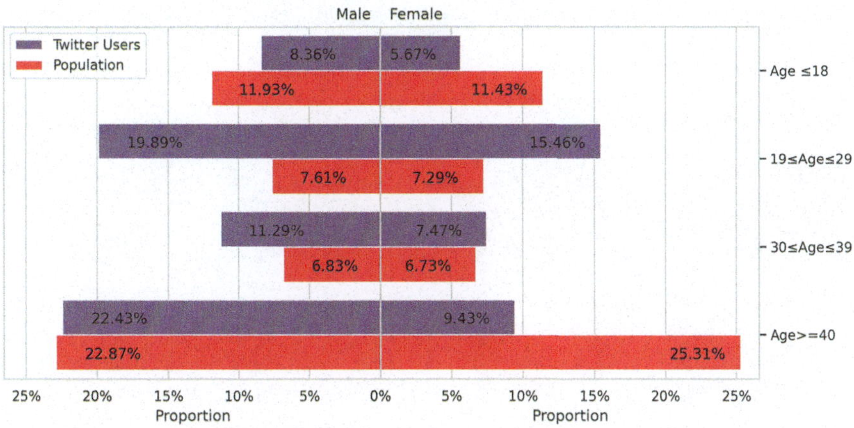

Fig. 4.5 Comparison of demographic composition differences between Twitter users discussing COVID-19 and the general US population

such biases in sentiment analysis to ensure an accurate representation of all social groups.

4.4.3 Sentiment Adjusted by Demographics (SAD) Index and Application

Figure 4.6 presents the spatial and temporal pattern of negative sentiment among different demographic groups before the adjustment. Figure 4.7 presents three scatter plots illustrating the original and adjusted percentages of Twitter/X users across three sentimental categories at the state level. The original state-level percentages of negative Twitter/X users ranged from 20.63% to 32.21%, averaging 28.82%. The state-level SAD index ranged from 22.01% to 33.12%, averaging 28.96%. Figure 4.8 compares the spatial patterns of public negative sentiment toward COVID-19 at the state level before and after the demographic adjustment. As depicted in Figs. 4.7 and 4.8, the SAD indexes closely mirrored the original percentages in most states. However, in 29 states and Washington D.C., the SAD index values exceeded the initial rates of negative Twitter/X users, with Vermont (VT) experiencing the largest increase from 20.63% to 22.01%. Conversely, in 21 states, the SAD index values were lower than the original percentages of negative Twitter/X users, with Texas showing the most significant decrease from 29.21% to 28.17%. The highest SAD index was observed in Wyoming (WY, 33.12%), followed by Montana (MT, 32.98%) and Alaska (AK, 32.35%). Vermont showed the lowest SAD index (22.00%), followed by Massachusetts (MA, 26.54%) and Nebraska (NE, 26.92%).

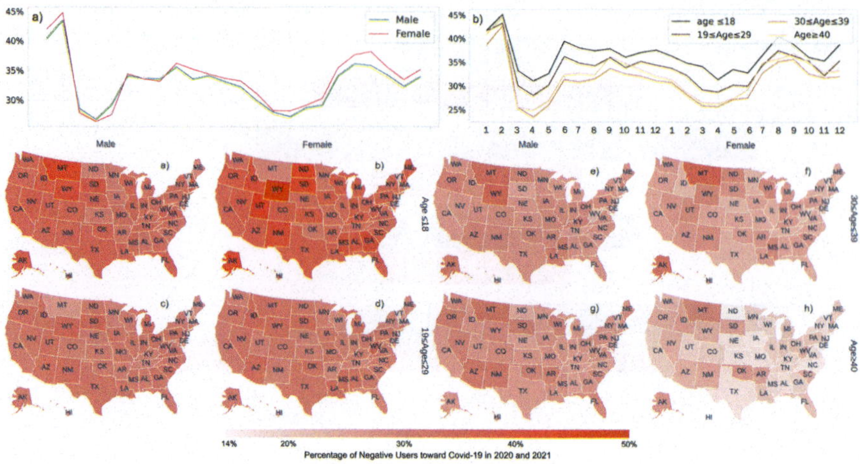

Fig. 4.6 Spatial and temporal patterns of negative sentiments among different demographic groups

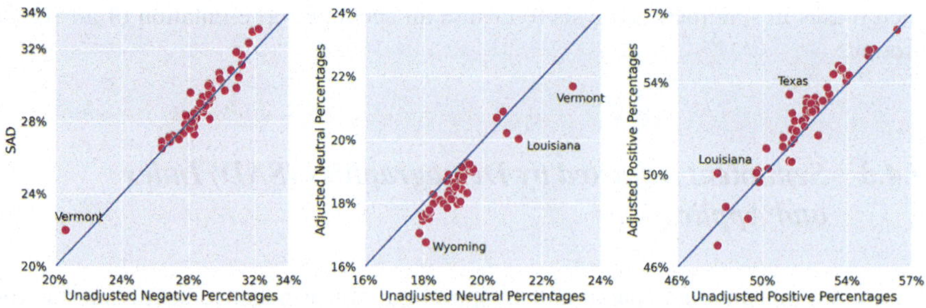

Fig. 4.7 Comparison at the state level between original and adjusted percentages of Twitter users across three sentiment levels toward COVID-19 (blue lines represent the y = x reference line)

This case study highlights the potential of NLP in enhancing the analysis and understanding of human behaviors and places semantics by extracting insights from geographically associated texts. The findings support governmental efforts to develop targeted, data-informed strategies to alleviate depressive emotions among various demographic groups. The proposed framework addresses demographic biases in social media analytics for similar studies.

4.5 Challenges and Discussion

The application of NLP in human geography presents a plethora of opportunities for gaining deeper insights into spatial–temporal phenomena through the analysis of

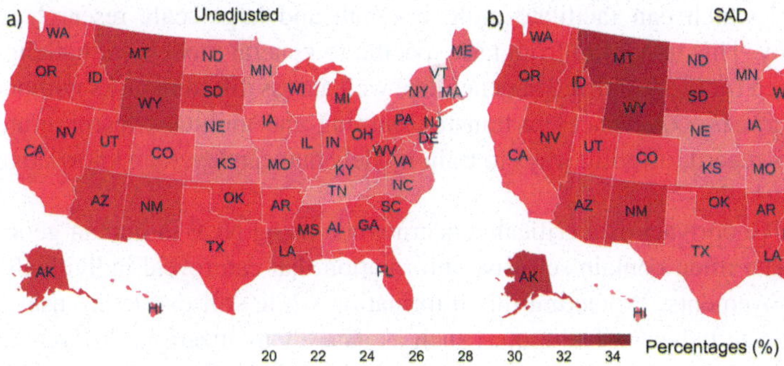

Fig. 4.8 State-level percentages of Twitter users expressing negative sentiment toward COVID-19 before and after demographic adjustments

textual data. However, it is fraught with several challenges that necessitate careful consideration and innovation.

Uncertainty of Textual Data: Textual data inherently carry a degree of uncertainty that complicates their use in human geography. Unlike numerical data, text is often ambiguous, context-dependent, and subjective. For instance, place names can have multiple spellings or refer to different locations depending on the context. Sentiment and tone in social media posts can be challenging to discern accurately due to diverse individual preference, differences in culture, and language inconsistencies when translated, leading to potential misinterpretations.

Data Quality Issues: The quality of textual data is another critical issue. Text data collected from sources like social media, news articles, and user-generated content can be noisy, biased, and incomplete. Social media data, for example, are often riddled with typographical errors that standard NLP tools struggle to process. Furthermore, these data sources may not represent all demographics equally, leading to biased insights. To mitigate these issues, rigorous data cleaning and preprocessing techniques are essential. Additionally, integrating multiple data sources can help in cross-verifying information and improving overall data quality.

Explainability of NLP Models: The explainability of NLP models is a significant challenge, particularly with the rise of deep learning techniques. Models like BERT and GPT-4, while powerful, are often considered "black boxes" due to their complexity and lack of transparency. This poses a problem for human geographers who need to understand how and why certain conclusions are drawn from the data. Explainable AI (XAI) techniques will continue playing crucial roles in enabling researchers to interpret model outputs and ensure that the models' decisions are based on understandable and justifiable criteria. Developing NLP models that are both accurate and interpretable needs further investigations.

Transition from Task-Specific to General Models: Most conventional NLP models are task-specific, designed to perform well on a particular task such as sentiment analysis, named entity recognition, or topic modeling. However, human geographers are seeking for more holistic approaches to handle various types of tasks seamlessly which can facilitate more in-depth and large-scale research in human geography. This transition from task-specific to general models is challenging due to the need for models that can generalize well across different tasks without sacrificing performance. Advances in transfer learning and multitask learning are promising in this regard, offering ways to train models that can leverage knowledge across various tasks.

Geoprivacy: Geoprivacy is a critical concern when using NLP in human geography. Textual data often contain sensitive information that can reveal individuals' locations and movements. Protecting this information while still extracting useful insights is a delicate balance. Techniques such as data anonymization, differential privacy, and secure multiparty computation are essential for safeguarding individuals' privacy. However, these techniques must be implemented carefully to ensure that the data remain useful for analysis without compromising privacy.

4.6 Future Directions

The integration of NLP with GIS offers exciting possibilities for human geography. Spatially aware NLP models that can understand and incorporate geographic context will be crucial for accurately interpreting text data related to specific locations. Moreover, the development of open-source, multilingual NLP tools will democratize access to these technologies, enabling researchers worldwide to apply NLP in diverse geographic contexts. Another promising direction is the use of NLP for real-time analysis and decision-making. For instance, during natural disasters, NLP can help analyze social media data to provide timely insights into affected areas, aiding in disaster response and management. However, this requires robust, scalable models that can handle large volumes of data quickly and accurately. Interdisciplinary collaboration will also be key to advancing NLP in human geography. Collaborations between geographers, computer scientists, linguists, and ethicists can lead to the development of more sophisticated models that are sensitive to the complexities of human language and spatial dynamics.

4.7 Conclusion

This chapter provides an overarching view of how advanced NLP can be leveraged to foster human geography research. It fulfilled three major objectives. First, it proposes two avenues through which NLP interacts tightly with human geography, namely geoparsing and the semantic analysis of an array of textual data. Second, two case studies are offered to demonstrate the application of NLP in studying human response among different population groups or communities in the context of the COVID-19 pandemic. The first case study scrutinizes the urgent needs of the underrepresented and vulnerable population under prompt and strict lockdown policies, while the second case study examines the demographic bias in public sentiment toward COVID-19 and proposes solutions to alleviate the bias issue. Finally, the challenges and future direction of integrating NLP and human geography are discussed. Overall, the data quality issue, uncertainty of textual data, explainability of black box models, limitations of task agnostic model training and application, and geoprivacy are the areas where scholars should spare effort to expand.

References

Arthur, R., & Williams, H. T. P. (2018). The Human Geography of Twitter (arXiv:1807.04107). *arXiv*. http://arxiv.org/abs/1807.04107

Blei, D. M. (2003). *Latent dirichlet allocation.*

Boulos, M. N. K., & Yang, S. P. (2013). Exergames for health and fitness: The roles of GPS and geosocial apps. *International Journal of Health Geographics, 12*(1), 18. https://doi.org/10.1186/1476-072X-12-18

Chowdhary, K. R. (2020). Natural language processing. In K. R. Chowdhary (Ed.), *Fundamentals of artificial intelligence* (pp. 603–649). Springer. https://doi.org/10.1007/978-81-322-3972-7_19

Dargatz, D. A., & Hill, G. W. (1996). Analysis of survey data. *Preventive Veterinary Medicine, 28*(4), 225–237. https://doi.org/10.1016/0167-5877(96)01052-5

Devlin, J., Chang, M.-W., Lee, K., & Toutanova, K. (2019). BERT: Pre-training of deep bidirectional transformers for language understanding (arXiv:1810.04805). *arXiv*. https://doi.org/10.48550/arXiv.1810.04805

Duffy, K. P., Cipparone, H. C., Johnson, E. S., Rickard, L. N., Beard, K., & Nascimento, F. (2020). Leveraging spatial dimensions of news media content analysis to explore place-based differences in natural resource issues. *Journal of Environmental Studies and Sciences, 10*(3), 303–309. https://doi.org/10.1007/s13412-020-00595-9

Garimella, B. S., Garlapati, H. S., Choul, S., Cherukuri, R., & Lanke, P. (2024). Advancing healthcare accessibility: Development of an AI-driven multimodal chatbot. In *2024 4th International conference on intelligent technologies (CONIT)* (pp. 1–10). https://doi.org/10.1109/CONIT61985.2024.10626795

Gelernter, J., & Balaji, S. (2013). An algorithm for local geoparsing of microtext. *GeoInformatica, 17*(4), 635–667. https://doi.org/10.1007/s10707-012-0173-8

Gregory, I. N., & Ell, P. S. (2007). *Historical GIS: Technologies, methodologies, and scholarship.* Cambridge University Press.

Hoggart, K. (2002). Researching human geography. In L. Lees & A. Davies (Eds.), *Co-published in the U.S.A.* Oxford University Press.

Hu, Y., Mao, H., & McKenzie, G. (2019). A natural language processing and geospatial clustering framework for harvesting local place names from geotagged housing advertisements. *International Journal of Geographical Information Science, 33*(4), 714–738. https://doi.org/10.1080/13658816.2018.1458986

Hu, X., Al-Olimat, H. S., Kersten, J., Wiegmann, M., Klan, F., Sun, Y., & Fan, H. (2022a). GazPNE: Annotation-free deep learning for place name extraction from microblogs leveraging gazetteer and synthetic data by rules. *International Journal of Geographical Information Science, 36*(2), 310–337. https://doi.org/10.1080/13658816.2021.1947507

Hu, X., Zhou, Z., Sun, Y., Kersten, J., Klan, F., Fan, H., & Wiegmann, M. (2022b). GazPNE2: A general place name extractor for microblogs fusing gazetteers and pretrained transformer models. *IEEE Internet of Things Journal, 9*(17), 16259–16271. https://doi.org/10.1109/JIOT.2022.3150967

Hu, Y., Mai, G., Cundy, C., Choi, K., Lao, N., Liu, W., Lakhanpal, G., Zhou, R. Z., & Joseph, K. (2023). Geo-knowledge-guided GPT models improve the extraction of location descriptions from disaster-related social media messages. *International Journal of Geographical Information Science, 37*(11), 2289–2318. https://doi.org/10.1080/13658816.2023.2266495

Hutto, C., & Gilbert, E. (2014). VADER: A parsimonious rule-based model for sentiment analysis of social media text. *Proceedings of the International AAAI Conference on Web and Social Media, 8*(1), Article 1. https://doi.org/10.1609/icwsm.v8i1.14550

Janstrup, K. H., Kostic, B., Møller, M., Rodrigues, F., Borysov, S., & Pereira, F. C. (2023). Predicting injury-severity for cyclist crashes using natural language processing and neural network modelling. *Safety Science, 164*, 106153. https://doi.org/10.1016/j.ssci.2023.106153

Kalton, G., & Flores-Cervantes, I. (2003). Weighting methods. *Journal of Official Statistics, 19*, 81–97.

Kitchin, R. (2013). Big data and human geography. *Dialogues in Human Geography, 3*, 262. https://doi.org/10.1177/2043820613513388

Lee, S.-S., Cha, S.-M., Ko, B., & Park, J. J. (2023). Extracting fallen objects on the road from accident reports using a natural language processing model-based approach. In *IEEE access* (Vol. 11, pp. 139521–139533). IEEE. https://doi.org/10.1109/ACCESS.2023.3339774

Li, H., & Yuan, B. (1998). Chinese Word Segmentation. In J. Guo, K. T. Lua, & J. Xu (Eds.), *Proceedings of the 12th Pacific Asia Conference on Language, Information and Computation* (pp. 212–217). Chinese and Oriental Languages Information Processing Society. http://hdl.handle.net/2065/12081

Li, T. J.-J., Sen, S., & Hecht, B. (2014). Leveraging advances in natural language processing to better understand Tobler's first law of geography. In *Proceedings of the 22nd ACM SIGSPATIAL international conference on advances in geographic information systems* (pp. 513–516). https://doi.org/10.1145/2666310.2666493

MacEachren, A. M., Jaiswal, A., Robinson, A. C., Pezanowski, S., Savelyev, A., Mitra, P., Zhang, X., & Blanford, J. (2011). SensePlace2: GeoTwitter analytics support for situational awareness. In *2011 IEEE Conference on Visual Analytics Science and Technology (VAST)* (pp. 181–190). IEEE. https://doi.org/10.1109/VAST.2011.6102456

Rubenstein, J. M. (2022). *Contemporary human geography* (5th ed.). Pearson.

Sit, M. A., Koylu, C., & Demir, I. (2019). Identifying disaster-related tweets and their semantic, spatial and temporal context using deep learning, natural language processing and spatial analysis: A case study of hurricane Irma. *International Journal of Digital Earth, 12*(11), 1205–1229. https://doi.org/10.1080/17538947.2018.1563219

Sodoge, J., Kuhlicke, C., & de Brito, M. M. (2023). Automatized spatio-temporal detection of drought impacts from newspaper articles using natural language processing and machine learning. *Weather and Climate Extremes, 41*, 100574. https://doi.org/10.1016/j.wace.2023.100574

Sui, D., Elwood, S., & Goodchild, M. (2012). *Crowdsourcing geographic knowledge: Volunteered geographic information (VGI) in theory and practice*. Springer.

Underwood, W. E. (2012). What can topic models of PMLA teach us about the history of literary scholarship? *Journal of Digital Humanities, 2*(1), 39–48.

Valcamonico, D., Baraldi, P., Amigoni, F., & Zio, E. (2022). A framework based on natural language processing and machine learning for the classification of the severity of road accidents from reports. *Proceedings of the Institution of Mechanical Engineers, Part O: Journal of Risk and Reliability, 238*, 957. https://doi.org/10.1177/1748006X221140196

Vaske, J., Wright, J. P., Boisvert, D., & Beaver, K. M. (2011). Gender, genetic risk, and criminal behavior. *Psychiatry Research, 185*(3), 376–381. https://doi.org/10.1016/j.psychres.2010.07.044

Vaswani, A., Shazeer, N., Parmar, N., Uszkoreit, J., Jones, L., Gomez, A. N., Kaiser, Ł. U., & Polosukhin, I. (2017). Attention is all you need. *Advances in Neural Information Processing Systems, 30*, 6000. https://proceedings.neurips.cc/paper_files/paper/2017/hash/3f5ee243547dee91fbd053c1c4a845aa-Abstract.html

Wang, J., Hu, Y., & Joseph, K. (2020). NeuroTPR: A neuro-net toponym recognition model for extracting locations from social media messages. *Transactions in GIS, 24*(3), 719–735. https://doi.org/10.1111/tgis.12627

Wang, S., Huang, X., Liu, P., Zhang, M., Biljecki, F., Hu, T., Fu, X., Liu, L., Liu, X., Wang, R., Huang, Y., Yan, J., Jiang, J., Chukwu, M., Reza Naghedi, S., Hemmati, M., Shao, Y., Jia, N., Xiao, Z ... Bao, S. (2024). Mapping the landscape and roadmap of geospatial artificial intelligence (GeoAI) in quantitative human geography: An extensive systematic review. *International Journal of Applied Earth Observation and Geoinformation, 128*, 103734. https://doi.org/10.1016/j.jag.2024.103734

Wang, Z., Hale, S., Adelani, D. I., Grabowicz, P., Hartman, T., Flöck, F., & Jurgens, D. (2019). Demographic inference and representative population estimates from multilingual social media data. *The World Wide Web Conference*, 2056–2067. https://doi.org/10.1145/3308558.3313684

Yum, S. (2020). Social Network Analysis for Coronavirus (COVID-19) in the United States. *Social Science Quarterly, 101*(4), 1642–1647. https://doi.org/10.1111/ssqu.12808

Zhou, B., Zou, L., Mostafavi, A., Lin, B., Yang, M., Gharaibeh, N., Cai, H., Abedin, J., & Mandal, D. (2022). VictimFinder: Harvesting rescue requests in disaster response from social media with BERT. *Computers, Environment and Urban Systems, 95*, 101824. https://doi.org/10.1016/j.compenvurbsys.2022.101824

Zhou, B., Zou, L., Hu, Y., Qiang, Y., & Goldberg, D. (2023). TopoBERT: A plug and play toponym recognition module harnessing fine-tuned BERT. *International Journal of Digital Earth, 16*(1), 3045–3064. https://doi.org/10.1080/17538947.2023.2239794

Zhu, Y., Li, Y., Yue, Y., Qiang, J., & Yuan, Y. (2020). A hybrid classification method via character embedding in Chinese short text with few words. *IEEE Access, 8*, 92120–92128. https://doi.org/10.1109/ACCESS.2020.2994450

Chapter 5
Explainable AI in Spatial Analysis

Ziqi Li

Abstract This chapter discusses the opportunities for eXplainable Artificial Intelligence (XAI) within the realm of spatial analysis. A key objective in spatial analysis is to model spatial relationships and infer spatial processes to generate knowledge from spatial data, which has been largely based on spatial statistical methods. More recently, machine learning offers scalable and flexible approaches that complement traditional methods and has been increasingly applied in spatial data science. Despite its advantages, machine learning is often criticized for being a black box, which limits our understanding of model behavior and output. Recognizing this limitation, XAI has emerged as a pivotal field in AI that provides methods to explain the output of machine learning models to enhance transparency and understanding. These methods are crucial for model diagnosis, bias detection, and ensuring the reliability of results obtained from machine learning models. This chapter introduces key concepts and methods in XAI with a focus on Shapley value-based approaches, which is arguably the most popular XAI method, and their integration with spatial analysis. An empirical example of county-level voting behaviors in the 2020 Presidential election is presented to demonstrate the use of Shapley values and spatial analysis with a comparison to multi-scale geographically weighted regression. The chapter concludes with a discussion on the challenges and limitations of current XAI techniques and proposes new directions.

Keywords Explainable AI (XAI) · Spatial Analysis · Geospatial Data Science

5.1 Introduction

Spatial modeling has long been a cornerstone of spatial analysis, which provides essential tools to understand spatial relationships underlying data and to predict

Z. Li (✉)
Department of Geography, The Spatial Data Science Center, Florida State University, Tallahassee, FL, USA
e-mail: Ziqi.Li@fsu.edu

© The Author(s), under exclusive license to Springer Nature Switzerland AG 2025
X. Huang et al. (eds.), *GeoAI and Human Geography*, Springer Geography,
https://doi.org/10.1007/978-3-031-87421-5_5

future spatial scenarios. Traditionally, spatial modeling has largely focused on statistical approaches. Widely applied methods include spatial econometric models (e.g., spatial lag model, spatial error model, spatial Durbin model), spatially varying coefficient models (such as geographically weighted regression (GWR) and eigenvector filtering), Kriging-based models, and Bayesian spatial models, among others (Anselin, 1988; Fotheringham et al., 2002; Griffith, 2003; Hengl et al., 2007; Lindgren & Rue, 2015; Lee, 2013). These statistical models are advantageous in which they can explicitly specify spatial effects such as spatial autocorrelation and spatial heterogeneity. They also offer inferential frameworks that help quantify uncertainties and infer underlying processes. However, most statistical approaches often come with strong data and model requirements, such as assumptions of data distribution, model linearity, and scale, stationarity and isotropy of spatial effects, which may simplify spatial relationships in the real world. Moreover, these models can be challenging to specify and are computationally intensive, particularly with moderate-to-large datasets, limiting their scalability and utilities where increasingly fine-scale, large-coverage spatial data are available. These limitations put challenges in discovering hidden combinations of spatial, nonlinear, and interaction effects that are more likely to occur in the real world but may go undetected due to intrinsic model specification, estimation, or computational issues. In fact, many social and natural behaviors and processes are highly nonlinear and are influenced by spatial contexts which include social interactions, economic activities, urbanization processes, climate systems, and ecological dynamics, among others (Helbing, 2010; Rial et al., 2004; Lek et al., 1996).

Over the past decade, machine learning has emerged as a powerful technique in spatial data science (Li, 2020; Janowicz et al., 2020; Hu et al., 2023; Gao et al., 2024). Machine learning presents significant opportunities for spatial data modeling by addressing many of the limitations inherent in traditional spatial and geostatistical approaches. These models excel at capturing nonlinear interactions, require fewer assumptions, and are highly efficient in processing high-dimensional and high-volume data. As universal approximators, machine learning models do not rely on explicit model specifications and data assumptions; instead, they learn functions and effects directly from the data (Sonoda & Murata, 2017). Additionally, they are scalable to large datasets and can seamlessly incorporate diverse data types (e.g., numerical and categorical) and distributions. The integration of machine learning in the geospatial domain has given rise to the emerging field of geospatial AI (GeoAI) that applies artificial intelligence (AI) to address spatial problems (Gao et al., 2023). GeoAI has shown significant potential and promise across a wide range of applications, including weather prediction (Lam et al., 2023), spatial interpolation (Zhu et al., 2020), spatial object classification and detection (Li, 2020), and many social and human geography studies (Wang et al., 2024a, b).

However, machine learning models are often considered black boxes because their internal workings are not easily understandable. This is primarily due to their complexity, which often involves nonlinear transformations, high-dimensional feature interactions, model ensembling, and iterative training processes. These mechanisms result in a vast number of parameters, making them difficult to interpret and posing challenges from both societal and scientific perspectives (Gunning & Aha, 2019). From a societal perspective, this opacity undermines trust and accountability, particularly in critical areas such as health, finance, criminal justice, and disaster response, where decisions can significantly impact individuals' lives. It also complicates efforts to ensure fairness and mitigate biases, potentially leading to discriminatory outcomes. In the context of spatial analysis, ensuring spatial fairness and reducing spatial inequalities have recently become a focus as AI is increasingly adopted in applications involving location information (e.g., Xie et al., 2022; Saxena et al., 2024). From a scientific perspective, understanding how a model makes predictions is essential for leveraging AI for scientific discovery beyond just predictive tasks. One of the central questions in spatial analysis has been to understand the spatial processes that generate observed spatial patterns, ultimately answering the question, "Why are things where they are?" Leveraging AI's capacity to handle large-scale data in a flexible way can enhance the discovery and understanding of spatial processes, but this cannot be achieved without AI explainability (Reichstein et al., 2019). Moreover, the inability to understand model decisions limits debugging and improvement efforts, hindering performance enhancement, which is especially important for the development of geospatial machine learning models. Addressing these issues requires the substantial development of eXplainable AI (XAI) techniques.

The next sections in this chapter will discuss the basics of XAI and its integration and application in spatial analysis. The chapter concludes by highlighting some caveats of XAI and future directions.

5.2 XAI Basics and Taxonomies

Machine learning has primarily been used for three types of data: text, image, and tabular data. The use of XAI varies significantly depending on the type of data, with each requiring specific techniques to uncover the reasoning behind model predictions. For text data, such as that used in natural language processing (NLP), attention mechanisms are often employed to highlight the words or phrases the model relies on, providing insights into its decision-making process. For image data, methods such as gradient-weighted class activation mapping (Grad-CAM) and occlusion sensitivity are used to identify key pixels in an image that influence the model's output. These methods are particularly valuable for modeling geospatial data obtained from remote sensing, street view images, and maps (Fu et al., 2024). Hsu and Li (2023) offer a detailed comparison and discussion of image-based XAI methods for geospatial applications. In addition, Cheng et al. (2024) present a review of XAI's geospatial applications, with a focus on image-related tasks. For tabular data, which is also commonly used in spatial analysis (e.g., census or georeferenced survey data), feature importance methods that identify important features in the model and partial dependence plots that visualize functional forms of relationships are widely used.

XAI methods can be divided into two main categories: model-based and model-agnostic approaches (Murdoch et al., 2019). Model-based methods are integrated into specific models, offering explanations that are inherently tied to the model's structure. Inherently interpretable models, such as linear regression and decision trees, provide clear explanations through regression coefficients or tree splits. Additionally, model-based methods, such as those used to explain neural networks, are often used in computer vision or NLP tasks. Some feature importance measures for tree-based models, such as those based on impurity or entropy, also fall into this category. On the other hand, model-agnostic methods can be applied to any model, regardless of its structure. The focus of these methods is on the relationship between the input data and the model's output, without relying on the model's internal workings. Examples include permutation feature importance, partial dependence plots, LIME, and Shapley value-based approaches.

In addition to being classified as model-based or model-agnostic, XAI methods can also be distinguished by the level of explanation they provide: global or local (Lundberg et al., 2020). Global XAI methods offer insights into a model's overall behavior, giving a broad view of its decision-making process. These methods are useful for identifying general patterns and rules across a dataset. For example, linear regression coefficients provide a global explanation by showing how each feature affects predictions overall. In more complex models, global explanations often involve examining feature importance or decision rules across the entire dataset such as various global feature importance measures. In recent years, more developments have been focused on local XAI methods which explain the model's behavior for individual predictions. Local XAI offers detailed, case-specific explanations that are particularly important when individuals are of concern rather than groups of the population. More details on local XAI will be introduced in the next section.

5.3 XAI and Spatial Analysis

The advancement from global to local interpretation provides opportunities for explaining models of geospatial data, where observations are georeferenced by coordinates, addresses, or administrative boundaries, allowing explanations to be visualized as maps (Li, 2022). Among the earliest prominent local XAI methods is Locally Interpretable Model-agnostic Explanations (LIME) by Ribeiro et al. (2016). LIME assumes that any complex model can be approximated as locally linear at a given observation. By creating a simple local surrogate model (e.g., linear regression) from perturbed data drawn from a neighborhood around each observation, it approximates relationships at the local level. However, one of LIME's major challenges is defining local proximity, as results can be highly sensitive to this choice (Molnar, 2020). Additionally, the accuracy of local approximations may vary across observations, complicating the comparison of local coefficients. In an effort to address these issues and make LIME more spatially explicit, Jin et al.

(2023) extended the LIME framework by developing the Geographically Localized Interpretable Model-agnostic Explanations (GLIME) method, which defines local neighborhoods based on geographic proximity. They applied GLIME to explain an LSTM model used to classify residential mobility types using both neighborhood- and individual-level characteristics. In this approach, the local explanation model is fixed at the state level, enabling the calculation of state-level feature importance for different feature groups. The results can be visualized on maps to highlight the features that contribute most to predicting residential mobility patterns, which vary across states. GLIME shows promise in capturing state-level effects by sampling data from geographic neighborhoods to account for spatial dependence in each local explanation model. However, it still inherits several limitations from LIME, including the arbitrary selection of neighborhood boundaries, which is subject to the modifiable areal unit problem, and the lack of guarantees for local accuracy and model fidelity. Despite these limitations, GLIME remains a useful tool when interpreting the effects of fixed geographic neighborhoods (e.g., based on administrative boundaries) is of interest.

Another key local XAI development is Shapley value-based approaches from game theory (Shapley, 1953). The Shapley value considers how to fairly distribute contribution among players participating in a coalition game, and the Shapley value for a player j (denoted as φ_j) in a game is given by

$$\varphi_j = \sum_{S \subseteq M \setminus \{j\}} \frac{s!\,(p-s-1)!}{p!} \left(f\,(S \cup \{j\}) - f(S)\right), \tag{5.1}$$

where p is the total number of players, $M\setminus\{j\}$ is a set of all possible combinations of players excluding j, S is a player set in $M\setminus\{j\}$ with a size of s, $f(S)$ is the outcome of S, and $f(S \cup \{j\})$ is the outcome with players in S plus player i. The interpretation of Eq. (5.1) is that the Shapley value of a player is the weighted average of its marginal contribution to the game outcome, taken over all possible combinations of players. In the context of machine learning, the players are features, the model is the game, and the prediction represents the outcome. Therefore, Shapley values provide an additive breakdown of individual prediction into marginal contributions from each feature (Štrumbelj and Kononenko, 2014) as shown by Eq. 5.2:

$$\hat{y}_i = \varphi_0 + \sum\nolimits_{j=1}^{p} \varphi_{ji}, \tag{5.2}$$

where φ_{ji} is the Shapley value for feature j of observation i, and φ_0 is a base value to adjust for the difference between the sum of Shapley values and the prediction value. The Shapley value-based approach has gained significant attention due to the development of Shapley Additive Explanations (SHAP) by Lundberg (2017) that unifies several XAI methods into one Shapley value framework. Lundberg (2017) introduced a Shapley-compliant kernel for LIME that enables the two methods to converge on the same solution. The Shapley-compliant kernel is theoretically stronger than LIME, as it maintains local accuracy across all observations. SHAP also incorporates DeepLIFT for image-related explanations and proposes several estimation methods, such as Tree SHAP, for scalable explanations in tree-based models (Lundberg et al., 2020). SHAP's popularity is enhanced by its user-friendly Python package, *shap*, which includes various Shapley value estimation algorithms and visualization techniques. SHAP is widely employed in leading AI platforms, including Google Vertex, IBM Watson OpenScale, Amazon SageMaker, and Microsoft Fabric, and has been applied in numerous domains. Its academic influence is also substantial, with tens of thousands of papers published based on SHAP so far. In spatial studies, SHAP has been widely applied to explain models involving geospatial data to understand how features determine spatial patterns, with applications in areas such as house price modeling (L. Chen et al., 2020), urban climate analysis (Yuan et al., 2022), species distribution (Song & Estes, 2023), travel behavior (Kashifi et al., 2022; Li, 2023; Zhang et al., 2024), accident analysis (Parsa et al., 2020), disaster prediction (Abdollahi and Pradhan, 2023; Zhang et al., 2023; Cilli et al., 2023; Dikshit et al., 2021), hydrology (Yang & Chui, 2021; Wang et al., 2022), and public health (Kim and Kim, 2022; Zou et al., 2024), among others.

Although the abovementioned examples demonstrate the utility of local interpretation methods in geospatial-related applications, it is important to note that most of existing works primarily focus on explaining nonspatial effects, such as global nonlinearity or variable interaction. As a result, the explanations or effects are not explicitly spatial. It is worth noting that several studies (e.g., Dou et al., 2023; Wang et al., 2024a, b) have attempted mapping SHAP values to explore the spatial distribution of feature contributions. These maps highlight areas where features have strong/weak and positive/negative contributions to predictions, providing spatial patterns of feature importance. However, it is important to note that SHAP values differ from regression coefficients and should not be interpreted in the same way. Unlike geographically weighted regression (GWR), in which local coefficients directly indicate the sign and strength of the relationship between a feature and the prediction, mapped SHAP values represent the marginal contribution of a feature to a specific prediction and do not imply a directional relationship in the same sense. For example, a positive SHAP value can be observed when a feature's value is below average and there is a negative relationship between feature and prediction. Li (2022) explored further the connection between SHAP and well-known spatial statistical models using simulated data and found that when coordinates are input as features, coordinate interactions and their interactions with other features can be analogous to spatial fixed effect and spatially varying effects that have similar coefficient values compared to the spatial lag model and multi-scale geographically weighted regression (MGWR) model. However, the off-the-shelf SHAP values have two limitations for spatial data: (1) interaction effects can only be calculated for tree-based models not for others such as neural networks; and (2) SHAP does not consider coordinates as a single joint feature which violates Shapley values' properties. To solve both issues, Li (2024) developed GeoShapley that extended Shapley interaction values and Joint Shapley values and proposed a kernel-based estimation method that is agnostic to model structures. More computational details can be found in Li (2024), but the final output of GeoShapley has four components that add up to the model prediction:

$$\hat{y} = \phi_0 + \boldsymbol{\phi}_{\mathbf{GEO}} + \sum\nolimits_{j=1}^{p} \boldsymbol{\phi}_j + \sum\nolimits_{j=1}^{p} \boldsymbol{\phi}_{(\mathbf{GEO},j)}, \tag{5.3}$$

where ϕ_0 is a constant base value representing the average prediction given the background data, serving as the global intercept. $\boldsymbol{\phi}_{\mathbf{GEO}}$ is a vector of size n that measures the intrinsic location effect in the model, analogous to the local intercept in geographically weighted regression models or spatial fixed effects. The intrinsic location effect captures unmeasurable contextual influences, which can stem from cultural factors, historical legacies, people's perceptions and attachments to places, and local social interactions, among other factors (Sampson, 2011; Fotheringham & Li, 2023). $\boldsymbol{\phi}_j$ is a vector of size n for each nonlocation feature j, representing location-invariant effects, such as global linear and nonlinear contributions to the model. $\boldsymbol{\phi}_{(\mathbf{GEO},j)}$ is a vector of size n for each nonlocation feature j, capturing the spatially varying interaction effects in the model. These effects can be further visualized as spatially varying coefficients, analogous to the coefficients in geographically weighted regression, by estimating the partial derivative between $\boldsymbol{\phi}_{(\mathbf{GEO},j)}$ and $\boldsymbol{X}_j$. If there are no spatial effects in the model, the terms $\boldsymbol{\phi}_{\mathbf{GEO}}$ and $\boldsymbol{\phi}_{(\mathbf{GEO},j)}$ will be zero, reducing Eq. (5.3) to Eq. (5.2), which corresponds to the normal SHAP values. GeoShapley thus provides a coherent bridge between traditional statistical approaches and modern machine learning models. Though recent, GeoShapley has been applied to measure spatial effects from machine learning models in empirical geospatial studies (e.g., Ke et al., 2025; Peng et al., 2025; Foroutan et al., 2025).

In addition to LIME and Shapley value-based methods, it is worth noting there are few other local XAI techniques that have been explored or developed within the literature. For instance, Masrur et al. (2022) introduced a spatiotemporally interpretable random forest (iST-RF) that incorporates a spatiotemporal sampling strategy for model training and weighted prediction. This model includes an impurity-based feature importance score to enable both local and global interpretations. The iST-RF outperformed aspatial random forest models in predicting wildfire occurrence. Its sampling strategies also facilitate the construction of local tree models that reveal local feature importance, aiding in the exploration of how environmental variables influence wildfire occurrence and their regional partial relationships. Liu et al. (2024) explored applying GNNExplainer to graph neural network (GNN) models for predicting traffic volume over street networks. GNNExplainer creates counterfactual explanations by perturbing the graph structure to assess changes in predictions. This approach produces local feature importance scores that highlight key node connections and demonstrate how features impact predictions. Given the growing use of GNNs in spatial applications, GNNExplainer offers significant potential for interpreting these complex networks. However, it is worth noting that the feature importance scores in both GNNExplainer and iST-RF are not tied to specific prediction units, making them less intuitive compared to Shapley value-based feature importance scores.

5.4 An Example of Predicting Voting Preferences in 2020 Presidential Election

This section demonstrates an empirical example application of XAI in spatial analysis. Given the focus of this book on human geography and GeoAI, voting behavior has been chosen as the example, as studying electoral patterns is a key area of interest in human geography. Here, we model the voting results from the 2020 Presidential election at the county level and examine its determinants using machine learning and GeoShapley. The data is sourced from Fotheringham and Li (2023), who originally obtained the voting data from the MIT Election Lab and socio-demographic data from the 2015–2019 ACS 5-year estimates. The dependent variable is the county-level percentage of votes for the Democratic Party in a two-party contest, excluding third-party votes. Seven key socio-demographic variables (percentage of Black, percentage of Hispanic, percentage of bachelor's degree or higher, percentage of foreign-born, log-transformed population density, median household income, and percentage of third-party vote) were selected to explain the spatial distribution of voting patterns. The analysis covers the contiguous USA, including 49 states (and Washington, D.C.), for a total of 3108 counties.

For model training, the Microsoft's Fast Library for Automated Machine Learning and Tuning (FLAML) python package (Wang et al., 2021) was used to train the machine learning models, with hyperparameters optimized through fivefold cross-validation. GeoShapley values were then calculated for the final trained model using the *geoshapley* python package. To estimate the uncertainty of these values, a bootstrap sampling process was applied: 500 bootstrap resamples of the dataset were used to retrain the model, and GeoShapley values were recalculated for each sample. This process generated a distribution of GeoShapley values to account for uncertainties within the machine learning model. To provide a benchmark, an multi-scale geographically weighted regression (MGWR) model was also estimated using the same data by the *mgwr* python package (Oshan et al., 2019), offering a comparison with a more established spatial model. Code and results reported in this book chapter can be found at this https://github.com/Ziqi-Li/XAI_spatial_analysis_chapter.

In terms of accuracy, the best-performing machine learning model was XGBoost, with a fivefold cross-validated out-of-sample R^2 value of 0.933. The MGWR model achieved the same value R^2, but it is an in-sample R^2. Due to fundamental differences in model estimation and training procedures, these two R^2 values cannot be directly compared, but they generally suggest that both models perform similarly well for this dataset. Now, we can first look at the results from XGBoost and GeoShapley. Figure 5.1 presents a ranking of global feature importance in the XGBoost model based on averaged absolute GeoShapley values across all counties. For each feature, the blue component represents the averaged primary effect, and the red component represents the feature's averaged location-varying effect. The figure shows that the percentage of Black residents and the percentage of bachelor's degrees have the highest total contribution to the model's prediction, followed by the

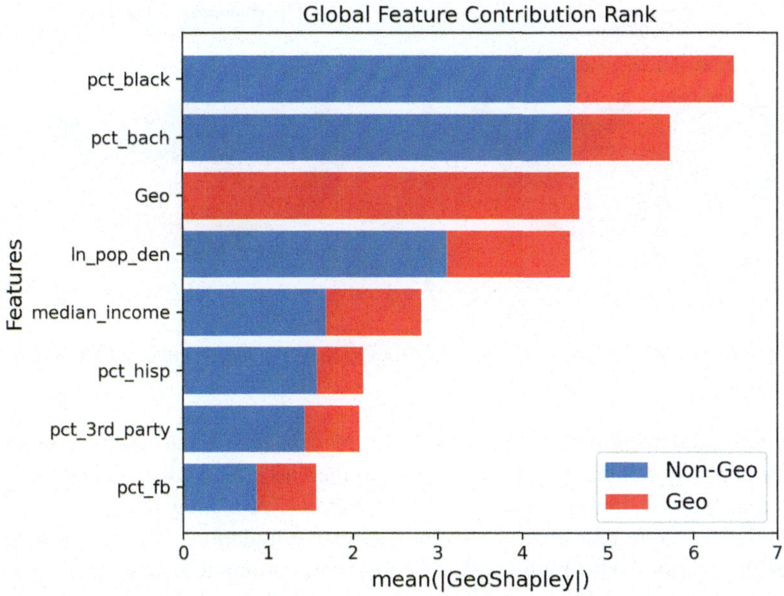

Fig. 5.1 Global feature importance ranking in the XGBoost model based on GeoShapley values

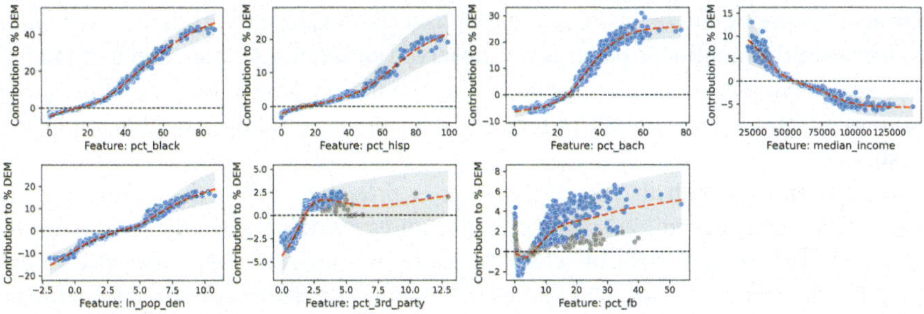

Fig. 5.2 Partial dependence plots of primary effects

intrinsic location effect and population density. For each feature, there is a varying amount of location-specific effect, but these effects are smaller than the location-invariant primary effects.

Figure 5.2 shows the primary effects using partial dependence plots. The y-axis in each plot represents the contribution of the corresponding feature to the prediction of the Democratic vote share. All features exhibit varying degrees of nonlinear association with Democratic vote share. For instance, the percentage of Black population and individuals with a bachelor's degree or higher show a positive relationship characterized by an S-shaped curve. As the values of these features increase, the Democratic vote share initially increases slowly, then accelerates, before flattening out as the feature values approach their maximum.

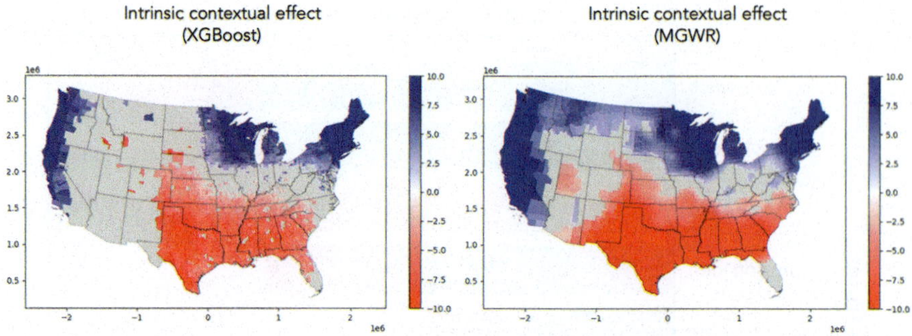

Fig. 5.3 Intrinsic contextual effect estimated from XGBoost + GeoShapley and MGWR

This indicates nonlinear dynamics: at lower percentages of bachelor's degrees and Black population, the Democratic vote share increases gradually. As feature values increase, the rise in Democratic votes accelerates, reflecting the strong political alignment of more educated and racially diverse populations with the Democratic Party. In counties with high levels of educational attainment and racial diversity, Democratic support grows rapidly, but at very high levels, the increase flattens out again, as the population becomes overwhelmingly Democratic. These nonlinear relationships suggest that the effect of education and race on Democratic voting is strongest at moderate levels, while the impact diminishes at very low or very high values. Similar nonlinear patterns are observed in other variables: median income has a negative nonlinear relationship, while population density (logged), third-party vote share, and the percentage of foreign-born population show a nonlinear positive association.

A comparison with the MGWR model is presented below. Li (2024) demonstrated that the contribution from location features is analogous to the local intercept in an MGWR model, both of which measure the intrinsic contextual effects that quantify the importance of location when holding other features constant. Estimates beyond their 95% confidence intervals are masked in gray. The two maps reveal a similar pattern: counties in the South are intrinsically leaning toward the Republican Party, while counties in the Pacific West, Midwest, and Northeast are intrinsically leaning toward the Democratic Party, with other demographic factors held constant. Such pattern aligns well with the general public perception of regional political ideology (Fig. 5.3).

The comparison between MGWR and XGBoost + GeoShapley spatially varying coefficients for the top three most important features is shown in Fig. 5.4. To obtain spatially varying coefficients from GeoShapley, we can use a spatial smoothing method, such as GWR used here, to regress the combined primary and location-varying effects against the respective feature values. These values are then comparable to MGWR's local coefficients. Figure 5.4 shows that both approaches reveal similar coefficient patterns. Nationally, for both models, the percentage of Black population, population density, and percentage of bachelor's degree have

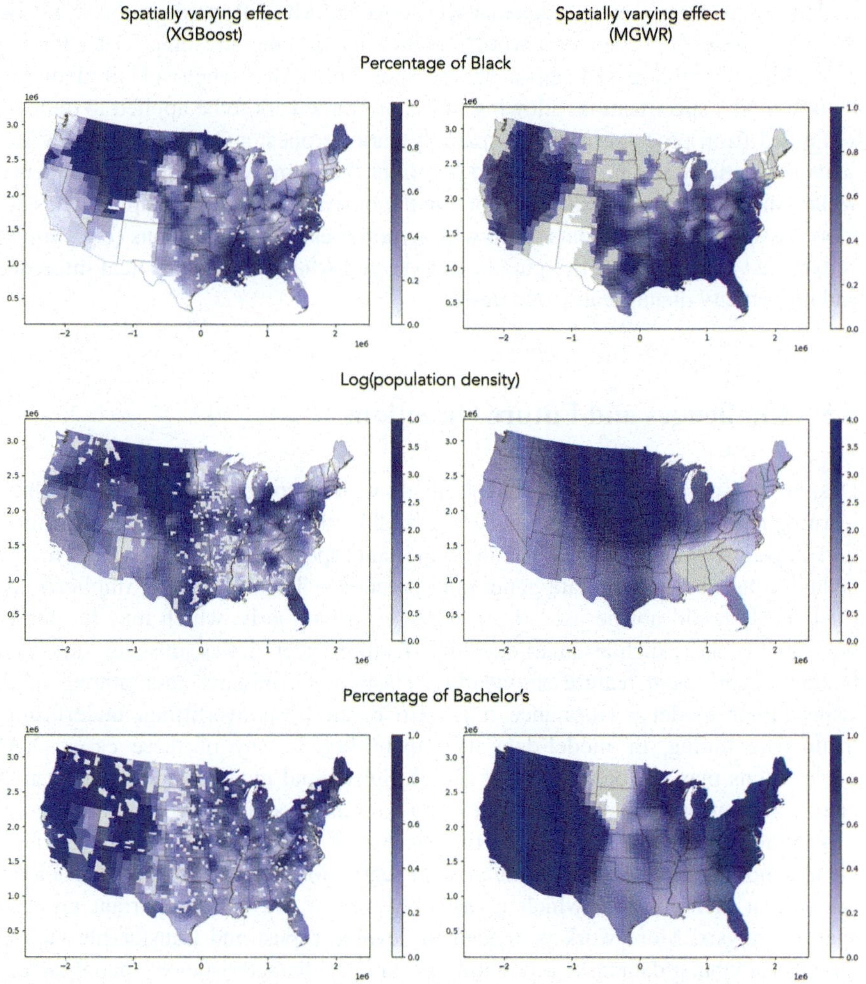

Fig. 5.4 Spatially varying coefficients estimated from XGBoost + GeoShapley and MGWR

a positive relationship with Democratic support. Locally and regionally, for the percentage of Black population, both maps highlight stronger associations with Democratic support in the Mountain West and Southeast, whereas states with smaller Black populations, such as Arizona, New Mexico, and Colorado, show little to no association. For population density, the Northwest Central Region shows a stronger positive association compared to the rest of the country. For the percentage of bachelor's degree feature, the West, Midwest, and Northeast display stronger positive associations compared to other regions, particularly in the Dakotas, where the association is low or insignificant. More detailed interpretations in the voting context can be found in Fotheringham et al. (2023) and Li and Fotheringham (2022).

The comparison demonstrates that XGBoost and MGWR yield similar results in this case study. However, it is worth mentioning that the advantages of adopting a machine learning-based approach are more favorable when: (1) it does not require model specification, allowing the same framework to be applied to datasets generated from processes beyond spatially varying ones; and (2) GeoShapley can separate spatially varying effects and highlight the primary nonlinear components, which linear statistical models cannot. On the other hand, statistical approaches are more favorable when the data align with hypotheses and assumptions (e.g., voting is expected to be affected by spatial context), and when formal statistical inference and uncertainty quantification are desired.

5.5 Challenges and Future Directions

Despite the success of various XAI methods which provide insights into machine learning models that were previously unavailable, there are several intrinsic limitations, caveats, or challenges. The most important aspect is that XAI describes model behavior but not the true data-generating processes. This has several implications. First, XAI should not be used if a model performs poorly, which may be due to poor-quality data, such as small data size, feature uncertainties, missing important features, insufficient feature engineering, biased data, missing data, among other issues. Poor model performance may also result from overfitting, underfitting, inadequate tuning, or model-data incompatibility. In any of these cases, XAI explanations may not be sensible or useful. Substantial model training and feature engineering efforts are required to develop a robust model before generating good explanations. Furthermore, despite the success of GeoAI in image-related tasks, GeoAI models for traditional tabular data regression and classification problems are still in their infancy, which is arguably one of the most important tasks in spatial analysis. More work is needed to develop robust and transferable GeoAI models for tabular data tasks accounting for unique characteristics of spatial tabular data. Finally, computing uncertainties for XAI explanations will help quantify the uncertainties in the model. For the example presented above, a bootstrap strategy can be adopted to generate a distribution of GeoShapley values, from which confidence intervals can be derived. The second implication is that model explanations can vary between different models, especially when features are correlated. For example, Shapley values for random forest and XGBoost can differ because of how each model handles feature correlations (Chen et al., 2020; Li, 2022). In random forest, feature importance is often spread more evenly across correlated features (Strobl et al., 2007), while in XGBoost, one feature may dominate the explanation due to its sequential learning process (Chen & Guestrin, 2016). Although Shapley values still reflect what a model considers important, special attention is needed when interpreting and reporting the results.

One caveat worth stressing is in interpreting the explanations derived from Shapley values. Shapley values represent the marginal contribution of features to the model prediction while holding all other features constant. Though there is a direct link between Shapley values and linear regression coefficients, Shapley values themselves do not have the same interpretation as regression coefficients. Positive Shapley values can result from lower-than-average feature values while having a negative relationship with prediction or higher-than-average feature values while having a positive relationship with prediction. This can cause confusion, especially when visualizing Shapley values spatially, as it involves visualizing four-dimensional spatial partial dependence, which is inherently more challenging than aspatial effects like those shown in Fig. 5.2. GeoShapley attempts to use GWR as a spatial smoother to generate spatially varying coefficients, but these are inevitably exploratory and deviate from the true GeoShapley values which adhere to game theory properties. Future work should focus on better and more formal methods for visualizing spatial contributions in models.

Additionally, although a significant body of literature in GeoAI has focused on improving model performance, equally important is the development of more spatially explicit interpretation methods to better understand geographic knowledge from machine learning models. GLIME by Jin et al. (2023) and GeoShapley by Li (2024) are good examples, and they inspire further developments to highlight spatial effects that spatial data scientists are interested in.

Lastly, there has been increasing interest in spatial causality to better understand the "why" question, going beyond correlation analysis (Akbari, Winter, and Tomko, 2023; Credit and Lehnert, 2023; Gao et al., 2023; Zhang and Wolf, 2023). Janzing, Minorics, and Blöbaum (2020) argued that Shapley values inherently have a causal interpretation when using interventional algorithms (such as kernel SHAP and GeoShapley), which examine how a model behaves when the dependence among features is deliberately disrupted. Shapley values also provide a way to investigate more complex causal effects within machine learning models and have been shown to be effective in measuring heterogeneous treatment effects using instrumental variables (Syrgkanis et al., 2019). Furthermore, Heskes et al. (2020) introduced causal Shapley values, a model-agnostic approach to measure the direct and indirect effects of each feature on the model's prediction. Extending and applying these causal ML methods to geospatial tasks would be a valuable direction for future research.

References

Anselin, L. (1988). *Spatial econometrics: Methods and models* (Vol. 4). Springer Science & Business Media.

Chen, T., & Guestrin, C. (2016, August). Xgboost: A scalable tree boosting system. In *Proceedings of the 22nd acm sigkdd international conference on knowledge discovery and data mining* (pp. 785–794).

Chen, H., Janizek, J. D., Lundberg, S., & Lee, S. I. (2020). True to the model or true to the data?. *arXiv preprint arXiv:2006.16234.*

Cheng, X., Vischer, M., Schellin, Z., Arras, L., Kuglitsch, M. M., Samek, W., & Ma, J. (2024). Explainability in GeoAI. In *Handbook of geospatial artificial intelligence* (pp. 177–200). CRC Press.

Dou, M., Gu, Y., & Fan, H. (2023). Incorporating neighborhoods with explainable artificial intelligence for modeling fine-scale housing prices. *Applied Geography, 158*, 103032.

Foroutan, E., Hu, T., & Li, Z. (2025). Revealing key factors of heat-related illnesses using geospatial explainable AI model: A case study in Texas, USA. *Sustainable Cities and Society, 122*, 106243.

Fotheringham, A. S., & Li, Z. (2023). Measuring the unmeasurable: Models of geographical context. *Annals of the American Association of Geographers, 113*(10), 2269–2286.

Fotheringham, A. S., Brunsdon, C., & Charlton, M. E. (2002). Geographically weighted regression. *The Sage handbook of spatial analysis, 1*, 243–254.

Fotheringham, A. S., Oshan, T. M., & Li, Z. (2023). *Multiscale geographically weighted regression: Theory and practice*. CRC Press.

Fu, C., Zhou, Z., Xin, Y., & Weibel, R. (2024). Reasoning cartographic knowledge in deep learning-based map generalization with explainable AI. *International Journal of Geographical Information Science, 38*, 1–22.

Gao, S., Hu, Y., & Li, W. (Eds.). (2023). *Handbook of geospatial artificial intelligence*. CRC Press.

Griffith, D. A. (2003). *Spatial filtering* (pp. 91–130). Springer.

Gunning, D., & Aha, D. (2019). DARPA's explainable artificial intelligence (XAI) program. *AI Magazine, 40*(2), 44–58.

Helbing, D. (2010). *Quantitative sociodynamics: Stochastic methods and models of social interaction processes*. Springer Science & Business Media.

Hengl, T., Heuvelink, G. B., & Rossiter, D. G. (2007). About regression-kriging: From equations to case studies. *Computers & Geosciences, 33*(10), 1301–1315.

Hsu, C. Y., & Li, W. (2023). Explainable GeoAI: Can saliency maps help interpret artificial intelligence's learning process? An empirical study on natural feature detection. *International Journal of Geographical Information Science, 37*(5), 963–987.

Jin, C., Park, S., Ha, H. J., Lee, J., Kim, J., Hutchenreuther, J., & Nara, A. (2023). Predicting households' residential mobility trajectories with geographically localized interpretable model-agnostic explanation (GLIME). *International Journal of Geographical Information Science, 37*(12), 2597–2619.

Ke, E., Zhao, J., & Zhao, Y. (2025). Investigating the influence of nonlinear spatial heterogeneity in urban flooding factors using geographic explainable artificial intelligence. *Journal of Hydrology, 648*, 132398.

Lee, D. (2013). CARBayes: An R package for Bayesian spatial modeling with conditional autoregressive priors. *Journal of Statistical Software, 55*(13), 1–24.

Lek, S., Delacoste, M., Baran, P., Dimopoulos, I., Lauga, J., & Aulagnier, S. (1996). Application of neural networks to modelling nonlinear relationships in ecology. *Ecological Modelling, 90*(1), 39–52.

Li, W. (2020). GeoAI: Where machine learning and big data converge in GIScience. *Journal of Spatial Information Science, 20*, 71–77.

Li, Z. (2022). Extracting spatial effects from machine learning model using local interpretation method: An example of SHAP and XGBoost. *Computers, Environment and Urban Systems, 96*, 101845.

Li, Z. (2023). Leveraging explainable artificial intelligence and big trip data to understand factors influencing willingness to ridesharing. *Travel Behaviour and Society, 31*, 284–294.

Li, Z. (2024). GeoShapley: A game theory approach to measuring spatial effects in machine learning models. *Annals of the American Association of Geographers, 114*, 1–21.

Li, Z., & Fotheringham, A. S. (2022). The spatial and temporal dynamics of voter preference determinants in four US presidential elections (2008–2020). *Transactions in GIS, 26*(3), 1609–1628.

Lindgren, F., & Rue, H. (2015). Bayesian spatial modelling with R-INLA. *Journal of Statistical Software, 63*(19), 1.

Liu, P., Zhang, Y., & Biljecki, F. (2024). Explainable spatially explicit geospatial artificial intelligence in urban analytics. *Environment and Planning B: Urban Analytics and City Science, 51*(5), 1104–1123.

Lundberg, S. (2017). A unified approach to interpreting model predictions. *arXiv preprint arXiv:1705.07874.*

Lundberg, S. M., Erion, G., Chen, H., DeGrave, A., Prutkin, J. M., Nair, B., & Lee, S. I. (2020). From local explanations to global understanding with explainable AI for trees. *Nature Machine Intelligence, 2*(1), 56–67.

Murdoch, W. J., Singh, C., Kumbier, K., Abbasi-Asl, R., & Yu, B. (2019). Definitions, methods, and applications in interpretable machine learning. *Proceedings of the National Academy of Sciences, 116*(44), 22071–22080.

Oshan, T. M., Li, Z., Kang, W., Wolf, L. J., & Fotheringham, A. S. (2019). Mgwr: A python implementation of multiscale geographically weighted regression for investigating process spatial heterogeneity and scale. *ISPRS International Journal of Geo-Information, 8*(6), 269.

Peng, Z., Ji, H., Yuan, R., Wang, Y., Easa, S. M., Wang, C., & Zhao, X. (2025). Modeling and spatial analysis of heavy-duty truck CO2 using travel activities. *Journal of Transport Geography, 124*, 104158.

Rial, J. A., Pielke, R. A., Beniston, M., Claussen, M., Canadell, J., Cox, P., & Salas, J. D. (2004). Nonlinearities, feedbacks and critical thresholds within the Earth's climate system. *Climatic Change, 65*, 11–38.

Ribeiro, M. T., Singh, S., & Guestrin, C. (2016, August). "Why should i trust you?" Explaining the predictions of any classifier. In *Proceedings of the 22nd ACM SIGKDD international conference on knowledge discovery and data mining* (pp. 1135–1144).

Sampson, R. J. (2011). 11 neighborhood effects, causal mechanisms and the social structure of the city. In *Analytical sociology and social mechanisms* (Vol. 227). Cambrigde University.

Saxena, N. A., Zhang, W., & Shahabi, C. (2024). Spatial fairness: The case for its importance, limitations of existing work, and guidelines for future research. *arXiv preprint arXiv:2403.14040.*

Song, L., & Estes, L. (2023). ITSDM: Isolation forest-based presence-only species distribution modelling and explanation in R. *Methods in Ecology and Evolution, 14*(3), 831–840.

Sonoda, S., & Murata, N. (2017). Neural network with unbounded activation functions is universal approximator. *Applied and Computational Harmonic Analysis, 43*(2), 233–268.

Strobl, C., Boulesteix, A. L., Zeileis, A., & Hothorn, T. (2007). Bias in random forest variable importance measures: Illustrations, sources and a solution. *BMC Bioinformatics, 8*, 1–21.

Wang, C., Wu, Q., Weimer, M., & Zhu, E. (2021). Flaml: A fast and lightweight automl library. *Proceedings of Machine Learning and Systems, 3*, 434–447.

Wang, S., Peng, H., & Liang, S. (2022). Prediction of estuarine water quality using interpretable machine learning approach. *Journal of Hydrology, 605*, 127320.

Wang, N., Zhang, H., Dahal, A., Cheng, W., Zhao, M., & Lombardo, L. (2024a). On the use of explainable AI for susceptibility modeling: Examining the spatial pattern of SHAP values. *Geoscience Frontiers, 15*(4), 101800.

Wang, S., Huang, X., Liu, P., Zhang, M., Biljecki, F., Hu, T., & Bao, S. (2024b). Mapping the landscape and roadmap of geospatial artificial intelligence (GeoAI) in quantitative human geography: An extensive systematic review. *International Journal of Applied Earth Observation and Geoinformation, 128*, 103734.

Xie, Y., He, E., Jia, X., Chen, W., Skakun, S., Bao, H., & Ravirathinam, P. (2022, June). Fairness by "where": A statistically-robust and model-agnostic bi-level learning framework. *Proceedings of the AAAI conference on artificial intelligence, 36*(11), 12208–12216.

Yang, Y., & Chui, T. F. M. (2021). Modeling and interpreting hydrological responses of sustainable urban drainage systems with explainable machine learning methods. *Hydrology and Earth System Sciences, 25*(11), 5839–5858.

Zhang, X., Zhou, Z., Xu, Y., & Zhao, X. (2024). Analyzing spatial heterogeneity of ridesourcing usage determinants using explainable machine learning. *Journal of Transport Geography, 114*, 103782.

Chapter 6
Human-Centered Computer Vision for Urban Sensing

Hanyu Yin and Fan Zhang

Abstract As urbanization continues to accelerate, the structural and functional complexity of cities grows, making the effective management and understanding of urban spaces increasingly critical. At the same time, advancements in the collection, sharing, and analysis have greatly improved our capacity to comprehend the impact of urban physical environments on human activities. Urban sensing technology, in particular, has emerged as an indispensable tool for studying urban dynamics and enhancing the quality of human life (Liu et al., *Annals of the Association of American Geographers, 105*:512–530, 2015). Computer vision technology has become increasingly important for understanding and representing the urban physical environment, making it a key tool for urban sensing (Zhang et al., *Annals of the American Association of Geographers, 114*:876–897, 2024). However, traditional computer vision methods often focus on technical improvement and data analysis, overlooking the most essential component of cities—people. Human-centered computer vision bridges this gap by not only analyzing physical spaces, but also focusing on residents' behavior patterns, social interactions, and living needs (Luca et al., *ACM Computing Surveys, 55(1)*:1–7:44, 2021; Salazar-Miranda et al., *Landscape and Urban Planning, 234*:104715, 2023). By gaining a deeper understanding of human-environment interactions, we can better investigate urban dynamics and formulate more sustainable development policies. This chapter discusses the role and applications of human-centered computer vision in urban sensing, focusing on how this technology can be used to enhance our understanding of city residents and improve their quality of life.

Keywords Human-centered research · Urban sensing · Computer vision

H. Yin · F. Zhang (✉)
Institute of Remote Sensing and Geographical Information System, School of Earth and Space Sciences, Peking University, Beijing, China
e-mail: fanzhanggis@pku.edu.cn

© The Author(s), under exclusive license to Springer Nature Switzerland AG 2025

X. Huang et al. (eds.), *GeoAI and Human Geography*, Springer Geography,
https://doi.org/10.1007/978-3-031-87421-5_6

6.1 Human-Centered Urban Sensing and Computer Vision

6.1.1 Overview of Urban Sensing and Computer Vision

Human activities and the urban physical environment are deeply interconnected (Stern, 1993). On the one hand, urban space and environmental quality affect people's lives, mobility, and interpersonal relationships (Mouratidis, 2018, 2021). On the other hand, humans continuously construct cities to promote sustainable urban development (Goudie, 2018; Xiao et al., 2024). To optimize the built environment, it is essential to explore human activities in various urban contexts and their interrelationship. Urban sensing addresses this need by employing technological means to collect, process, and analyze data from human activities, thus providing valuable insights into urban dynamics and structures (Luca et al., 2021; Salazar-Miranda et al., 2023). This approach facilitates sustainable urban management and planning, ultimately improving the quality of life for residents (Liu et al., 2015; Long & Thill, 2015).

In the field of urban sensing, computer vision has become a crucial technology. It enables computers to "see" and interpret visual information by processing and analyzing images and videos. Depending on the tasks and observational perspectives, computer vision can be categorized into different aspects. Its primary tasks for urban sensing applications include object detection, semantic segmentation, and image restoration (Ibrahim et al., 2020; Marasinghe et al., 2024). Object detection identifies specific objects within an image and determines their locations, facilitating real-time monitoring and analysis of dynamic changes in urban environments (Zou et al., 2023). Semantic segmentation divides an image into regions with specific meanings, aiding in the analysis of urban morphological features and the composition of scene elements (Zhang et al., 2018a). Image Restoration generates high-quality, complete images from partial or low-quality inputs, which is useful for restoring and analyzing historical images and simulating missing datasets (He et al., 2022).

In urban analysis using computer vision, different perspectives such as top-down, oblique, and horizontal views can offer diverse insights into the spatial characteristics of the environment (Lefèvre et al., 2017; Luo et al., 2022). The top-down view, commonly used in remote sensing imagery such as satellite images and vertical drone photos, is effective for capturing large-scale data. However, this perspective usually lacks the resolution to detail internal urban spaces (Herold et al., 2004; Van de Voorde et al., 2011). The oblique view employs oblique photogrammetry techniques to capture three-dimensional information about urban areas, though data collection using drones can be relatively expensive (Colomina & Molina, 2014; H. Yao et al., 2019). The horizontal view, typically seen in street view imagery, covers most global cities and provides detailed representations of the urban environment from a human perspective, making it a valuable data source for human-centered studies (Zhang et al., 2018b).

6.2 Overview and Applications of Street View Imagery

6.2.1 Overview of Street View Imagery

Street view imagery consists of images that depict urban streetscapes, providing human-perspective visual information, and is easily accessible through online map services. This data enables users to navigate real urban environments remotely via interactive panoramic images, offering a highly immersive experience. In addition, street view imagery presents several significant advantages, including ease of access, extensive spatiotemporal coverage, reduced data bias, and low cost. Furthermore, it provides a consistent and objective representation of the built environment from a standardized viewpoint, making it a valuable resource for urban studies and human-centered research (Huang et al., 2023).

Street view imagery can be categorized into two main types based on its source: commercial providers and volunteered geographic information (VGI). Commercial street view imagery is primarily provided by companies such as Google, Bing, Here, Baidu, Amap, and Tencent, typically requiring paid access (Anguelov et al., 2010). This type of data is renowned for its superior quality, achieved through significant investments by suppliers in data collection, processing, and maintenance. These providers set high industry standards in terms of image quality, resolution, coverage, and update frequency, making their data invaluable for urban planning, transportation, tourism, and geographic information systems (GIS) (Anguelov et al., 2010). On the other hand, VGI-based street view imagery is collected and updated through public participation, with individuals taking and uploading photos using their smartphone applications. Users can upload street view images through open-source platforms like Mapillary and OpenStreetView, or social media channels such as Twitter and Weibo. With the development of VGI technology, such crowdsourced street view imagery has become an important supplementary source for street view imagery, although it is generally less standardized compared to commercial street view imagery.

Street view imagery offers significant advantages, including extensive coverage, easy accessibility, and a human-perspective presentation. This data encompasses a wide range of geographic areas, reflecting the environmental characteristics of various cities and regions. With the support of existing platforms and technologies, researchers can conveniently access large volumes of street view imagery. Moreover, since these images are captured from a human perspective, they closely align with the actual perceptions of city residents, making them valuable for human-centered urban studies (Hou et al., 2024; Kang et al., 2020). However, street view imagery also presents some limitations, such as low temporal resolution and uneven spatial sampling. The infrequent updates of street view images make it difficult to capture real-time urban dynamics. Additionally, the varying coverage density across different regions can result in uneven spatial data distribution, potentially neglecting some remote and underdeveloped areas(Juhász & Hochmair, 2016; Miranda et al., 2020).

6.2.2 Combining Computer Vision and Street View for Urban Sensing

Places serve as spatial carriers of human activities, and exploring the characteristics of different places helps to understand human activity patterns. Street view images are valuable data sources for urban sensing studies, while computer vision technology plays an important role in urban data mining (Zhang et al., 2024). Applying various computer vision tasks on street view imagery can enhance our understanding of the characteristics of places where people live (Biljecki & Ito, 2021). For example, semantic segmentation of street view images can answer the question, "How are urban scenes composed?" This helps to distinguish areas rich in natural elements like sky and trees from those dominated by artificial elements such as roads and buildings (Zhang et al., 2018a). Comparing street view features across different cities and classifying images can address the question, "What makes a city unique?" Using deep convolutional networks and classification algorithms, distinctive landmarks in each city, such as Tiananmen Square in Beijing or the Eiffel Tower in Paris, can be identified (Zhang et al., 2019b). This method of image comparison also helps explore questions like, "Is there a trend of cities looking increasingly similar?" and "What are the differences in architectural styles among cities?"

Additionally, combining street view imagery with other multi-source data helps to better understand urban scene characteristics and analyze their impact on people. For instance, integrating street view imagery with vehicle flow data can help identify urban villages (Huang et al., 2023) and predict changes in traffic patterns (Zhang et al., 2019a). Merging street view imagery with environmental monitoring data allows researchers to calculate environmental noise (Huang et al., 2024) and air quality (O'Regan et al., 2022) from an image perspective. By combining street view imagery with points of interest (POI) data, we can analyze functional area types (Ye et al., 2021), street quality (Wang et al., 2022), and commercial vitality (Li & Long, 2024). Integrating street view imagery with resident health data facilitates studies on various health aspects, such as physiological health (e.g., obesity (X. Xiao et al., 2021), diabetes (Keralis et al., 2020), myocardial infarction (Yao et al., 2022a, 2022b) and psychological health (e.g., depression (Helbich et al., 2019)) Combining street view imagery with socioeconomic data allows for the analysis of the relationship between the physical environment and crime cases (Yao et al., 2023; Zhang et al., 2021). Additionally, integrating street view images with remote sensing images enables the study of urban spatial morphology (Hu et al., 2020; Middel et al., 2019), urban heat islands (Dirksen et al., 2019; Shaker et al., 2019), and urban disasters (Xing et al., 2023).

In conclusion, street view imagery provides detailed geographic and visual information. Leveraging computer vision technology to analyze this data greatly enhances our understanding of urban spaces and human activities. Moreover, integrating street view imagery with multi-source data further broadens and deepens research insights, providing robust support for scientific urban planning and management.

6.3 Human-Centered Place Sensing Case Studies

Street view imagery reflects the urban physical environment from a human perspective, but relying solely on images limits our understanding of how people experience and interact with these environments. Different urban physical environments can influence human perceptions and behaviors. Therefore, by estimating emotional perceptions and behavioral responses from the environments shown in street view imagery, we can conduct more human-centered urban research. This approach not only enhances our understanding of the impact of urban environments on residents' lives but also uncovers the complexity of human-environment interactions, offering new insights and methods for scientific urban planning and management. The following sections present two specific case studies to demonstrate how this human-centered sensing approach can be applied in practical research.

6.3.1 AI Perception Map

With the rapid pace of global urbanization, creating more livable and human-centered urban environments has become a key challenge in urban planning and management. Residents' perceptions and sentimental responses to the urban environment significantly reflect the city's livability. However, traditional survey and data collection methods are often time-consuming and labor-intensive, making it difficult to capture residents' feelings on different urban physical environments. By combining computer vision with people's sentimental perceptions of street views, we can automate the evaluation of residents' perceptions of urban spaces and efficiently generate AI perception maps. This section will introduce research published in the journal "*Landscape and Urban Planning*" titled "Measuring human perceptions of a large-scale urban region using machine learning," which demonstrates how to create perception maps using AI methods.

To quantify people's sentimental perceptions of street environments, corresponding street view imagery and people's environmental perception data are needed. In 2013, the MIT Media Lab launched the "Place Pulse 2.0" project. This was an online data collection platform designed to gather human perception ratings of urban landscapes (Fig. 6.1). On the website, participants are shown two side-by-side street view images, and are asked to choose one image in response to the question, "Which place looks more like X?" where X can be one of six human perception dimensions: "safer," "livelier," "more boring," "wealthier," "more depressing," and "more beautiful." Participants are required to choose one of three options—the left image, the right image, or "equal"—to indicate their perceptions for the pair of street view images. The image dataset includes 110,988 street view images taken between 2007 and 2012, covering 56 cities in 28 countries across six continents. As of October 2016, 81,630 online participants had contributed 1,169,078 pairwise comparisons. Since volunteers came from diverse regions and each street view

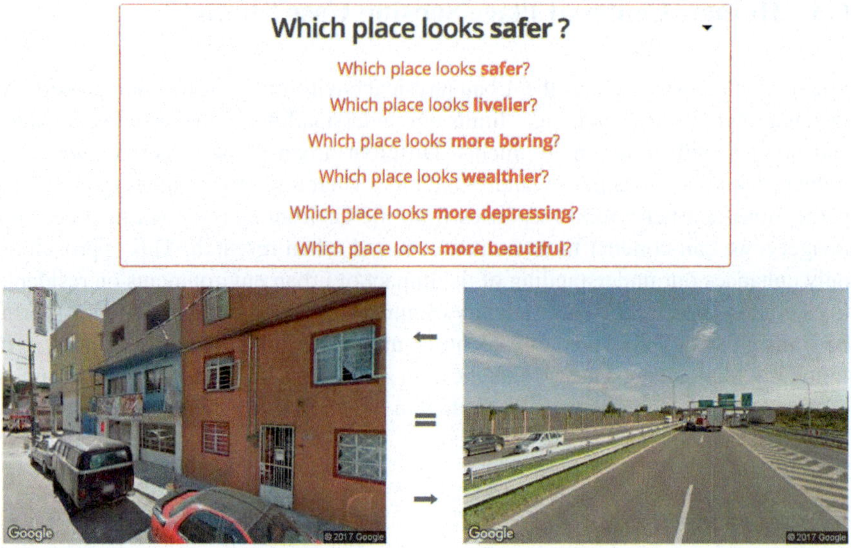

Fig. 6.1 Street view and user perception rating platform. (this figure is adapted from (Zhang et al., 2018b))

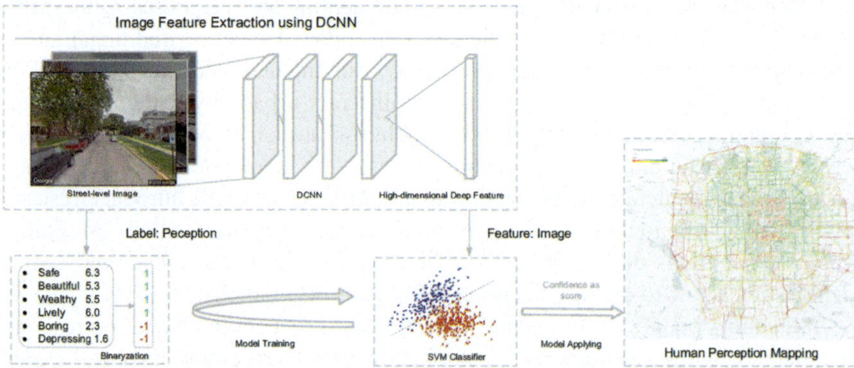

Fig. 6.2 Predicting human's perceptions of road environments using street view imagery. (this figure is adapted from (Zhang et al., 2018b))

image was compared multiple times, this process greatly reduced perception differences among different groups. As a result, it provides relatively representative views of the general public's perception of the urban physical environment.

The study trained a deep learning model, as shown in Fig. 6.2, to evaluate peoples' perception scores for the street environments of Beijing. The process included the following steps: (1) Perception scores were categorized as either 1 or − 1 based on their numerical values to minimize the impact of minor score differences on the model. (2) An SVM model was used to explore the nonlinear relationship of street view features extracted by DCNN and perception categories. (3) Street views of Beijing were used as input data, and the positive confidence levels (probability of being classified as 1) predicted by the SVM model were used as perception scores for each image. (4) These perception scores were then mapped onto the street scale for spatial analysis.

6.3.2 Perception Bias Between Humans and AI

Researchers can efficiently and automatically assess people's emotional perceptions of places using AI. This data-driven and computer vision-based approach provides relatively universal conclusions. However, local residents' perceptions of their living environments are often more authentic, and utilizing local census data can better reflect community characteristics. Additionally, the results of AI-based perception analysis may differ from people's actual perceptions. Examining these differences helps to better understand the causes behind human-machine perception bias. The following cases will introduce two relevant studies: "Assessing differences in safety perceptions using GeoAI and survey across neighborhoods in Stockholm, Sweden" and "Perception bias: Deciphering a mismatch between urban crime and perception of safety." These studies demonstrate how to use perception bias to conduct research on community-scale characteristics.

Local residents have a deeper understanding of the characteristics and socioeconomic activities of their community. Evaluating residents' perceptions of their familiar environment through surveys can provide valuable insights into community economic characteristics. Kang et al. (2023) recruited volunteers to assess the safety of their living environment and compared these assessments with AI-based perception analysis results (Fig. 6.3). They then explored the factors contributing to differences in safety perceptions by incorporating community economic attributes.

Figure 6.4 displays the safety perceptions of places in Stockholm (Sweden) as assessed by computer vision, the safety perceptions of residents based on social surveys, and the differences between the two. By integrating community socioeconomic attributes (such as population proportions and POIs), the relationship between these attributes and perception bias can be explored. For instance, the study found that AI assessments might overestimate the safety of areas with bars, commercial centers, and well-maintained residential areas because these places appear more economically active. However, local residents may perceive these areas as less safe, as they believe such areas are more likely to attract criminals. Additionally, residents may label certain places as "unsafe" based on their life experiences. For example, they might view gas stations as more prone to crime, a nuance that AI-based assessments could overlook.

On the other hand, residents' perceptions of community safety do not always align with actual crime rates. This means there may be areas where residents feel safe but crime rates are high, and areas where residents feel unsafe but crime rates are low. Analyzing the spatial distribution of these discrepancies helps residents

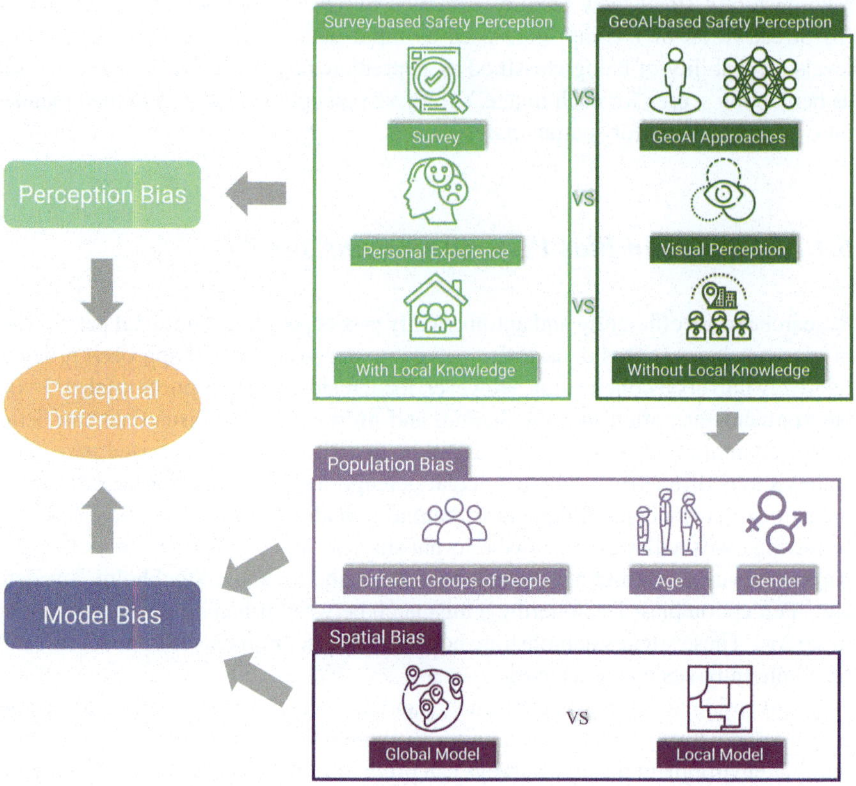

Fig. 6.3 Concept and framework of perception bias. (this figure is adapted from Kang et al. (2023))

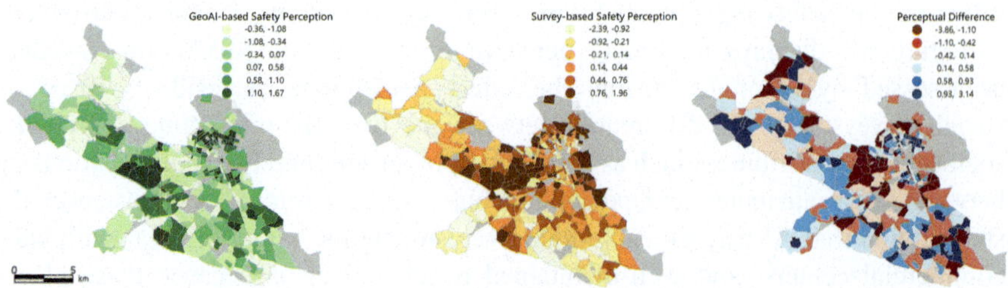

Fig. 6.4 AI-based safety perception scores (Left); Community survey-based safety perception scores (Middle); Perception bias between the two (Right). (this figure is adapted from Kang et al. (2023))

better understand safety and aids the government in efficiently deploying police resources. Zhang et al. (2021) explored the differences between people's safety perceptions and actual crime rates in Houston (USA) to identify the socioeconomic factors contributing to these discrepancies. Figure 6.5 shows some mismatches

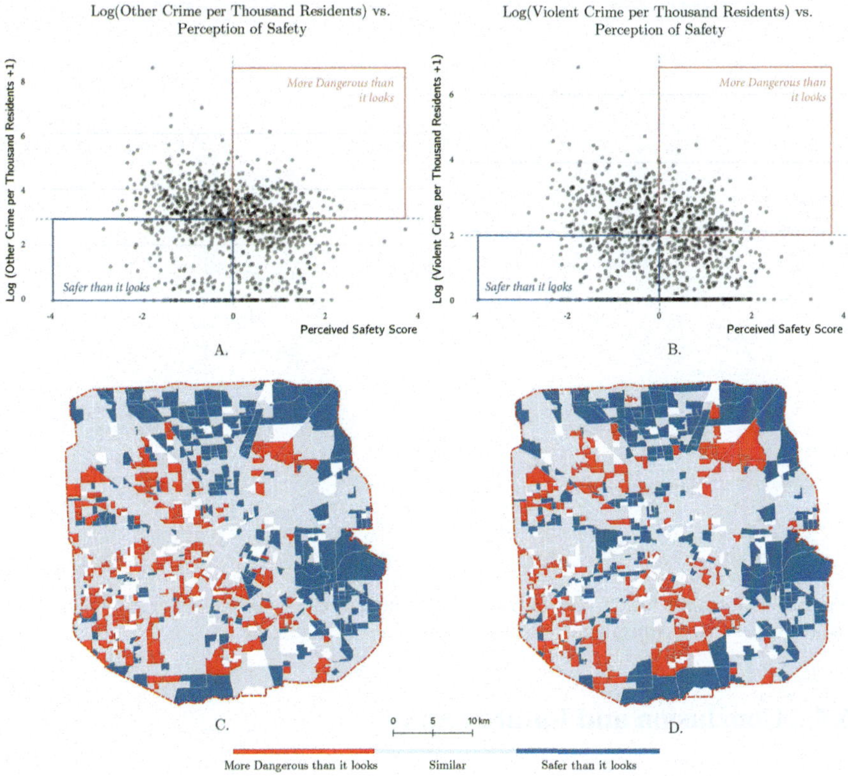

Fig. 6.5 (**a**, **b**) Classification of the relationship between safety perception and crime rate into three categories; (**c**, **d**) Spatial distribution of the three relationship categories. (this figure is adapted from Zhang et al. (2021))

between perceived safety and crime rates, as well as the spatial distribution of these perception bias areas. The study categorized areas with high safety perception scores and high crime rates as "More Dangerous than it looks" (red), areas with low safety perception scores and low crime rates as "Safer than it looks" (blue), and the remaining areas with aligned perceptions and crime rates as "Similar" (gray).

Additionally, the study examined how community attributes like POIs, population flow, and demographic distribution relate to perception bias. Figure 6.6 shows the community factors linked to perception bias for both violent and other crimes. The findings reveal that areas with high visitation at 6 PM and higher population density are more associated with being "Safer Than it looks," indicating that some busy daytime areas appear less safe but actually have lower crime rates. In contrast, areas with high visitation at 2 AM, diverse POIs, higher unemployment rates, and greater racial diversity are more associated with being "More Dangerous than it looks," suggesting that these communities are perceived as more dangerous than they really are.

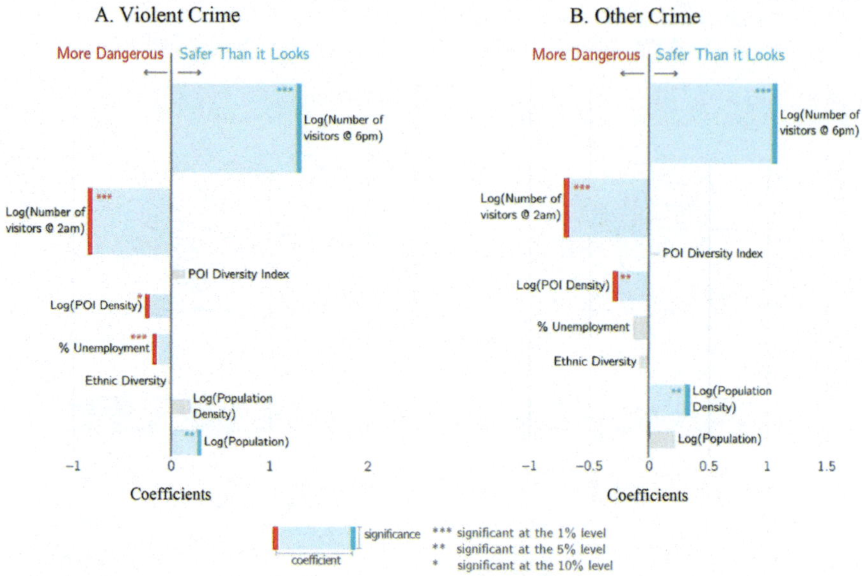

Fig. 6.6 Community factors contributing to differences in safety perception for (**a**) Violent Crime and (**b**) Other Crime. (this figure is adapted from Zhang et al. (2021))

6.4 Conclusion and Future Outlook

The application of human-centered computer vision technology in urban sensing has greatly enhanced our understanding of urban physical environments and residents' lives. This chapter introduces the concept of human-centered urban sensing and related technologies, providing examples of how computer vision and street view imagery can be used to study urban characteristics, residents' emotional perceptions, and human-machine perception biases. By integrating local residents' perception data with AI-based evaluation results, we can gain deeper insights into community characteristics and their impact on residents' lives.

In the future, with the continuous advancement of data collection and processing technologies, urban sensing technology will become more precise and comprehensive. Especially as the integration of multi-source data becomes more widespread, we will be able to conduct urban research on a larger scale and with greater detail. For example, combining real-time updated street view imagery with high-resolution remote sensing images will enable real-time monitoring of urban dynamics. Behavior data collected from smartphones and IoT devices will more accurately reflect residents' lifestyle patterns and needs. Additionally, with the ongoing development of deep learning and other advanced algorithms, the application of computer vision technology in urban sensing will become more intelligent and automated. This will help urban planners and managers to formulate better policies and measures, improving the livability and sustainability of cities.

In conclusion, human-centered computer vision offers powerful tools and methods for urban sensing studies. Future research should further explore its applications across different cities and environments, providing richer theoretical and practical support for scientific urban management and planning.

References

Anguelov, D., Dulong, C., Filip, D., Frueh, C., Lafon, S., Lyon, R., Ogale, A., Vincent, L., & Weaver, J. (2010). Google street view: Capturing the world at street level. *Computer, 43*(6), 32–38.

Biljecki, F., & Ito, K. (2021). Street view imagery in urban analytics and GIS: A review. *Landscape and Urban Planning, 215*, 104217.

Colomina, I., & Molina, P. (2014). Unmanned aerial systems for photogrammetry and remote sensing: A review. *ISPRS Journal of Photogrammetry and Remote Sensing, 92*, 79–97.

Dirksen, M., Ronda, R. J., Theeuwes, N. E., & Pagani, G. A. (2019). Sky view factor calculations and its application in urban heat Island studies. *Urban Climate, 30*, 100498.

Goudie, A. S. (2018). *Human impact on the natural environment*. John Wiley & Sons.

He, K., Chen, X., Xie, S., Li, Y., Dollár, P., & Girshick, R. (2022). *Masked autoencoders are scalable vision learners*. 16000–16009. In *Proceedings of the IEEE/CVF conference on computer vision and pattern recognition* (pp. 16000–16009).

Helbich, M., Yao, Y., Liu, Y., Zhang, J., Liu, P., & Wang, R. (2019). Using deep learning to examine street view green and blue spaces and their associations with geriatric depression in Beijing, China. *Environment International, 126*, 107–117.

Herold, M., Roberts, D. A., Gardner, M. E., & Dennison, P. E. (2004). Spectrometry for urban area remote sensing—Development and analysis of a spectral library from 350 to 2400 nm. *Remote Sensing of Environment, 91*(3), 304–319.

Hou, C., Li, Y., & Zhang, F. (2024). Sensing urban physical environment with GeoAI and street-level imagery. In *Handbook of geospatial approaches to sustainable Cities*. CRC Press.

Hu, C.-B., Zhang, F., Gong, F.-Y., Ratti, C., & Li, X. (2020). Classification and mapping of urban canyon geometry using Google street view images and deep multitask learning. *Building and Environment, 167*, 106424.

Huang, Y., Zhang, F., Gao, Y., Tu, W., Duarte, F., Ratti, C., Guo, D., & Liu, Y. (2023). Comprehensive urban space representation with varying numbers of street-level images. *Computers, Environment and Urban Systems, 106*, 102043.

Huang, J., Fei, T., Kang, Y., Li, J., Liu, Z., & Wu, G. (2024). Estimating urban noise along road network from street view imagery. *International Journal of Geographical Information Science, 38*(1), 128–155.

Ibrahim, M. R., Haworth, J., & Cheng, T. (2020). Understanding cities with machine eyes: A review of deep computer vision in urban analytics. *Cities, 96*, 102481.

Juhász, L., & Hochmair, H. H. (2016). User contribution patterns and completeness evaluation of Mapillary, a crowdsourced street level photo service. *Transactions in GIS, 20*(6), 925–947.

Kang, Y., Zhang, F., Gao, S., Lin, H., & Liu, Y. (2020). A review of urban physical environment sensing using street view imagery in public health studies. *Annals of GIS, 26*(3), 261–275.

Kang, Y., Abraham, J., Ceccato, V., Duarte, F., Gao, S., Ljungqvist, L., Zhang, F., Näsman, P., & Ratti, C. (2023). Assessing differences in safety perceptions using GeoAI and survey across neighbourhoods in Stockholm, Sweden. *Landscape and Urban Planning, 236*, 104768.

Keralis, J. M., Javanmardi, M., Khanna, S., Dwivedi, P., Huang, D., Tasdizen, T., & Nguyen, Q. C. (2020). Health and the built environment in United States cities: Measuring associations using Google street view-derived indicators of the built environment. *BMC Public Health, 20*(1), 215.

Lefèvre, S., Tuia, D., Wegner, J. D., Produit, T., & Nassar, A. S. (2017). Toward seamless Multiview scene analysis from satellite to street level. *Proceedings of the IEEE, 105*(10), 1884–1899.

Li, Y., & Long, Y. (2024). Inferring storefront vacancy using mobile sensing images and computer vision approaches. *Computers, Environment and Urban Systems, 108*, 102071.

Liu, Y., Liu, X., Gao, S., Gong, L., Kang, C., Zhi, Y., Chi, G., & Shi, L. (2015). Social sensing: A new approach to understanding our socioeconomic environments. *Annals of the Association of American Geographers, 105*(3), 512–530.

Long, Y., & Thill, J.-C. (2015). Combining smart card data and household travel survey to analyze jobs–housing relationships in Beijing. *Computers, Environment and Urban Systems, 53*, 19–35.

Luca, M., Barlacchi, G., Lepri, B., & Pappalardo, L. (2021). A survey on deep learning for human mobility. *ACM Computing Surveys, 55*(1), 1–7:44.

Luo, J., Zhao, T., Cao, L., & Biljecki, F. (2022). Semantic Riverscapes: Perception and evaluation of linear landscapes from oblique imagery using computer vision. *Landscape and Urban Planning, 228*, 104569.

Marasinghe, R., Yigitcanlar, T., Mayere, S., Washington, T., & Limb, M. (2024). Computer vision applications for urban planning: A systematic review of opportunities and constraints. *Sustainable Cities and Society, 100*, 105047.

Middel, A., Lukasczyk, J., Zakrzewski, S., Arnold, M., & Maciejewski, R. (2019). Urban form and composition of street canyons: A human-centric big data and deep learning approach. *Landscape and Urban Planning, 183*, 122–132.

Miranda, F., Hosseini, M., Lage, M., Doraiswamy, H., Dove, G., & Silva, C. T. (2020). Urban mosaic: Visual exploration of streetscapes using large-scale image data. In *Proceedings of the 2020 CHI conference on human factors in computing systems* (pp. 1–15).

Mouratidis, K. (2018). Built environment and social Well-being: How does urban form affect social life and personal relationships? *Cities, 74*, 7–20.

Mouratidis, K. (2021). Urban planning and quality of life: A review of pathways linking the built environment to subjective Well-being. *Cities, 115*, 103229.

O'Regan, A. C., Byrne, R., Hellebust, S., & Nyhan, M. M. (2022). Associations between Google street view-derived urban greenspace metrics and air pollution measured using a distributed sensor network. *Sustainable Cities and Society, 87*, 104221.

Salazar-Miranda, A., Zhang, F., Sun, M., Leoni, P., Duarte, F., & Ratti, C. (2023). Smart curbs: Measuring street activities in real-time using computer vision. *Landscape and Urban Planning, 234*, 104715.

Shaker, R. R., Altman, Y., Deng, C., Vaz, E., & Forsythe, K. W. (2019). Investigating urban heat Island through spatial analysis of new York City streetscapes. *Journal of Cleaner Production, 233*, 972–992.

Stern, P. C. (1993). A second environmental science: Human-environment interactions. *Science, 260*(5116), 1897–1899.

Van de Voorde, T., Jacquet, W., & Canters, F. (2011). Mapping form and function in urban areas: An approach based on urban metrics and continuous impervious surface data. *Landscape and Urban Planning, 102*(3), 143–155.

Wang, M., He, Y., Meng, H., Zhang, Y., Zhu, B., Mango, J., & Li, X. (2022). Assessing street space quality using street view imagery and function-driven method: The case of Xiamen, China. *ISPRS International Journal of Geo-Information, 11*(5), 282.

Xiao, X., Wang, R., Knibbs, L. D., Jalaludin, B., Heinrich, J., Markevych, I., Gao, M., Xu, S.-L., Wu, Q.-Z., Zeng, X.-W., Chen, G.-B., Hu, L.-W., Yang, B.-Y., Yu, Y., & Dong, G.-H. (2021). Street view greenness is associated with lower risk of obesity in adults: Findings from the 33 Chinese community health study. *Environmental Research, 200*, 111434.

Xiao, R., Yin, H., Liu, R., Zhang, Z., Chinzorig, S., Qin, K., Tan, W., Wan, Y., Gao, Z., Xu, C., Liu, L., & Jia, T. (2024). Exploring the relationship between land use change patterns and variation in environmental factors within urban agglomeration. *Sustainable Cities and Society, 108*, 105447.

Xing, Z., Yang, S., Zan, X., Dong, X., Yao, Y., Liu, Z., & Zhang, X. (2023). Flood vulnerability assessment of urban buildings based on integrating high-resolution remote sensing and street view images. *Sustainable Cities and Society, 92*, 104467.

Yao, H., Qin, R., & Chen, X. (2019). Unmanned aerial vehicle for remote sensing applications—A review. *Remote Sensing, 11*(12), 1443.

Yao, Y., Xu, C., Yin, H., Shao, L., & Wang, R. (2022a). More visible greenspace, stronger heart? Evidence from ischaemic heart disease emergency department visits by middle-aged and older adults in Hubei, China. *Landscape and Urban Planning, 224*, 104444.

Yao, Y., Yin, H., Xu, C., Chen, D., Shao, L., Guan, Q., & Wang, R. (2022b). Assessing myocardial infarction severity from the urban environment perspective in Wuhan, China. *Journal of Environmental Management, 317*, 115438.

Yao, Y., Dong, A., Liu, Z., Jiang, Y., Guo, Z., Cheng, J., Guan, Q., & Luo, P. (2023). Extracting the pickpocketing information implied in the built environment by treating it as the anomalies. *Cities, 143*, 104575.

Ye, C., Zhang, F., Mu, L., Gao, Y., & Liu, Y. (2021). Urban function recognition by integrating social media and street-level imagery. *Environment and Planning B: Urban Analytics and City Science, 48*(6), 1430–1444.

Zhang, F., Zhang, D., Liu, Y., & Lin, H. (2018a). Representing place locales using scene elements. *Computers, Environment and Urban Systems, 71*, 153–164.

Zhang, F., Zhou, B., Liu, L., Liu, Y., Fung, H. H., Lin, H., & Ratti, C. (2018b). Measuring human perceptions of a large-scale urban region using machine learning. *Landscape and Urban Planning, 180*, 148–160.

Zhang, F., Wu, L., Zhu, D., & Liu, Y. (2019a). Social sensing from street-level imagery: A case study in learning spatio-temporal urban mobility patterns. *ISPRS Journal of Photogrammetry and Remote Sensing, 153*, 48–58.

Zhang, F., Zhou, B., Ratti, C., & Liu, Y. (2019b). Discovering place-informative scenes and objects using social media photos. *Royal Society Open Science, 6*(3), 181375.

Zhang, F., Fan, Z., Kang, Y., Hu, Y., & Ratti, C. (2021). "Perception bias": Deciphering a mismatch between urban crime and perception of safety. *Landscape and Urban Planning, 207*, 104003.

Zhang, F., Salazar-Miranda, A., Duarte, F., Vale, L., Hack, G., Chen, M., Liu, Y., Batty, M., & Ratti, C. (2024). Urban visual intelligence: Studying cities with artificial intelligence and street-level imagery. *Annals of the American Association of Geographers, 114*(5), 876–897.

Zou, Z., Chen, K., Shi, Z., Guo, Y., & Ye, J. (2023). Object detection in 20 years: A survey. *Proceedings of the IEEE, 111*(3), 257–276.

Chapter 7
Social Network Analysis in Human Geography

Xi Gong, Yujian Lu, and Yan Lin

Abstract Social media data offers unprecedented opportunities for human geography research. Compared to traditional data collection methods, such as surveys and interviews, data collected via social media is more cost-effective, georeferenced, timelier, and available in greater volumes which therefore complement traditional approaches. As a form of spatial-temporal big data, social media provides unique insights into spatial-temporal patterns in peoples, places, and environments as well as their relationships and social dynamics—key research areas in human geography. This chapter introduces four key categories of social network analysis commonly conducted to address research questions in human geography—topic/event detection and tracking, public perspective exploration, collaboration

---

X. Gong
Department of Biobehavioral Health, The Pennsylvania State University, University Park, PA, USA

Institute for Computational and Data Sciences (ICDS), The Pennsylvania State University, University Park, PA, USA

Department of Geography & Environmental Studies, UNM Center for the Advancement of Spatial Informatics Research and Education (ASPIRE), University of New Mexico, Albuquerque, NM, USA
e-mail: xigong@psu.edu

Y. Lu (✉)
Department of Geography & Environmental Studies, UNM Center for the Advancement of Spatial Informatics Research and Education (ASPIRE), University of New Mexico, Albuquerque, NM, USA
e-mail: yujianlu@unm.edu

Y. Lin
Department of Geography, The Pennsylvania State University, University Park, PA, USA

Social Science Research Institute (SSRI), The Pennsylvania State University, University Park, PA, USA

Department of Geography & Environmental Studies, UNM Center for the Advancement of Spatial Informatics Research and Education (ASPIRE), University of New Mexico, Albuquerque, NM, USA
e-mail: yanlin@psu.edu

© The Author(s), under exclusive license to Springer Nature Switzerland AG 2025
X. Huang et al. (eds.), *GeoAI and Human Geography*, Springer Geography,
https://doi.org/10.1007/978-3-031-87421-5_7

and communication network analysis, and human mobility dynamics analysis—and provides a review of example applications for each. While numerous opportunities are available, challenges persist, including sample representativeness, sentiment analysis misclassification, geotag inaccuracies, and rising costs of data collection. The field remains dynamic and invites researchers to collaboratively address human geography questions through social network analysis.

Keywords Social network analysis · Human geography · Social media data · Sentiment analysis · Spatial-temporal patterns

7.1 Introduction

Social media refers to a broad range of interactive Web 2.0 applications that enable users to create and share user-generated contents in virtual communities (Kaplan & Haenlein, 2010; Kietzmann et al., 2011). Social media platforms are generally categorized into six types, including microblogging sites (e.g., Twitter/X, Sina Weibo), social networking sites (e.g., Facebook, LinkedIn), content-sharing platforms (e.g., Instagram, YouTube), product and service review sites (e.g., Yelp, TripAdvisor), collaborative knowledge-sharing platforms (e.g., Wikipedia, OpenStreetMap), and others (e.g., Airbnb, Craigslist) (Gong & Yang, 2020). By July 2024, the number of active social media users worldwide reached 5.17 billion (Kepios, 2024), with many individuals incorporating social media visits into their daily routines. Therefore, on social media platforms, users generate massive amounts of digital footprints, including time-stamped text, graphic content, as well as network information (e.g., following, reposting, and at-mentioning others). These are innovative and valuable resources for observing, analyzing, and even predicting human behavior and interactions. Notably, many social media platforms also allow the sharing of georeferenced information, such as coordinates, place names, and points of interest (Gong & Yang, 2020). Therefore, as a novel form of spatial-temporal big data, the georeferenced social media data offers unprecedented opportunities to reveal spatial-temporal patterns in human and social dynamics. Compared with traditional data collection methods such as surveys and interviews, data collected via social media are more cost-effective, georeferenced, timelier, and available in greater volumes (Rizwan et al., 2020). With the georeferenced social media big data from different platforms, various types of social network analysis in human geography topics are enabled, including topic/event detection and tracking, public perspective exploration, collaboration and communication network analysis, and human mobility dynamics analysis (Fig. 7.1). This chapter will first review these four categories of social network analysis for human geography research, each supported by some detailed case studies. This chapter will then discuss challenges and related future research directions. The chapter does not seek to provide a comprehensive examination of each social network analysis case study in human geography; instead, it offers a concise overview of the current state of this dynamic research field for both researchers and the interested public.

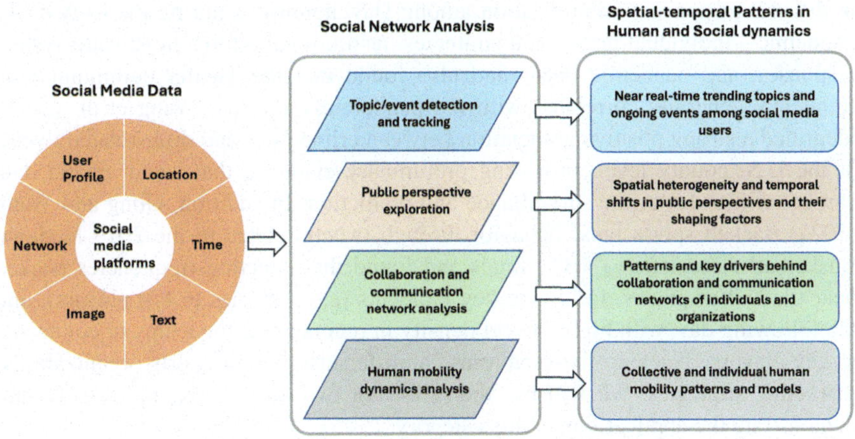

Fig. 7.1 Types of social network analysis in human geography

7.2 Social Network Analysis in Human Geography

7.2.1 Category 1: Topic/Event Detection and Tracking

Since social media has the potential to generate widespread discussions and sharing on trending topics among large user groups, it offers valuable opportunities and near real-time information for detecting and tracking hot topics and ongoing events, such as natural disasters, health-related incidents, crimes, and sports events. Eilander et al. (2016) utilized flood-related geotagged tweets to extract flood depth and location information, creating a near real-time flood probability map for Jakarta, Indonesia. This approach enabled the identification of affected areas and individuals in need of assistance during and after the disaster. Similarly, by integrating geotagged social media data from Twitter and Facebook with volunteered geographic information (VGI) from the Crowdmap platform, as well as authoritative data including Public Fire Department of Split (PFDS) data, satellite images, and official data from the Natural Protection and Rescue Directorate (NPRD), a wildfire crisis map was developed to monitor the wildfire and enhance awareness during the forest fire in Split, Croatia, in July 2017 (Tavra et al., 2021). During the COVID-19 pandemic, social media data was also used to track and monitor severity of pandemic and related policies. Hasan et al. (2022) developed a natural language processing pipeline to analyze patient posts related to COVID-19, providing valuable information on symptoms, severity, and prevalence for public health practitioners and researchers. Shen et al. (2020) used COVID-19 symptoms and diagnosis-related Weibo posts to accurately predict daily case count 2 weeks ahead of official report in China. Boon-Itt and Skunkan (2020) tracked and explored public discussion topics regarding COVID-19 over time through 107,990 related English tweets worldwide from December 13 to March 9, 2020. Gong et al. (2022) identified four main topics in disaster response communication among U.S. nonprofits during the COVID-19 pandemic—individual prevention strategies, institutional efforts, healthcare system responses, and pandemic data—and also found that the Twitter communication frequency generally correlated with the pandemic's severity. Wang et al. (2019) identified a strong positive association between crime rates and drug-related tweets at the U.S. county level, providing preliminary evidence that social media data can be utilized for crime surveillance and prediction. In addition, Gong and Wang (2021) tracked sports fans' behavior through tweets related to the 2019 National Basketball Association (NBA) finals and found that tweeting frequencies peaked near the end of games, following key incidents (e.g., injuries to top players), and the following day, with higher tweet density in regions near the teams or connected to key players. Tracking these patterns can offer insights for sports organizations' marketing strategies. We will use the following two detailed examples to further demonstrate the applications in this category.

Many institutions use social media as a tool for outreach. Based on the analysis of 65,325 posts from 106 Twitter accounts affiliated with U.S. geography departments across 39 states, Gong and Lane (2020) identified and tracked these departments' institutional outreach activities. The study found a steady increase in departmental Twitter accounts over the past decade, with more frequent tweets during full academic months, particularly in April (typical AAG month), and on Thursdays. By text mining of the hashtags and topical keywords in U.S. geography departments' outreach communication, the study identified four common categories of topics, including research topics (e.g., map, geography, and climate change), activities (e.g., GIS day and Geography Awareness Week), conferences (e.g., AAG annual meetings and ESRI user conferences), and organizations (e.g., ESRI and NASA). By further comparing the identified patterns with the suggested criteria for effective outreach from the literature, the study offered suggestions to improve departmental outreach performance through Twitter, such as tweeting more frequently even during weekends and winter/summer breaks, and incorporating community buzzwords and suggested hashtags related to popular geographical research, events, organizations, and conference when drafting tweets.

During the COVID-19 pandemic, U.S. state governors led efforts to curb the virus's spread within their states. Given the extensive use of Twitter for crisis communication by these governors, Gong and Ye (2021) analyzed Twitter data from 94 official and personal accounts of all 50 governors to track how state governments responded to the crisis. The study found that the overall daily tweet volume aligned with the severity of the pandemic in the United States, and individual governors' daily tweeting patterns positively correlated with daily case growth rates in their states. Additionally, most governors acted preemptively, mentioning COVID-19 before the first reported case in their states to mitigate public fear and uncertainty. Content analysis revealed that tweets containing topical keywords, such as virus names, information updates, and public health strategies, received greater public attention. A co-occurrence network of high-frequency terms in governors' tweets highlighted four key topics of public interest: public health orders, live-streamed news conferences, COVID-19 testing, and other updates on government responses.

To better inform the public, governors should focus more on tweeting about these topics.

7.2.2 Category 2: Public Perspective Exploration

As individuals share, comment on, and discuss social issues and events on social media, they often express personal opinions through texts and/or graphics. Sentiment analysis of these social media posts, collected from a large number of users across extensive geographic areas and over extended time frames, can offer valuable insights into the spatial heterogeneity and temporal changes of public perspectives on the discussed topics. Sentiment analysis serves as an automated and efficient tool for extracting and interpreting summarized opinions and detailed emotions from large volumes of textual data (Lombardo et al., 2019; Xu et al., 2022). Typical sentiment analysis approaches include lexicon-based methods (e.g., dictionary-based and corpus-based approaches), traditional machine learning methods (e.g., Naive Bayes and decision trees), deep learning methods (e.g., convolutional neural networks and recurrent neural networks), hybrid methods (combining the aforementioned categories), and other methods (e.g., Bidirectional Encoder Representations from Transformers (BERT) and transfer learning) (Mao et al., 2024). Mapping sentiment analysis results across space and time allows researchers to investigate and even predict public opinions on a given topic. By classifying COVID-19-related tweets in Indonesia during the pandemic into positive, neutral, and negative categories, Rahmanti et al. (2021) found that most Twitter users expressed a positive view of the government's partial lockdown policy. Wang et al. (2016) analyzed over 700,000 U.S. tweets to assess public sentiment on gun control following the 2021 Sandy Hook shooting and found that while both pro-gun and anti-gun sentiments increased, pro-gun sentiment remained elevated for several days, with varying responses across states. Li et al. (2024) examined public attention and attitudes toward shared bikes in China using over 721,000 Sina Weibo posts from January 2017 to August 2023. Sentiment analysis reveals an overall positive attitude, largely due to discounted monthly passes, but also highlights negative sentiments related to improper parking management.

Here is a detailed study showcasing how social network analysis can be employed to explore public perspectives. People's attitudes toward hydraulic fracturing (fracking) are diverse and influenced by various factors, but existing research studying the attitudes often relies on limited survey and interview samples, which may introduce bias. In contrast, Gong et al. (2023) used georeferenced tweets related to fracking from 2018 to 2019 across the entire United States to provide a more comprehensive view of public perspectives on the issue. A Naïve Bayes text classifier was trained using 2000 manually classified fracking-related tweets, categorized as positive, neutral, or negative sentiment. This classifier was then applied to assess the sentiments of all other fracking-related tweets. Subsequently, multiscale geographically weighted regression (MGWR) was used to analyze the associations at the county level between the percentage of negative tweets toward fracking and potential shaping factors across five categories: socio-demographics, economic development, social equity and politics, environmental impacts, and fracking-related information. The results reveal spatial heterogeneity and varying scales of these associations, which can be utilized to predict future public attitudes toward fracking. For example, a larger African American population and/or lower educational levels are associated with reduced opposition to fracking across all counties, demonstrating global stationarity. In contrast, fracking is more likely to be opposed in eastern and central U.S. counties with higher unemployment rates and in counties east of the Great Plains with fewer nearby fracking sites, indicating regional associations. The methodology employed in this study can also be applied to investigate public perspective toward other controversial topics.

In addition to identifying and tracking public perspectives toward a given topic, sentiment information derived from social network analysis can also be utilized to validate theories of human behaviors. Sports fans' tweets offer valuable big data for analyzing their attitudes toward team performance and game results, as demonstrated by Gong and Wang's (2021) case study using Twitter data collected during the 2019 NBA Finals between the Toronto Raptors and the Golden State Warriors. The study first generated a sentiment polarity score (indicating positive, neutral, and negative sentiments) for each tweet using a Naïve Bayes text classifier trained on the movie reviews corpus from the Natural Language Toolkit (NLTK). The results revealed significant local hotspots with higher percentages of positive tweets about the Warriors near California, clearly highlighting the team's strong fan base in the region. Across all games, the relative percentage of positive tweets for a team increased following a win, decreased after a loss, and was moderated by key incidents. This study also validated theoretical fan behavior tendencies by analyzing trends in fans' attitudes related to game results in Ontario, Canada. Specifically, the quantity and percentage of positive tweets toward the Raptors increased after each victory, aligning with the Basking in Reflected Glory (BIRGing) phenomenon, where sports fans feel a sense of personal success through their team's achievements. Conversely, when Raptors fans received negative information about the Raptors (a loss), they exhibited a tendency to derogate the opponent (the Warriors), as evidenced by an increase in negative tweets directed at the Warriors, reflecting the blasting tendency.

7.2.3 Category 3: Collaboration and Communication Network Analysis

Social media platforms inherently function as networks, where users act as nodes (e.g., individuals or organizations) and interactions among users serve as edges (e.g., follows, reposts, and at-mentions). These interactions can be complex, being either directed or undirected (Wasserman & Faust, 1994), and may also be weighted according to the strength, intensity, or capacity of the connection between nodes (Tabassum et al., 2018). Although social media interactions occur in virtual spaces, they still reflect the geography of communication and collaboration. Human interactions in both cyberspace and the physical world are influenced by similar social, economic, cultural, and linguistic constraints, leading social media networks to largely mirror existing real-world patterns (Bild et al., 2015; Stephens & Poorthuis, 2015). For instance, the distance decay effect is also evident in social media networks, although the slope and fluctuation of the decay curves may vary (Han et al., 2018). Twitter networks, in particular, exhibit stronger and more effective information transmission within local geographies (Stephens & Poorthuis, 2015; Takhteyev et al., 2012). Therefore, analyzing the patterns and driving factors within networks of individuals and/or organizations on social media can provide insights into the geography of complex collaboration and communication relationships among users. Several studies have utilized social media data to analyze networks of institutional users, such as the communication network among U.S. Geography departments on Twitter (Gong & Lane, 2020) and the collaboration network among U.S. nonprofits and government agencies for mitigating the effects of the COVID-19 pandemic (Gong et al., 2022). Other research focuses on networks of individual users, including collaboration network among U.S. governors in crisis communications during the pandemic (Gong & Ye, 2021) and social interaction network among members within a volunteered geographic information (VGI) community (Zhang et al., 2024). The following paragraphs will examine a case study from each group to illustrate the analysis of collaboration and communication networks.

The National Voluntary Organizations Active in Disaster (NVOAD) is a coalition of over 70 national nonprofit organizations, including faith-based, community-based, and other groups, that have been responding to disasters across the United States and globally for more than 50 years (NVOAD, 2021). The COVID-19 pandemic has introduced complex challenges, necessitating collaboration between these organizations and government agencies within and across sectors to address them collectively. Gong et al. (2022) analyzed 19,343 COVID-19-related tweets from 74 NVOAD nonprofits and three government agencies (HHS, FEMA, and CDC), offering insights into nonprofit communication and collaboration during the pandemic from both network and geographical perspectives. Network connections among these organizations were primarily established through account follows or mentioning/retweeting/quoting in tweets. Analysis of the following network and interaction network revealed that three government agencies played key leadership roles in providing COVID-19 guidelines to NVOAD nonprofits, while nonprofits in Human Services, International and Foreign Affairs, and Public and Societal Benefit sectors, particularly the American Red Cross, were central in nonprofit communications. Geographically, nonprofit connections and interactions are more common within the same city or neighboring states. Both mission homophily and heterophily influence these network patterns, depending on the nonprofit subsector. Despite this, these interactions highlight significant opportunities for increased collaboration among nonprofits.

The iNaturalist citizen science is a volunteered geographic information (VGI) project where participants identify or vote on species in observations submitted by others to achieve consensus species identification. Contributors can be modeled as nodes and their interactions as edges in a social network. Zhang et al. (2024) analyzed this social network, comprising 17,935 nodes and 49,006 edges, focusing on characteristics such as driving factors, edge formation, edge strength, and community formation using correlation analysis, random simulations, and community detection, respectively. The results indicate that half of the interactions occurred between observers and identifiers within 590 km of each other. The median similarity in taxon composition and land cover composition between observers and identifiers involved in species identification is 0.86 and 0.73, respectively. Thus, geographic proximity, shared species taxon interests, and common observation environments preferences are key drivers of inter-contributor species identification interactions on iNaturalist. Moreover, the study results also validate theories regarding the social forces influencing social link formation. Specifically, shorter geographic distance reflects physical proximity, while high similarities in species taxon composition and land cover illustrate homophily effects, which contribute to greater interconnectedness among social actors and the formation of communities or clusters within social networks (Bedi & Sharma, 2016; McCulloh et al., 2013).

7.2.4 Category 4: Human Mobility Dynamics Analysis

Geotagged social media data, especially posts from the same user over time, offer publicly accessible, high-resolution location histories of individuals that can serve as proxies for human mobility. With the rapid expansion of social media datasets covering large populations across extensive geographic areas over prolonged periods, scientists can quantitatively analyze collective mobility patterns. This facilitates the development of models that capture and replicate spatiotemporal trends in human movement. Researchers have leveraged this type of social network analysis to address a variety of real-world applications. Jurdak et al. (2015) analyzed over six million geotagged tweets in Australia to examine the diversity of movement patterns among individuals within and between cities. They found that both short- and long-distance movers predominantly spend their time in large metropolitan areas, whereas intermediate-distance movers tend to travel within a broader region around key cities and along major roads connecting them. By analyzing 2.3 million geotagged tweets in Great London Area during the COVID-19 pandemic, Zhong et al. (2023) extracted daily trips, origin-destination matrices, and spatial networks, and further generated mobility indices to monitoring changes in people's travel patterns. They observed a shift to fewer but longer trips starting in March 2020, a reduced interest in central and sub-central activity locations in 2020 compared to outer areas, and a sign of return to pre-pandemic patterns in 2021. The location histories extracted from geotagged social media data are also feasible to delineate users' activity spaces. Hu et al. (2020) mapped users' activity spaces using 20 million geotagged tweets from over 560,000 unique Twitter users in the Los Angeles metropolitan area in 2015. Their results indicate a significant relationship between the spatial measures of activity spaces (average distance and area covered between users' home locations and activity locations) and neighborhood spatial and socioeconomic characteristics. Comito (2019) extracted human trajectories using 7,424,112 geotagged tweets from 292,195 mobile users in London in 2013. Based on these historic trajectory patterns, the study developed a method integrating frequent pattern mining, trajectory similarity, and supervised classification to predict users' future movements, thereby supporting urban computing applications.

The following case study provides a detailed example of how social media data can uncover human mobility dynamics. During the COVID-19 pandemic, there was a need for an efficient and less privacy-concerning method to monitor human mobility, which is closely associated with viral transmission. Huang et al. (2020) analyzed over 580 million geotagged tweets worldwide in 2020 to trace the location histories of more than 9.5 million users, aiming to illustrate reductions in human mobility at global, national, and U.S. state levels during the pandemic. Two types of distance metrics were employed to assess Twitter-based mobility: single-day distance for daily travel behavior and cross-day distance for displacement between consecutive days. The analysis revealed significant deviations from baseline in both metrics globally, with a marked reduction in periodicity of weekly routines following the WHO's declaration of COVID-19 as a pandemic on March 11, 2020. At the country level, discrepancies in mobility responsiveness were evident, reflecting varying patterns across different epidemic phases. Additionally, Twitter-based mobility changes closely aligned with the timing of announced mitigation measures. In the United States, Twitter-based mobility declined in most states following the National Emergency declaration on March 13, 2020, though the extent of impact varied widely across states. These findings suggest that mobility patterns derived from social media data effectively capture and quantitatively reflect human movement.

7.3 Challenges and Future Research Directions

Although social network analysis offers numerous opportunities and is widely used in human geography research, several challenges also require attention. Firstly, most existing research has focused on a single social media platform, such as Twitter/X (Gong & Ye, 2021), Instagram (Le Busque et al., 2022), and Facebook (De Jong, 2015). Each platform serves distinct user demographics in terms of age, gender, language, etc. Integrating datasets from diverse social media platforms can enhance population coverage and sample diversity. Future studies should consider this approach to obtain a more comprehensive view of human behavior, attitudes, and networks. However, issues related to compatibility among different datasets—such as data structure, sampling methods, and spatiotemporal resolution—must be carefully addressed.

Secondly, sampling bias may arise as only a fraction of social media data is accessible via platform APIs; for instance, Twitter's streaming APIs capture only about 1% of the public tweet stream (Pfeffer et al., 2018). Twitter's undisclosed sampling methods and lack of guaranteed sampling stability prevent accurate estimation of this bias (Wang et al., 2015). Although Wang et al. (2015) asserted that a small sampling ratio via Twitter streaming APIs can adequately preserve information from the full Twitter streams for research and applications, including sentiment analysis, user activity characterization, and event detection, it is important to note that sampling methods may vary with updated API versions. Researchers must therefore remain vigilant regarding the completeness and potential sampling biases of their data.

Thirdly, most sentiment analysis models are trained on specific datasets, and no model is universally perfect. Some researchers use models trained on general datasets, like product reviews, which may not align well with the data being studied. Others train their own classifiers on manually labeled, study-specific samples, which can enhance sentiment prediction but is time-consuming and prone to misclassification. Future research could involve training multiple classifiers on manually labeled social media samples and selecting the best-performing model based on cross-validation accuracy. Moreover, current sentiment analysis models largely concentrate on a single language, primarily English, which excludes data from other languages on social media. Future advancements should incorporate sentiment analysis models for multiple languages. Additionally, the training data for these models must be regularly updated to capture evolving popular culture and accurately reflect the most current social dynamic patterns.

Fourthly, location spoofing poses a significant challenge in social network analysis by potentially introducing bias and misleading conclusions due to the intentional fabrication of actual social media data locations (Zhao & Sui, 2017). However, it may also serve to protect individuals' privacy, particularly concerning sensitive locations like their homes. Researchers must balance the need for accurate data with the ethical responsibility to respect privacy. To enhance the reliability of geotagged data, future studies should develop methods for detecting location spoofing, such as Bayesian time geographic approaches (Zhao & Sui, 2017), to strengthen the credibility of research findings.

Finally, the rising costs of social media data collection have become a significant concern, as some platforms have begun charging fees based on data volume, even for scientific research. For example, Twitter/X began charging for API access in February 2023 (X Developer Platform, 2024), followed by Reddit implementing similar fees (Binder, 2023). Consequently, the high cost of APIs may impose a financial burden on researchers, making social media analysis both unaffordable and difficult to replicate.

It is important to note that social media data collection and analysis should not replace traditional approaches used in human geography research. Instead, these methods provide complementary resources and should be integrated with conventional approaches when possible to produce meaningful results and more accurately reveal the relationships between humans, places, and environments—key concerns in human geography. As the field evolves toward more hybrid methodologies, social media approaches add plurality and depth to human geography research.

7.4 Conclusion

Social media data, as a cost-effective source of big data, provides near real-time spatial, temporal, textual, visual, and network information, along with user profile details, thereby opening new avenues for addressing research questions in human geography. This chapter does not aim to exhaustively cover all existing social network analysis methods in human geography research; instead, it highlights the most commonly used approaches to demonstrate their applications. Researchers typically detect and track trending topics and ongoing events through text mining of posts and discussions among social media users, explore spatial heterogeneity and temporal shifts in public perspectives and their shaping factors via sentiment analysis, examine patterns and key drivers behind collaboration and communication networks of individuals and organizations using network analysis, and quantitatively analyze collective and individual mobility dynamics by extracting and reconstructing historical trajectories from geotagged social media posts. While social network analysis offers valuable insights, it may not fully address complex human geography research problems on its own. It is often complemented by other methodologies like spatial statistics and time geography, and requires the integration of theories and domain knowledge from other disciplines like public administration and sports management to effectively filter raw data and interpret the resulting patterns. Despite the opportunities and challenges, the field remains dynamic and invites researchers to collaboratively address human geography questions through social network analysis.

References

Bedi, P., & Sharma, C. (2016). Community detection in social networks. *Wiley Interdisciplinary Reviews: Data Mining and Knowledge Discovery, 6*(3), 115–135.

Bild, D. R., Liu, Y., Dick, R. P., Mao, Z. M., & Wallach, D. S. (2015). Aggregate characterization of user behavior in twitter and analysis of the retweet graph. *ACM Transactions on Internet Technology (TOIT), 15*(1), 1–24.

Binder, M. (2023). *Twitter and Reddit's high-priced APIs are bad news for the internet's future*. Mashable. https://mashable.com/article/social-media-paid-api-internet-future

Boon-Itt, S., & Skunkan, Y. (2020). Public perception of the COVID-19 pandemic on twitter: Sentiment analysis and topic modeling study. *JMIR Public Health and Surveillance, 6*(4), e21978.

Comito, C. (2019). Mining human mobility from social media to support urban computing applications. In *2019 15th international conference on distributed computing in sensor systems (DCOSS)* (pp. 514–521). https://doi.org/10.1109/DCOSS.2019.00099

De Jong, A. (2015). Using Facebook as a space for storytelling in geographical research. *Geographical Research, 53*(2), 211–223.

Eilander, D., Trambauer, P., Wagemaker, J., & van Loenen, A. (2016). Harvesting social media for generation of near real-time flood maps. *Procedia Engineering, 154*, 176–183. https://doi.org/10.1016/j.proeng.2016.07.441

Gong, X., & Lane, K. M. D. (2020). Institutional twitter usage among U.S. geography departments. *Professional Geographer, 72*(2), 219–237. https://doi.org/10.1080/00330124.2019.1653770

Gong, X., & Wang, Y. (2021). Exploring dynamics of sports fan behavior using social media big data—a case study of the 2019 National Basketball Association Finals. *Applied Geography, 129*, 102438. https://doi.org/10.1016/J.APGEOG.2021.102438

Gong, X., & Yang, X. (2020). Social media platforms. *Geographic Information Science & Technology Body of Knowledge, 2020*(Q3), 1. https://doi.org/10.22224/gistbok/2020.3.2

Gong, X., & Ye, X. (2021). Governors fighting crisis: Responses to the COVID-19 pandemic across U.S. states on twitter. *Professional Geographer, 73*, 1–19. https://doi.org/10.1080/00330124.2021.1895850

Gong, X., Peng, S., Lu, Y., Wang, S., Huang, X., & Ye, X. (2022). Social network analysis of nonprofits in disaster response: The case of twitter during the COVID-19 pandemic in the United States. *Social Science Computer Review, 41*(6), 2029–2054. https://doi.org/10.1177/08944393221130674

Gong, X., Lu, Y., Beene, D., Li, Z., Hu, T., Morgan, M., & Lin, Y. (2023). Understanding public perspectives on fracking in the United States using social media big data. *Annals of GIS, 29*(1), 21–35. https://doi.org/10.1080/19475683.2022.2121856

Han, S. Y., Tsou, M.-H., & Clarke, K. C. (2018). Revisiting the death of geography in the era of big data: The friction of distance in cyberspace and real space. *International Journal of Digital Earth, 11*(5), 451–469. https://doi.org/10.1080/17538947.2017.1330366

Hasan, A., Levene, M., Weston, D., Fromson, R., Koslover, N., & Levene, T. (2022). Monitoring COVID-19 on social media: Development of an end-to-end natural language processing pipeline using a novel triage and diagnosis approach. *Journal of Medical Internet Research, 24*(2), e30397. https://doi.org/10.2196/30397

Hu, L., Li, Z., & Ye, X. (2020). Delineating and modeling activity space using geotagged social media data. *Cartography and Geographic Information Science, 47*(3), 277–288. https://doi.org/10.1080/15230406.2019.1705187

Huang, X., Li, Z., Jiang, Y., Li, X., & Porter, D. (2020). Twitter reveals human mobility dynamics during the COVID-19 pandemic. *PLoS One, 15*(11), e0241957.

Jurdak, R., Zhao, K., Liu, J., AbouJaoude, M., Cameron, M., & Newth, D. (2015). Understanding human mobility from twitter. *PLoS One, 10*(7), e0131469. https://doi.org/10.1371/journal.pone.0131469

Kaplan, A. M., & Haenlein, M. (2010). Users of the world, unite! The challenges and opportunities of social media. *Business Horizons, 53*(1), 59–68. https://doi.org/10.1016/j.bushor.2009.09.003

Kepios, M. (2024). We Are Social. In *Global social media statistics*. DATAREPORTAL. https://datareportal.com/social-media-users

Kietzmann, J. H., Hermkens, K., McCarthy, I. P., & Silvestre, B. S. (2011). Social media? Get serious! Understanding the functional building blocks of social media. *Business Horizons, 54*(3), 241–251. https://doi.org/10.1016/j.bushor.2011.01.005

Le Busque, B., Mingoia, J., & Litchfield, C. (2022). Slow tourism on Instagram: An image content and geotag analysis. *Tourism Recreation Research, 47*(5–6), 623–630.

Li, Y., Wang, S., Tang, J. H. C. G., Peng, Z., & Zhuge, C. (2024). Public attention and attitudes towards bike-sharing in China: A text mining approach. *Transportation Research Part D: Transport and Environment, 134*, 104348. https://doi.org/10.1016/j.trd.2024.104348

Lombardo, G., Fornacciari, P., Mordonini, M., Sani, L., & Tomaiuolo, M. (2019). A combined approach for the analysis of support groups on facebook-the case of patients of hidradenitis suppurativa. *Multimedia Tools and Applications, 78*, 3321–3339.

Mao, Y., Liu, Q., & Zhang, Y. (2024). Sentiment analysis methods, applications, and challenges: A systematic literature review. *Journal of King Saud University—Computer and Information Sciences, 36*(4), 102048. https://doi.org/10.1016/j.jksuci.2024.102048

McCulloh, I., Armstrong, H., & Johnson, A. (2013). *Social network analysis with applications*. Wiley.

NVOAD. (2021). *National Voluntary Organizations Active in Disaster*. National Voluntary Organizations Active in Disaster. https://www.nvoad.org/

Pfeffer, J., Mayer, K., & Morstatter, F. (2018). Tampering with Twitter's sample API. *EPJ Data Science, 7*(1), 50.

Rahmanti, A. R., Ningrum, D. N. A., Lazuardi, L., Yang, H.-C., & Li, Y.-C. (2021). Social media data analytics for outbreak risk communication: Public attention on the "new Normal" during the COVID-19 pandemic in Indonesia. *Computer Methods and Programs in Biomedicine, 205*, 106083. https://doi.org/10.1016/j.cmpb.2021.106083

Rizwan, M., Wan, W., & Gwiazdzinski, L. (2020). Visualization, spatiotemporal patterns, and directional analysis of urban activities using geolocation data extracted from LBSN. *ISPRS International Journal of Geo-Information, 9*(2), 137. https://doi.org/10.3390/ijgi9020137

Shen, C., Chen, A., Luo, C., Zhang, J., Feng, B., & Liao, W. (2020). Using reports of symptoms and diagnoses on social media to predict COVID-19 case counts in mainland China: Observational infoveillance study. *Journal of Medical Internet Research, 22*(5), e19421.

Stephens, M., & Poorthuis, A. (2015). Follow thy neighbor: Connecting the social and the spatial networks on twitter. *Computers, Environment and Urban Systems, 53*, 87–95. https://doi.org/10.1016/j.compenvurbsys.2014.07.002

Tabassum, S., Pereira, F. S. F., Fernandes, S., & Gama, J. (2018). Social network analysis: An overview. *Wiley Interdisciplinary Reviews: Data Mining and Knowledge Discovery, 8*(5), e1256.

Takhteyev, Y., Gruzd, A., & Wellman, B. (2012). Geography of twitter networks. *Social Networks, 34*(1), 73–81.

Tavra, M., Racetin, I., & Peroš, J. (2021). The role of crowdsourcing and social media in crisis mapping: A case study of a wildfire reaching Croatian City of Split. *Geoenvironmental Disasters, 8*(1), 10. https://doi.org/10.1186/s40677-021-00181-3

Wang, Y., Callan, J., & Zheng, B. (2015). Should we use the sample? Analyzing datasets sampled from Twitter's stream API. *ACM Transactions on the Web (TWEB), 9*(3), 1–23.

Wang, N., Varghese, B., & Donnelly, P. D. (2016). A machine learning analysis of twitter sentiment to the Sandy hook shootings. In *2016 IEEE 12th international conference on E-science (e-science)* (pp. 303–312). https://doi.org/10.1109/eScience.2016.7870913

Wang, Y., Yu, W., Liu, S., & Young, S. D. (2019). The relationship between social media data and crime rates in the United States. *Social Media+ Society, 5*(1), 2056305119834585.

Wasserman, S., & Faust, K. (1994). *Social network analysis: Methods and applications*. Cambridge University Press.

X Developer Platform. (2024). *About the X API*. X Developer Platform. https://developer.x.com/en/docs/twitter-api/getting-started/about-twitter-api#item0

Xu, Q. A., Chang, V., & Jayne, C. (2022). A systematic review of social media-based sentiment analysis: Emerging trends and challenges. *Decision Analytics Journal, 3*, 100073. https://doi.org/10.1016/j.dajour.2022.100073

Zhang, G., Gong, X., & Zhu, D. (2024). Geographic proximity and homophily effects drive social interactions within VGI communities: An example of iNaturalist. *International Journal of Digital Earth, 17*(1), 2297948. https://doi.org/10.1080/17538947.2023.2297948

Zhao, B., & Sui, D. Z. (2017). True lies in geospatial big data: Detecting location spoofing in social media. *Annals of GIS, 23*(1), 1–14.

Zhong, C., Morphet, R., & Yoshida, M. (2023). Twitter mobility dynamics during the COVID-19 pandemic: A case study of London. *PLoS One, 18*(4), e0284902. https://doi.org/10.1371/journal.pone.0284902

Chapter 8
Converging GeoAI and CyberGIS for Human Geography

Fangzheng Lyu

Abstract This chapter examines the convergence of geospatial artificial intelligence (GeoAI) and cyberinfrastructure-based geographic information systems (cyberGIS) in advancing human geography research. Leveraging the wide availability of geospatial big data, this chapter illustrates how the integration of GeoAI and cyberGIS can tackle the data, modeling, and computational challenges in various human geography sectors. In particular, the convergence of GeoAI's advanced modeling capabilities with CyberGIS's high-performance computing capabilities is highlighted, enabling sophisticated modeling, near-real-time analysis, and high-resolution, high-accuracy, and large-scale processing of diverse geospatial data to reveal the complex relationship between human and environment. A case study on analyzing human sentiments of heat exposure in Mecklenburg County, North Carolina, demonstrates the application of this convergence in urban geography. Using location-based social media data, natural language processing model, and cyberGIS capability, the study reveals spatial patterns of human perception of urban heat. This case study shows the potential of GeoAI and cyberGIS in enabling more sophisticated, timely, and scalable analysis in human geography, while also addressing ethical concerns and computational challenges associated with big data.

Keywords GeoAI · CyberGIS · High-performance computing · Location-based social media data

8.1 Introduction

The availability of large-scale geospatial data, artificial intelligence (AI) advancements, and high-performance computing (HPC) capabilities has significantly transformed the geospatial landscape (Janowicz et al., 2020). Driven by the wide

F. Lyu (✉)
Department of Geography, Virginia Tech, Blacksburg, VA, USA
e-mail: fangzheng@vt.edu

© The Author(s), under exclusive license to Springer Nature Switzerland AG 2025

X. Huang et al. (eds.), *GeoAI and Human Geography*, Springer Geography,
https://doi.org/10.1007/978-3-031-87421-5_8

availability of large-scale geospatial data (e.g., high-resolution remote sensing images, real-time social sensing data), human geography, which focuses on spatial relationships between human communities, cultures, economies, and their environmental interactions, faces significant opportunities, challenges, and risks with the emergence of geospatial big data (Kitchin, 2013). Specifically, the field of human geography benefits from unprecedented access to rich, detailed, and timely data, enabling more sophisticated analysis and broader insights into spatial and temporal relationships between human and environments (Lazer et al., 2009). However, handling, analyzing, and interpreting the large-scale geospatial data poses major challenges in new methodological and computational approaches and frameworks (Batty et al., 2012). Meanwhile, the sensitivity of human-related data raises critical ethical concerns and privacy issues regarding its collection, use, and storage (Nelson et al., 2022). And major questions are raised on how data-intensive human geography research can be conducted in a responsible way. In the context of big data, the geospatial artificial intelligence (GeoAI) and cyberinfrastructure-based geographic information systems (cyberGIS) have been identified as major enablers on the modeling and computational perspectives to not only addressing challenges and opportunities but also mitigating the ethical concerns in the field of human geography.

Recent advancements in artificial intelligence, particularly large language models and explainable AI, have significantly enhanced the capability to address complex problems and leverage big data while simultaneously improving the interpretability of the AI model (Angelov et al., 2021). In the geospatial domain, GeoAI develops intelligent computer programs to mimic the processes of human perception, spatial reasoning, and discovery about geographical phenomena and dynamics to solve problems in human environmental systems and their interactions (Gao, 2021). It provides unique modeling capability for geospatial knowledge discovery and is widely applied in fields such as urban geography (Liu & Biljecki, 2022), health geography (Sahana et al., 2023), transport geography (Wang et al., 2024a), and beyond. However, a significant constraint for GeoAI is its high demand for computing resources, especially for large-scale models like generative AI or geospatial foundation models, where computing capability is crucial for large-scale, high-resolution, near-real-time responses, and advanced AI modeling. To address this challenge, cyberGIS has emerged as a new generation of GIS, bridging the gap between advanced cyberinfrastructure (e.g., supercomputers) and modern geographic information science (Shaowen Wang et al., 2013), serving as an enabler from the computing side to enable the analysis of large-scale geospatial data, and computationally intensive spatial analysis and modeling, and collaborative geographic problem solving (Wang, 2010).

While the integration of GeoAI and CyberGIS has the potential to yield synergistic benefits by combining the advanced modeling and computing capability, these fields currently operate relatively independently at the current stage, each addressing challenges in human geography and other geospatial domains. And the amount of scholarly works that effectively converges GeoAI and cyberGIS to tackle geospatial challenges remains limited.

The author believes that the convergence of GeoAI and cyberGIS will be crucial, as they can complement each other by bringing unique advantages from both sides, providing significant opportunities for advancing human geography research in the future. In this chapter, we first review the existing literature on the convergence of GeoAI and cyberGIS in advancing various subdomains of human geography, followed by a case study that applies GeoAI and cyberGIS to human geography, specifically analyzing human sentiment of heat exposure in major urban areas using location-based social media (LBSM) data.

8.2 Converging GeoAI and CyberGIS for Advancing Human Geography

There are three major layers for integrating GeoAI and cyberGIS for human geography research (see Fig. 8.1).

CyberGIS and geospatial big data are situated in the bottom layer providing computational and infrastructure support and geospatial data input for GeoAI models. In particular, cyberGIS provides advanced computing resources and infrastructure to not only support data- and computational-intensive models, but also support geospatial middleware (e.g., CyberGIS-Compute, CyberGIS-Jupyter) that

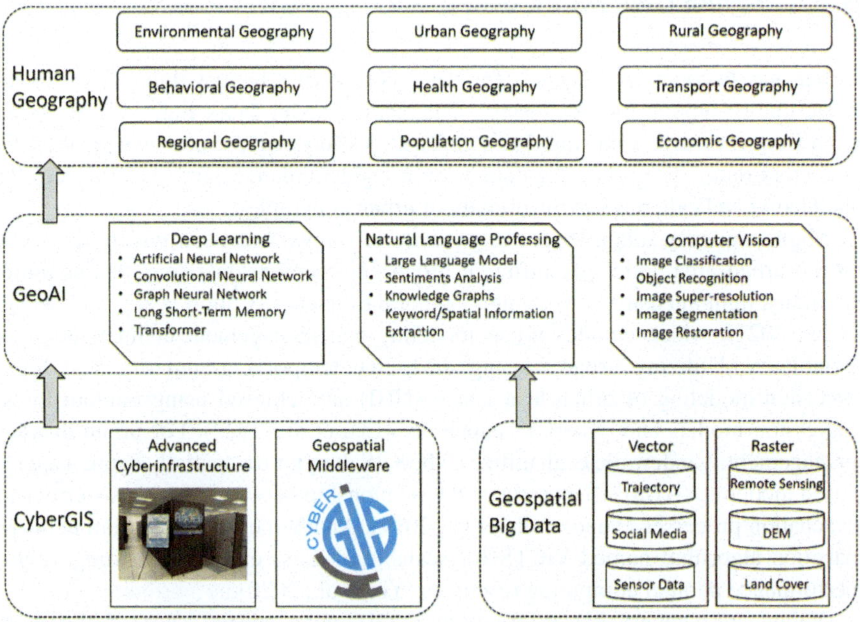

Fig. 8.1 Converging GeoAI and cyberGIS for advancing human geography

allows users with little background in programming background to access advanced computing resources, enabling broader community engagement, and facilitating collaboration (Lyu et al., 2019; Michels et al., 2024b; Yin et al., 2017a). Geospatial big data (e.g., high-resolution remote sensing images, social media data, sensor data) refers to spatial datasets that exceed the storage and computational capacity of current systems (Lee & Kang, 2015). While geospatial big data presents significant opportunities for research and knowledge discovery, extensive computing resources and sophisticated models are required to extract meaningful information from these large-scale spatial datasets.

GeoAI is the middle layer providing robust modeling capability. Deep learning models, natural language processing (NLP), computer vision (CV) have been widely adopted in the geospatial domain. In particular, the NLP has been widely used for geospatial information extracted from text corpus (e.g., social media posts, news outlets). Geospatial knowledge hypercube, for example, extracts spatial-related text data and constructs knowledge hypercube with NLP (Wang et al., 2023). And CV models (e.g., image classification, object recognition) are widely used in remote sensing images.

The top layer includes the different subdomains of human geography. Detailed reviews of the existing literature on selected human geography research areas enabled by GeoAI and cyberGIS will be further illustrated in the following sections.

8.2.1 Urban Geography

Urban has become an epicenter for data representing large volume (e.g., large-scale fine-resolution remote sensing data, social media data), high velocity (e.g., high-frequency urban sensing data), and wide variety (e.g., street view data, traffic trajectory data). The urban big data have brought challenges and opportunities for the GeoAI and cyberGIS in the domain of urban geography.

High-frequency urban sensor (e.g., array of things) has been widely deployed in the urban area collecting different urban environment data (e.g., temperature, humidity, light intensity) at very high temporal frequency (Catlett et al., 2017; Wang et al., 2021), which enables urban modeling and management at finer temporal granularity. Utilizing urban sensing data, fine temporal granularity, and high-precision modeling of urban heat island (UHI) are achieved using random forest regression models and cyberGIS capability (Lyu et al., 2022). For urban disaster management, Vandewalle et al. utilize CyberGIS-Jupyter for spatially explicit agent-based modeling (ABM) to investigate the relationship between disaster severity and evacuation processes (Vandewalle et al., 2019). And Yin et al. proposed a network partition algorithm named ViCTS for scalable of mass evacuation through high-performance parallel computing resources (Yin et al., 2020).

Volunteered geographic information (VGI), as a special case of the user-generated content, has been widely used in domain such as urban planning and management (Goodchild, 2007; Steiniger et al., 2016; Sui et al., 2013). Using social media data, studies have been conducted to understand the exposure of heat based using NLP and high-performance geospatial computing techniques (Lyu et al., 2024). Meanwhile, an urban environmental monitoring system is achieved by leveraging CyberGIS capabilities to enable citizens to collect pollution data using smartphones, employing advanced data handling techniques and GPU processing to quickly generate pollution maps (Yin et al., 2017b). For urban disaster management and response, Zhang et al. propose a cyberGIS-enabled multi-criteria spatial decision support system that integrates cyberGIS and multi-criteria decision analysis models, and social vulnerability indicators, along with social media data, to facilitate rapid decision-making during flood emergencies in Austin, Texas (Zhang et al., 2019). Leveraging Twitter data and urban infrastructure registry data, Zhang et al. present cyberGIS-enabled spatiotemporal population model to estimate human activity dynamics in the urban area, utilizing an object class-oriented space-time composite approach and a temporal weight matrix derived from social media activity (Zhang et al., 2021). And Soliman et al. analyze individual-level mobility patterns from geo-located tweet data using density-based spatial clustering of applications with noise (DBSCAN) supported by advanced cyberinfrastructure (Soliman et al., 2017).

8.2.2 Health Geography

Computational and data sciences are revolutionizing GIS through cyberGIS and advanced modeling, which offer a transformative, non-parametric, scalable, and stochastic approach to spatial analysis in health studies (Shi & Wang, 2015). Agent-based Model (ABM), a microscale model that simulates the simultaneous operations and interactions of multiple agents, has been widely adopted in epidemiology research (Tracy et al., 2018). As simulation process requires extensive computing resources, Song et al. combine fuzzy cognitive maps, ABM, and cyberGIS to simulate the COVID-19 spread at Bengaluru Urban District in India, revealing how individual behavior influences broader epidemic patterns (Song et al., 2024). Another major challenge for ABM analysis lies in parameter uncertainty and calibration. To address this challenge, Kang et al. utilized spatiotemporally explicit uncertainty, global sensitivity analysis, and high-performance computers to optimize the ABM calibration for disease transmission simulation (Kang et al., 2020a; Kang et al., 2022). And Michels et al. proposed using Particle Swarm Optimization (PSO) to efficiently calibrate ABM for model influenza transmission in Miami (Michels et al., 2022).

Spatial accessibility to healthcare resources (e.g., healthcare personnel, testing resources, hospital beds, and ventilators) is critical people's health, especially during pandemic (e.g., COVID-19) (Kang et al., 2020b). Michels and Wang introduce AgentAccess, an agent-based model combined with cyberGIS for spatial accessibility analysis, demonstrating its effectiveness in simulating individual decision-making for healthcare resource access through a case study of hospital bed accessibility in Cook County, IL (Michels & Wang, 2023).

8.2.3 Transport Geography

New data sources (e.g., mobile phone records, smart cards, social media) enable more detailed mobility pattern analysis than conventional transport models. And machine learning approaches have been used extensively to address data uncertainty and integrate diverse datasets for transportation research (Anda et al., 2017). Traffic trajectory and origin-destination (OD) data, for example, have been utilized with different AI models for traffic prediction, urban planning, and management (Sun et al., 2019; Wang et al., 2022).

While AI models offer robust modeling capabilities for transportation analysis, the convergence of GeoAI and CyberGIS is particularly effective in data- and computation-intensive analysis. Michels et al. propose SPASTC, a spatial partitioning algorithm for efficient travel-time computation in large transportation networks. SPASTC addresses memory and computational challenges by using spatial domain decomposition, preserving spatial relationships while respecting user-specified memory limits (Michels et al., 2024a). In addition, using graph neural networks and cyberGIS capability, Wang et al. proposed EAST-Net, an enhanced spatiotemporal network for modeling mobility dynamics in Mobility-as-a-Service (MaaS) scenarios (Wang et al., 2024b). It incorporates a heterogeneous mobility information network to represent multimodal transport interactions and a memory-augmented dynamic filter generator to adapt to open-world events.

8.3 Case Study on Human Sentiments of Heat Exposure

As heat-related illness contributes to a significant portion of mortality, combined with the huge challenge of climate change and urban heat island (UHI) effect, it is critical to find the locations of vulnerable communities that are exposure to extreme heat, especially in the densely populated urban area (Shindell et al., 2020). This study integrates GeoAI and cyberGIS to analyze social media (LBSM) data, examining human perceptions of heat exposure in urban areas. This case study investigates the human sentiments of heat in the urban context leveraging GeoAI and cyberGIS capability. The methodology employs NLP and advanced computing techniques to collect, process, and visualize LBSM data, providing dynamic insights into heat-related sentiments across urban locations.

Mecklenburg County, North Carolina, is the study area for this research. Home to the city of Charlotte, it is a rapidly growing urban center with over one million residents. The county experienced significant urban heat island effects due to its dense population and rapid expansion (Li et al., 2023).

8.3.1 Location-Based Social Media Data

LBSM data (e.g., Twitter/X data) has become a versatile data source for solving urban problems due to its large volume, real time, low cost, and wide availability (McKitrick et al., 2022). Sentiment analysis, for example, has been widely studied by scholar by applying NLP algorithms to LBSM data contents and studies people's view on different locations. Sentiment analysis, for example, has been extensively applied to location-based LBSM. By utilizing NLP algorithms to examine people's perceptions of various locations, sentiment analysis enables researchers to extract and analyze sentiment patterns from user-generated content, providing insights into public opinions about specific places or geographical areas (Paolanti et al., 2021). Urban geographers have been leveraging the wealth of LBSM data, generated by dense populations, to examine mobility patterns (Luo et al., 2016), identify place of interest (Van Canneyt et al., 2012), study collective behaviors (Lee et al., 2013), and detect events (Béjar et al., 2016).

8.3.2 Understanding Social Media Content with GeoAI

In order to understand the content of social media data, a keyword-based NLP model is developed and a heat dictionary to understand the tweet posts. Figure 8.2 shows the workflow for generating the heat dictionary. In the first step, the word2vec tool is widely used in sentiment analysis and urban region identification (Xue et al., 2014), combined with a text corpus named One Billion Words Benchmark corpus, transforming words/phases into 300-dimension vector representation. And the distance (*d*) between different words is defined as the cosine similarity between words to measure their relatedness. Then, to evaluate the heat index in tweets, we identified synonyms for hot weather (*hwords*) and cold weather (*cwords*) based on the Oxford dictionary, where *hwords* included terms like "warm," "scorching," and "humid," and "*cwords*" included terms like "chilly," "frosty," and "frigid." The summation of distances between different words and *hwords* and *cwords* was used to determine their relatedness to weather. After normalizing this value between 0 and 1, a higher value for weather relatedness indicates stronger weather relevance. To filter out words that are irrelevant to weather from the heat dictionary, we kept the top 20,000 weather-related words/phrases and calculated their polarization toward either hot or cold weather by comparing their similarity distances to *hwords* and *cwords*. And in the last step for heat dictionary construction, we multiplied the word's relatedness to weather and its polarization toward hot weather and then normalized the value on a scale from -1 to 1, with -1 indicates a cold weather word and 1 toward hot weather.

The construction of the heat dictionary represents a method for keyword-based NLP algorithm, and it can be applied to the contents of tweets. After retrieving the key information from the LBSM data including timestamps, geospatial location,

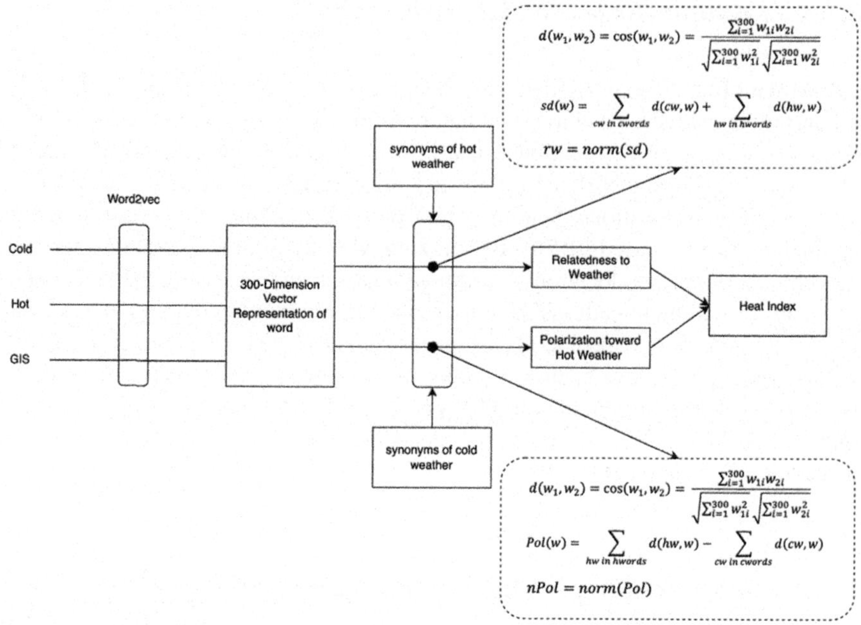

Fig. 8.2 Heat dictionary generation pipeline

and contents, we calculated heat exposure for each tweet by summing the heat index values of all the words in the textual content using the heat dictionary.

$$he = \sum_{w \text{ in twitter\_content}} hd(w)$$

8.3.3 CyberGIS for Geospatial Data Management and Processing

The cyberGIS capabilities serve two main purposes in this case study. First, cyberGIS help ensure geoprivacy and protect user information. A scientific workflow was developed using CyberGIS-Jupyter to allow indirect access to Twitter data. This data is collected and stored on a geospatial high-performance computer named Resourcing Open Geospatial Education and Research (ROGER). Users with the necessary credentials can request access to the Twitter data, which is retrieved via a REST-based API that pulls tweets based on user-initiated search queries, which prevent users from directly accessing the raw data. Second, cyberGIS supports data collection, management, and rapid analysis through advanced cyberinfrastructure. Given the large scale of social media data, we implemented a tweet collection workflow in Python3 using the *tweepy* library on ROGER, which continuously collects Twitter data from streaming endpoints in JSON format. For processing this large-scale location-based social media data, particularly for generating near-real-time analyses of human sentiments regarding heat exposure in urban areas, cyberGIS is crucial in providing the necessary computing power to provide a near-real-time solution to urban management.

8.3.4 Result

This case study explores human sentiments related to heat exposure in Mecklenburg County, North Carolina, focusing on the case study date of September 25, 2021. Although Twitter has a large user base, only about 1% of the data is accessible via the *tweepy* API due to user privacy concerns, and roughly 3% of all tweets include geospatial information (Leetaru et al., 2013). Out of 4202 tweets collected from Mecklenburg County on September 25, 2021, 142 were identified as weather-related, containing at least one word that matched a word in the heat dictionary. For this study, the census tract was chosen as the spatial unit, and the Census Tract boundary data was obtained from NC OneMap (https://www.nconemap.gov/). Figure 8.3 presents a histogram of human sentiments of heat exposure in Mecklenburg County in the study day. The normalized heat exposure values

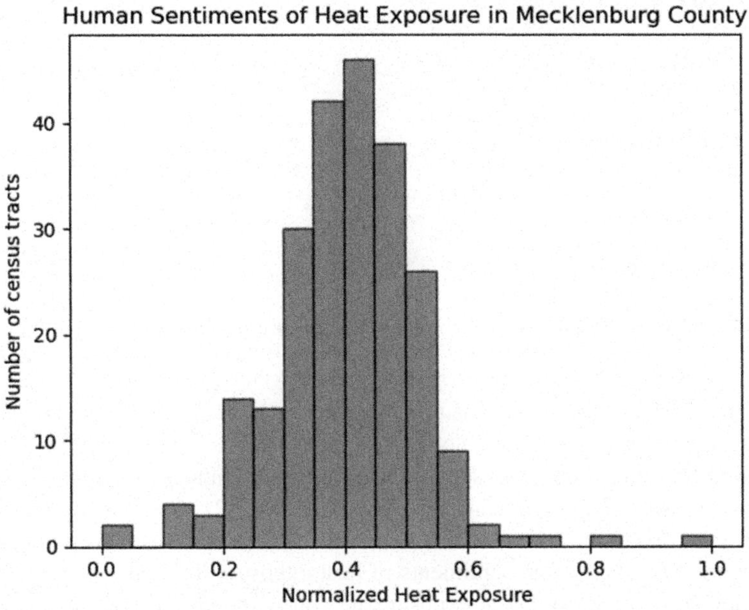

Fig. 8.3 Histogram of human sentiments of heat exposure in Mecklenburg County on September 25, 2021

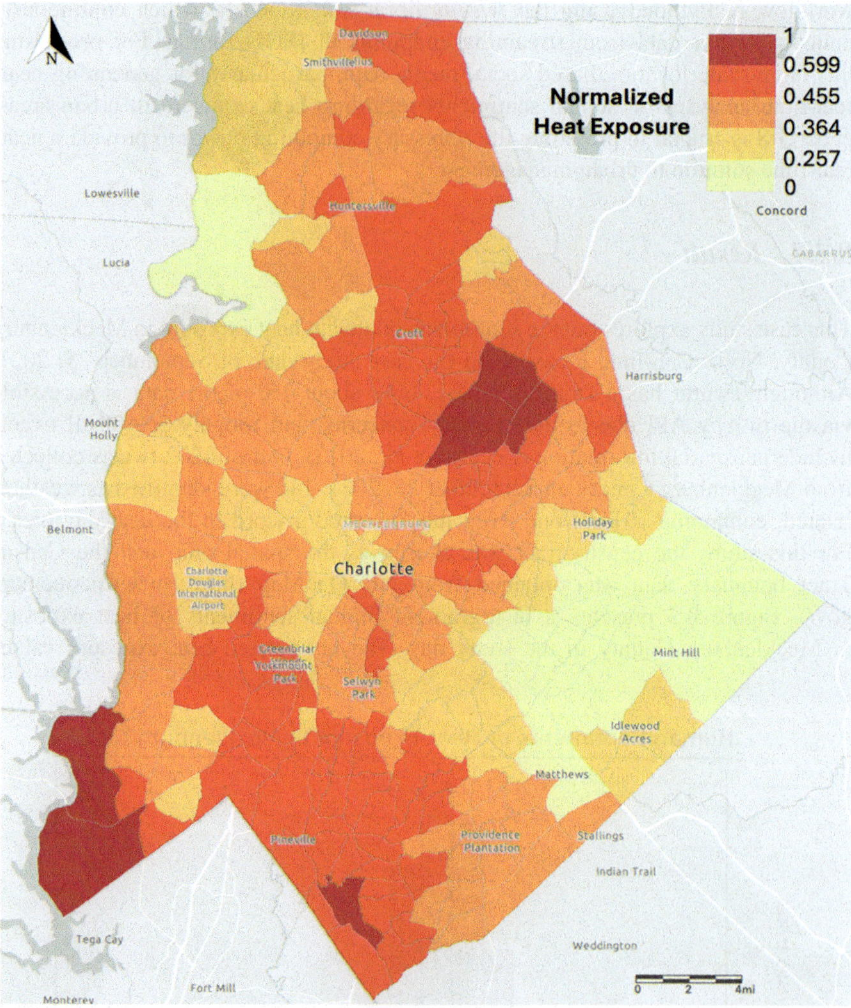

Fig. 8.4 Human sentiments of heat exposure in Mecklenburg County on September 25

generally follow a normal distribution, with the ranges between 0.35–0.4 and 0.4–0.45 encompassing more than 40 of the 233 census tracts in Mecklenburg County.

Figure 8.4 illustrates the spatial distribution of human sentiments of heat exposure in Mecklenburg County on September 25, 2021. Contrary to what might be expected due to the urban heat island effect, the downtown area is not identified as a hotspot for the human sentiments of heat exposure. Instead, clusters of high-level human sentiments of heat exposure are concentrated in the southwestern and northeastern parts of the county. In the northeastern region, areas such as Mineral Springs/Rumple Road and Hidden Valley, primarily residential zones surrounding downtown Charlotte, report high levels of heat-related sentiments. Similarly, in the southeastern part of the county, areas like Palisades and Ballantyne West, located on the periphery of the county, also show relatively high heat-related sentiments based on LBSM data. In contrast, the eastern and northwestern parts of Mecklenburg County exhibit relatively lower levels of heat-related sentiments according to the spatial distribution.

Human sentiments of heat exposure are highly complex, as they are influenced not only by environmental factors but also by the characteristics (e.g., socioeconomic status) of the people living in the area. Evaluating human sentiments of heat exposure is crucial, especially due to the correlation between heat and people's health, particularly in densely populated urban areas that experience from UHI effects. Although this case study focuses on a single county in the United States and a specific date, the combination of GeoAI and cyberGIS, along with large-scale LBSM data, offers the potential to scale this analysis to other major cities across the world due to the wide availability of Twitter. This approach could also enable continuous monitoring of daily or even hourly changes in human sentiments of heat exposure in near real time, which could be invaluable for identifying sudden heat or cold events and helping with resource allocation to mitigate risks from extreme weather.

8.4 Conclusion

The convergence of GeoAI and cyberGIS offers significant opportunities for human geography research, enabling analysis of complex spatial relationships and large-scale geospatial data. This integration addresses key challenges in data management (e.g., data collection, data storage, data ethics), modeling (e.g., heterogeneous data fusion, complex spatial analysis), and computing (e.g., fine-resolution, large-scale, data-and computational-intensive analysis). This chapter first reviews the recent literatures on GeoAI and cyberGIS applications in human geography, with specific focus on urban geography, health geography, and transport geography domains. In the following section, we present a case study applying GeoAI and cyberGIS to urban geography, examining human sentiments of heat exposure in Mecklenburg County, North Carolina. Utilizing LBSM data, natural language processing, and high-performance computing, the study reveals the spatial patterns of heat exposure perceptions by understanding the contents of geo-tagged Twitter posts. This demonstrates the capacity of integrated GeoAI and cyberGIS approaches to provide unique insights into urban problems.

Rapid advancements in AI, high-performance computing, and data availability (e.g., social media, social sensing) facilitate the convergence of GeoAI and cyberGIS in geographic analysis, enabling sophisticated modeling, near-real-time analysis, and high-resolution processing of diverse geospatial data for knowledge discovery in human geography. And the integration of GeoAI and cyberGIS could provide new opportunities addressing the ethical concerns regarding data privacy and responsible AI in human geography research. In conclusion, the convergence of GeoAI and CyberGIS holds potential in human geography research, providing advanced modeling and computing tools to enhance our understanding of the interactions between humans and the environment.

References

Anda, C., Erath, A., & Fourie, P. J. (2017). Transport modelling in the age of big data. *International Journal of Urban Sciences, 21*(sup1), 19–42. https://doi.org/10.1080/12265934.2017.1281150

Angelov, P. P., Soares, E. A., Jiang, R., Arnold, N. I., & Atkinson, P. M. (2021). Explainable artificial intelligence: An analytical review. *Wiley Interdisciplinary Reviews Data Mining and Knowledge Discovery, 11*(5), e1424. https://doi.org/10.1002/widm.1424

Batty, M., Axhausen, K. W., Giannotti, F., Pozdnoukhov, A., Bazzani, A., Wachowicz, M., Ouzounis, G., & Portugali, Y. (2012). Smart cities of the future. *The European Physical Journal. Special Topics, 214*(1), 481–518. https://doi.org/10.1140/epjst/e2012-01703-3

Béjar, J., Álvarez, S., García, D., Gómez, I., Oliva, L., Tejeda, A., & Vázquez-Salceda, J. (2016). Discovery of spatio-temporal patterns from location-based social networks. *Journal of Experimental & Theoretical Artificial Intelligence: JETAI, 28*(1–2), 313–329. https://doi.org/10.1080/0952813x.2015.1024492

Catlett, C. E., Beckman, P. H., Sankaran, R., & Galvin, K. K. (2017). Array of things: A scientific research instrument in the public way: Platform design and early lessons learned. In *Proceedings of the 2nd international workshop on science of smart city operations and platforms engineering* (pp. 26–33).

Gao, S. (2021). Geospatial artificial intelligence (GeoAI). In *Geography*. Oxford University Press. https://doi.org/10.1093/obo/9780199874002-0228

Goodchild, M. F. (2007). Citizens as sensors: The world of volunteered geography. *GeoJournal, 69*(4), 211–221. https://doi.org/10.1007/s10708-007-9111-y

Janowicz, K., Gao, S., McKenzie, G., Hu, Y., & Bhaduri, B. (2020). GeoAI: Spatially explicit artificial intelligence techniques for geographic knowledge discovery and beyond. *Geographical Information Systems, 34*(4), 625–636. https://doi.org/10.1080/13658816.2019.1684500

Kang, J.-Y., Aldstadt, J., Vandewalle, R., Yin, D., & Wang, S. (2020a). A CyberGIS approach to spatiotemporally explicit uncertainty and global sensitivity analysis for agent-based modeling of vector-borne disease transmission. *Annals of the American Association of Geographers, 110*(6), 1855–1873. https://doi.org/10.1080/24694452.2020.1723400

Kang, J.-Y., Michels, A., Lyu, F., Wang, S., Agbodo, N., Freeman, V. L., & Wang, S. (2020b). Rapidly measuring spatial accessibility of COVID-19 healthcare resources: A case study of Illinois, USA. *International Journal of Health Geographics, 19*(1), 36. https://doi.org/10.1186/s12942-020-00229-x

Kang, J.-Y., Michels, A., Crooks, A., Aldstadt, J., & Wang, S. (2022). An integrated framework of global sensitivity analysis and calibration for spatially explicit agent-based models. *Transactions in GIS: TG, 26*(1), 100–128. https://doi.org/10.1111/tgis.12837

Kitchin, R. (2013). Big data and human geography. *Dialogues in Human Geography, 3*(3), 262–267. https://doi.org/10.1177/2043820613513388

Lazer, D., Pentland, A., Adamic, L., Aral, S., Barabási, A.-L., Brewer, D., Christakis, N., Contractor, N., Fowler, J., Gutmann, M., Jebara, T., King, G., Macy, M., Roy, D., & Van Alstyne, M. (2009). Computational social science. *Science (New York, N.Y.), 323*(5915), 721–723. https://doi.org/10.1126/science.1167742

Lee, J.-G., & Kang, M. (2015). Geospatial big data: Challenges and opportunities. *Big Data Research, 2*(2), 74–81. https://doi.org/10.1016/j.bdr.2015.01.003

Lee, K., Tamilarasan, P., & Caverlee, J. (2013). Crowdturfers, campaigns, and social media: Tracking and revealing crowdsourced manipulation of social media. *Proceedings of the International AAAI Conference on Web and Social Media, 7*, 331–340.
Leetaru, K., Wang, S., Cao, G., Padmanabhan, A., & Shook, E. (2013). Mapping the global twitter heartbeat: The geography of twitter. *First Monday, 18*(5), 4. https://doi.org/10.5210/fm.v18i5.4366
Li, X., Ma, X., Lyu, F., & Song, Y. (2023). Examine the environmental inequity impact of urban heat mitigation on redlining legacy: Case study of Charlotte's retrofitting, 2001–2020. *Frontiers in Environmental Science, 11*, 1218819. https://doi.org/10.3389/fenvs.2023.1218819
Liu, P., & Biljecki, F. (2022). A review of spatially-explicit GeoAI applications in urban geography. *International Journal of Applied Earth Observation and Geoinformation: ITC Journal, 112*(102936), 102936. https://doi.org/10.1016/j.jag.2022.102936
Luo, F., Cao, G., Mulligan, K., & Li, X. (2016). Explore spatiotemporal and demographic characteristics of human mobility via twitter: A case study of Chicago. *Applied Geography (Sevenoaks, England), 70*, 11–25. https://doi.org/10.1016/j.apgeog.2016.03.001
Lyu, F., Yin, D., Padmanabhan, A., Choi, Y., Goodall, J. L., Castronova, A., Tarboton, D., & Wang, S. (2019, July 28). Reproducible hydrological modeling with CyberGIS-Jupyter. In *Proceedings of the practice and experience in advanced research computing on rise of the machines (learning). PEARC '19: Practice and experience in advanced research computing, Chicago IL USA*. https://doi.org/10.1145/3332186.3333052
Lyu, F., Wang, S., Han, S. Y., Catlett, C., & Wang, S. (2022). An integrated cyberGIS and machine learning framework for fine-scale prediction of urban Heat Island using satellite remote sensing and urban sensor network data. *Urban Informatics, 1*(1), 6. https://doi.org/10.1007/s44212-022-00002-4
Lyu, F., Zhou, L., Park, J., Baig, F., & Wang, S. (2024). Mapping dynamic human sentiments of heat exposure with location-based social media data. *International Journal of Geographical Information Science, 38*, 1291.
McKitrick, M. K., Schuurman, N., & Crooks, V. A. (2022). Collecting, analyzing, and visualizing location-based social media data: Review of methods in GIS-social media analysis. *GeoJournal, 88*(1), 1035–1057. https://doi.org/10.1007/s10708-022-10584-w
Michels, A., & Wang, S. (2023). *An agent-based modeling approach to spatial accessibility*. Purdue University. https://doi.org/10.5703/1288284317670
Michels, A., Kang, J.-Y., & Wang, S. (2022). Particle swarm optimization for calibration in spatially explicit agent-based modeling. *Journal of Artificial Societies and Social Simulation: JASSS, 25*(2), 8. https://doi.org/10.18564/jasss.4796
Michels, A. C., Park, J., Kang, J.-Y., & Wang, S. (2024a). SPASTC: A spatial partitioning algorithm for scalable travel-time computation. *Geographical Information Systems, 38*(5), 803–824. https://doi.org/10.1080/13658816.2024.2326445
Michels, A. C., Padmanabhan, A., Xiao, Z., Kotak, M., Baig, F., & Wang, S. (2024b). CyberGIS-compute: Middleware for democratizing scalable geocomputation. *SoftwareX, 26*(101691), 101691. https://doi.org/10.1016/j.softx.2024.101691
Nelson, T. A., Goodchild, M. F., & Wright, D. J. (2022). Accelerating ethics, empathy, and equity in geographic information science. *Proceedings of the National Academy of Sciences of the United States of America, 119*(19), e2119967119. https://doi.org/10.1073/pnas.2119967119
Paolanti, M., Mancini, A., Frontoni, E., Felicetti, A., Marinelli, L., Marcheggiani, E., & Pierdicca, R. (2021). Tourism destination management using sentiment analysis and geo-location information: A deep learning approach. *Information Technology & Tourism, 23*(2), 241–264. https://doi.org/10.1007/s40558-021-00196-4
Sahana, K. S., Karthik, C. B., Sree, M. M., Chaithra, M., Abhinandithe, S. K., Snehalatha, B., Sridhar, R., Balasubramanian, S., & Madhu, B. (2023). Geospatial artificial intelligence (GeoAI): Applications in health care. *International Journal of Health & Allied Sciences, 11*(4), 4. https://doi.org/10.55691/2278-344x.1044
Shi, X., & Wang, S. (2015). Computational and data sciences for health-GIS. *Annals of GIS, 21*(2), 111–118. https://doi.org/10.1080/19475683.2015.1027735

Shindell, D., Zhang, Y., Scott, M., Ru, M., Stark, K., & Ebi, K. L. (2020). The effects of heat exposure on human mortality throughout the United States. *GeoHealth, 4*(4), e2019GH000234. https://doi.org/10.1029/2019GH000234

Soliman, A., Soltani, K., Yin, J., Padmanabhan, A., & Wang, S. (2017). Social sensing of urban land use based on analysis of twitter users' mobility patterns. *PLoS One, 12*(7), e0181657. https://doi.org/10.1371/journal.pone.0181657

Song, Z., Zhang, Z., Lyu, F., Bishop, M., Liu, J., & Chi, Z. (2024). From individual motivation to geospatial epidemiology: A novel approach using fuzzy cognitive maps and agent-based modeling for large-scale disease spread. *Sustainability, 16*(12), 5036. https://doi.org/10.3390/su16125036

Steiniger, S., Poorazizi, M. E., & Hunter, A. J. S. (2016). Planning with citizens: Implementation of an e-planning platform and analysis of research needs. *Urban Planning, 1*(2), 46–64. https://doi.org/10.17645/up.v1i2.607

Sui, D., Goodchild, M., & Elwood, S. (2013). Volunteered geographic information, the exaflood, and the growing digital divide. In *Crowdsourcing geographic knowledge* (pp. 1–12). Springer. https://doi.org/10.1007/978-94-007-4587-2_1

Sun, S., Chen, J., & Sun, J. (2019). Traffic congestion prediction based on GPS trajectory data. *International Journal of Distributed Sensor Networks, 15*(5), 155014771984744. https://doi.org/10.1177/1550147719847440

Tracy, M., Cerdá, M., & Keyes, K. M. (2018). Agent-based modeling in public health: Current applications and future directions. *Annual Review of Public Health, 39*, 77–94. https://doi.org/10.1146/annurev-publhealth-040617-014317

Van Canneyt, S., Schockaert, S., Van Laere, O., & Dhoedt, B. (2012, December). Detecting places of interest using social media. In *2012 IEEE/WIC/ACM International Conferences on Web Intelligence and Intelligent Agent Technology. 2012 IEEE/WIC/ACM international joint conferences on web intelligence (WI) and intelligent agent technologies (IAT), Macau, China.*https://doi.org/10.1109/wi-iat.2012.19

Vandewalle, R., Kang, J.-Y., Yin, D., & Wang, S. (2019, November 5). Integrating CyberGIS-Jupyter and spatial agent-based modelling to evaluate emergency evacuation time. In *Proceedings of the 2nd ACM SIGSPATIAL international workshop on GeoSpatial simulation. SIGSPATIAL '19: 27th ACM SIGSPATIAL international conference on advances in geographic information systems, Chicago Illinois*. https://doi.org/10.1145/3356470.3365530

Wang, S. (2010). A CyberGIS framework for the synthesis of cyberinfrastructure, GIS, and spatial analysis. *Annals of the Association of American Geographers, 100*(3), 535–557. https://doi.org/10.1080/00045601003791243

Wang, S., Anselin, L., Bhaduri, B., Crosby, C., Goodchild, M. F., Liu, Y., & Nyerges, T. L. (2013). CyberGIS software: A synthetic review and integration roadmap. *Geographical Information Systems, 27*(11), 2122–2145. https://doi.org/10.1080/13658816.2013.776049

Wang, S., Lyu, F., Wang, S., Catlett, C. E., Padmanabhan, A., & Soltani, K. (2021). Integrating CyberGIS and urban sensing for reproducible streaming analytics. In *Urban informatics* (pp. 663–681). Springer. https://doi.org/10.1007/978-981-15-8983-6_36

Wang, C., Lyu, F., Wu, S., Wang, Y., Xu, L., Zhang, F., Wang, S., Wang, Y., & Du, Z. (2022). A deep trajectory clustering method based on sequence-to-sequence autoencoder model. *Transactions in GIS: TG, 26*(4), 1801–1820. https://doi.org/10.1111/tgis.12905

Wang, Z., Jin, B., Hu, W., Jiang, M., Kang, S., Li, Z., Zhou, S., Han, J., & Wang, S. (2023, November 13). Geospatial knowledge hypercube. In *Proceedings of the 31st ACM international conference on advances in geographic information systems. SIGSPATIAL '23: 31st ACM international conference on advances in geographic information systems, Hamburg Germany*. https://doi.org/10.1145/3589132.3625629

Wang, S., Huang, X., Liu, P., Zhang, M., Biljecki, F., Hu, T., Fu, X., Liu, L., Liu, X., Wang, R., Huang, Y., Yan, J., Jiang, J., Chukwu, M., Reza Naghedi, S., Hemmati, M., Shao, Y., Jia, N., Xiao, Z., et al. (2024a). Mapping the landscape and roadmap of geospatial artificial intelligence (GeoAI) in quantitative human geography: An extensive systematic review. *International Journal of Applied Earth Observation and Geoinformation: ITC Journal, 128*(103734), 103734. https://doi.org/10.1016/j.jag.2024.103734

Wang, Z., Jiang, R., Xue, H., Salim, F. D., Song, X., Shibasaki, R., Hu, W., & Wang, S. (2024b). Learning spatio-temporal dynamics on mobility networks for adaptation to open-world events. *Artificial Intelligence, 335*(104120), 104120. https://doi.org/10.1016/j.artint.2024.104120

Xue, B., Fu, C., & Shaobin, Z. (2014, June). A study on sentiment computing and classification of Sina Weibo with Word2vec. In *2014 IEEE International Congress on Big Data. 2014 IEEE International Congress on Big Data (BigData congress), Anchorage, AK, USA*. https://doi.org/10.1109/bigdata.congress.2014.59

Yin, D., Liu, Y., Padmanabhan, A., Terstriep, J., Rush, J., & Wang, S. (2017a). A CyberGIS-Jupyter framework for geospatial analytics at scale. *Proceedings of the Practice and Experience in Advanced Research Computing 2017 on Sustainability, Success and Impact, 18*, 1–18:8. https://doi.org/10.1145/3093338.3093378

Yin, J., Gao, Y., & Wang, S. (2017b). CyberGIS-enabled urban sensing from volunteered citizen participation using mobile devices. In *Springer geography* (pp. 83–96). Springer International Publishing. https://doi.org/10.1007/978-3-319-40902-3_5

Yin, D., Wang, S., & Ouyang, Y. (2020). ViCTS: A novel network partition algorithm for scalable agent-based modeling of mass evacuation. *Computers, Environment and Urban Systems, 80*(101452), 101452. https://doi.org/10.1016/j.compenvurbsys.2019.101452

Zhang, Z., Hu, H., Yin, D., Kashem, S., Li, R., Cai, H., Perkins, D., & Wang, S. (2019). A cyberGIS-enabled multi-criteria spatial decision support system: A case study on flood emergency management. *International Journal of Digital Earth, 12*(11), 1364–1381. https://doi.org/10.1080/17538947.2018.1543363

Zhang, Z., Yin, D., Virrantaus, K., Ye, X., & Wang, S. (2021). Modeling human activity dynamics: An object-class oriented space–time composite model based on social media and urban infrastructure data. *Computational Urban Science, 1*(1), 7. https://doi.org/10.1007/s43762-021-00006-x

Chapter 9
High-Performance Geospatial Computing and Computationally Reproducible Human Geography

Xiao Huang and Jeon-Young Kang

Abstract The exponential growth of geospatial data has transformed human geography, presenting both new opportunities and significant computational challenges. This chapter explores how high-performance geospatial computing (HPGC) addresses these challenges by leveraging advanced computational resources to efficiently process and analyze large-scale geospatial datasets. HPGC enables complex simulations, real-time analytics, and intricate spatial modeling, expanding the capabilities of human geography research. Key technological advancements discussed include parallel processing, graphics processing units (GPU) integration, and cloud-based infrastructures. The chapter also emphasizes the importance of computationally reproducible human geography, highlighting the need for results that can be independently verified using original data and code. HPGC facilitates reproducibility by supporting the sharing of comprehensive analytical workflows and enhancing transparency and collaboration. By examining the synergy between HPGC and reproducible practices, this chapter underscores how these advancements empower researchers to address complex spatial challenges effectively and advance the discipline.

Keywords High-performance geospatial computing (HPGC) · Geospatial big data · Computational reproducibility

9.1 Introduction

In an era defined by the exponential growth of data, the field of human geography stands at the cusp of a transformative evolution. The proliferation of geospatial

X. Huang (✉)
Department of Environmental Sciences, Emory University, Atlanta, GA, USA
e-mail: xiao.huang2@emory.edu

J.-Y. Kang
Department of Geography, Kyung Hee University, Seoul, South Korea
e-mail: geokang@khu.ac.kr

© The Author(s), under exclusive license to Springer Nature Switzerland AG 2025
X. Huang et al. (eds.), *GeoAI and Human Geography*, Springer Geography,
https://doi.org/10.1007/978-3-031-87421-5_9

data—from satellite imagery and global positioning system (GPS) traces to social media check-ins and urban sensor networks—has unlocked unprecedented opportunities to explore the complex spatial dynamics of human societies. However, this deluge of data also presents significant computational challenges (Huang, 2024). Traditional methods of data processing and analysis are often inadequate for handling the volume, velocity, and variety of contemporary geospatial datasets. To address these challenges, high-performance geospatial computing (HPGC) has emerged as a paradigm that leverages advanced computational resources to efficiently process and analyze large-scale geospatial data. HPGC is not merely a technological advancement; it is a catalyst that enables deeper insights into human geographical phenomena by facilitating complex simulations, real-time analytics, and intricate spatial modeling.

The development of HPGC has been instrumental in pushing the boundaries of human geography. From the advent of parallel processing and distributed computing systems to the integration of graphics processing units (GPUs) and cloud-based infrastructures, the evolution of computational technologies has significantly enhanced our ability to handle and interpret vast geospatial datasets. These advancements have empowered geographers to tackle previously intractable problems, such as real-time urban traffic modeling (Meduri et al., 2023), large-scale environmental impact assessments (Liang et al., 2020; Feng et al., 2022), infectious disease modeling (Kang et al., 2020), and the spatial analysis of social networks (Huang et al., 2020; Li et al., 2021). By accelerating data processing and enabling more sophisticated analytical techniques, HPGC has become an indispensable tool in the geographer's arsenal.

Parallel to these technological advancements is the growing emphasis on computationally reproducible human geography. In the scientific community, reproducibility is a cornerstone of credible and reliable research (McNutt, 2014). Computational reproducibility refers to the ability of researchers to reproduce the results of a study using the original data and code. In human geography, where analyses often involve complex models and large datasets, ensuring computational reproducibility is crucial for validating findings, fostering collaboration, and advancing the discipline (Kedron et al., 2021; Nüst & Pebesma, 2021). Computationally Reproducible Human Geography thus embodies the principles of transparency, repeatability, and rigor in spatial analysis. High-performance geospatial computing (HPGC) plays a pivotal role in facilitating computational reproducibility in human geography. By providing the computational power necessary to handle extensive datasets and complex models, HPGC enables researchers to share their results as well as the computational workflows that produced them. This capacity for sharing complete analytical pipelines—including data preprocessing, modeling, and visualization steps—enhances the transparency and reproducibility of geographical research. Moreover, the integration of open-source platforms, standardized data formats, and collaborative tools within HPGC frameworks further support reproducible practices.

This chapter delves into the intersection of high-performance geospatial computing and computationally reproducible human geography. We will explore the historical development of HPGC, highlighting key technological milestones and their impact on geographical research. We will define computational reproducibility within the context of human geography and discuss its importance in the contemporary research landscape. Finally, we will examine how HPGC facilitates reproducible exploration in human geography, emphasizing a synergy between advanced computational capabilities and the principles of open, transparent science.

9.2 The Evolution of High-Performance Geospatial Computing

9.2.1 Early Computational Methods in Geography

The integration of computational tools into geography began in the mid-twentieth century, marking a significant shift from traditional qualitative approaches to more quantitative and analytical methods. Early geographers relied heavily on manual cartography and statistical calculations to analyze spatial phenomena. The introduction of computers in the 1950s and 1960s opened new avenues for data processing and spatial analysis, albeit with limited capabilities. During this period, the first geographic information systems (GIS) emerged, such as the Canada geographic information system (CGIS) developed in the 1960s for land inventory management (Maliene et al., 2011). These early GIS platforms allowed for the digital storage and basic manipulation of spatial data, setting the foundation for modern geospatial analysis.

Despite these advancements, traditional geospatial computing methods faced significant limitations. The hardware of the time was constrained by minimal processing power and memory capacity, restricting the size and complexity of datasets that could be analyzed (Breunig et al., 2020). Data input and output were often cumbersome processes, involving punch cards and magnetic tapes, which were prone to errors and data loss (Orlik & Orlikova, 2014). Software options were limited and frequently proprietary, hindering widespread adoption and collaboration among researchers (Kiehle et al., 2007). Additionally, the lack of standardized data formats made data sharing difficult, isolating studies and preventing the accumulation of collective knowledge. These constraints meant that early computational geography was accessible only to those with significant resources, limiting its impact on the broader field of human geography.

The limitations extended to the analytical capabilities of early systems. Spatial analyses were often simplistic, focusing on basic overlay operations and lacking the sophistication needed to model complex human and environmental interactions. Visualization tools were rudimentary, producing static maps that failed to capture the dynamic nature of geographical processes. The computational inefficiency also meant that running even simple analyses could take an inordinate amount of time, discouraging iterative exploration and hypothesis testing. Consequently, while early computational methods in geography represented a significant step forward, they were insufficient for addressing the increasingly complex and data-intensive questions that would arise in the following decades.

9.2.2 Technological Advancements in HPGC

The evolution of HPGC has been propelled by significant technological advancements aimed at overcoming the limitations of earlier methods. One of the most pivotal developments has been the advent of parallel processing and distributed computing systems. Parallel processing allows multiple computations to be carried out simultaneously by dividing tasks across multiple processors (Asanovic et al., 2009). This approach significantly reduces processing time for large datasets and complex models. Distributed computing extends this concept by linking together a network of computers to share computational tasks (Kshemkalyani & Singhal, 2011). Technologies such as cluster computing and grid computing have been instrumental in enabling geospatial analyses that were previously computationally prohibitive.

The integration of graphics processing units (GPUs) into geospatial analysis represents another major leap forward. Originally designed to handle the intense graphical computations for rendering images and videos, GPUs are adept at performing a high number of simple calculations in parallel. This makes them ideal for handling the large matrices and raster data common in geospatial analyses. Software frameworks like NVIDIA's CUDA and OpenCL have made it possible to harness GPU power for general-purpose computing tasks, including spatial statistics, image processing, and simulation models (Li et al., 2020). The result is a substantial acceleration of computational tasks, enabling real-time processing and more complex analyses.

The emergence of cloud-based infrastructures has further revolutionized HPGC by providing scalable and flexible computational resources. Cloud platforms such as Amazon Web Services (AWS), Microsoft Azure, and Google Cloud offer specialized services for geospatial data storage, processing, and analysis. These services eliminate the need for significant upfront investment in hardware and allow researchers to access high-performance computing resources on demand. Cloud computing facilitates collaboration by enabling multiple users to access and work on the same datasets and computational environments from different locations. It also supports the integration of various tools and platforms, promoting interoperability and efficiency in geospatial research.

9.2.3 Impact on Human Geography Research

The advancements in high-performance geospatial computing have had a profound impact on human geography research, fundamentally changing how spatial data is processed and analyzed. One of the most significant impacts is the enhanced capability to process large-scale geospatial data. Researchers can now handle datasets encompassing millions of data points, such as global positioning system (GPS) traces, satellite imagery, and social media geotags. This capability allows for more comprehensive studies on urbanization patterns, migration flows, and environmental change, providing insights at scales that were previously unattainable (Ferreira & Vale, 2024).

The ability to perform complex simulations and real-time analytics has also been greatly enhanced. High-performance computing enables the modeling of intricate systems, such as urban growth dynamics, transportation networks, and disaster response scenarios. For instance, agent-based models can simulate the behaviors of individuals within a spatial context, offering valuable predictions about traffic congestion (Cuevas, 2020) or disease spread (Kang & Aldstadt, 2019). Real-time analytics have become increasingly important in applications like emergency management, where immediate analysis of spatial data can inform critical decision-making during natural disasters or public health crises (Huang et al., 2021; Hu et al., 2021).

These technological advancements have facilitated the expansion of research frontiers in human geography. Researchers are now exploring interdisciplinary approaches that integrate geospatial analysis with fields like economics, sociology, and environmental science. The capacity to process and analyze big data has opened new avenues for studying phenomena such as the impacts of climate change on human populations, the spatial dimensions of social networks, and the patterns of resource distribution and consumption. High-performance geospatial computing has also democratized access to advanced analytical tools, enabling researchers from various backgrounds and institutions to contribute to the field.

In summary, the evolution of HPGC has addressed the limitations of traditional methods and unlocked new possibilities for exploration and discovery in human geography. By providing the tools necessary to handle vast and complex datasets, perform sophisticated analyses, and collaborate across disciplines, HPGC has become an indispensable component of contemporary geographical research.

9.3 Computationally Reproducible Human Geography

9.3.1 Definition and Importance

Computational reproducibility refers to the ability of independent researchers to obtain the same results using the original data and computational methods used in a study (Piccolo & Frampton, 2016). In the context of research, it ensures that findings are not only verifiable but also reliable, fostering trust within the scientific community. Computational reproducibility goes beyond simply sharing data; it encompasses the entire computational workflow, including data processing, analysis algorithms, and visualization techniques. This level of transparency allows others to validate results, identify potential errors, and build upon previous work without unnecessary duplication of effort. In human geography, computational reproducibility holds significant importance due to the complex and data-intensive nature of the field. Human geography often involves analyzing large geospatial datasets, running intricate models, and interpreting multifaceted spatial relationships. The reproducibility of such studies is crucial for several reasons. First, it allows for the validation of findings, ensuring that conclusions drawn about human spatial behaviors and patterns are sound and reliable. Second, it facilitates cumulative knowledge building by enabling researchers to build upon each other's work, refining models, and expanding analyses to new contexts or scales. Third, reproducibility enhances the credibility of the geography discipline, demonstrating a commitment to rigorous scientific standards.

Moreover, computational reproducibility in human geography is essential for informing policy and decision-making. Geographic research frequently influences urban planning, environmental management, and social policy. Reproducible studies provide policymakers with confidence in the evidence base, knowing that the results have been thoroughly vetted and can be independently verified. In an era where data-driven decisions are increasingly prominent, the reproducibility of geospatial analyses ensures that policies are grounded in robust and trustworthy research (Hwang et al., 2021; Miller & Goodchild, 2015).

Finally, computational reproducibility aligns with the broader movement toward open science and transparency in research. By making data and computational methods openly available, human geographers contribute to a more collaborative and inclusive scientific community. This openness is expected to accelerate innovation, as researchers from diverse backgrounds bring new perspectives and techniques to existing problems.

9.3.2 Challenges in Achieving Reproducibility

Despite its importance, achieving computational reproducibility in human geography faces several challenges. One of the primary obstacles is the inherent complexity of geospatial models and analyses. Geospatial data often come in various formats and scales, requiring sophisticated processing techniques to manage and analyze effectively (Kedron et al., 2021). Models may incorporate numerous variables, intricate algorithms, and custom code, which can be difficult to document comprehensively. The complexity increases when models are stochastic or involve machine learning algorithms, where results can vary based on initial conditions or random seed values (Gundersen et al., 2022). This intricacy makes it challenging for other researchers to replicate studies precisely, especially if detailed documentation and code are not provided.

Another significant challenge is issues related to data accessibility. Geospatial data can be subject to licensing restrictions, confidentiality agreements, or proprietary formats, limiting the ability of researchers to share datasets openly. Even when data are publicly available, they may be too large to distribute conveniently or require specialized software to interpret. Additionally, the dynamic nature of certain geospatial datasets, such as real-time sensor data or social media feeds, means that data snapshots used in original studies may no longer be accessible (Huang et al., 2022). This temporal aspect can hinder efforts to reproduce analyses that depend on specific data points captured at particular times.

Software dependencies pose further hurdles to reproducibility. Geospatial analyses often rely on a combination of software tools, libraries, and platforms that may have version-specific features or compatibility issues. Over time, software updates or deprecations can render original code unusable or produce different results due to changes in algorithms or default settings. Researchers attempting to reproduce a study may struggle to recreate the exact computational environment, especially if the original setup was not thoroughly documented. Proprietary software adds another layer of complexity, as access may be limited by licensing costs or institutional availability, preventing some researchers from replicating studies that depend on these tools.

9.3.3 Principles of Reproducible Research

To overcome the challenges of reproducibility, researchers in human geography can adopt several key principles of reproducible research. Foremost among these is transparency in methodology and data usage. This involves providing detailed documentation of all steps in the research process, from data collection and preprocessing to analysis and visualization (Leonelli et al., 2015). Researchers should include descriptions of the algorithms used, parameter settings, and any assumptions made during the study. Being transparent about methods allows other researchers to understand exactly how results were obtained and identify any potential sources of error or bias.

Repeatability and validation of results are also critical components. Repeatability refers to the ability of the original researchers to obtain consistent results using the same data and methods, while reproducibility extends this capability to independent researchers (Tullis & Kar, 2021). To facilitate this, sharing the original code and computational workflows is essential. Utilizing version control systems like GIT can help track changes in code over time and allow others to access the exact codebase used in the study. Providing test datasets or subsets of data can enable others to validate specific parts of the analysis without needing access to the entire dataset.

The adoption of open science practices further supports reproducibility (Brunsdon & Comber, 2021). Open science advocates for making research outputs, including data, code, and publications, freely available to the public. Using open-source software and standard data formats enhances accessibility and reduces barriers related to proprietary tools. Platforms like GitHub, GitLab, or repositories like Zenodo and Figshare allow researchers to share code and data openly. Pre-registering studies and publishing in open-access journals also contribute to transparency and allow for broader dissemination of research findings. Additionally, documenting the computational environment is important. Tools like containerization (e.g., Docker) or virtualization can encapsulate the software environment, ensuring that all dependencies and software versions are preserved. This approach enables other researchers to recreate the exact environment in which the original analysis was conducted, minimizing issues related to software updates or incompatibilities.

9.4 Conclusion

The convergence of HPGC and computationally reproducible human geography marks a transformative moment in geographic research. HPGC provides the computational muscle needed to process and analyze vast, complex geospatial datasets, unlocking insights into spatial phenomena that were previously beyond reach. This advancement has expanded the horizons of human geography, enabling sophisticated modeling of urban dynamics, environmental impacts, and social networks. As data volumes continue to grow, HPGC becomes ever more crucial in managing and interpreting this information effectively. Simultaneously, the emphasis on computational reproducibility ensures that these powerful analyses are transparent, verifiable, and trustworthy. By adhering to principles such as open data, shared methodologies, and detailed documentation, researchers enhance the credibility of their findings. This is particularly vital in human geography, where studies often inform policy decisions and resource management. Reproducibility allows for validation and fosters collaboration, enabling the scientific community to build upon each other's work confidently. In essence, embracing both high-performance geospatial computing and computational reproducibility positions human geography to address the pressing spatial challenges of our time more effectively. It empowers researchers to delve deeper into complex issues while maintaining rigorous standards of scientific integrity. As the field progresses, maintaining this dual focus will be essential for advancing knowledge and ensuring that research contributes meaningfully to society.

References

Asanovic, K., Bodik, R., Demmel, J., Keaveny, T., Keutzer, K., Kubiatowicz, J., & Yelick, K. (2009). A view of the parallel computing landscape. *Communications of the ACM, 52*(10), 56–67.

Breunig, M., Bradley, P. E., Jahn, M., Kuper, P., Mazroob, N., Rösch, N., & Jadidi, M. (2020). Geospatial data management research: Progress and future directions. *ISPRS International Journal of Geo-Information, 9*(2), 95.

Brunsdon, C., & Comber, A. (2021). Opening practice: Supporting reproducibility and critical spatial data science. *Journal of Geographical Systems, 23*(4), 477–496.

Cuevas, E. (2020). An agent-based model to evaluate the COVID-19 transmission risks in facilities. *Computers in Biology and Medicine, 121*, 103827.
Feng, S., Li, W., Xu, J., Liang, T., Ma, X., Wang, W., & Yu, H. (2022). Land use/land cover mapping based on GEE for the monitoring of changes in ecosystem types in the upper Yellow River basin over the Tibetan plateau. *Remote Sensing, 14*(21), 5361.
Ferreira, D., & Vale, M. (2024). Big data and geography. In *The encyclopedia of human geography* (pp. 1–4). Springer International Publishing.
Gundersen, O. E., Shamsaliei, S., & Isdahl, R. J. (2022). Do machine learning platforms provide out-of-the-box reproducibility? *Future Generation Computer Systems, 126*, 34–47.
Hu, T., Wang, S., She, B., Zhang, M., Huang, X., Cui, Y., & Li, Z. (2021). Human mobility data in the COVID-19 pandemic: Characteristics, applications, and challenges. *International Journal of Digital Earth, 14*(9), 1126–1147.
Huang, X. (2024). Computational urban science needs to go beyond computational. *Computational Urban Science, 4*(1), 18.
Huang, X., Li, Z., Jiang, Y., Li, X., & Porter, D. (2020). Twitter reveals human mobility dynamics during the COVID-19 pandemic. *PLoS One, 15*(11), e0241957.
Huang, X., Li, Z., Jiang, Y., Ye, X., Deng, C., Zhang, J., & Li, X. (2021). The characteristics of multi-source mobility datasets and how they reveal the luxury nature of social distancing in the US during the COVID-19 pandemic. *International Journal of Digital Earth, 14*(4), 424–442.
Huang, X., Wang, S., Zhang, M., Hu, T., Hohl, A., She, B., & Li, Z. (2022). Social media mining under the COVID-19 context: Progress, challenges, and opportunities. *International Journal of Applied Earth Observation and Geoinformation, 113*, 102967.
Hwang, S., Nam, T., & Ha, H. (2021). From evidence-based policy making to data-driven administration: Proposing the data vs. value framework. *International Review of Public Administration, 26*(3), 291–307.
Kang, J. Y., & Aldstadt, J. (2019). Using multiple scale spatio-temporal patterns for validating spatially explicit agent-based models. *International Journal of Geographical Information Science, 33*(1), 193–213.
Kang, J. Y., Aldstadt, J., Vandewalle, R., Yin, D., & Wang, S. (2020). A CyberGIS approach to spatiotemporally explicit uncertainty and global sensitivity analysis for agent-based modeling of vector-borne disease transmission. *Annals of the American Association of Geographers, 110*(6), 1855–1873.
Kedron, P., Frazier, A. E., Trgovac, A. B., Nelson, T., & Fotheringham, A. S. (2021). Reproducibility and replicability in geographical analysis. *Geographical Analysis, 53*(1), 135–147.
Kiehle, C., Greve, K., & Heier, C. (2007). Requirements for next generation spatial data infrastructures-standardized web based geoprocessing and web service orchestration. *Transactions in GIS, 11*(6), 819–834.
Kshemkalyani, A. D., & Singhal, M. (2011). *Distributed computing: Principles, algorithms, and systems*. Cambridge University Press.
Leonelli, S., Spichtinger, D., & Prainsack, B. (2015). Sticks and carrots: Encouraging open science at its source. *Geo: Geography and Environment, 2*(1), 12–16.
Li, Z., Gui, Z., Hofer, B., Li, Y., Scheider, S., & Shekhar, S. (2020). Geospatial information processing technologies. In *Manual of digital earth* (pp. 191–227). Springer.
Li, Z., Huang, X., Hu, T., Ning, H., Ye, X., Huang, B., & Li, X. (2021). ODT FLOW: Extracting, analyzing, and sharing multi-source multi-scale human mobility. *PLoS One, 16*(8), e0255259.
Liang, J., Xie, Y., Sha, Z., & Zhou, A. (2020). Modeling urban growth sustainability in the cloud by augmenting Google earth engine (GEE). *Computers, Environment and Urban Systems, 84*, 101542.
Maliene, V., Grigonis, V., Palevičius, V., & Griffiths, S. (2011). Geographic information system: Old principles with new capabilities. *Urban Design International, 16*, 1–6.
McNutt, M. (2014). Reproducibility. *Science, 343*(6168), 229–229.
Meduri, K., Nadella, G. S., Gonaygunta, H., & Meduri, S. S. (2023). Developing a fog computing-based AI framework for real-time traffic management and optimization. *International Journal of Sustainable Development in Computing Science, 5*(4), 1–24.

Miller, H. J., & Goodchild, M. F. (2015). Data-driven geography. *GeoJournal, 80*, 449–461.
Nüst, D., & Pebesma, E. (2021). Practical reproducibility in geography and geosciences. *Annals of the American Association of Geographers, 111*(5), 1300–1310.
Orlik, A., & Orlikova, L. (2014). Current trends in formats and coordinate transformations of geospatial data—Based on MyGeoData converter. *Central European Journal of Geosciences, 6*, 354–362.
Piccolo, S. R., & Frampton, M. B. (2016). Tools and techniques for computational reproducibility. *GigaScience, 5*(1), s13742–s13016.
Tullis, J. A., & Kar, B. (2021). Where is the provenance? Ethical replicability and reproducibility in GIScience and its critical applications. *Annals of the American Association of Geographers, 111*(5), 1318–1328.

Part III
Branches of Human Geography Intersected With GeoAI

Chapter 10
GeoAI and Cultural Geography

Gunnar Mallon and Daniella Vos

... until statistical methods become much more sophisticated, they cannot handle the multiple strands that compose any culturalist's interests in even a simple topic (Shurmer-Smith, 2002, 97)

Abstract This chapter explores the emerging potential of geospatial artificial intelligence (GeoAI) in cultural geography, a subfield traditionally driven by qualitative methods. Cultural geographers focus on the meanings, identities, beliefs, and practices that shape places, landscapes, and material culture. While established approaches rely heavily on contextualized qualitative investigations, recent advances in AI and computational methods offer tools to enhance and expand the scale of cultural inquiries. GeoAI, encompassing machine learning, deep learning, and spatial analysis techniques, can uncover subtle patterns and relationships that might remain undetected through conventional methods alone. For example, computational approaches can identify emerging cultural trends in social media data, classify features of cultural landscapes in images, or predict shifts in cultural practices under changing environmental conditions. However, the integration of GeoAI into cultural geography also raises ethical considerations regarding data privacy, bias, and the risk of oversimplifying complex cultural phenomena. This chapter argues that cultural geographers need not abandon traditional qualitative methods. Instead, by thoughtfully combining GeoAI techniques with established approaches and community engagement, they can deepen understanding, broaden analytical scales, and discover new research frontiers in the study of cultural processes.

Keywords GeoAI · Cultural geography · Qualitative methods · Spatial analysis · Ethical considerations

---

G. Mallon (✉) · D. Vos
Department of Cultural Geography, Faculty of Spatial Sciences, University of Groningen, Groningen, The Netherlands
e-mail: g.mallon@rug.nl

© The Author(s), under exclusive license to Springer Nature Switzerland AG 2025

X. Huang et al. (eds.), *GeoAI and Human Geography*, Springer Geography,
https://doi.org/10.1007/978-3-031-87421-5_10

10.1 Overview of Methods in Cultural Geography

While the definition of cultural geography has been widely debated and may be as difficult to define as culture itself (Anderson, 2015; Jackson, 2000), some of the more general notions about this discipline can be taken as a starting point to describe what cultural geographers do. This is what we will be focusing on in this introduction, and not only to avoid diving into decades of debate. This chapter aims to provide a reflective overview of applications of geospatial artificial intelligence (GeoAI) in cultural geography, and the necessary context for such an overview is a methodological one.

From this perspective, cultural geography, or what cultural geographers do, can be defined as the study of the meaning, ideas, customs, beliefs, identities, and other aspects of culture that are bestowed upon places, practices, and material culture. In other words—cultural geography seeks to understand how the cultural "products" of human communities are tied to experience and identity in relation to a broader natural and social context. Since many, if not all, of those cultural interactions and relations are linked by a shared spatial context, this is often a key aspect of their study.

While spatial context is likely to remain an important consideration in cultural geography, the key themes of the discipline have defined the nature of methods being applied in such investigations. Exploring sense of place in relation to identity (Myers, 2005), reading cultural landscapes (Matless, 2014), or investigating cultural practices and rituals (Saint-Blancat & Cancellieri, 2014), often require a contextualized (and therefore local) qualitative approach. While culture can take both abstract (i.e., practices and verbal communication) and tangible (i.e., places and materials) forms, it has been argued that the meaning and perception of either is most directly captured in a qualitative way, and this approach has been typical in human geography so far (Heinrich et al., 2024; Shurmer-Smith, 2002).

However, to understand cultural processes at a larger scale, quantitative methods may offer a more suitable means to investigate a broader cultural context. Traditionally, studies of diffusion and migration patterns tend to apply spatial quantitative analysis techniques (Zhang et al., 2020 and references therein). The need for such methods becomes more apparent when considering the effects of a proposed "death of distance" (Couclelis, 1996) in an increasingly digital era, when culture can spread and be shared more easily across physical boundaries. Contextualizing, and applying a local focus, then get a different meaning than a traditional spatial one. Studies of social media are a good example of this, some even allowing for definitions of space and place to remain vague by using quantitative spatial methods (Clasper, 2017), a task some may associate with qualitative methods. In addition to a larger scale and broader cultural context being addressed, the application of computational quantitative methods in particular makes sense when one considers the pivotal role that computers play in the lives of geographers today (Burt, 2010). While interviews remain an important qualitative method in the cultural geographer's toolkit, interview data is increasingly coded and analyzed using specialized software. Our current reliance on computers, and the (not unrelated) increase in data, have made the use of quantitative methods more common across all academic disciplines in the past decades, including cultural geography.

10.2 Methods of GeoAI for Cultural Geography

As has been touched on in previous chapters, GeoAI spreads across a wide variety of human geography topics, reaching well beyond just mapping features or crunching numbers. In the following sections, we will show how different GeoAI methods help us analyze and quantify cultural dynamics. This isn't about ticking a box or just categorizing places; it's about understanding how people's stories, identities, and practices move and transform within the cultural landscapes they inhabit.

These days, researchers handle immense volumes of information, from old archival documents gathering dust in storerooms to digital social media posts reacting to a local music festival, or layers of remote sensing data capturing changes in a city's physical form. Trying to make sense of the sheer complexity of the data and picking out subtle nuances can feel like an overwhelming task, and this is a perfect application of GeoAI technology. When applied correctly and ethically (see below), GeoAI algorithms can expose underlying patterns that would otherwise go unnoticed and consequently open new research avenues for exploration.

GeoAI is an umbrella term for many geospatially focussed artificial intelligence algorithms, ranging from classification to clustering, dimensionality reduction, and deep learning. By using these advanced quantitative methods together with more traditional qualitative ones, we can start to explore cultural phenomena at a much deeper level. For example, spatial clustering has the potential to highlight neighborhoods with unusually high linguistic diversity. Hot-spot analysis easily identifies neighborhoods with diverse cultural activities from a large dataset. Even simple techniques such as geographically weighted regression can help to tease out subtle differences between areas that might look identical at first glance.

Out of the available techniques, machine learning and deep learning models form the foundation of a lot of GeoAI integration in cultural geographical research. The methods do not need to be complicated or "black boxes." Relatively straightforward techniques, such as random forests or k-nearest neighbors, can easily identify meaningful patterns in cultural data. Deep learning systems (e.g., convolutional neural networks), on the other hand, are better suited to tease out fine detail or more nuances connections from the dataset. When combined with qualitative methods, analyses can provide cultural geographers with deeper insights than ever before and allow them to connect with communities on a more meaningful level.

Predictive modeling can add another dimension to the cultural geographical analysis. It does not just show us what is happening right now but makes inferences and predictions about changes that might come. This could include how environmental policies will reshape local traditions, or demographic changes that influence how a community celebrates its heritage. By building a temporal element into these models, we can estimate how cultural practices might evolve under climate change or imagine how heritage tourism might change as neighborhoods become more gentrified. This forward-looking perspective can guide stakeholders to preserve and protect their community's cultural values.

GeoAI can also play a crucial role in dealing with visual data. Techniques drawn from computer vision, such as image classification or image segmentation, can allow cultural geographers to quickly pick out architectural styles, spot cultural clothing in large photographic datasets, or map the subtle shifts in how a landscape looks over time. By automating what used to be months of manual work, these methods free up researchers to collect more targeted data or to interpret the findings instead of spending valuable time organizing data.

Just as computer vision models can play a crucial role in visual data, Natural Language Processing (NLP) methods, such as sentiment analysis or topic modeling, can help us sift through large volumes of documents and transcripts. They can let us see how cultural narratives evolve. By deep diving into large volumes of text, NLP models can identify subtle shifts in cultural dimensions that might be emerging over time. Rather than relying on a handful of representative texts, thousands or even millions of words can be analyzed, and patterns that shape how communities understand themselves and each other can be discovered.

However, one crucial point remains. It is vital to remember that communities are not just data sources. The core of cultural geography is the people that make up the cultures. The use of GeoAI, therefore, needs to be in tandem with human-centered approaches. This ensures that local knowledge and perspectives are not ignored but that they help to inform how we collect and interpret cultural information.

Although we've listed many techniques here, this list is by no means exhaustive. New tools and applications keep surfacing, and each one adds another string to the human-and cultural geographer's bow. When used together with traditional data collection instruments, we can start to see cultural patterns and spot trends that we might have overlooked otherwise. GeoAI tools have the great potential to enhance research and allow us to engage with data at a level of detail we could not have imagined a couple of decades ago.

10.3 Applications of GeoAI in Cultural Geography

Despite the range of possible techniques that fall within GeoAI and their early application in other fields of geography (Li & Hsu, 2022), the applications of GeoAI have so far remained limited in cultural geography.

The investigation of some topics in cultural geography shows methodological continuity from earlier quantitative analyses, where datasets containing large amounts of digital information were previously explored using mapping and other GIS techniques, such topics are now approached using, for example, spatial machine learning and clustering algorithms. These types of data lend themselves to the use of GeoAI techniques since they allow us to explore a wide range of cultural perceptions at a large scale. Social media is a classic example of digital big data suitable for such applications, providing not only a wealth of cultural information but also embedded geographic locations. It is therefore not surprising that social media studies are common research topics being explored through AI in cultural geography.

Lansley and Longley (2016), for example, applied an unsupervised learning algorithm to classify Twitter tweets in London to contextualize the topics and attitudes expressed through social media across time and place. By overlaying tweets with land use data, they were able to characterize the environments in which Twitter activities took place; residential areas, non-domestic ones, and public green spaces. This enabled the researchers to link the topics of tweets to the type of urban environment, linking human perception to place. It could have been the effect of lockdowns increasing our reliance on digital means, or simply the advancements in AI at the time, but the COVID-19 pandemic resulted in several AI applications in social media studies (Hantrais et al., 2020). In parallel, research into cultural ecosystem services (CES), previously typically involving hours of analyses of image and text information, has also benefited from GeoAI tooling. Cardoso et al. (2022), for example, used convolutional neural networks on images posted to Flickr and Wikiloc platforms to identify and classify human and natural elements relevant to CES.

The study of cultural landscapes, which previously relied on many hours of manpower through qualitative interpretation and GIS analysis, was one of the first themes in cultural geography to incorporate GeoAI methods for identification and conservation purposes. Hong (2020), for example, shows how computer vision and machine learning techniques can be integrated to study the linguistic urban landscape of a neighborhood in Seoul. GeoAI has also been increasingly used for the identification of cultural heritage sites and for detecting human and natural threats to cultural heritage landscapes (Ben-Romdhane et al., 2023; Cuca et al., 2023). Casillo et al. (2022) illustrate how deep learning can be applied to the conservation of cultural buildings. They applied a deep convolutional generative adversarial network to predict the rising of dampness in masonry structures through a correlation between weather conditions, the humidity of the surrounding soil, and the moisture within building walls. In addition to its utilization for the preservation of cultural heritage, AI has also been used to create virtual simulations that enable museum visitors to experience cultural (heritage) landscapes and interact with museum artifacts (Kiourt et al., 2016).

While applications discussed above relate to research questions that lend themselves to the use of quantitative tools, (Geo)AI is also being applied to topics that have traditionally been explored entirely by qualitative methods. Pattern recognition and augmentation AI techniques have been applied to correct and interpret handwritten texts, facilitating further qualitative analysis (Guha et al., 2020; Suissa et al., 2022). Zhang et al. (2019) experimented with human ratings of street-level imagery to predict human perceptions of large-scale urban regions. It is important to note at this point that GeoAI techniques for image analysis, while promising, still need to be further applied and assessed to fully understand their potential. Analyses of images with the aim of predicting human perceptions of urban or natural environments in particular, have enjoyed varying degrees of success and are difficult to verify due to the large scale of analysis and required data (which is why such techniques were applied in the first place). During the validation of a methodology for predicting socio-cultural preferences of CES across different stakeholders, a machine learning technique relying on maximum entropy (MaxEnt software) was not able to fully predict the real demand of CES (Marine et al., 2022). This having been said, the applications and value of GeoAI applications to facilitate qualitative research in cultural geography can reach further than image analysis. The use of immersive technologies to aid social impact assessment, for example, allows different stakeholders to explore the consequences of alterations to an environment without having to make changes in the real world (Miranda, 2024).

10.4 Methodological Integration of GeoAI in Cultural Geography

So where does the application of GeoAI in cultural geography stand? Incorporating GeoAI into cultural geography shouldn't necessarily mean tossing out everything we already know about understanding places and their cultural significance. Instead, it offers a great opportunity to pick out nuances from datasets that can often be overlooked. As cultural geographers, we rely heavily on extensive fieldwork, recorded interviews, reading of landscapes and cultural values, and the study of archives to decipher how people attach meanings and identities to the spaces they inhabit (Anderson, 2015; Norton, 2006). These methods largely, but not exclusively, form the core of the discipline. The challenge now lies in finding a way to incorporate computational tools without losing that personal, grounded perspective, which is deeply rooted at the core of cultural geography.

With a strong emphasis on qualitative research methods, the step to integrating deeply quantitative methods, which sometimes include black-box elements (e.g., neural networks) can seem like a large step, involving a shift in research thinking. However, these steps need not be as large as sometimes perceived. GeoAI techniques should not replace existing research but rather augment it and add an additional dimension to analyses. The core of cultural geography should retain its anthropocentric focus.

For example, research may start with traditional qualitative approaches—talking to community members, visiting key sites, and reading local histories—to establish key cultural parameters of interest. Once the broad strokes and key cultural features in the data have been identified, GeoAI techniques handle data that would otherwise feel overwhelmingly complex to untangle intricate relationships. This might mean analyzing thousands of social media posts to spot unexpected cultural patterns that can later be investigated in person. Or perhaps feeding historical photographs into an image analysis model, looking for subtle shifts in architecture over time, before going back to the field to explore how these temporal changes relate to human narratives of urban change.

In other words, GeoAI can help identify patterns that may have gone unnoticed otherwise. It's not an "either/or" proposition, but rather more like an iterative process: starting with a qualitative sense of what's meaningful, using GeoAI to explore large or complex datasets, and then returning to human-centered methods to make sense of what was found. This keeps the research anchored in real-world cultural complexity rather than reducing it to neat lines on a map.

Machine learning can therefore be a highly useful and beneficial tool when you're trying to detect patterns in complex heterogeneous cultural data. Let's say you're looking at images of urban murals from different neighborhoods. After an initial data collection phase, a supervised learning model (like a random forest) could help you classify murals by style or theme, maybe revealing that one particular visual motif spreads along certain migrant communities' settlement patterns (Ben-Romdhane et al., 2023). The data might tell you something interesting—perhaps murals near a historical market feature symbolism related to agricultural traditions, while those near newer shopping complexes do not. Once subtle patterns have been identified, more targeted data can be collected. Local residents can be interviewed and asked about their interpretation of the murals and what they mean to them. Without the intermediate machine learning analysis step, the subtle connections between distance from the historical market and mural motive may have gone entirely undetected.

However, machine learning and GeoAI algorithms are merely analytical tools that, on their own, cannot explain why something matters. They can identify that something, interesting or not, is going on. It's ultimately still up to us, as a cultural geographer, to interpret the results, dive deeper into the data, and understand why the patterns that emerged exist.

GeoAI also opens doors for more nuanced spatial and network analysis. As we know, cultural practices do not exist in isolation, but rather connect and evolve through trade routes, migration streams, religious pilgrimages, and countless other forms of social exchange. Using spatial models can help trace how a particular craft tradition spreads from one region to another over several decades. Or perhaps, you construct a network of language use across social media platforms, to investigate hidden clusters of dialect speakers. Similarly, as with patterns in murals, research starts with data collection, in this case with social media posts before GeoAI techniques are applied. However, this is not where the analysis of data should stop. The most important part of the work lies in digging deeper and interpreting the identified patterns.

One could even apply predictive models to guess how cultural practices might shift if, for example, climate change affects agriculture in a particular region. The inferential potential of GeoAI could help address subtle questions about whether scarcity of specific food items as a result of climate change will alter a community's yearly festival traditions. It cannot tell you with certainty, of course, but it can help you consider multiple scenarios and figure out which ones deserve closer attention. This approach encourages a kind of forward-looking cultural geography, where you are not just documenting what has happened but also thinking about what could happen next—and how communities might adapt.

As promising and exciting as the incorporation of GeoAI methods sounds, it's never as straightforward and without hurdles. Imperfections in cultural data, which is inherently scattered and often incomplete or biased, will continue to persist. For example, historical records might overrepresent the voices of certain groups (e.g., wealthy sections of society), while underrepresenting the voices of others. Equally, social media posts might reflect younger, more tech-savvy populations than older generations. When we feed uneven data into AI models, without careful preprocessing of data, we run the risk of having these imbalances amplified. We might end up "discovering" patterns that reflect nothing more than gaps or biases in the data. And this returns us to the main concern about the use of GeoAI in cultural geography. The potential of the available tools is great, however, it requires careful quantitative considerations and insights alongside the more common qualitative methodologies. This is nowhere more important in data preprocessing and in addressing data biases.

There's also the issue of interpretability. Deep learning models, for instance, can often feel like opaque black boxes even to experts in the field of artificial intelligence. They might identify interesting patterns, but it's not always clear how these have emerged. However, cultural geography is not just about detecting patterns—it's about understanding them. When addressed under a Popperian lens, researchers need to take a step back, perhaps use simpler models, and rely on methods that help them explain the reasoning behind the results: "If you can't explain why your model grouped certain neighbourhoods, what does that mean for your analysis?" This leads to the philosophical question of whether it's worth sacrificing a bit of predictive power for the sake of transparency and reproducibility.

A useful approach would be to involve the local community. After a computational framework has identified patterns in the cultural data, work with local residents and cultural stakeholders can help to make sense of these, or to put the results into question. Local cultural knowledge, such as the local paint factory closing or stricter government monitoring of social media posts (Nguyen Van Quoc et al., 2024) may not have been collected during the initial data collection phase but can offer tremendous insight into the unearthed patterns. Incorporating the local community in the interpretation of the GeoAI model outputs can add a significant step in interpreting and validating these.

10.5 Ethical Considerations in GeoAI for Cultural Geography

As we incorporate GeoAI more and more into our research, it becomes crucial to think carefully about the ethical implications. Although advanced algorithms can reveal subtle patterns, as we have seen, they also run the risk of flattening cultural complexity. Cultures rarely fit neatly into algorithmic categories. Relying too heavily on automated classification can very easily overlook the human meanings woven into everyday life. Seeing and using GeoAI as a mere tool therefore remains paramount. We have to acknowledge its limitations and still do the thinking ourselves. Furthermore, the question of data privacy looms large, and it remains unclear how we should best protect individuals whose daily expressions end up in a dataset they never agreed to join. Take our example of using social media data. Researchers might harvest geolocated posts through the social media's programmable application interface without fully considering whether the people creating that content ever consented to its academic use.

GeoAI models further run into issues of cultural representation. AI models are only as good as the data that they are fed. If there is bias, intentional or not, in the source data that AI models are trained on, their output will reflect these and in some cases amplify them. As we know, social science data needs to be carefully collected as it can easily skew toward dominant groups. Without careful checking and preprocessing of data, subtle voices that may not be the focus of the study risk being drowned out. Instead of gaining new insights, there's a danger of reproducing existing hierarchies of power and cultural biases. This might involve overemphasizing readily digitized cultural elements. Furthermore, activities or narratives that are less likely to appear online may be neglected as a result. Language-based analyses can misunderstand indigenous dialects or reduce them to generic categories, reducing or in extreme cases erasing local identities. While GeoAI can be a powerful tool, it can also end up presenting a distorted map of cultural life, one that privileges what's easy to measure at the expense of what's actually meaningful to the people who live it.

We must further consider the risk that new technologies could drive the discipline toward a more mechanistic and deterministic approach. There's a potential risk that cultural geographers, seduced by deep data-driven insights, might downplay the significance of qualitative traditions. This could produce a kind of technological determinism, suggesting that algorithms alone could untangle the complexities of cultural interactions. In reality, cultural phenomena do not conform to neat categorization. It takes a mix of methods, both computational and human-centered, to fully understand why particular cultural phenomena exist.

Charting a responsible path forward calls for genuine collaboration. Cultural geographers need to work closely with community members, but also with data scientists, and ethicists to review how datasets are collected and ultimately interpreted. Being transparent about the methods used is one step toward building trust and minimizing harm.

10.6 Conclusion

While GeoAI methods have been embraced in many domains of geography, the field of cultural geography does not seem to have fully taken advantage of recent developments and applications of these. This probably has a lot to do with the discipline's history, in particular, earlier critique on quantitative thinking and methods as adequate means of investigating the meaning, ideas, customs, beliefs, identities, and other aspects of culture that are bestowed upon places, practices, and material culture.

In this overview, we hope that we were able to demonstrate that GeoAI techniques may carry much value for studying even the most qualitative aspects of cultural practices, identities, and places. Rather than applying such techniques in a quantitative framework, one could utilize them to aid qualitative analyses and help expand the scope and scale of what can be done by only using human eyes and language.

In order to achieve this, a better understanding of the potential of GeoAI tools is needed, which could be achieved through collaboration between cultural geographers with domain expertise and researchers trained in quantitative methods. In order to bridge gaps between disciplines, it is always necessary to familiarize yourself with the "other," and keep an open mind. While every discipline has its own toolkit, the latter can always be expanded. Referring back to Shurmer-Smith's statement about waiting for statistical methods to become sophisticated enough to handle the multiple strands composing culturalists' interests in even simple topics—we think it's safe to say that we have finally reached that point. Cultural geographers need not wait further for the development of more sophisticated statistical tools. All they have to do is reach out and explore.

References

Anderson, J. (2015). *Understanding cultural geography: Places and traces* (2nd ed.). Routledge.

Ben-Romdhane, H., Francis, D., Cherif, C., Pavlopoulos, K., Ghedira, H., & Griffiths, S. (2023). Detecting and predicting archaeological sites using remote sensing and machine learning—Application to the Saruq Al-Hadid site, Dubai, UAE. *Geosciences, 13*, 179.

Burt, W. (2010). Structuralism, attitude and the computer: Questioning the notion of "cultural computing". *IFIP Advances in Information & Communication Technology, 333*, 128–136.

Cardoso, A. S., Renna, F., Moreno-Llorca, R., Alcaraz-Segura, D., Tabik, S., Ladle, R. J., & Vaz, A. S. (2022). Classifying the content of social media images to support cultural ecosystem service assessments using deep learning models. *Ecosystem Services, 54*, 101410.

Casillo, M., Colace, F., Gupta, B. B., Lorusso, A., Marongiu, F., & Santaniello, D. (2022). A deep learning approach to protecting cultural heritage buildings through IoT-based systems. In *IEEE international conference on smart computing (SMARTCOMP), Helsinki, Finland, 2022* (pp. 252–256).

Clasper, L. T. (2017). Exploring vernacular perceptions of spatial entities: Using twitter data and R for delimiting vague, informal neighbourhood units in inner London, UK. *Proceedings of VGI-analytics, 2017*, 9.

Couclelis, H. (1996). The death of distance. *Environment and Planning B: Planning and Design, 23*(4), 387–389.

Cuca, B., Zaina, F., & Tapete, D. (2023). Monitoring of damages to cultural heritage across Europe using remote sensing and earth observation: Assessment of scientific and Grey literature. *Remote Sensing, 15*, 3748.

Guha, R., Das, N., Kundu, M., Nasipuri, M., & Santosh, K. C. (2020). DevNet: An efficient CNN architecture for handwritten Devanagari character recognition. *International Journal of Pattern Recognition and Artificial Intelligence, 34*, 12.

Hantrais, L., Allin, P., Kritikos, M., Sogomonjan, M., Anand, P. B., Livingstone, S., Williams, M., & Innes, M. (2020). Covid-19 and the digital revolution. *Contemporary Social Science, 16*, 256–270.

Heinrich, A. J., Marguin, S., Million, A., & Stollmann, J. (2024). *Handbook of qualitative and visual methods in spatial research*. Transcript Verlag.

Hong, S.-Y. (2020). Linguistic landscapes on street-level images. *International Journal of Geo-Information, 9*, 57.

Jackson, P. (2000). Rematerializing social and cultural geography. *Social & Cultural Geography, 1*(1), 9–14.

Kiourt, C., Koutsoudis, A., Markantonatou, S., & Pavlidis G. (2016). The 'Syntheis' virtual museum. *Mediterranean Archaeology and Archaeometry, 16*, 1–9.

Lansley, G., & Longley, P. A. (2016). The geography of twitter topics in London. *Computers, Environment and Urban Systems, 58*, 85–96.

Li, W., & Hsu, C.-Y. (2022). GeoAI for Large-scale image analysis and machine vision: Recent Progress of Artificial Intelligence in Geography. *International Journal of Geo-Information, 11*, 385.

Marine, N., Arnaiz-Schmitz, C., Santos-Cid, L., & Schmitz, M. F. (2022). Can we Forsee landscape interest? Maximum entropy applied to social media photographs: A case study in Madrid. *Land, 11*, 715.

Matless, D. (2014). *In the nature of landscape : Cultural geography on the Norfolk broads*. Wiley.

Miranda, J. C. (2024). Immersive technologies: New ways to visualise a possible future world. In *Handbook of social impact assessment and management*. Edward Elgar Publishing.

Myers, G. A. (2005). Place and humanistic African cultural geography: A Tanzanian case. *Journal of Cultural Geography, 22*(2), 1–26.

Nguyen Van Quoc, T., Trell, E.-M., & Mallon, G. (2024). More than keyboard heroes? #ichoosefish, disaster farming, and environmental protest in Vietnam. *Media, Culture & Society, 46*, 1141–1161.

Norton, W. (2006). *Cultural geography : Environments, landscapes, identities, inequalities* (2nd ed.). Oxford University Press.

Saint-Blancat, C., & Cancellieri, A. (2014). From invisibility to visibility? The appropriation of public space through a religious ritual: The Filipino procession of *Santacruzan* in Padua, Italy. *Social & Cultural Geography, 15*, 645–663.

Shurmer-Smith, P. (2002). *Doing cultural geography* (p. 248). Sage Publications.

Suissa, O., Elmalech, A., & Zhitomirsky-Geffet, M. (2022). Text analysis using deep learning neural networks in digital humanities and information science. *JASIST. Special Issue on Digital Humanities, 73*, 268–287.

Zhang, F., Wu, L., Zhu, D., & Liu, Y. (2019). Social sensing from street-level imagery: A case study in learning spatio-temporal urban mobility patterns. *ISPRS Journal of Photogrammetry and Remote Sensing, 153*, 48–58.

Zhang, H., Zhou, X., Tang, G., Ji, G., Zhang, X., & Xiong, L. (2020). Inference method for cultural diffusion patterns using a field model. *Transactions in GIS, 24*, 1578–1601.

Chapter 11
GeoAI and Economic Geography

Siyuan Kang, Lingfeng He, Zezheng Lin, and Wei Luo

Abstract This chapter examines the transformative integration of Geographic Artificial Intelligence (GeoAI) into economic geography, focusing on its innovative contributions and potential applications. GeoAI combines geospatial technologies and artificial intelligence to address core challenges in economic geography, including spatial-temporal distributions, network relationships, and dynamic transformations. The chapter explores how GeoAI facilitates nuanced analyses of international trade, supply chain resilience, geopolitical impacts, and real-time economic dynamics through advanced methodologies. Despite its potential, GeoAI faces challenges, such as data quality, model limitations, and ethical concerns, which highlight the need for rigorous frameworks and interdisciplinary collaboration. The chapter concludes by outlining future directions, emphasizing the expansion of generative AI, the advancement of real-time analytics, and the integration of GeoAI into policy-making, aiming to establish its role as an indispensable tool for modern economic systems.

Keywords GeoAI · Economic geography · Spatiotemporal analysis · Supply chain resilience · Generative AI

S. Kang · W. Luo (✉)
GeoSpatialX Lab, Geography Department, National University of Singapore, Singapore, Singapore
e-mail: siyuan_kang@u.nus.edu; geowl@nus.edu.sg

L. He
Institute for Empirical Social Science Research, Xi'an Jiaotong University, Xi'an, China
e-mail: helingfeng@stu.xjtu.edu.cn

Z. Lin
Steering Committee of ESG & Carbon Neutral Investment, United Nations Industrial Development Organization, Beijing, China
e-mail: jasonlinzezheng@gmail.coml

© The Author(s), under exclusive license to Springer Nature Switzerland AG 2025
X. Huang et al. (eds.), *GeoAI and Human Geography*, Springer Geography,
https://doi.org/10.1007/978-3-031-87421-5_11

11.1 Introduction

Economic geography has traditionally focused on the spatial organization of economic activities and the mechanisms underpinning their distribution and evolution (Aoyama et al., 2010). However, the increasing globalization and interconnectedness of economies introduce complexities that challenge conventional analytical tools (Balland et al. 2022). Traditional approaches often fall short in capturing the non-linear interactions, feedback loops, and temporal dynamics that characterize modern economic systems. GeoAI addresses these limitations by integrating predictive and prescriptive modeling capabilities (Li, 2022). These advancements enable researchers and policymakers not only to anticipate economic shifts but also to optimize resource allocation and assess policy impacts in real time, equipping them with tools essential for navigating today's multifaceted economic challenges.

The integration of Geographic Artificial Intelligence (GeoAI) into economic geography represents a paradigm shift in addressing the spatial and temporal complexities inherent in global economic systems. This interdisciplinary innovation combines geospatial technologies with artificial intelligence, creating a robust framework for analyzing dynamic, high-dimensional, and multilayered datasets that were previously inaccessible through conventional methodologies (Xie, 2020). GeoAI transcends the static and linear models of traditional approaches, offering researchers an unprecedented ability to explore the evolving interactions within global economic landscapes (Queiroz et al., 2024). By bridging gaps in data synthesis and analytical precision, GeoAI unveils new pathways for understanding and predicting economic phenomena that are deeply rooted in spatiotemporal intricacies.

The potential of GeoAI extends beyond enhancing analytical capacity; it offers new methods for analyzing economic phenomena, enabling more precise insights into complex spatial and temporal dynamics. By prioritizing dynamic processes over static snapshots, GeoAI has the potential to shift the focus from merely describing spatial phenomena to modeling their evolution and projecting future scenarios. For instance, machine learning, when applied to international trade data, could reveal emerging trade clusters and help forecast potential shifts in global supply chains (Breinlich et al., 2022). Similarly, neural network models could advance the study of economic resilience by simulating the potential impacts of external shocks—such as geopolitical conflicts or environmental crises—on economic networks, potentially providing insights that are difficult to access through traditional approaches (Chen et al., 2021; Lei et al., 2023; Mittal & Panchal, 2023).

Moreover, GeoAI has the potential to serve as a methodological bridge between micro-and macro-level analyses. At the micro-level, it could facilitate granular studies of urban economies, infrastructure investments, and localized trade dynamics (Agbaje et al., 2024; Sachit et al., 2024). At the macro-level, GeoAI may empower the investigation of complex global systems such as multinational supply chains and international trade networks, offering the possibility of unprecedented insights into cross-scale economic interactions (Cao et al., 2024). This dual capacity is particularly promising for addressing global challenges, such as climate change and economic inequality, which demand integrated, multi-scalar approaches.

Despite its transformative potential, the adoption of GeoAI also raises critical questions about data quality, algorithmic bias, and ethical considerations. As reliance on AI-driven insights grows, ensuring the transparency, equity, and accountability of these models becomes imperative. Addressing these challenges is essential to realizing GeoAI's promise as a transformative tool in economic geography.

This chapter examines the role of GeoAI in economic geography, focusing on its contributions to addressing core spatial-economic issues, such as spatial and temporal distribution, network relationships, and dynamic transformations. It highlights how GeoAI enables innovative analyses of complex systems through applications in spatiotemporal trade analysis, supply chain modeling, geopolitical impact assessments, and real-time decision support. The discussion also acknowledges the challenges and limitations of GeoAI, including data quality, model reliability, and ethical considerations, while pointing to future directions, such as advancing generative AI, bridging theory with practice, and fostering policy integration.

11.2 GeoAI-Driven Innovations in Economic Geography: Applications and Insights

11.2.1 Core Issues in Economic Geography

Economic geography delves into the spatial organization of economic activities, investigating the underlying mechanisms that dictate their distribution, evolution, and interaction (Aoyama et al., 2010). At its core, the discipline addresses three interrelated themes: **spatial and temporal distribution, network relationships, and dynamic transformations** (Grabher, 2006; Hess, 2004; Schwanen, 2018). These dimensions underscore the complexity of economic systems, which are marked by interdependencies that traverse spatial scales and evolve over time.

11.2.1.1 Spatial and Temporal Distribution

The spatial and temporal distribution of economic activities forms the foundation of economic geography. This theme examines questions such as: Why do industries cluster in specific locations? How do trade hubs emerge and persist? What factors influence the siting of infrastructure critical to global commerce? Spatial clustering often reflects the cumulative effects of resource endowments, institutional frameworks, and agglomeration economies (Spencer et al., 2010; Wei, 2015). For instance, regions like Silicon Valley exemplify how spatial concentrations of innovation and capital generate enduring economic advantages (Engel, 2015; Lee, 2000).

Temporal dimensions, on the other hand, reveal the shifting dynamics of these distributions. Cyclical trade flows, technological adoption timelines, and policy-driven market fluctuations illustrate how economic landscapes evolve over time. For example, the diffusion of renewable energy technologies follows distinct temporal trajectories shaped by policy incentives, market demands, and regional energy capacities (Bai et al., 2020; Essletzbichler, 2016; Norberg-Bohm, 2000). Traditional models in economic geography, such as spatial econometrics and location theory, often struggle to fully capture the interplay of these temporal and spatial processes, leading to oversimplified interpretations of complex economic realities.

11.2.1.2 Network Relationships

Economic systems operate within networks characterized by interconnected nodes—nations, cities, enterprises—and the flows of goods, capital, and information that link them (Luo & MacEachren, 2014; Malecki, 2002). International trade and global supply chains epitomize these networks, whose structures and dynamics are influenced by geopolitical, technological, and market forces. Economic geography scrutinizes the topology of these networks to understand resilience, efficiency, and adaptability (Liu et al., 2024; Luo et al., 2014; Shi et al., 2022; Wu et al., 2024; Yu et al., 2024).

For instance, trade agreements like NAFTA or RCEP reshape trade routes, influencing the prominence of specific nodes within the network (Baladad, 2021; Datta & Misra, 2024; Datta & Satghare, 2024). During disruptions such as pandemics or geopolitical conflicts, these networks reveal their vulnerabilities; supply chain bottlenecks, transport delays, and localized economic shocks highlight the fragility of interdependent systems. Analyzing these phenomena requires advanced methodologies that integrate spatial configurations with temporal dependencies, enabling a nuanced understanding of how networks evolve under stress.

11.2.1.3 Dynamic Transformations

Dynamic transformations encompass the processes through which economic systems adapt to technological innovation, policy changes, and external shocks (Dosi & Nelson, 2010; Schot & Edward Steinmueller, 2018). These transformations are neither linear nor predictable; they involve feedback loops, path dependencies, and emergent phenomena (Alatriste-Contreras et al., 2019; Albin, 2021). For example, the global transition toward renewable energy has triggered profound shifts in trade flows, industrial geographies, and investment patterns (Blondeel et al., 2021; Hassan et al., 2024). The declining reliance on fossil fuels in favor of green energy resources alters both the supply and demand landscapes, impacting regions differently based on their resource endowments and industrial capacities (Nacke et al., 2022; Vakulchuk et al., 2020).

Dynamic transformations are further illustrated by the rise of digital economies and e-commerce platforms, which reconfigure traditional notions of economic space (Song et al., 2022; Zhang et al., 2022). The proliferation of remote work and digital marketplaces has blurred the boundaries between local and global economic interactions. These developments necessitate a process-oriented approach that captures the fluidity and interconnectedness of contemporary economic systems.

11.2.2 Current Applications of GeoAI in Economic Geography

GeoAI has significantly reshaped economic geography by enabling a deeper, more nuanced understanding of complex spatial and temporal economic phenomena (Huang et al., 2024; Janowicz et al., 2022). By integrating artificial intelligence with geospatial technologies, it enhances the ability to analyze, model, and predict a range of economic processes that involve dynamic and interconnected systems (Rane et al., 2024). Its applications span critical areas, including international trade, supply chain modeling, the assessment of geopolitical and environmental impacts, and innovative uses of generative AI for policy simulation and scenario analysis.

11.2.2.1 Spatiotemporal Analysis of International Trade

International trade is inherently dynamic, shaped by shifting spatial flows and temporal patterns influenced by market demands, geopolitical policies, and global events. Traditional econometric models often fail to capture the intricate interdependencies that drive these changes. GeoAI, particularly through tools like Self-Organizing Maps (SOMs) and advanced neural networks, addresses these limitations by incorporating spatial and temporal dimensions into economic analysis.

Luo and Yuan (2021) demonstrated the utility of SOM–trajectory for analyzing over a century of multilateral trade data across 207 countries. By applying this model, the study successfully classified complex trade interactions into distinct clusters, revealing temporal shifts linked to globalization and regional economic integration. For instance, SOM–trajectory uncovered the emergence of trade clusters in East Asia, driven by regional trade agreements like the ASEAN Free Trade Area (AFTA), contrasting with stagnation in some developed regions. These patterns highlight the significance of trade agreements in shaping regional trade dynamics and the uneven pace of globalization across regions. SOM–trajectory also reveals the realignment of trade partnerships during economic crises or policy disruptions, such as Brexit, offering valuable insights for countries seeking to diversify trade routes or strengthen regional alliances to mitigate vulnerabilities.

Long Short-Term Memory (LSTM) networks, a type of recurrent neural network, are particularly suited for analyzing time-series data. LSTM networks have been employed to predict future trade flows by modeling historical trade patterns, examining tariff structures, and analyzing trade agreements (Yang et al., 2023). For example, the potential impacts of Brexit on EU-UK trade were modeled using LSTM networks, which provided early warnings about sector-specific vulnerabilities and opportunities for alternative partnerships. By understanding these temporal shifts, policymakers can take proactive measures to safeguard their economies.

GeoAI also aids policymakers by simulating the effects of trade policy changes, such as tariff adjustments or new environmental regulations. Through scenario modeling, GeoAI can estimate the spatial impacts of such policies on global trade flows and regional economies (von Cramon-Taubadel et al., 2010; der Mensbrugghe & Dominique, 2013). By incorporating multiple economic variables and policy parameters, researchers can provide data-driven insights to inform decision-making and optimize policy outcomes.

11.2.2.2 Supply Chain Modeling and Resilience Analysis

Supply chains are the backbone of global commerce, interlinking production, transportation, and consumption across regions (Ponte et al., 2019; Sanders, 2024). However, their complexity and dependence on various external factors make them highly vulnerable to disruptions, whether due to pandemics, geopolitical tensions, or natural disasters. GeoAI offers advanced modeling and simulation tools that enhance supply chain resilience and performance.

GeoAI employs dynamic network analysis to model supply chain structures and identify their vulnerabilities. Through machine learning algorithms, researchers analyze network metrics such as node centrality and edge weights, which highlight critical components in global supply chains. For example, during the COVID-19 pandemic, such models identified high-risk nodes in the global supply chain, enabling real-time adjustments to minimize disruptions (Luo et al., 2023a). These models also provided insights into potential rerouting strategies and alternative suppliers, proving invaluable for maintaining supply chain stability in times of crisis.

Generative Adversarial Networks (GANs) have been increasingly used to simulate potential supply chain disruptions, such as port closures or shortages of raw materials (Habibi et al., 2023; Rezki & Mansouri, 2024). These models generate synthetic data to evaluate how supply chains would adapt under stress. For example, GANs can simulate the impact of environmental events, like flooding or earthquakes, on transportation networks and port operations, providing stakeholders with insights into optimal inventory management and contingency plans. Such simulations allow businesses to prepare for worst-case scenarios and implement more effective resilience strategies.

The integration of GeoAI with Internet of Things (IoT) technologies further enhances supply chain management by enabling real-time monitoring. IoT sensors embedded in logistics infrastructure provide data on environmental conditions, transportation delays, and warehouse capacities (Joel et al., 2024). GeoAI analyzes this data to dynamically adjust supply chain operations, reducing costs and improving efficiency. For example, real-time traffic data can inform optimal routing decisions, ensuring the timely delivery of goods while avoiding congestion and delays.

11.2.2.3 Impact Assessment of Geopolitical and Environmental Events

The global economy is increasingly influenced by geopolitical events, such as trade wars and conflicts, as well as environmental challenges, such as climate change and natural disasters. GeoAI provides robust methodologies for quantifying the spatial and temporal impacts of these disruptions on economic systems.

The geopolitical conflict has underscored the vulnerabilities within world's several economic sectors' trade networks. Using GeoAI, researchers were able to map the reconfiguration of energy trade routes and identify alternative suppliers in response to sanctions and shifting geopolitical alliances. By integrating Difference-in-Differences (DID) econometric models with spatial analysis, GeoAI quantifies changes in trade volumes and assesses how alternative hubs or transit points have emerged in the wake of the conflict (Luo et al., 2023b). These models offer policymakers insights into the cascading effects of geopolitical tensions and help identify strategies to mitigate economic disruptions caused by trade embargoes or sanctions.

GeoAI also integrates climate models with economic data to assess the long-term impacts of environmental changes on global trade. For instance, studies have used GeoAI to simulate how rising sea levels and extreme weather events might disrupt port operations and trade routes (Durlik et al., 2024; Mittal & Panchal, 2023). By modeling the economic consequences of environmental disruptions, GeoAI helps identify vulnerable regions and suggests adaptive strategies, such as resilient infrastructure investments and the diversification of trade networks. This proactive approach supports policymakers in planning for a climate-resilient future.

11.2.2.4 Real-Time Analytics and Decision Support

GeoAI's integration with real-time data sources, such as IoT sensors, satellite imagery, and social media feeds, opens new exploratory pathways for decision-making in economic geography (Chen et al., 2023a; Franchi et al., 2024; Mortaheb & Jankowski, 2023). Although these applications are still in their infancy, they demonstrate significant potential in areas like supply chain management, disaster response, and market trend analysis.

During natural disasters, GeoAI can process real-time satellite imagery to assess infrastructure damage and prioritize resource allocation (Agbaje et al., 2024; Cristian, 2024). For instance, AI-driven maps have been experimentally used to identify the most affected areas during flooding, helping planners determine optimal routes for aid delivery while avoiding damaged infrastructure. While these applications have shown promise in controlled scenarios, scaling them for large-scale disaster management remains a challenge requiring further research and validation.

Another area of exploration involves leveraging GeoAI for sentiment analysis using social media data. By analyzing trends in consumer behavior and political sentiment, GeoAI may help businesses and governments adjust trade strategies, predict market shifts, and prepare for potential social unrest or political volatility (Papic, 2020; Polinsky, 2021). However, this approach depends heavily on the quality and representativeness of the data, as well as the reliability of the analytical models, which are still under development.

An emerging exploratory application is using GeoAI for operational optimization, such as inventory management and logistics planning (Cao et al., 2024; El Makhloufi, 2023; Joel et al., 2024). As IoT and AI technologies evolve, GeoAI systems could potentially suggest dynamic adjustments to resource allocations or propose route changes to mitigate inefficiencies. Preliminary trials in small-scale settings have demonstrated initial success, but ensuring scalability and security for high-frequency, high-volume data streams remains a pressing issue for future research.

11.3 Challenges and Limitations of GeoAI

Despite its transformative potential, GeoAI faces several challenges and limitations that constrain its broader application in economic geography. These challenges arise from issues related to data, algorithms, ethical considerations, and technical integration.

GeoAI relies on diverse datasets, ranging from satellite imagery to trade logs and social media feeds (Lunga et al., 2019; Zhou, 2019). However, data sources often differ in structure, format, and granularity. For instance, international trade data may be aggregated by different regions, time periods, or product classifications, leading to inconsistencies when integrated into a single analytical framework (Zipkin et al., 2021). These variations necessitate rigorous preprocessing and standardization, which can be resource intensive. In many developing countries, the lack of reliable and comprehensive datasets impedes the application of GeoAI. Trade records, logistics data, and supply chain networks are often underreported or inaccessible due to political, economic, or technological constraints (Rao et al., 2021). Moreover, proprietary datasets held by corporations further limit access, creating an uneven playing field for researchers and policymakers (Korinek & Vipra, 2024). Economic systems are inherently dynamic, requiring real-time data to capture temporal variations. However, the lack of reliable mechanisms for continuous data collection and updating poses challenges to GeoAI applications (Alastal & Shaqfa, 2022). Without timely updates, models risk becoming obsolete, reducing their relevance for decision-making.

Many GeoAI models, such as neural networks and GANs, are often criticized for their "black box" nature (Kang et al., 2024). While these models excel in pattern recognition and predictive accuracy, their complexity makes it difficult to interpret the underlying mechanisms driving their outputs (Amiri et al., 2023). This opacity limits their applicability in policy-oriented research, where interpretability and transparency are crucial. GeoAI models often perform well on specific datasets but may fail to generalize across diverse contexts. For example, a model trained on trade data from developed countries may not accurately predict patterns in emerging economies. Overfitting becomes a concern, particularly when working with limited or skewed datasets. GeoAI models, especially those involving deep learning, require significant computational resources for training and deployment. High-performance hardware and cloud computing infrastructure are often prerequisites, making GeoAI applications inaccessible for institutions with limited budgets (Das & Cetl, 2021; Dou et al., 2023).

GeoAI models inherit biases from their training data, which can perpetuate inequalities (Chen et al., 2023b). For example, trade networks involving underrepresented regions may be mischaracterized due to limited data availability. Ensuring fairness requires proactive measures such as dataset balancing and algorithmic audits (Patel & Uddin, 2022). The integration of IoT and social media data into GeoAI systems raises concerns about privacy and data security (Biu et al., 2024). Trade-sensitive information, when improperly handled, can lead to economic espionage or unintended consequences. Generative AI, while powerful, carries risks of misuse. The creation of synthetic datasets or simulated scenarios can be manipulated to mislead policymakers or skew public opinion(Sharma & Sharma, 2024). Ensuring ethical guidelines and transparency in generative AI applications is paramount.

11.4 Future Directions for GeoAI

To overcome existing limitations and unlock its transformative potential, GeoAI must advance through technological innovations, interdisciplinary collaborations, and strategic integration into policy and industry. This section outlines promising future directions for GeoAI, emphasizing its potential rather than current achievements.

Generative AI holds significant promise for addressing persistent data gaps in economic geography. By creating synthetic datasets that approximate real-world scenarios, this technology could enable researchers to conduct analyses in underrepresented regions where reliable data are lacking (Smart et al., 2024). For instance, synthetic trade data might be generated to simulate the potential impacts of trade liberalization in regions without detailed historical records (Zhou & Li, 2022). Such advancements could provide researchers with valuable tools to analyze economic systems in ways previously constrained by data availability. In addition, generative AI could potentially streamline policy evaluation by simulating various economic and policy scenarios (Akhavan & Jalali, 2024; Capraro et al., 2024). For example, AI-driven models might predict how different carbon pricing mechanisms could influence global trade flows, offering policymakers actionable insights (Akter et al., 2024; Khallaf, 2024). These models, by reducing the time and resources required for traditional policy analysis, could significantly enhance decision-making processes in contexts requiring rapid evaluation. While these applications remain largely exploratory, they represent a compelling direction for future development. Research should focus on enhancing the reliability and interpretability of generative AI models to ensure their outputs are robust and useful for real-world applications.

Integrating GeoAI with Internet of Things (IoT) technologies could enable real-time monitoring and modeling of economic networks (Andronie et al., 2023; Miranda et al., 2024). While still in the developmental stages, such systems have the potential to provide dynamic insights into supply chains, trade routes, and production systems. IoT sensors embedded in logistics networks, production hubs, and transportation systems could continuously feed real-time data into GeoAI platforms, allowing stakeholders to better understand and respond to disruptions (Bibri et al., 2024; Chowdhury et al., 2024). For example, during geopolitical conflicts or natural disasters, real-time GeoAI systems might analyze satellite imagery, traffic updates, and social media feeds to identify disrupted trade hubs and propose alternative routes (Agbaje et al., 2024; Bibri et al., 2024). Although this integration is not yet widely implemented, it could eventually provide businesses and policymakers with immediate, data-driven strategies to minimize economic losses. Additionally, real-time GeoAI could optimize routine operations, such as inventory management and logistics planning. As IoT and AI technologies evolve, GeoAI systems might dynamically adjust resource allocations or suggest route changes to mitigate inefficiencies. Future research should explore the scalability and security of such systems to ensure they can handle high-frequency, high-volume data streams effectively.

Theoretical advancements in GeoAI remain in their infancy, offering significant opportunities for future exploration. Currently, GeoAI applications often operate independently of established economic geography theories, such as spatial interaction models or agglomeration economies (Jin et al., 2018; Liu, 2019; Mortaheb & Jankowski, 2023). Developing a unified framework that integrates GeoAI methodologies with these theories could enhance the rigor and relevance of GeoAI-based studies. For instance, GeoAI could be used to test existing theories by uncovering previously undetectable patterns in large spatial datasets. In particular, theories related to economic resilience and adaptation could benefit from GeoAI's ability to model dynamic, multi-scalar interactions. However, these possibilities remain largely untapped, requiring further interdisciplinary collaboration between geographers, data scientists, and policymakers to align GeoAI tools with theoretical needs. Future efforts should prioritize creating frameworks that guide the design of GeoAI models in ways that are both scientifically grounded and practically applicable. By fostering collaboration across disciplines, GeoAI could become a critical tool for bridging gaps between theoretical understanding and real-world challenges.

GeoAI's potential for supporting evidence-based policy-making is immense, particularly in the areas of regional development, trade optimization, and sustainability

(He & Chen, 2024; Koliousis et al., 2024). While applications are still under development, GeoAI-driven models could eventually assist policymakers in identifying growth hubs, analyzing trade dependencies, and proposing infrastructure investments. For example, future GeoAI models might optimize the placement of logistics hubs to enhance regional connectivity and economic competitiveness, especially in rapidly urbanizing regions (Bibri et al., 2024; Naqvi et al., 2021). In the private sector, GeoAI could enhance supply chain resilience and efficiency. Applications under exploration include predictive models for identifying bottlenecks, optimizing inventory levels, and improving route planning (Mahmood, 2022). For instance, businesses might leverage GeoAI insights to anticipate potential disruptions, such as extreme weather events or geopolitical tensions, and preemptively adjust their operations. GeoAI's role in achieving sustainability goals is another promising area for future development. By analyzing the spatial impacts of green policies, such as renewable energy adoption or carbon offset programs, GeoAI could help policymakers design strategies that balance environmental and economic objectives. While such applications are not yet widely realized, they represent an important direction for aligning GeoAI tools with global sustainability initiatives.

References

Agbaje, T. H., Abomaye-Nimenibo, N., Ezeh, C. J., Bello, A., & Olorunnishola, A. (2024). Building damage assessment in aftermath of disaster events by leveraging Geoai (geospatial artificial intelligence). *World Journal of Advanced Research and Reviews, 23*(1), 667–687.

Akhavan, A., & Jalali, M. S. (2024). Generative AI and simulation modeling: How should you (not) use large language models like CHATGPT. *System Dynamics Review, 40*(3), e1773. https://doi.org/10.1002/sdr.1773

Akter, S., Babu, M. M., Hani, U., Sultana, S., Bandara, R., & Grant, D. (2024). Unleashing the power of artificial intelligence for climate action in industrial markets. *Industrial Marketing Management, 117*, 92–113.

Alastal, A. I., & Shaqfa, A. H. (2022). Geoai technologies and their application areas in urban planning and development: Concepts, opportunities and challenges in Smart City (Kuwait, study case). *Journal of Data Analysis and Information Processing, 10*(2), 110–126.

Alatriste-Contreras, M. G., Brida, J. G., & Anyul, M. P. (2019). Structural change and economic dynamics: Rethinking from the complexity approach. *Journal of Dynamics & Games, 6*(2), 87.

Albin, P. (2021). *Barriers and bounds to rationality: Essays on economic complexity and dynamics in interactive systems*. Princeton University Press.

Amiri, Z., Heidari, A., Navimipour, N. J., Unal, M., & Mousavi, A. (2023). Adventures in data analysis: A systematic review of deep learning techniques for pattern recognition in cyber-physical-social systems. *Multimedia Tools and Applications, 83*(8), 22909–22973. https://doi.org/10.1007/s11042-023-16382-x

Andronie, M., Lăzăroiu, G., Iatagan, M., Hurloiu, I., Ştefănescu, R., Dijmărescu, A., & Dijmărescu, I. (2023). Big data management algorithms, deep learning-based object detection technologies, and geospatial simulation and sensor fusion tools in the internet of robotic things. *ISPRS International Journal of Geo-Information, 12*(2), 35.

Aoyama, Y., Murphy, J. T., & Hanson, S. (2010). *Key concepts in economic geography*. Sage.

Bai, C., Feng, C., Kerui, D., Wang, Y., & Gong, Y. (2020). Understanding spatial-temporal evolution of renewable energy technology innovation in China: Evidence from convergence analysis. *Energy Policy, 143*, 111570.

Baladad, J. A. (2021). Of thesis: Mapping National Interests: Semantic network analysis of USA. *Brookings Papers on Economic Activity, 2*, 381–476.

Balland, P. A., Broekel, T., Diodato, D., Giuliani, E., Hausmann, R., o'Clery, N., & Rigby, D. (2022). Reprint of the new paradigm of economic complexity. *Research Policy, 51*(8), 104568.

Bibri, S. E., Huang, J., Jagatheesaperumal, S. K., & Krogstie, J. (2024). The synergistic interplay of artificial intelligence and digital twin in environmentally planning sustainable Smart cities: A comprehensive systematic review. *Environmental Science and Ecotechnology, 20*, 100433.

Biu, P. W., Oliha, J. S., & Obi, O. C. (2024). The evolving role of geospatial intelligence in enhancing urban security: A review of applications and outcomes. *Engineering Science & Technology Journal, 5*(2), 483–495.

Blondeel, M., Bradshaw, M. J., Bridge, G., & Kuzemko, C. (2021). The geopolitics of energy system transformation: A review. *Geography Compass, 15*(7), e12580. https://doi.org/10.1111/gec3.12580

Breinlich, H., Corradi, V., Rocha, N., Michele Ruta, J. M. C., Silva, S., & Zylkin, T. (2022). *Machine learning in international trade research-evaluating the impact of trade agreements*. World Bank Group.

Cao, K., Zhou, C., Church, R., Li, X., & Li, W. (2024). Revisiting spatial optimization in the era of geospatial big data and GeoAI. *International Journal of Applied Earth Observation and Geoinformation, 129*, 103832.

Capraro, V., Lentsch, A., Acemoglu, D., Akgun, S., Akhmedova, A., Bilancini, E., Bonnefon, J.-F., Brañas-Garza, P., Butera, L., & Douglas, K. M. (2024). The impact of generative artificial intelligence on socioeconomic inequalities and policy making. *PNAS Nexus, 3*(6), pgae191.

Chen, S., Song, Y., Ding, Y., Zhang, M., & Nie, R. (2021). Using Long short-term memory model to study risk assessment and prediction of China's oil import from the perspective of resilience theory. *Energy, 215*, 119152.

Chen, M., Claramunt, C., Çöltekin, A., Liu, X., Peng, P., Robinson, A. C., Wang, D., Strobl, J., Wilson, J. P., & Batty, M. (2023a). Artificial intelligence and visual analytics in geographical space and cyberspace: Research opportunities and challenges. *Earth-Science Reviews, 241*, 104438.

Chen, P., Linna, W., & Wang, L. (2023b). AI fairness in data management and analytics: A review on challenges, methodologies and applications. *Applied Sciences, 13*(18), 10258.

Chowdhury, M., Dey, K., & Apon, A. (2024). *Data analytics for intelligent transportation systems*. Elsevier.

Cristian, Alexandru. 2024. 'Post-disaster recovery and environmental monitoring using GIS and GeoAI: Building damage assessment insights from McDonough County'.

Das, A., & Cetl, V. (2021). *Study on opportunities and challenges of collaboration for geospatial services*. European Commission.

Datta, B., & Misra, S. (2024). Port efficiency and infrastructure development: Catalysts for the India-Middle East-European Union economic corridor. In *Global cargo industry: Resilience of Asia-Pacific shipping industries* (pp. 1–34). IGI Global.

Datta, B., & Satghare, H. R. (2024). *Global cargo industry: Resilience of Asia-Pacific shipping industries: Resilience of Asia-Pacific shipping industries*. IGI Global.

der Mensbrugghe, V., & Dominique. (2013). Modeling the global economy–forward-looking scenarios for agriculture. In *Handbook of computable general equilibrium modeling* (Vol. 1, pp. 933–994). Elsevier.

Dosi, G., & Nelson, R. R. (2010). Technical change and industrial dynamics as evolutionary processes. *Handbook of the Economics of Innovation, 1*, 51–127.

Dou, F., Ye, J., Yuan, G., Lu, Q., Niu, W., Sun, H., Guan, L., Lu, G., Mai, G., Liu, N., Lu, J., Liu, Z., Wu, Z., Tan, C., Xu, S., Wang, X., Li, G., Chai, L., Li, S., Sun, J., Sun, H., Shao, Y., Li, C., Liu, T., & Song, W. (2023). *Towards artificial general intelligence (AGI) in the internet of things (IoT): Opportunities and challenges*. Cornell University.

Durlik, I., Miller, T., Kostecka, E., & Tuński, T. (2024). Artificial intelligence in maritime transportation: A comprehensive review of safety and risk management applications. *Applied Sciences, 14*(18), 8420.

El Makhloufi, A. (2023). *AI application in transport and logistics: Opportunities and challenges (An exploratory study)*. Amsterdam University of Applied Sciences.

Engel, J. S. (2015). Global clusters of innovation: Lessons from Silicon Valley. *California Management Review, 57*(2), 36–65. https://doi.org/10.1525/cmr.2015.57.2.36

Essletzbichler, J. (2016). Renewable energy technology and path creation: A multi-scalar approach to energy transition in the UK. In *Path dependence and new path creation in renewable energy technologies* (pp. 63–88). Routledge.

Franchi, F., Galassi, A., & Graziosi, F. (2024). AI-assisted GIS toward GEO-AI: Trends and innovations overview. In *2024 IEEE 8th forum on research and Technologies for Society and Industry Innovation (RTSI)* (pp. 149–153). IEEE.

Grabher, G. (2006). Trading routes, bypasses, and risky intersections: Mapping the travels of 'networks' between economic sociology and economic geography. *Progress in Human Geography, 30*(2), 163–189. https://doi.org/10.1191/0309132506ph600oa

Habibi, F., Chakrabortty, R. K., & Abbasi, A. (2023). Towards facing uncertainties in biofuel supply chain networks: A systematic literature review. *Environmental Science and Pollution Research, 30*(45), 100360–100390. https://doi.org/10.1007/s11356-023-29331-w

Hassan, Q., Viktor, P., Al-Musawi, T. J., Ali, B. M., Algburi, S., Alzoubi, H. M., Al-Jiboory, A. K., Sameen, A. Z., Salman, H. M., & Jaszczur, M. (2024). The renewable energy role in the global energy transformations. *Renewable Energy Focus, 48*, 100545.

He, W., & Chen, M. (2024). Advancing urban life: A systematic review of emerging technologies and artificial intelligence in Urban Design and planning. *Buildings, 14*(3), 835.

Hess, M. (2004). "Spatial" relationships? Towards a reconceptualization of embedded ness. *Progress in Human Geography, 28*(2), 165–186. https://doi.org/10.1191/0309132504ph479oa

Huang, X., Wang, S., Tianjun, L., Liu, Y., & Serrano-Estrada, L. (2024). Crowdsourced geospatial data is reshaping urban sciences. *International Journal of Applied Earth Observation and Geoinformation, 127*, 103687.

Janowicz, K., Sieber, R., & Crampton, J. (2022). GeoAI, counter-AI, and human geography: A conversation. *Dialogues in Human Geography, 12*(3), 446–458. https://doi.org/10.1177/20438206221132510

Jin, R., Gong, J., Deng, M., Wan, Y., & Yang, X. (2018). A framework for spatiotemporal analysis of regional economic agglomeration patterns. *Sustainability, 10*(8), 2800.

Joel, O. S., Oyewole, A. T., Odunaiya, O. G., & Soyombo, O. T. (2024). Leveraging artificial intelligence for enhanced supply chain optimization: A comprehensive review of current practices and future potentials. *International Journal of Management & Entrepreneurship Research, 6*(3), 707–721.

Kang, Y., Gao, S., & Roth, R. E. (2024). Artificial intelligence studies in cartography: A review and synthesis of methods, applications, and ethics. *Cartography and Geographic Information Science, 51*(4), 599–630. https://doi.org/10.1080/15230406.2023.2295943

Khallaf, A. N. (2024). Using Ai to help reduce the effect of global warming. *Power System Technology, 48*(1), 1927–1947.

Koliousis, I., Al-Surmi, A., & Bashiri, M. (2024). Artificial intelligence and policy making; can small municipalities enable digital transformation? *International Journal of Production Economics, 274*, 109324.

Korinek, A., & Vipra, J. (2024). Concentrating intelligence: Scaling and market structure in artificial intelligence. *Economic Policy, 225*, 225.

Lee, C.-M. (2000). *The Silicon Valley edge: A habitat for innovation and entrepreneurship*. Stanford University Press.

Lei, Y., Liang, Z., & Ruan, P. (2023). Evaluation on the impact of digital transformation on the economic resilience of the energy industry in the context of artificial intelligence. *Energy Reports, 9*, 785–792.

Li, W. (2022). GeoAI in social science. In *Handbook of spatial analysis in the social sciences* (pp. 291–304). Elger.

Liu, X. (2019). *Quantifying cities through user-generated geosocial content with GeoAI*. The Pennsylvania State University.

Liu, Q., Chen, R., & Qian, Y. (2024). Improving sustainable performance of China's new energy industry through collaborative innovation network resilience. *Renewable and Sustainable Energy Reviews, 201*, 114625.

Lunga, D. D., Alemohammad, H., Liu, Y., Newsam, S., Pacifici, F., Santos-Villalobos, H., Shook, E., Stewart, R. N., Voisin, S., & Yang, L. (2019). *The trillion pixel GeoAI challenge workshop*. Oak Ridge National Lab.(ORNL).

Luo, W., & MacEachren, A. M. (2014). Geo-social visual analytics. *Journal of Spatial Information Science, 2014*(8), 27–66.

Luo, W., & Yuan, M. (2021). Identifying temporal patterns of multilateral spatial interactions: Using international trades as an example. *Transactions in GIS, 25*(4), 1888–1909. https://doi.org/10.1111/tgis.12745

Luo, W., Yin, P., Di, Q., Hardisty, F., & MacEachren, A. M. (2014). A Geovisual analytic approach to understanding geo-social relationships in the international trade network. *PLoS One, 9*(2), e88666.

Luo, W., He, L., Yang, Z., Zhang, S., Wang, Y., Liu, D., Sheng, H., He, L., Xia, J., & Chen, M. (2023a). Spatio-temporal heterogeneity in the international trade resilience during COVID-19. *Applied Geography, 154*, 102923.

Luo, W., Kang, S., Hu, S., Lixian, S., and Dai, R. (2023b). '*Dual effects of the US-China trade war and COVID-19 on United States imports: Transfer of China's industrial chain?*'

Mahmood, H. (2022). *Strategic foresight to applications of geospatial artificial intelligence (GeoAI) to achieve disaster-related sustainable development goals*. ESCAP.

Malecki, E. J. (2002). The economic geography of the Internet's infrastructure*. *Economic Geography, 78*(4), 399–424. https://doi.org/10.1111/j.1944-8287.2002.tb00193.x

Miranda, R., Alves, C., Sousa, R., Chaves, A., Montenegro, L., Peixoto, H., Durães, D., Machado, R., Abelha, A., & Novais, P. (2024). Revolutionising the quality of life: The role of real-time sensing in Smart cities. *Electronics, 13*(3), 550.

Mittal, U., & Panchal, D. (2023). AI-based evaluation system for supply chain vulnerabilities and resilience amidst external shocks: An empirical approach. *Reports in Mechanical Engineering, 4*(1), 276–289.

Mortaheb, R., & Jankowski, P. (2023). Smart City re-imagined: City planning and GeoAI in the age of big data. *Journal of Urban Management, 12*(1), 4–15.

Nacke, L., Cherp, A., & Jewell, J. (2022). Phases of fossil fuel decline: Diagnostic framework for policy sequencing and feasible transition pathways in resource dependent regions. *Oxford Open Energy, 1*, oiac002.

Naqvi, Asad, S. A., & Naqvi, R. Z. (2021). Geographical information systems (GIS) in industry 4.0: Revolution for sustainable development. In C. M. Hussain & P. Di Sia (Eds.), *Handbook of Smart materials, technologies, and devices* (pp. 1–27). Springer International Publishing.

Norberg-Bohm, V. (2000). Creating incentives for environmentally enhancing technological change: Lessons from 30 years of US energy technology policy. *Technological Forecasting and Social Change, 65*(2), 125–148.

Papic, M. (2020). *Geopolitical alpha: An investment framework for predicting the future*. Wiley.

Patel, P., & Uddin, M. N. (2022). AI for algorithmic auditing: Mitigating bias and improving fairness in big data systems. *International Journal of Social Analytics, 7*(12), 39–48.

Polinsky, J. M. (2021). *Utilizing economic indicators in a machine learning platform to create an interdisciplinary forecasting framework for Natural Resource Management and Policy Planning*. Oregon State University.

Ponte, S., Gereffi, G., & Raj-Reichert, G. (2019). Introduction to the handbook on global value chains. In *Handbook on global value chains* (pp. 1–27). Edward Elgar Publishing.

Queiroz, L. P., Nogueira, I. B. R., & Ribeiro, A. M. (2024). Flavor engineering: A comprehensive review of biological foundations, AI integration, industrial development, and socio-cultural dynamics. *Food Research International, 115100*, 196.

Rane, N. L., Paramesha, M., Rane, J., & Kaya, O. (2024). Artificial intelligence, machine learning, and deep learning for enabling Smart and sustainable cities and infrastructure. *Artificial Intelligence and Industry in Society, 5*, 2–25.

Rao, S., Gulley, A., Russell, M., & Patton, J. (2021). On the quest for supply chain transparency through Blockchain: Lessons learned from two serialized data projects. *Journal of Business Logistics, 42*(1), 88–100. https://doi.org/10.1111/jbl.12272

Rezki, N., & Mansouri, M. (2024). Machine learning for proactive supply chain risk management: Predicting delays and enhancing operational efficiency. *Management Systems in Production Engineering, 32*(3), 345–356. https://doi.org/10.2478/mspe-2024-0033

Sachit, M. S., Shafri, H. Z. M., Abdullah, A. F., Rafie, A. S. M., Mohamed, B. A., & Gibril. (2024). A Novel GeoAI-Based Multidisciplinary Model for SpatioTemporal Decision-Making of Utility-Scale Wind–Solar Installations: To Promote Green Infrastructure in Iraq. *The Egyptian Journal of Remote Sensing and Space Sciences, 27*(1), 120–136.

Sanders, N. R. (2024). *Supply chain management: A global perspective*. John Wiley & Sons.

Schot, J., & Edward Steinmueller, W. (2018). Three frames for innovation policy: R&D, Systems of Innovation and Transformative Change. *Research Policy, 47*(9), 1554–1567.

Schwanen, T. (2018). Thinking complex interconnections: Transition, nexus and geography. *Transactions of the Institute of British Geographers, 43*(2), 262–283. https://doi.org/10.1111/tran.12223

Sharma, A. K., & Sharma, R. (2024). Generative artificial intelligence and legal frameworks: Identifying challenges and proposing regulatory reforms. *Kutafin Law Review, 11*(3), 415–451.

Shi, J., Wang, X., Wang, C., Liu, H., Miao, Y., & Ci, F. (2022). Evaluation and influencing factors of network resilience in Guangdong-Hong Kong-Macao Greater Bay Area: A structural perspective. *Sustainability, 14*(13), 8005.

Smart, A., Hutchinson, B., Amugongo, L. M., Dikker, S., Zito, A., Ebinama, A., Wudiri, Z., Wang, D., van Liemt, E., Sedoc, J., Olojo, S., Uwakwe, S., Wornyo, E., Schmer-Galunder, S., and Smith-Loud, J. (2024). '*Socially responsible data for large multilingual language models*'.

Song, Y., Escobar, O., Arzubiaga, U., & De Massis, A. (2022). The digital transformation of a traditional market into an entrepreneurial ecosystem. *Review of Managerial Science, 16*(1), 65–88. https://doi.org/10.1007/s11846-020-00438-5

Spencer, G. M., Vinodrai, T., Gertler, M. S., & Wolfe, D. A. (2010). Do clusters make a difference? Defining and assessing their economic performance. *Regional Studies, 44*(6), 697–715. https://doi.org/10.1080/00343400903107736

Vakulchuk, R., Overland, I., & Scholten, D. (2020). Renewable energy and geopolitics: A review. *Renewable and Sustainable Energy Reviews, 122*, 109547.

von Cramon-Taubadel, S., Hess, S., & Brummer, B. (2010). A preliminary analysis of the impact of a Ukraine-EU free trade agreement on agriculture. In *World Bank Policy Research Working Paper* (Vol. 5264). Elsevier.

Wei, Y. D. (2015). Spatiality of regional inequality. *Applied Geography, 61*, 1–10.

Wu, Z., Cheng, S., Keyi, X., & Qian, Y. (2024). Ecological network resilience evaluation and ecological strategic space identification based on complex network theory: A case study of Nanjing City. *Ecological Indicators, 158*, 111604.

Xie, Y. (2020). '*GeoAI: Challenges and opportunities*'. PhD Thesis, University of Minnesota.

Yang, C.-H., Lee, C.-F., & Chang, P.-Y. (2023). Export-and import-based economic models for predicting global trade using deep learning. *Expert Systems with Applications, 218*, 119590.

Yu, C., Zhou, Z., & Gao, J. (2024). Rural network resilience: A new tool for exploring the mechanisms and pathways of rural sustainable development. *Sustainability, 16*(14), 5850.

Zhang, Y., Long, H., Ma, L., Shuangshuang, T., Li, Y., & Ge, D. (2022). Analysis of rural economic restructuring driven by E-commerce based on the space of flows: The case of Xiaying Village in Central China. *Journal of Rural Studies, 93*, 196–209.

Zhou, X. (2019). *GeoAI-enhanced techniques to support geographical knowledge discovery from big geospatial data*. Arizona State University.

Zhou, A., & Li, J. (2022). How do trade liberalization and human capital affect renewable energy consumption? Evidence from the panel threshold model. *Renewable Energy, 184*, 332–342.

Zipkin, E. F., Zylstra, E. R., Wright, A. D., Saunders, S. P., Finley, A. O., Dietze, M. C., Itter, M. S., & Tingley, M. W. (2021). Addressing data integration challenges to link ecological processes across scales. *Frontiers in Ecology and the Environment, 19*(1), 30–38. https://doi.org/10.1002/fee.2290

Chapter 12
From Space to Context: Contextualizing the Food Environment with GeoAI

Xiang Chen, Congcong Miao, Weixuan Lyu, Gengchen Mai, and Ran Xu

Abstract This chapter advocates a paradigm shift from traditional spatial modeling to context-rich approaches in examining the food environment and its impact on dietary health. While previous studies focus heavily on spatial distribution—proximity, density, and variety of stores—the underlying contexts that shape food choices often remain overlooked. By extending beyond simple spatial proxies, we emphasize the importance of environmental contexts, including nutrition quality, store attributes, cultural relevance, and people's perceptions, in influencing food behaviors. Emerging geospatial artificial intelligence (GeoAI) provides powerful tools to identify, interpret, and visualize these contextual factors. Drawing from a case study in Hartford, Connecticut, we illustrate how AI-based nutrition assessment of geotagged food images generates fine-grained environmental insights that transcend the limitations of traditional space-based models. This approach not only advances the conceptual understanding of how contextual factors guide food choices but also supports policy interventions, such as menu labeling, that are sensitive to the complex interplay between place and perception. Our findings highlight the need for integrative frameworks that combine spatial analysis with contextual intelligence, paving the way for more nuanced, human-centered inquiries into the link between the food environment and public health.

Keywords GeoAI · Food environment · Nutritional context · Spatial analysis · Dietary health

X. Chen (✉) · C. Miao · W. Lyu
Department of Geography, Sustainability, Community, and Urban Studies, University of Connecticut, Storrs, CT, USA
e-mail: xiang.chen@uconn.edu

G. Mai
Department of Geography and the Environment, University of Texas at Austin, Austin, TX, USA

R. Xu
Department of Allied Health Sciences, University of Connecticut, Storrs, CT, USA

© The Author(s), under exclusive license to Springer Nature Switzerland AG 2025
X. Huang et al. (eds.), *GeoAI and Human Geography*, Springer Geography,
https://doi.org/10.1007/978-3-031-87421-5_12

12.1 Introduction

Where people choose to procure food can influence their daily diets. Large supermarkets tend to offer a wide variety of healthy food options, such as fresh fruits and vegetables, which are generally low in calories and high in fiber. Purchasing fresh produce from supermarkets and maintaining homemade diets are healthy practices encouraged by nutrition scientists (Caspi et al., 2012). Conversely, frequenting fast-food restaurants can lead to the overconsumption of fat, sodium, and sugar, increasing the risk of developing obesity and obesity-related comorbidities (Fleischhacker et al., 2011).

Examining the geographies of food is crucial for identifying the environmental influences that contribute to health-promoting or health-damaging dietary behaviors (Matthews et al., 2009). A direct and measurable environmental influence is space, including the spatial separation between a person's home and food provisioning nearby, which, to some extent, influences the likelihood of healthy food procurement. The past two decades have witnessed a flourishing spatial evolution in research on the food environment, emphasizing the role of space in facilitating or deterring food access. A wide array of spatial models (e.g., density, variety, and proximity) were proposed as proxies for spatial deterrence and were realized by recourse to Geographic Information Systems (GIS) (Charreire et al. 2010). These endeavors, further substantiated by policy tools such as the Food Access Research Atlas (USDA, 2024a) and the Food Environment Atlas (USDA, 2024b), put spatial metaphors such as "food deserts" (i.e., areas lacking healthy foods) (Cummins & Macintyre, 2002) into practice.

While this spatial revolution portrays the landscape of food (i.e., foodscape), unraveling the linkage between the food environment and dietary health is still far from conclusive. For example, although the density of fast-food restaurants is intuitively seen as a risk factor for poor dietary health (e.g., higher obesity rates), there is no consistent statistical evidence supporting this conclusion. In a systematic review of 15 studies connecting the fast-food environment with obesity indicators, 8 studies (53%) showed no significant associations (Fleischhacker et al., 2011). While many geographical uncertainties, such as the modifiable areal unit problem (MAUP) (Chen, Ye, et al., 2022a) and the ecological fallacy (Schwartz, 1994), can affect the strength of the associations, the contexts in which food environmental influences are translated into food choices and subsequently dietary health are beyond simple spatial models.

We argue that the food environmental contexts are missing in these spatial models. Specifically, the aggregation and interpretation of spatial data can result in the loss of important contextual information. This information, such as the nutritional quality of foods and the perception of the built environment, can potentially influence the likelihood of consumers' store patronage and even change their health behavior in the long run. These situational influences, however, do not exist in traditional spatial data. They can be extracted from additional environmental and human-generated datasets and then interpreted through model-assisted analyses.

We thus call for a paradigm shift from space to context, meaning that future studies should focus on examining the different environmental contexts that induce health behavioral changes, rather than developing complex spatial models to reiterate the concept of space.

We believe that emerging technologies in geospatial artificial intelligence (GeoAI) can assist with this transition. Rooted in the convergence of GIScience and computer science, GeoAI has ushered in a new phase of spatial analysis and inquiries by supporting data-intensive research and enhancing empirical, theoretical, and methodological explorations within the urban context (Mortaheb & Jankowski, 2023). GeoAI applies spatially explicit AI techniques to address spatial problems, such as identifying patterns and predicting trends based on spatiotemporal attributes in the data (Janowicz et al., 2020; Mai et al., 2022). However, we found very few use cases of GeoAI for food studies. In this chapter, we first review the methodological development in this unique subdiscipline of health geography. Then, we propose a conceptual framework to illustrate how food environmental contexts can be structured to influence food behaviors. Following this framework, we demonstrate a case study using GeoAI for mapping food nutrition as an important food environmental context. Lastly, we highlight the potential and limitations of GeoAI for future research.

12.2 Space and People in the Food Environment

How the food environment shapes food behaviors, eventually leading to long-term dietary health outcomes is a widely discussed topic in health geography (Suarez et al., 2015; Haslam et al., 2022; Huang et al., 2019). As food access plays a marked role in individuals' consumption patterns and can eventually shape their physical well-being, the majority of food environmental studies spiral around the problem of "food security," a concept defined by the U.S. Department of Agriculture (USDA) as "access by all people at all times to enough food for an active, healthy life" (USDA, 2023). As such, the last two decades have seen growing attention to food security that explored the problem primarily from two different angles: space and people.

A space-oriented perspective emphasizes studying the physical environment where food is procured or consumed (McKinnon et al., 2009). Traditional food environmental studies are focused on the in-store environment, using standardized store audit tools (e.g., market basket) to evaluate the availability, pricing, and signage of food items (Donkin et al., 2000). The emergence of GIS has revolutionized this assessment method, shifting in situ observations to large-scale spatial assessments. GIS enables the delineation of food distribution patterns, such as the proximity, diversity, and variety of stores, which are often demarcated by administrative units (e.g., census tracts) (Apparicio et al., 2007). Then, these area-based geographical measures are employed to correlate with residential socioeconomic variables (e.g., household incomes) or dietary health variables (e.g., obesity rate) to reveal issues of health inequities (Fleischhacker et al., 2011). These space-health associations, however, are not consistent across different case studies and remain subject to debate (Fleischhacker et al., 2011; Caspi et al., 2012).

This disconnection can be attributed to several methodological fallacies relating to the effects of boundaries and scales. The first fallacy is the MAUP, referring to that statistical analyses using different geographical scales or areal units would generate very different observations (Fotheringham & Wong, 1991). The MAUP is a common issue in food environmental studies and can moderate the strength of statistical associations (Chen, Ye, et al., 2022a). A related issue to the MAUP is the edge effect, which means that food stores tend to be located near census boundaries or edges. This proximity to edges can obscure the statistical significance of pattern identification by census unit (Chen, 2017). The second fallacy is the ecological fallacy, referring to the fact that drawing evidence from aggregated census data may misinterpret the specifics of individuals (Schwartz, 1994), and as a result, macro-scale correlations are less meaningful at the individual level. Moreover, examining the food environment from a generalized spatial perspective falls short in identifying individual differences in socioeconomics and the contexts in which food choices are made. In reality, people's mobility, environmental perception, and lived experience can concurrently influence their food behaviors and induce unique dietary health outcomes (Chen & Kwan, 2015). Ignoring these contexts makes any identified food disparity patterns less interpretable. In a related vein, the generalization of space in studying the food environment echoes the criticism of GIS as a tool for measuring spatial patterns in a positivist approach (Taylor, 1990), while the underlying contexts justifying these patterns are often ignored.

These criticisms sparked a paradigm shift by bringing people to the spotlight. Traditionally, the people-oriented perspective focuses on individuals or households by using food and nutrition assessment tools, such as structured surveys and interviews (Carletto et al., 2013). These individual-based evaluations not only aid in verifying food intakes that differ by household but also tap into measures of the food environment near a household, such as proximity to nearby food stores. More significantly, data collected on an individual scale can capture contextual influences, including educational attainment and cultural relevance, which are indirectly linked to food behaviors.

These people-oriented examinations, however, have an obvious flaw that deserves closer scrutiny. The flaw is a focus on the static food environment, such as using an individual's home address to represent food environmental exposure. The everyday effort to procure food is rather dynamic and can occur in various environments, including home, worksites, and schools. Drawing evidence solely from food procurement near the household or in the residential community ignores other environments where food is purchased or consumed (Lucan, 2015). This multi-environmental exposure is fundamentally driven by human mobility. Mobility resources (e.g., personal cars, public transit) can significantly help individuals overcome spatial barriers to food access (Jin et al., 2023). Advances in geospatial science now accommodate the need to capture dynamic food environmental exposure, using location-based technologies such as the Global Positioning System (GPS) (Christian, 2012; Chaix et al., 2013) to track individual food activities.

Recently, the utilization of large-scale mobility data has facilitated the identification of dynamic food environmental exposure, revealing that the majority of people in the United States procure foods beyond their residential communities (Xu et al., 2023).

12.3 Conceptualizing Food Environmental Contexts

While endeavors to model space and people have been progressing rapidly, the pathway from the environment to health remains unclear. A void in these modeling approaches is the lack of discourse on food environmental contexts. For example, there has always been an oversimplification in defining the food environment, presuming grocery stores as healthy sources and fast-food restaurants and convenience stores as unhealthy. Even the Centers for Disease Control and Prevention (CDC) adopts this dichotomy in labeling the community food environment (CDC 2012). In reality, ultra-processed, nutrient-poor foods are very common in "healthy" grocery stores. Other contextual layers, such as the ethnicity of food, food price, and store service, play equally important roles in shaping the likelihood of store patronage. These contexts, which largely differ from place to place, are rarely discussed in spatial modeling approaches. Beyond space, they interact with individual variables in the making of food choices. To this end, we propose a conceptual framework to illustrate how these environmental contexts mechanistically influence food behaviors, as shown in Fig. 12.1.

Based on the definitions proposed by Glanz et al. (2005), we identify two different environmental contexts: the consumer nutrition environment and the community nutrition environment. The consumer nutrition environment refers to the dietary contexts within a store that potentially drive a consumer's food behavior. Examples of these contexts include food nutrition, price and promotion, in-store environment, and service and amenities. The community nutrition environment, on the other hand, refers to the physical environment of a store that influence a consumer's food choices. Examples of these contexts include store type, location, operating hours, and the presence of nearby transport infrastructure. While the community nutrition environment has been widely modeled by GIS as mentioned earlier, the study of the consumer nutrition environment, especially on a broad scale, remains scant.

These food environmental contexts interact with individual variables to influence food behaviors. Differences in individual variables such as access to a car, disposable income, home location, available time, and nutrition education can lead to disparities in the ways individuals gather and process information about the environmental contexts. For example, a grocery store may not be considered a shopping destination due to financial constraints or the lack of car access at the individual level (Payán et al., 2021). When these individual variables interact with environmental contexts, the interaction allows the individual to generate perceptual criteria based on which a food place is considered accessible, given their available

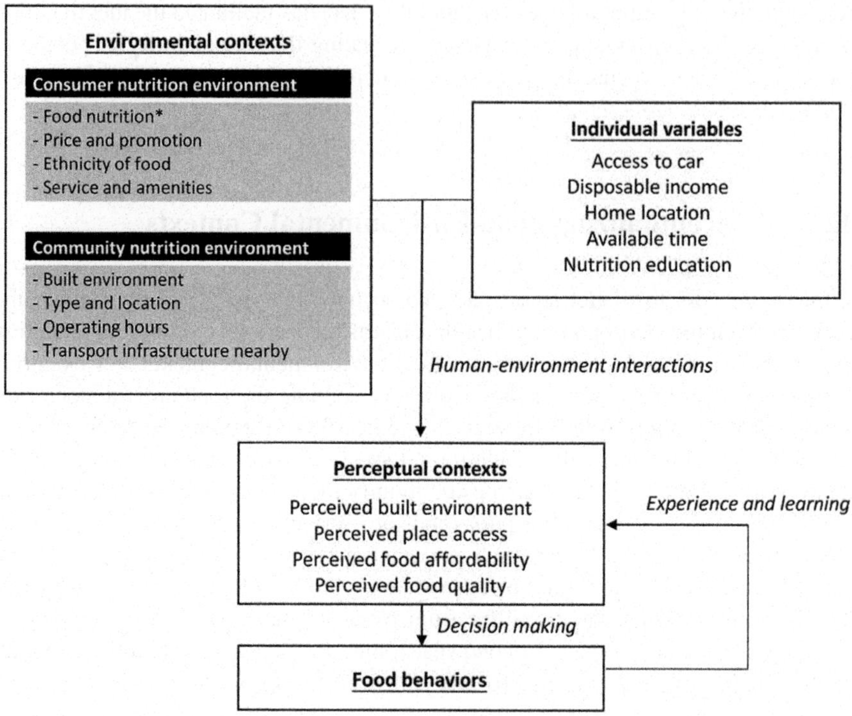

Fig. 12.1 A conceptual framework of food environmental contexts. The asterisk (*) indicates the GeoAI application in this chapter

resources and socio-cultural interpretation of the food place. We call these criteria the perceptual contexts of the food environment. The perceptual contexts can thus largely influence the decision-making process about food behaviors, including where to shop, how to get there, and what to buy and consume. These perceptual contexts are rather dynamic. Through interacting with people and places, individual perceptions will change over time, and their attitudes toward others and their sense of place will constantly evolve. This experience and learning process can further shape their perceptual contexts. Lastly, it should be noted that this framework is structured towards the retail food environment (e.g., restaurants, cafes, grocery stores, and convenience stores), while other organizational food environments (e.g., workplace) are excluded because of the lack of generality to all community members.

Some variables in the environmental contexts, such as food nutrition, cannot be easily acquired on a broad scale. Also, variables in the perceptual contexts cannot be adequately evaluated when human perception is involved. These evaluations can be assisted by GeoAI, which allows for batch processing, learning, and reinterpretation of large-scale, geotagged information regarding environmental contexts. In the subsequent sections, we present how GeoAI can be employed as a promising tool in the nutrition assessment of the food environment.

12.4 AI for Nutrition Assessment

The ambiguous linkage between the food environment and dietary health, as stated earlier, can be attributed to the mislabeling of food places. Specifically, existing studies have followed a dichotomous approach, categorizing fast-food restaurants as health-damaging and chained grocery stores as health-promoting. In reality, fresh-made, nutrient-balanced meals can be ordered at fast-food restaurants, while ultra-processed, nutrient-poor foods are common in grocery stores. This nutritional granularity of the food environment, which directly impacts dietary health, remains unsolved in the existing paradigm. While it is necessary to adopt a quality-based, fine-grained approach to assess the nutritional context, acquiring large-scale food nutrition data is considerably challenging.

One potential method to overcome this data limitation is through crowdsourcing online food images and analyzing them by AI-based nutrition assessment models. These food images on business review platforms (e.g., Google Places, Yelp, Facebook, and TripAdvisor), which are examples of volunteered geographic information (VGI) (Chen & Yang, 2014; Goodchild & Li, 2012), provide a fruitful set of geotagged nutrition data. Then, AI-based nutrition assessment models can be applied to these food images to estimate the nutritional information (e.g., calories, sugar, carbohydrates, etc.). While this technology is relatively new, it has a historical root in nutrition science. Before delving into the application, we would like to review its development in two different phases.

The first phase, which we coin as "the digitalization phase," appeared in the early 2000s. This phase enabled the conversion of traditional dietary data (e.g., 24HRs, food records, and food frequency questionnaire) to a digital form. One of the earliest attempts was the Remote Food Photography Method (RFPM). The RFPM was about utilizing digital photos to collect individual food intakes. These photos were then analyzed by dietitians in terms of food contents and portion sizes (Williamson et al., 2004; Martin et al., 2009). Later, the emergence of mobile devices and communication technologies facilitated the data transfer. One example was the Wellnavi (Matsushita Electric Works, Ltd., Japan), which was a handheld personal digital assistant (PDA) with a mobile phone card designed to take food images and transfer them via cellular networks (Wang et al., 2002). However, these methods still required participants to have the necessary training in photo-taking and also relied on nutrition assessments in research and clinical settings.

The second phase, which we term "the AI phase," appeared in the early 2020s and is currently underway. The AI-based image recognition models have revolutionized the way nutrition information can be derived from food images. Thanks to the widespread use of smartphones and social media, food-related images and videos are continuously being generated, enabling the creation of large-scale, well-annotated image databases (Kawano & Yanai, 2015). Meanwhile, researchers have developed various targeted deep convolutional network structures to enhance model performance in image recognition (Liu et al., 2016). Models trained using these image databases and state-of-the-art network structures can accurately recognize the type and specificity of food items without the need for human raters (Sahoo et al., 2019). Then, the image recognition models are further linked to a nutrient composition database to estimate the nutrition information, including calories, macronutrients (i.e., carbohydrates, fats, and proteins), and micronutrients (i.e., vitamins and minerals), thereby making the transition from image recognition to nutrition assessment. These AI-based nutrition assessment models are primarily used for individual dietary assessments, where users can take a picture of their meals and become informed of their nutritional intake. Examples of these models include Calorie Mama, Cheat Day, Mealviser, and Foodzilla.

In a recent assessment of Calorie Mama, it was found the model has achieved a top-1 accuracy of 63% and a top-5 accuracy of 88% (Van Asbroeck & Matthys, 2020). However, the adoption of these AI models for studying the food environment on a broad scale remains extremely limited. To date, there have been only two case studies, one in Columbus, Ohio (Chen et al., 2022b) and the other in Hartford, Connecticut (Chen et al., 2021; Lyu et al., 2023), both studying the restaurant environment. These studies interpret the "geo" component of the GeoAI, by identifying the nutritional landscape of food provisioning in a city. These applications fill the missing environmental context about food nutrition and can elaborate on the pathway to health inequities. Following this discussion, we would like to exemplify how this technology unfolds in the Hartford study.

12.5 Case Study: Mapping Nutritional Landscape

The city of Hartford is the capital of Connecticut, United States. Hartford has a total population of approximately 1.21 million (U.S. Census Bureau, 2024). The city is characterized by high levels of racial segregation and concentrated poverty, especially when compared to neighboring cities. According to the 2022 census data, Hartford has a low-income rate of 51%, with considerable minorities (46% being Latino and 34% being non-Hispanic Black) (DataHaven, 2024). This socioeconomic landscape raised potential issues of food injustice, where access to healthy food sources could be limited for low-income, minority groups.

In this example, we employed Calorie Mama to assess the nutritional quality of restaurants in Hartford. The workflow is shown in Fig. 12.2. Specifically, we collected our first dataset, the restaurant directory (e.g., name, address, and operating hours), from Yelp by using its Fusion application programming interface (API) (Yelp Fusion, 2024). We further cross-validated the data with Google Place, as some business locations were permanently closed or mislabeled. These additional cross-validation ensured restaurants under study were up-to-date and accurate.

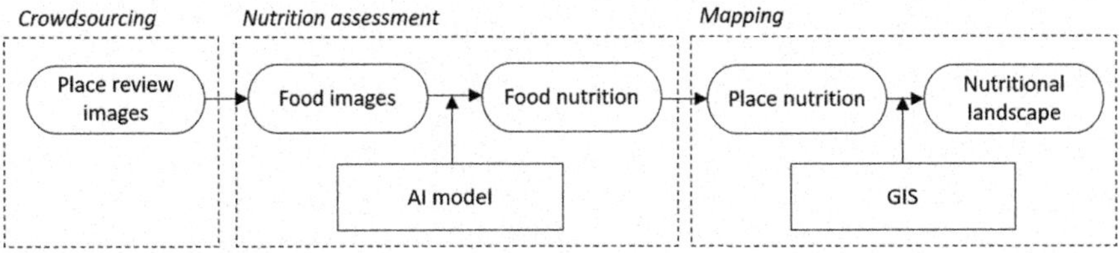

Fig. 12.2 A workflow of GeoAI for nutrition assessment of the food environment

Ultimately, 487 restaurants met the inclusion criteria. We further geocoded them based on their business addresses in ESRI ArcGIS Pro.

Our next dataset was the food review images. They were crowdsourced from Google Places and TripAdvisor and then consolidated by linking to the restaurant directory database. The initial food image data had considerable noise (e.g., staged photos, beverages, buildings, and food consumers), so we refined them by extracting only the food items. Our final food image dataset included 15,908 images derived from 470 restaurants, with 17 restaurants dropped from the analysis due to insufficient data points (i.e., fewer than five food images). The selected food images were further cropped to a standard size, which was a requirement for input into the AI model.

We then employed Calorie Mama's API for batch processing and deriving the nutritional information from each food image. The output for each image was the standard nutritional information. In addition, two trained raters validated the accuracy of the model in food item identification. The validation showed a relatively high accuracy level of 75.1% with an inter-rater reliability of 94.7%. In a follow-up study, we found that these online food images are representative of the popular dishes from the restaurant's menus (Lyu et al., 2023).

Based on this AI-derived nutrition information, we performed a nutrition assessment of each restaurant. Specifically, we adopted one of the nutritional measures: calorie density, expressed in a standardized SI unit (i.e., average calories per 1 kg of food). The calorific level of each restaurant was derived from the average of the estimated calorie densities of all food images for the restaurant. This measure conveys the nutritional context of the food place, as restaurants with a calorie-dense menu tend to be obesogenic. We then color-coded this information on a map to portray the nutritional landscape of the restaurants in the city (Fig. 12.3).

The case study demonstrates how an AI-based image recognition model can be deployed to estimate the nutritional information of food images and become expandable to the nutrition assessment of a city. In this case, the power of AI to support effective analyses of image data is further amplified by geographical methods, including crowdsourcing, geocoding, and geovisualization. These methods, which represent the "geo" component, empower AI models with fruitful geographical data for input and cartographical tools for output. More importantly, they transform the way in which food environmental data can be decoded, contextualized, and interpreted. The AI-derived nutritional landscape, as an important environmental driver

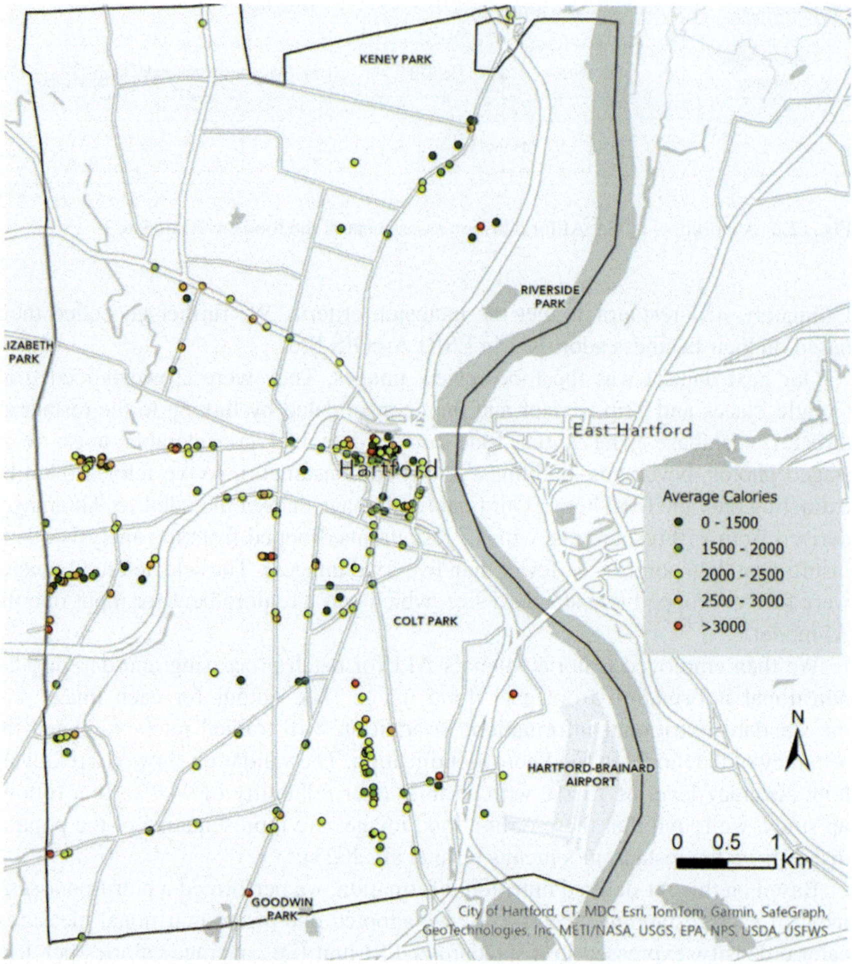

Fig. 12.3 The nutritional landscape of restaurants in the city of Hartford

of community health, can be used to understand the underlying food provisioning patterns on a broad scale and help explain many previously unexplained contexts leading to health inequities.

12.6 Food for Thought: GeoAI for Future Research

Beyond the recognition of food nutrition by image recognition, we believe other food environmental contexts can be derived, assessed, and visualized in a similar approach. We thus identify four emerging areas in the realm that can be potentially revolutionized by GeoAI.

The first study area is the identification of key environmental contexts that drive food behaviors. Many studies have explored how different structural, social, and physical environmental variables shape individuals' food behaviors (Story et al., 2008). Most of these studies have employed traditional statistical models (e.g., linear regression), which makes the identification of statistically relevant relationships difficult when there are many tiers of variables at play, when the effects are nonlinear, or when the relationships are subject to spatial non-stationarity (Stewart Fotheringham et al., 1996). Traditional models also fall short in identifying "what works for whom" when there are nuanced interactions between individual characteristics and the food environment (e.g., the effect of household's vehicle access on grocery shopping might vary by the communities' public transportation system). GeoAI has the potential to overcome these methodological limitations. For example, machine learning (ML) models, combined with interpretable measures (e.g., XGBoost (Chen & Guestrin, 2016) combined with GeoShapley values) (Li, 2024), can select and test important predictors from a large number of spatial and non-spatial variables. These models can account for various nonlinear and interactive effects, allowing researchers to evaluate the overall importance of a variable as well as a variable's contribution in any given local context (Abdel Abdel Magid et al., 2024). Other ML models, such as the generalized random forest or the causal (Wager & Athey, 2018; Athey et al., 2016), allow researchers to statistically evaluate the human–environment interactions and explore how the effect of individual characteristics varies across environmental contexts. These spatially explicit AI tools can offer valuable insights and allow researchers to generate a more nuanced understanding of the intricate dynamics among individuals, environments, and the health behaviors they collectively induce.

The second study area is sentiment analysis of food experiences. Food consumers' visits to a food store can be influenced by online reviews. These reviews provide a rich source of qualitative data that characterize place-based food experiences. However, traditional qualitative methods for analyzing this data can be subjective and labor-intensive. Instead of manually reading and scoring large volumes of text, natural language processing (NLP) technologies can be used to understand people's sentiments (e.g., perceptions and opinions) about places, including points of interest and neighborhoods (Hu et al., 2019). Many classic ML models, such as naïve Bayes, maximum entropy, and support vector machines (SVM), have been employed for sentiment analysis. These models typically analyze sentiment based on text features extracted from original documents, such as unigrams, bigrams, part-of-speech tags, and term frequency. However, these methods may not be effective in all situations. For instance, the sentence "What is this restaurant?" carries a negative connotation without using overtly negative terms (e.g., bad, terrible, worst), which can be challenging for traditional ML models to interpret. Recently, with the development of transformer-based language models like bidirectional encoder representations from transformers (BERT) (Devlin et al.,

2018) and RoBERT (Masala et al., 2020), as well as large language models, such as ChatGPT and Gemini, it has become possible to use these pretrained deep language representation models to uncover the hidden semantics in reviews and achieve higher accuracy in sentiment analysis. We believe these pretrained language models offer a new paradigm for understanding food experiences and feedback, thereby making food behaviors more explainable.

The third study area is the evaluation of food environmental perception. The association between the built environment and people's dietary behaviors has been studied in public health and consumer research (Rahmanian et al., 2014). However, existing measurements of the built environment are focused on the objective evaluation of the built environment, such as the accessibility of transportation facilities and the coverage of green space (Barthel et al., 2015). Yet, people's perceptions of the built environment can be closely linked to their dietary behaviors. For instance, perceptions of safety, aesthetics, and liveliness of a neighborhood can directly influence people's likelihood to visit local businesses (Zhang et al., 2018). These perceptual attributes relating to the built environment are difficult to measure due to the challenges in collecting broad-scale human subject data. In recent years, the abundance of geotagged street view images, combined with advancements in computer vision techniques, has provided a feasible solution to this problem. Crowdsourced environmental datasets (e.g., Google Street View, storefront images) and deep learning models have made it possible to evaluate people's perceptions of urban environments on a large scale (Zhang et al., 2018). This approach has already been applied in various fields related to human health (Larkin et al., 2022), and it holds promise for providing new insights into the assessment of human-perceived food environments and how such perceptions shape store patronization and eating patterns.

The fourth study area is the policy relevance, particularly relating to the menu labeling movement in the United States. There has been a growing shift in the American food culture, with more and more people choosing to dine out than cook at home (Jin et al., 2023). However, foods catered at restaurants are comparably calorie-rich and nutrient-poor than those prepared at home (Lin & Guthrie, 2012). To raise consumers' nutrition awareness and encourage healthy diets at restaurants, the U.S. Food and Drug Administration (FDA) proposed the menu labeling rule in December 2014, requiring restaurants and related food retail franchises with 20 or more establishments to disclose calorie information on food menus and menu boards (VanEpps et al., 2016). With rounds of political disputes, the final rule did not take full effect until May 7, 2018 (FDA, 2024). While the menu labeling rule is now implemented nationwide, it has two notable limitations: it only requires the disclosure of calorie content, while other nutritional information, including macronutrients and micronutrients, is not mandated; the rule does not apply to small, privately owned restaurants, making information disclosure at these establishments particularly challenging. In this context, the AI-based nutrition assessment has the potential to complement the calorie labeling policy by filling in the missing nutritional details on menus and estimating calorie information for non-franchised restaurants. Further justification and validation of the method will lead to the inception of a scalable and interactive nutrition assessment tool. It could encourage restaurant owners to adopt a dynamic and personalized approach to nutrition labeling, ultimately enabling consumers to make more informed, nutrition-conscious decisions.

12.7 Beyond the Hype: GeoAI at Its Limits

These opportunities come with certain limitations that we should be fully aware of.

Firstly, the performance of AI and ML models can vary greatly depending on the datasets, tasks, and spatial and temporal scopes involved. More advanced and complex AI models do not always guarantee better performance. Selecting the appropriate model to contextualize the food environment requires a deep understanding of the specific task and the underlying principles of the model. For example, the AI-based food image recognition model (i.e., Calorie Mama) used in our analysis is incapable of identifying portion sizes and multiple food items in an image. Comparing models and selecting the most suitable one for assessment is always necessary before implementation.

Secondly, biases in the training data, such as representation bias, sampling bias, and aggregation bias (Liu et al., 2022), can significantly impact model predictions and the conclusions. Even widely used pretrained ML models like ChatGPT and GPT-4V can exhibit biases related to race, gender, socioeconomic status, and geography (Mai et al., 2023; Manvi et al., 2024), leading to poor prediction results. For example, Calorie Mama is unable to detect many ethnic cuisines (e.g., Asian foods), likely due to a lack of training data and limited entries in the nutrient composition database for these cuisines. Additionally, online reviewers tend to favor high-end food establishments and stores in socially affluent neighborhoods. These biases must be carefully considered and addressed when employing AI models for assessment.

Thirdly, while there have been pioneering efforts to integrate spatial principles and address spatial inductive bias in GeoAI model development, such as spatial representation learning (Mai et al., 2021) and spatial-heterogeneity-aware deep neural networks (Xie et al., 2023), most existing GeoAI research still relies on general AI models to address geospatial problems. For example, the given case study is fundamentally the geovisualization of AI-derived nutrition assessment results; however, the "geo" component is not integrated into the development of a proprietary model to predict the outcome. Given the complexity and heterogeneity of geospatial datasets, understanding when and how to integrate spatial principles into AI model design to better grasp the complex relationships among environment, food, and health is a pressing issue.

12.8 Conclusions

The emergence of GeoAI signals a paradigm shift for geographers and public health researchers in unraveling the nexus of environment, food, and health—from a spatially oriented perspective to a context-driven, human-centered perspective. We argue that to deepen the understanding of the health effects permeating the food environment, it is essential to decipher the multiple environmental contexts in which a particular foodscape is formed. This exploration goes beyond a static food environment in space and must involve delving into the interaction between contexts and humans. Only by facilitating this transition can we fully elucidate the true effects of the food environment on community health.

References

Abdel Magid, H. S., Desjardins, M. R., & Yingjie, H. (2024). Opportunities and shortcomings of AI for spatial epidemiology and health disparities research on aging and the life course. *Health & Place, 89*(September), 103323. https://doi.org/10.1016/j.healthplace.2024.103323

Apparicio, P., Cloutier, M.-S., & Shearmur, R. (2007). The case of Montréal's missing food deserts: Evaluation of accessibility to food supermarkets. *International Journal of Health Geographics, 6*(1), 4. https://doi.org/10.1186/1476-072X-6-4

Athey, S., Tibshirani, J., & Wager, S. (2016). Generalized random forests. *arXiv*. https://doi.org/10.48550/ARXIV.1610.01271

Barthel, S., Parker, J., & Ernstson, H. (2015). Food and green space in cities: A resilience lens on gardens and urban environmental movements. *Urban Studies, 52*(7), 1321–1338. https://doi.org/10.1177/0042098012472744

Carletto, C., Zezza, A., & Banerjee, R. (2013). Towards better measurement of household food security: Harmonizing indicators and the role of household surveys. *Global Food Security, 2*(1), 30–40. https://doi.org/10.1016/j.gfs.2012.11.006

Caspi, C. E., Sorensen, G., Subramanian, S. V., & Kawachi, I. (2012). The local food environment and diet: A systematic review. *Health & Place, 18*(5), 1172–1187. https://doi.org/10.1016/j.healthplace.2012.05.006

CDC. (2012). Census Tract Level State Maps of the Modified Retail Food Environment Index (mRFEI). National Center for Chronic Disease Prevention and Health Promotion. *Division of Nutrition, Physical Activity, and Obesity*. Available online: https://stacks.cdc.gov/view/cdc/61367. Accessed on 25 March 2025.

Chaix, B., Méline, J., Duncan, S., Merrien, C., Karusisi, N., Perchoux, C., Lewin, A., Labadi, K., & Kestens, Y. (2013). GPS tracking in neighborhood and health studies: A step forward for environmental exposure assessment, a step backward for causal inference? *Health & Place, 21*, 46–51. https://doi.org/10.1016/j.healthplace.2013.01.003

Charreire, H., Casey, R., Salze, P., Simon, C., Chaix, B., Banos, A., & Oppert, J. M. (2010). Measuring the food environment using geographical information systems: A methodological review. *Public Health Nutrition, 13*(11), 1773–1785.

Chen, X. (2017). Take the edge off: A hybrid geographic food access measure. *Applied Geography, 87*, 149–159. https://doi.org/10.1016/j.apgeog.2017.07.013

Chen, T., & Guestrin, C. (2016). XGBoost: A scalable tree boosting system. In *Proceedings of the 22nd ACM SIGKDD international conference on knowledge discovery and data mining* (pp. 785–794). ACM. https://doi.org/10.1145/2939672.2939785

Chen, X., & Kwan, M.-P. (2015). Contextual uncertainties, human mobility, and perceived food environment: The uncertain geographic context problem in food access research. *American Journal of Public Health, 105*(9), 1734–1737. https://doi.org/10.2105/AJPH.2015.302792

Chen, Xiang, & Xining Yang. (2014). Does food environment influence food choices? A geographical analysis through 'Tweets.'" Applied Geography 51 (July):82–89. doi:https://doi.org/10.1016/j.apgeog.2014.04.003.

Chen, X., Johnson, E., Kulkarni, A., Ding, C., Ranelli, N., Chen, Y., & Ran, X. (2021). An exploratory approach to deriving nutrition information of restaurant food from crowdsourced food images: Case of Hartford. *Nutrients, 13*(11), 4132. https://doi.org/10.3390/nu13114132

Chen, X., Ye, X., Widener, M. J., Delmelle, E., Kwan, M.-P., Shannon, J., Racine, E. F., Aaron Adams, L., & Liang, and Peng Jia. (2022a). A systematic review of the modifiable areal unit problem (MAUP) in community food environmental research. *Urban Informatics, 1*(1), 22. https://doi.org/10.1007/s44212-022-00021-1

Chen, X., Zhao, B., & Yang, X. (2022b). The Obesogenity of restaurant food: Mapping the nutritional foodscape of Franklin County, Ohio using food review images. *Applied Geography, 144*, 102717. https://doi.org/10.1016/j.apgeog.2022.102717

Christian, W. J. (2012). Using geospatial technologies to explore activity-based retail food environments. *Spatial and Spatio-temporal Epidemiology, 3*(4), 287–295. https://doi.org/10.1016/j.sste.2012.09.001

Cummins, S., & Macintyre, S. (2002). Food deserts'—Evidence and assumption in health policy making. *BMJ (Clinical Research Ed.), 325*(7361), 436–438. https://doi.org/10.1136/bmj.325.7361.436

DataHaven. 2024. Community & neighborhood profiles. 2024. https://www.ctdatahaven.org/communities.

Devlin, J., Chang, M.-W., Lee, K., & Toutanova, K. (2018). BERT: Pre-training of deep bidirectional transformers for language understanding. *arXiv*. https://doi.org/10.48550/ARXIV.1810.04805

Donkin, A. J., Dowler, E. A., Stevenson, S. J., & Turner, S. A. (2000). Mapping access to food in a deprived area: The development of Price and availability indices. *Public Health Nutrition, 3*(1), 31–38. https://doi.org/10.1017/s1368980000000057

FDA. (2024). Menu labeling requirements. *U.S. Food & Drug Administration. FDA. March, 27*, 2024. https://www.fda.gov/food/food-labeling-nutrition/menu-labeling-requirements

Fleischhacker, S. E., Evenson, K. R., Rodriguez, D. A., & Ammerman, A. S. (2011). A systematic review of fast food access studies. *Obesity Reviews: An Official Journal of the International Association for the Study of Obesity, 12*(5), e460–e471. https://doi.org/10.1111/j.1467-789X.2010.00715.x

Fotheringham, A. S., & Wong, D. W. S. (1991). The modifiable areal unit problem in multivariate statistical analysis. *Environment and Planning A: Economy and Space, 23*(7), 1025–1044. https://doi.org/10.1068/a231025

Glanz, K., Sallis, J. F., Saelens, B. E., & Frank, L. D. (2005). Healthy nutrition environments: Concepts and measures. *American Journal of Health Promotion: AJHP, 19*(5), 330–333, ii. https://doi.org/10.4278/0890-1171-19.5.330

Goodchild, M. F., & Li, L. (2012). Assuring the quality of volunteered geographic information. *Spatial Statistics, 1*, 110–120. https://doi.org/10.1016/j.spasta.2012.03.002

Haslam, A., Nikolaus, C. J., & Sinclair, K.'. I. A. (2022). Association of Food Environment Characteristics with health outcomes in counties with a high proportion of native American residents. *Journal of Hunger & Environmental Nutrition*, 1–13. https://doi.org/10.1080/19320248.2022.2101414

Hu, Y., Deng, C., & Zhou, Z. (2019). A semantic and sentiment analysis on online neighborhood reviews for understanding the perceptions of people toward their living environments. *Annals of the American Association of Geographers, 109*(4), 1052–1073. https://doi.org/10.1080/24694452.2018.1535886

Huang, Y., Huang, D., & Nguyen, Q. C. (2019). Census tract food tweets and chronic disease outcomes in the U.S., 2015⁻2018. *International Journal of Environmental Research and Public Health, 16*(6), 975. https://doi.org/10.3390/ijerph16060975

Janowicz, K., Gao, S., McKenzie, G., Yingjie, H., & Bhaduri, B. (2020). GeoAI: Spatially explicit artificial intelligence techniques for geographic knowledge discovery and beyond. *International Journal of Geographical Information Science, 34*(4), 625–636. https://doi.org/10.1080/13658816.2019.1684500

Jin, A., Chen, X., Huang, X., Li, Z., Caspi, C. E., & Ran, X. (2023). Selective daily mobility bias in the community food environment: Case study of Greater Hartford, Connecticut. *Nutrients, 15*(2), 404. https://doi.org/10.3390/nu15020404

Kawano, Y., & Yanai, K. (2015). FoodCam: A real-time food recognition system on a smartphone. *Multimedia Tools and Applications, 74*(14), 5263–5287. https://doi.org/10.1007/s11042-014-2000-8

Larkin, A., Krishna, A., Chen, L., Amram, O., Avery, A. R., Duncan, G. E., & Hystad, P. (2022). Measuring and modelling perceptions of the built environment for epidemiological research using crowd-sourcing and image-based deep learning models. *Journal of Exposure Science & Environmental Epidemiology, 32*(6), 892–899. https://doi.org/10.1038/s41370-022-00489-8

Li, Z. (2024). GeoShapley: A game theory approach to measuring spatial effects in machine learning models. *Annals of the American Association of Geographers, 114*(7), 1365–1385. https://doi.org/10.1080/24694452.2024.2350982

Lin, Biing-Hwan, and Joanne Guthrie. 2012. "Nutritional quality of food prepared at home and away from home, 1977–2008."

Liu, C., Cao, Y., Luo, Y., Chen, G., Vokkarane, V., & Ma, Y. (2016). DeepFood: Deep learning-based food image recognition for computer-aided dietary assessment. In C. K. Chang, Y. Lorenzo Chiari, H. J. Cao, M. Mokhtari, & H. Aloulou (Eds.), *Inclusive smart cities and digital health* (Lecture notes in computer science) (Vol. 9677, pp. 37–48). Springer. https://doi.org/10.1007/978-3-319-39601-9_4

Liu, Z., Janowicz, K., Cai, L., Zhu, R., Mai, G., & Shi, M. (2022). Geoparsing: Solved or biased? An evaluation of geographic biases in geoparsing. *AGILE: GIScience Series, 3*(June), 1–13. https://doi.org/10.5194/agile-giss-3-9-2022

Lucan, S. C. (2015). Concerning limitations of food-environment research: A narrative review and commentary framed around obesity and diet-related diseases in youth. *Journal of the Academy of Nutrition and Dietetics, 115*(2), 205–212. https://doi.org/10.1016/j.jand.2014.08.019

Lyu, W., Seok, N., Chen, X., & Ran, X. (2023). Using crowdsourced food image data for assessing restaurant nutrition environment: A validation study. *Nutrients, 15*(19), 4287. https://doi.org/10.3390/nu15194287

Mai, G., Janowicz, K., Yingjie, H., Gao, S., Yan, B., Zhu, R., Cai, L., & Lao, N. (2021). A review of location encoding for GeoAI: Methods and applications. https://doi.org/10.48550/ARXIV.2111.04006

Mai, G., Yingjie, H., Gao, S., Cai, L., Martins, B., Scholz, J., Gao, J., & Janowicz, K. (2022). Symbolic and subsymbolic GEOAI : Geospatial knowledge graphs and spatially explicit machine learning. *Transactions in GIS, 26*(8), 3118–3124. https://doi.org/10.1111/tgis.13012

Mai, G., Huang, W., Sun, J., Song, S., Mishra, D., Liu, N., Gao, S., et al. (2023). On the opportunities and challenges of foundation models for geospatial artificial intelligence. *arXiv*. https://doi.org/10.48550/ARXIV.2304.06798

Manvi, Rohin, Marshall Burke, Samar Khanna, David Lobell, Gengchen Mai, and Stefano Ermon. 2024. "GeoLLM: Extracting geospatial knowledge from large language models."

Martin, C. K., Han, H., Coulon, S. M., Raymond Allen, H., Champagne, C. M., & Anton, S. D. (2009). A novel method to remotely measure food intake of free-living individuals in real time: The remote food photography method. *The British Journal of Nutrition, 101*(3), 446–456. https://doi.org/10.1017/S0007114508027438

Masala, M., Ruseti, S., & Dascalu, M. (2020). RoBERT—A Romanian BERT Model. In *Proceedings of the 28th international conference on computational linguistics* (pp. 6626–

6637). International Committee on Computational Linguistics. https://doi.org/10.18653/v1/2020.coling-main.581

Matthews, S. A., Moudon, A. V., & Daniel, M. (2009). Work Group II: Using geographic information systems for enhancing research relevant to policy on diet, physical activity, and weight. *American Journal of Preventive Medicine, 36*(4 Suppl), S171–S176. https://doi.org/10.1016/j.amepre.2009.01.011

McKinnon, R. A., Reedy, J., Morrissette, M. A., Lytle, L. A., & Yaroch, A. L. (2009). Measures of the food environment: A compilation of the literature, 1990-2007. *American Journal of Preventive Medicine, 36*(4 Suppl), S124–S133. https://doi.org/10.1016/j.amepre.2009.01.012

Mortaheb, R., & Jankowski, P. (2023). Smart city re-imagined: City planning and GeoAI in the age of big data. *Journal of Urban Management, 12*(1), 4–15. https://doi.org/10.1016/j.jum.2022.08.001

Payán, D. D., Rios, L. K. D., Ramírez, A. S., & Young, M.-E. D. T. (2021). Structural barriers influencing food insecurity, malnutrition, and health among Latinas during and after COVID-19: Considerations and recommendations. *Journal of the Academy of Nutrition and Dietetics, 121*(5), 837–843. https://doi.org/10.1016/j.jand.2021.01.005

Rahmanian, E., Gasevic, D., Vukmirovich, I., & Lear, S. A. (2014). The association between the built environment and dietary intake—A systematic review. *Asia Pacific Journal of Clinical Nutrition, 23*(2), 183–196. https://doi.org/10.6133/apjcn.2014.23.2.08

Sahoo, D., Hao, W., Ke, S., Xiongwei, W., Le, H., Achananuparp, P., Lim, E.-P., & Hoi, S. C. H. (2019). FoodAI: Food image recognition via deep learning for smart food logging. In *Proceedings of the 25th ACM SIGKDD international conference on knowledge discovery & data mining* (pp. 2260–2268). ACM. https://doi.org/10.1145/3292500.3330734

Schwartz, S. (1994). The fallacy of the ecological fallacy: The potential misuse of a concept and the consequences. *American Journal of Public Health, 84*(5), 819–824. https://doi.org/10.2105/ajph.84.5.819

Stewart Fotheringham, A., Charlton, M., & Brunsdon, C. (1996). The geography of parameter space: An investigation of spatial non-stationarity. *International Journal of Geographical Information Systems, 10*(5), 605–627. https://doi.org/10.1080/02693799608902100

Story, M., Kaphingst, K. M., Robinson-O'Brien, R., & Glanz, K. (2008). Creating healthy food and eating environments: Policy and environmental approaches. *Annual Review of Public Health, 29*(1), 253–272. https://doi.org/10.1146/annurev.publhealth.29.020907.090926

Suarez, J. J., Isakova, T., Cheryl, A. M., Anderson, L. E., Boulware, M. W., & Scialla, J. J. (2015). Food access, chronic kidney disease, and hypertension in the U.S. *American Journal of Preventive Medicine, 49*(6), 912–920. https://doi.org/10.1016/j.amepre.2015.07.017

Taylor, P. J. (1990). Editorial comment GKS. *Political Geography Quarterly, 9*(3), 211–212. https://doi.org/10.1016/0260-9827(90)90023-4

U.S. Census Bureau. (2024). QuickFacts: Hartford City, Connecticut. 2024. https://www.census.gov/quickfacts/fact/table/hartfordcityconnecticut,CT/PST045223.

USDA. (2023). Food security in the U.S. Economic Research Service U.S. Department of Agriculture. 2023. https://www.ers.usda.gov/topics/food-nutrition-assistance/food-security-in-the-u-s/.

USDA. (2024a). *Food access research atlas*. Economic Research Service U.S. Department of Agriculture. 2024. https://www.ers.usda.gov/data-products/food-access-research-atlas/.

USDA. (2024b). *Food environment atlas*. Economic Research Service U.S. Department of Agriculture. 2024. https://www.ers.usda.gov/data-products/food-environment-atlas/.

Van Asbroeck, S., & Matthys, C. (2020). Use of different food image recognition platforms in dietary assessment: Comparison study. *JMIR Formative Research, 4*(12), e15602. https://doi.org/10.2196/15602

VanEpps, E. M., Roberto, C. A., Park, S., Economos, C. D., & Bleich, S. N. (2016). Restaurant menu labeling policy: Review of evidence and controversies. *Current Obesity Reports, 5*(1), 72–80. https://doi.org/10.1007/s13679-016-0193-z

Wager, S., & Athey, S. (2018). Estimation and inference of heterogeneous treatment effects using random forests. *Journal of the American Statistical Association, 113*(523), 1228–1242. https://doi.org/10.1080/01621459.2017.1319839

Wang, D.-H., Kogashiwa, M., Ohta, S., & Kira, S. (2002). Validity and reliability of a dietary assessment method: The application of a digital camera with a mobile phone card attachment. *Journal of Nutritional Science and Vitaminology, 48*(6), 498–504. https://doi.org/10.3177/jnsv.48.498

Williamson, D. A., Allen, H. R., Davis Martin, P., Alfonso, A., Gerald, B., & Hunt, A. (2004). Digital photography: A new method for estimating food intake in cafeteria settings. *Eating and Weight Disorders: EWD, 9*(1), 24–28. https://doi.org/10.1007/BF03325041

Xie, Y., Chen, W., He, E., Jia, X., Bao, H., Zhou, X., Ghosh, R., & Ravirathinam, P. (2023). Harnessing heterogeneity in space with statistically guided meta-learning. *Knowledge and Information Systems, 65*(6), 2699–2729. https://doi.org/10.1007/s10115-023-01847-0

Xu, R., Huang, X., Zhang, K., Lyu, W., Ghosh, D., Li, Z., & Chen, X. (2023). Integrating human activity into food environments can better predict cardiometabolic diseases in the United States. *Nature Communications, 14*(1), 7326. https://doi.org/10.1038/s41467-023-42667-8

Yelp Fusion. (2024). Build standout consumer experiences with Yelp Data. 2024. https://fusion.yelp.com/.

Zhang, F., Zhou, B., Liu Liu, Y., Liu, H. H., Fung, H. L., & Ratti, C. (2018). Measuring human perceptions of a large-scale urban region using machine learning. *Landscape and Urban Planning, 180*, 148–160. https://doi.org/10.1016/j.landurbplan.2018.08.020

Chapter 13
GeoAI and Political Geography

Margath Walker and Jamie Winders

Abstract This chapter examines the growing, yet bifurcated, engagement in human geography with artificial intelligence (AI) as a research topic and with GeoAI as a research tool. Across its sections, it asks: How we can find bridges between these two efforts to integrate geography and AI, and what adjustments must geographers make now to ensure that scholarly conversations around GeoAI and around the geographies of AI can converge? In its main, the chapter reflects on what it means, on the one hand, to approach AI as a tool that helps us answer geographic questions and address geographic problems and, on the other, to frame AI as a phenomenon that transforms the focus of a given subfield (political geography, in this case). We draw attention to two areas where scholars of GeoAI and political geographers studying AI can find common ground and a common language: first, approaches to theories and second, understandings of our scholarship's impacts. To conclude, the chapter makes the case for getting comfortable with the uncomfortable and for being willing to work along and across concepts that might not translate fully across subfields, if we are to have a discipline-wide dialogue about AI, its applications, and its impacts.

Keywords GeoAI · Political geography · Ontology · Spatial analysis · Transdisciplinary collaboration

---

M. Walker
Departments of Geography and Geosciences and Urban and Public Affairs, University of Louisville, Louisville, KY, USA
e-mail: margath.walker@louisville.edu

J. Winders (✉)
Department of Geography and the Environment, Syracuse University, Syracuse, NY, USA
e-mail: jwinders@syr.edu

© The Author(s), under exclusive license to Springer Nature Switzerland AG 2025

X. Huang et al. (eds.), *GeoAI and Human Geography*, Springer Geography,
https://doi.org/10.1007/978-3-031-87421-5_13

13.1 Introduction

GeoXAI is social as well as technical. (Xing & Sieber, 2023, p. 637)

When we were asked to contribute a chapter to this edited collection on GeoAI and human geography, we sat down to reflect on what we knew about the topic. We had written about the geographies and spatialities of artificial intelligence (AI), particularly how a geographic perspective can bring new insights to the interdisciplinary literature on AI (Walker & Winders, 2021, 2023; Walker et al., 2021). One of us (Walker) oversees a research team pulling together case studies around the application of AI in various creative and extractive industries, including the urban bourbon industry and rural crypto-mining. Through these case studies, Walker's project investigates how AI is mediated by locational factors and how it transforms spatial landscapes. The other (Winders) founded an interdisciplinary research institute on the design/technology, policy/governance, and societal implications of autonomous systems and AI. Through this work, Winders spent time working with university faculty, industry leaders, and government officials around the thorny questions of how to think about and plan for the futures that AI might help create and shape.

After brainstorming about this chapter, however, we realized that despite our involvement with scholarship, teaching, and policy/industry work around AI, we knew very little about GeoAI as an emerging body of scholarship. Thus, we did what good scholars do and began a wider search for what geography journals had recently published on the topic. That effort identified a pattern on which the remainder of this chapter reflects—namely, a proliferation of publications on GeoAI in journals and outlets dedicated to GIS, urban analytics, and spatial information science and almost no publications on GeoAI in flagship human geography journals like *Progress in Human Geography, Annals of the American Association of Geographers*, or *Transactions of the Institute of British Geographers*. Three key journals in human geography—*Urban Geography, Antipode*, and *Political Geography*—had no publications at all on GeoAI. While these human geography outlets had little to say about GeoAI, all were home to a growing number of publications on the ethics, politics, spatialities, and geographies of AI.

What explained this pattern, and what does it mean for human geography's wider engagement with all things AI? Are we at risk, we wondered, of reproducing the divide observed during the early days of GIS, when geographers writing about the design and applications of GIS and geographers theorizing the assumptions and frameworks built into GIS published in different venues, undertook different research projects, and charted different courses through and around GIS as a topic (e.g., Cinnamon & Schuurman, 2013; O'Sullivan, 2006; Pickles, 1999)? Is human geography in danger of approaching AI, particularly GeoAI, in the same way? What might be the consequences of reproducing this pattern? More to the point, what would it take not to repeat those same patterns we saw (and often lamented) around GIS? What adjustments must we make now to ensure that scholarly conversations around GeoAI and around the geographies of AI, at best, converge and, at minimum, are in dialogue?

The remainder of this chapter considers these questions, asking what is required to create more meaningful and sustained conversations around GeoAI and geographies of AI and how we can create opportunities for these two growing fields to speak to each other and find points of overlap. After a brief discussion of GeoAI and its applications, we reflect on what it means, on the one hand, to approach AI as a tool that helps us answer geographic questions and address geographic problems and, on the other, to frame AI as an object of analysis that transforms the focus of a given subfield (political geography, in this case). What difference does it make, we ask, to treat AI as a tool that we use or as a thing that we study? From there, we draw attention to two key areas where, we suggest, scholars of GeoAI and political geographers studying AI can and should find common ground and a common language: first, approaches to theories and, second, understandings of our scholarship's impacts. To conclude, we make the case for getting comfortable with the uncomfortable and for being willing to work along and across concepts that might not translate fully across subfields, if we are to have a discipline-wide dialogue about AI, its applications, and its impacts.

13.2 GeoAI and Political Geography

There are far greater experts on GeoAI in this edited collection than we, so in this section, we highlight the aspects of GeoAI that resonate the most with us, as political and cultural geographers. As Li (2020) notes, GeoAI "sits at the junction of AI, geospatial big data, and high-performance computing (HPC) to provide a promising solution technology for data or compute-intensive geospatial problems" (p. 72). We return to the framing of GeoAI as a "solution technology" later in this chapter, but a key starting point for our understanding of GeoAI, and especially its relationship to political geography's engagement with AI, is acknowledging the difference between framing AI as a research tool and AI as an object of analysis.

Geographers whose scholarship focuses on GIS, urban analytics, and spatial science typically frame AI's merging with geography through the lens of GeoAI. GeoAI, for these scholars, is seen as a new technology that can help them address and potentially solve a series of urban and environmental problems, from traffic forecasting and urban planning (Liu et al., 2024) to public health challenges (Li, 2020) and crime behavior (Papadakis et al., 2022). Some geographers in this vein of scholarship raise key questions about not only the ethical issues that GeoAI prompts (Xing & Sieber, 2023; De Sabbata et al., 2023; Li et al., 2024) but also the complex geographic concepts that are hard to build into GeoAI systems (De Sabbata et al., 2023; Xing & Sieber, 2023; Janowicz et al., 2020). Nonetheless, many geographers writing about GeoAI remain optimistic about its value as a technology or tool for solving environmental and even social problems, with the right attention to geographic concepts and with a participatory framework sensitive to all communities and constituencies impacted by this technology.

Political geographers, by contrast, along with many human geographers in urban, social, and cultural geography, have taken a more circumspect approach to AI, approaching it less as a tool and more as a disruptive technology that has the potential to profoundly transform human life and livelihoods, not always for the better. A burgeoning literature on AI in political geography, for example, seeks to problematize AI's objectivity by signaling how AI systems are themselves immersed in spatial and societal structures (e.g., Goode, 2020; Lavallin & Downs, 2021). Political geographic perspectives on AI have usefully grounded emerging technologies in context to highlight how they might "touch down" across space and place and with what consequences for different groups, particularly marginalized communities (e.g., Alvarez Leon, 2024; McDuie-Ra & Gulson, 2020). When AI becomes part of the complex equation of managing territories and interactions across scales, this work argues, our understandings of the nature of political geography and geopolitical processes are potentially open to reinterpretation, suggesting that AI technologies themselves transfigure how we know the world.

Of course, part of making sense of GeoAI as an emerging area of scholarship and of its relationship to other streams of geographic scholarship on AI is figuring out whether GeoAI is similar to previous areas of geographic scholarship or previous waves of technological transformation. Our introduction, for example, posed the question of whether GeoAI may follow the path laid out by the emergence of GIS as a technology and of various strands of geographic scholarship around GIS. Xing and Sieber (2023) see potential parallels between Public Participation/Participatory Geographic Systems (P/PGIS) and the integration of explainable AI into GeoAI. There are rich opportunities here, we suggest, to spend more time thinking through how GeoAI does and does not seem similar to both previous technologies drawn into geographic analyses and previous areas of scholarship within our discipline. Along similar lines, there is the question of whether GeoAI is becoming its own subdiscipline or even discipline or whether it is better seen as a development within a given subfield. As Li et al. (2024) phrased it, "what may make it a long-lasting and high-impact scientific discipline instead of a 'local hotspot' in GIScience research" (p. 5)?

These questions around what, if any, past technology or phenomena AI is most like, as well as whether geographic engagement with AI might become a stand-alone subfield, are much more prevalent with the literature on GeoAI than in reflections on AI in political geography or other human geography subdisciplines. This difference in attention to where AI fits in the genealogy or projected future of a given subfield merits further reflection, as it suggests not only an uneven incorporation of AI into geographic scholarship across subfields but also different understandings of how AI may, or may not, transform future geographic scholarship more broadly. Mapping AI's trajectory through our discipline can tell us a great deal about how we understand emerging technologies as geographers, and this chapter provides a small step in that wider effort.

13.3 Same Words, Same Meanings?

Reading through the growing body of scholarship on GeoAI as political and cultural geographers felt both familiar and unfamiliar to us. On the one hand, this scholarship used geographic terms and concepts with which we are familiar and which we both have written about—scale, place, region, urban, theory, and so on. Scholars of GeoAI also grapple with some of the same foundational approaches seen in political geography. They, for example, frequently frame geography's defining feature as a perspective, not a particular topic of study, and point to the discipline's productive tension between nomothetic and ideographic approaches to knowledge production. Janowicz and company (2020) stress that "successful GeoAI research will have to address why (geo-)spatial matters by making a case for spatially explicit models" (p. 628), and similar arguments about the value of a geographic perspective can be found across political geography and other subdisciplines within human geography (e.g., Daniels, 2011; McFarlane, 2004; Rashidi et al., 2020).

Equally important, both political geographers engaging AI and GeoAI scholars are committed to articulating how and why AI itself cannot be applied or studied as an aspatial technology. For Li (2020), "the study of GeoAI should not merely involve a simple import of AI to geography" (p. 75). Political and social geographers writing about AI offer a similar sentiment. Vincent Del Casino and his co-authors (2020), for example, stress the geographies embedded in and produced by AI, arguing that AI shifts our perceptions and helps "create, our complex, continuously evolving, and contingent socio-spatial realities" (p. 611). In this way, both those studying GeoAI and political geographers writing about AI converge in their strong sense that AI cannot simply be added to either geographic analyses or geographic scholarship but, instead, must be approached through and within an explicitly spatial perspective.

These points of convergence aside, there are also key differences between scholarship on GeoAI and political geographic engagement with AI. While both fields use similar words and concepts, those terms often take on very different meanings in each area of research. Here, we highlight two such instances of differences between these areas: first, definitions of and approaches to theory and second, understandings of the impacts of our scholarship. We selected these two topics, both because they help us see where these two fields are currently pulling in different directions and because they point to what we see as potentially productive engagements between GeoAI experts and political geographers. As both groups grapple with how AI is transforming their respective subfields, geography, as an overall field, will be stronger if we can find ways to speak and work across the differences we describe below.

First, GeoAI scholars and political geographers think about and operationalize theory, or theories, in different ways. This distinction is not unique to these two areas of research and, in many ways, reflects much wider, deeper, and older debates within human geography around what theory means, how we do and should work with theory, and how geographers contribute to theory creation and refinement (e.g.,

Dixon & Jones, 1998; Tambassi, 2021; Ash, 2020). In the context of AI, however, a disruptive technology that has the potential to profoundly transform so many aspects of our individual and collective lives, the stakes around these conversations concerning theory seem particularly high, and the need to better understand these different uses of theory seems particularly urgent.

At a high level, GeoAI orients around what Li (2020) describes as "two major method classes: knowledge-driven, known as the top-down approach, and data-driven, known as the bottom-up approach" (p. 72). While the data-driven approach, associated most directly with machine learning, remains common and reflects the bottom-up approach to identifying patterns and solving problems that big data has long promised (e.g., Kitchin & McArdle, 2016; Resnyansky, 2019), the top-down, or ontological, approach relies on "a knowledge base to provide semantic definitions of real-world entities and their interrelationships" (Li, 2020, p. 73) and is growing in popularity, particularly vis-à-vis the knowledge graph.

We want to pause here to reflect in more detail on this use of "ontological." For GeoAI scholars, an ontological approach is one that begins with an established knowledge base or framework, which is then applied to a given situation and used to make sense of and order whatever problem, question, or situation is being studied. Here, ontology refers to the knowledge (about how the world works) that creates a shared set of ideas around a given topic (here, geographic data) for a given set of users and applications. Something that is ontological, in other words, orders and gives structure to knowledge about a topic in a way that is portable across domains or contexts. Ontologies, as Claramunt (2020) argues, "should abstract the world as it is" such that it offers "a meaningful representation of a given real-world domain: from fields to objects, from events to processes as well as causal to qualitative spatial and temporal relation" (p. 36).

This understanding of ontologies as a form of "abstracting" from the world "as it is" is different from how political geography, as well as other parts of critical human geography, understand and operationalize ontology as a concept. While there are political geographers whose work is grounded in a form of critical realism that aligns with the GeoAI approach to ontologies as a form of abstraction that is portable from situation to situation (e.g., Flint, 2010; Modelski, 1996), other strands of political geography, borrowing from wider veins of critical social theories, begin from the starting point of multiple ontologies, defined here as ideas about how the world must be in order for us to know it. From this perspective, the ontology used in GeoAI would be one of many potential ontological frameworks operationalized in political geography, each of which might be associated with different epistemological (how we come to know the world), methodological (how we study that world), and analytical (how we interpret that world) frameworks.

We, of course, are not the first to call attention to different approaches to theory and to baseline understandings of the world across subdisciplines or fields in geography. In the context of AI, however, these differences begin to carry more weight. Since some of AI's goals include approximating human intelligence and enabling AI systems to be able to make decisions, interpretations, and predictions with no or only limited human input, the question of which definition of ontology will shape GeoAI becomes more pressing. Particularly given conversations around counter AI, indigenous AI, and other efforts to create AI systems that reflect multiple ontologies and baseline assumptions about the world "as it is" (Mhlambi & Tiribelli, 2023; Janowicz et al., 2022; McQuillan, 2022), GeoAI scholars must be able to ask questions about and explain to other human geographers which ontologies are shaping the knowledge produced and consumed by AI systems and why those ontologies make the most sense for the question at hand. At the same time, political geographers must be able to think and speak across different understandings of ontology itself and find ways to translate GeoAI ontological approaches into political geography's engagement with the concept.

Second, although both GeoAI scholars and political geographers have strong commitments to engaging the "wicked" problems and complicated challenges seen in the world around us, the two bodies of scholarship diverge in how they think about and understand both the goals and impacts of their scholarly contributions. As already mentioned, GeoAI scholars often frame GeoAI as a "solution technology"—something poised not only to help us better understand problems but also to help us solve those problems through its predictive capability and its capacity to draw on often-disparate data sources that can proxy for the complex spatial relationships and dynamics that geographers study (De Sabbata et al., 2023). As Li (2020) notes, through the application of AI and deep learning, "complex hidden patterns can be revealed, new questions can be asked and answered, and new knowledge can be discovered" (p. 72). Through such work, the GeoAI model can be made "smarter at solving geospatial problems and developing artificial geospatial intelligence" (Li, 2020, p. 75).

Political geography, by contrast, while also focused on complex issues and the search for new questions and new knowledge, rarely positions that new knowledge as solving the problems or challenges that political geographers study—war, political struggles, violence, border conflicts, security, and so on. In most cases, political geography tackles what might be seen as intractable challenges and positions the arguments and "findings" of political geographic scholarship as advancing, transforming, refining, or challenging established ideas about those problems or challenges, rather than "solving" the problems or challenges themselves. For example, political geographic scholarship often centers on how power works at multiple scales and across territory and territoriality (e.g., Allen, 2016; Elden, 2007). Political geographers typically think about the relationships between people and their actions, which can take many forms including inclusion, exclusion, surveillance, and place-making itself, and the implication of such actions for the production of space (e.g., Mezzadra & Neilson, 2012; Smith et al., 2016). In recent years, the subdiscipline's theoretical turn has meant increased attention to the socially constructed nature of place and region in political geographers' reflection on the world around them (e.g., Reynolds, 1994; Paasi, 2022).

Undoubtedly, part of the distinction we are drawing here is the line between applied and basic research—a distinction that traces its way through both human and physical geography. Across the whole of geography, much ink has been spilled in normative arguments, debates, and programmatic statements about what kind of scholarship (with what kind of solutions) our field should offer (e.g., Aitken & Valentine, 2014; Sheppard, 2004; Smith, 1992; Tadaki et al., 2015; Warf & Sui, 2010). Again, however, the incorporation of AI into both political geography and GIS scholarship raises the stakes in these debates. The kinds of problems for which GeoAI is identified as a "solution technology" merit critical reflection by all parts of our discipline, including political geography. Could, or should, GeoAI, for example, be used to solve the problem of border conflicts? Of homelessness in cities? Of political polarization? Along similar lines, what would it mean for political geographers to reflect on problems or challenges that might be effectively addressed by GeoAI? What would that reflection look like, and what kinds of translations, conversations, and interactions across the lines between basic and applied research, between AI as a tool and AI as an object of study, would this work require?

13.4 Conclusion

> Looking ahead, the research agenda for GeoAI should aim to accompany its focus on spatially-explicit AI models – which is how GeoAI is most commonly understood, as a shorthand for geospatial artificial intelligence – with a broader purposeful aim to do geography with AI, a geographical artificial intelligence . . . Such an aim will require closer collaboration with our colleagues in other branches of geography. (De Sabbata et al., 2023, p. 2458–2459)

This chapter has attempted to think through what kinds of issues the "closer collaboration with our colleagues in other branches of geography" for which De Sabbata and co-authors call might require, in an effort to better link GeoAI scholars and political geographers. In doing so, it has focused on two aspects of our scholarship that, at the moment, may push in different directions, even while using similar language—theoretical terms and the desired impacts of our research. To close, we want to briefly highlight some areas where potential bridges are already in place between GeoAI and political geography, before closing with some recommendations on next steps.

A first potential bridge or pathway between GeoAI and political geography involves work on social sensing—"the use of (user-generated) digital content to better understand human dynamics" (Janowicz et al., 2020, p. 629). With links to the concept of place, a foundational idea across human geography, social sensing, along with critical attention to our "digital doubles" and digital footprints, has strong connections to key themes in political geography concerning, for example, surveillance, the state, smart borders, and biometrics (e.g., Everuss & Hsu, 2024; Andrejevic, 2019; Smith, 2016), as well as to other subfields like digital geographies. Such works could be productively linked to scholarship on user-generated data (Wu et al., 2019), to name just one example. Here, we see strong potential for generative partnerships between GeoAI and political geography.

A second potential bridge between GeoAI and political geographic work on AI involves the sentiment, shared by scholars in both areas, that geographers should not be "mere users of these models" but, instead, active contributors to both the use cases for AI and the wider study of AI and its impacts. Being active contributors to the study of AI's proliferation, as well as its ever-growing list of applications, means that we, as geographers, must foreground "fundamental geospatial research questions" (Li et al., 2024, p. 11) in our work. It also means, however, not only finding ways to speak across divides within our own field but also actively seeking ways to speak to scholars in the wide range of fields addressing this emerging technology.

Being part of and helping shape that wider interdisciplinary conversation means, in our view, getting comfortable with the uncomfortable in our own scholarship and getting comfortable with the unfamiliar from scholarship in other fields. Although Li (2020) notes a move toward "convergence research which requires the seamless integration of theories, knowledge, tools, and expertise across traditional discipline boundaries" to address "complex problems in a collaborative manner" (p. 75), our own foray into GeoAI did not feel seamless. Reading in a new field with new terms, new techniques, new authors, and new ways of producing knowledge can be exhilarating, but it can also be very uncomfortable. We definitely learned a great deal about GeoAI and have new questions and ideas for our own work. We also undoubtedly missed many of the key points, and certainly the nuances, of what we read, and we remain far from experts on the topic. If, however, we, as geographers, are to be contributors to, not just consumers of, AI technologies and if we are to be active participants in shaping how AI, as a technology, a set of practices, and a collection of world-changing possibilities, is understood, approached, created, managed, and sometimes resisted, we all have to get comfortable with being pushed out of our comfort zones as scholars.

Deepening scholarly work on GeoAI and, we would add, geographies of AI requires, as Li and his co-authors (2024) note, "a community effort" (p. 11). We are excited to be part of that community effort and look forward to seeing where it leads us. We hope that this chapter is a first step in truly embracing GeoAI as both technical and social, as Xing and Sieber (2023) suggest, and in finding and creating bridges and pathways across our discipline's engagement with this new technology.

References

Aitken, S. C., & Valentine, G. (Eds.). (2014). *Approaches to human geography: Philosophies, theories, people and practices*. Sage.

Allen, J. (2016). *Topologies of power: Beyond territory and networks*. Routledge.

Alvarez Leon, L. F. (2024). AI and global climate change: The political economy of data and energy in geographic perspective. *Geo: Geography and Environment, 11*(1), e00134.

Andrejevic, M. (2019). *Automating surveillance. Surveillance & Society, 17*(1/2), 7–13.

Ash, J. (2020). Flat ontology and geography. *Dialogues in Human Geography, 10*(3), 345–361.

Cinnamon, J., & Schuurman, N. (2013). Confronting the data-divide in a time of spatial turns and volunteered geographic information. *GeoJournal, 78*, 657–674.
Claramunt, C. (2020). Ontologies for geospatial information: Progress and challenges ahead. *Journal of Spatial Information Science, 20*, 35–41.
Daniels, S. (2011). Geographical imagination. *Transactions of the Institute of British Geographers, 36*(2), 182–187.
De Sabbata, S., Ballatore, A., Miller, H. J., Sieber, R., Tyukin, I., & Yeboah, G. (2023). GeoAI in urban analytics. *International Journal of Geographical Information Science, 37*(12), 2455–2463.
Del Casino Jr, V. J., House-Peters, L., Crampton, J. W., & Gerhardt, H. (2020). The social life of robots: The politics of algorithms, governance, and sovereignty. *Antipode, 52*(3), 605–618.
Dixon, D. P., & Jones, J. P. (1998). My dinner with Derrida, or spatial analysis and poststructuralism do lunch. *Environment and Planning A: Economy and Space, 30*(2), 247–260.
Elden, S. (2007). Governmentality, calculation, territory. *Environment and Planning D: Society and Space, 25*(3), 562–580.
Everuss, L., & Hsu, E. L. (2024). 12 modern bordering systems and digital identities: Data doubles, enrolment exclusion and the creation of digital polities. In *The De Gruyter handbook of artificial intelligence, identity and technology studies* (Vol. 1, p. 211). De Gruyter.
Flint, C. (2010). Geographic perspectives on world-systems theory. In R. Denemark (Ed.), *International Studies Association Compendium* (pp. 2828–2845). Wiley-Blackwell.
Goode, P. J. (2020). Artificial intelligence and the future of nationalism. *Nations and Nationalism*, 1–14. https://doi.org/10.1111/nana.12684
Janowicz, K., Gao, S., McKenzie, G., Hu, Y., & Bhaduri, B. (2020). GeoAI: Spatially explicit artificial intelligence techniques for geographic knowledge discovery and beyond. *International Journal of Geographical Information Science, 34*(4), 625–636.
Janowicz, K., Sieber, R., & Crampton, J. (2022). GeoAI, counter-AI, and human geography: A conversation. *Dialogues in Human Geography, 12*(3), 446–458.
Kitchin, R., & McArdle, G. (2016). What makes big data, big data? Exploring the ontological characteristics of 26 datasets. *Big Data & Society, 3*(1), 2053951716631130.
Lavallin, A., & Downs, J. A. (2021). Machine learning in geography: Past, present, and future. *Geography Compass*, 15.
Li, W. (2020). GeoAI: Where machine learning and big data converge in GIScience. *Journal of Spatial Information Science, 20*, 71–77.
Li, W., Arundel, S. T., Gao, S., Goodchild, M. F., Hu, Y., Wang, S., & Zipf, A. (2024). GeoAI for science and the science of GeoAI. *Journal of Spatial Information Science, 29*, 1–17.
Liu, P., Zhang, Y., & Biljecki, F. (2024). Explainable spatially explicit geospatial artificial intelligence in urban analytics. *Environment and Planning B: Urban Analytics and City Science, 51*(5), 1104–1123.
McDuie-Ra, D., & Gulson, K. (2020). The backroads of AI: The uneven geographies of artificial intelligence and development. *Area, 52*(3), 626–633.
McFarlane, C. (2004). Geographical imaginations and spaces of political engagement: examples from the Indian Alliance. *Antipode, 36*(5), 890–916.
McQuillan, D. (2022). *Resisting AI: An anti-fascist approach to artificial intelligence*. Policy Press.
Mezzadra, S., & Neilson, B. (2012). Between inclusion and exclusion: On the topology of global space and borders. *Theory, Culture & Society, 29*(4-5), 58–75.
Mhlambi, S., & Tiribelli, S. (2023). Decolonizing AI ethics: Relational autonomy as a means to counter AI harms. *Topoi, 42*(3), 867–880.
Modelski, G. (1996). Evolutionary paradigm for global politics. *International Studies Quarterly, 40*(3), 321–342.
O'Sullivan, D. (2006). Geographical information science: Critical GIS. *Progress in human geography, 30*(6), 783–791.
Paasi, A. (2022). Examining the persistence of bounded spaces: remarks on regions, territories, and the practices of bordering. *Geografiska Annaler: Series B, Human Geography, 104*(1), 9–26.

Papadakis, E., Adams, B., Gao, S., Martins, B., Baryannis, G., & Ristea, A. (2022). Explainable artificial intelligence in the spatial domain (X-GeoAI). *Transactions in GIS, 26*(6), 2413–2414.
Pickles, J. (1999). Arguments, debates, and dialogues: the GIS-social theory debate and the concern for alternatives. *Geographical information systems, 1*, 49–60.
Rashidi, Y., Ahmadypoor, Z., Alemi, A., & Bayat, M. (2020). The role of geographical imagination and geopolitical representation in dividing space/place into. *Geopolitics Quarterly, 16*(60), 79–100.
Resnyansky, L. (2019). Conceptual frameworks for social and cultural Big Data analytics: Answering the epistemological challenge. *Big Data & Society, 6*(1), 2053951718823815.
Reynolds, D. R. (1994). Political geography: the power of place and the spatiality of politics. *Progress in Human Geography, 18*(2), 234–247.
Sheppard, E. (2004). Practicing geography. *Annals of the Association of American Geographers, 94*(4), 744–747.
Smith, N. (1992). History and philosophy of geography: Real wars, theory wars. *Progress in human geography, 16*(2), 257–271.
Smith, G. J. (2016). Surveillance, data and embodiment: On the work of being watched. *Body & Society, 22*(2), 108–139.
Smith, S., Swanson, N. W., & Gökarıksel, B. (2016). Territory, bodies and borders. *Area, 48*(3), 258–261.
Tadaki, M., Brierley, G., Dickson, M., Le Heron, R., & Salmond, J. (2015). Cultivating critical practices in physical geography. *The Geographical Journal, 181*(2), 160–171.
Tambassi, T. (2021). *The philosophy of geo-ontologies: Applied ontology of geography*. Springer.
Walker, M., & Winders, J. (2021). Where is artificial intelligence? Geographies, ethics, and practices of AI. *Space and Polity, 25.2*, 163–166. With Margath Walker.
Walker, M., & Winders, J. (2023). Geographies of Artificial Intelligence: Labor, Surveillance, and Activism. *Human Geography*. https://journals.sagepub.com/doi/full/10.1177/19427786231208458
Walker, M., Winders, J., & Boamah, E. F. (2021). Locating artificial intelligence: A research agenda. *Space and Polity, 25*(2), 202–219.
Warf, B., & Sui, D. (2010). From GIS to neogeography: ontological implications and theories of truth. *Annals of GIS, 16*(4), 197–209.
Wu, X., Wang, J., Shi, L., Gao, Y., & Liu, Y. (2019). A fuzzy formal concept analysis-based approach to uncovering spatial hierarchies among vague places extracted from user-generated data. *International Journal of Geographical Information Science, 33*(5), 991–1016. https://doi.org/10.1080/13658816.2019.1566550
Xing, J., & Sieber, R. (2023). The challenges of integrating explainable artificial intelligence into GeoAI. *Transactions in GIS, 27*, 626–645.

Chapter 14
GeoAI and Health Geography

Changzhen Wang and Mengxi Zhang

Abstract This chapter offers a comprehensive overview of how geographic artificial intelligence (GeoAI) has been employed within health geography to address a wide spectrum of issues. First, it examines how GeoAI techniques have been used to estimate environmental exposures—such as air pollution and green spaces—and to measure human perceptions of the built environment, shedding light on their associations with various health outcomes. Second, it highlights GeoAI's role in analyzing social and environmental determinants of health, especially when dealing with large, complex datasets that traditional statistical methods find challenging. Third, it covers the integration of GeoAI into disease surveillance and forecasting systems, enabling earlier detection of outbreaks and improved resource allocation. Throughout the chapter, we discuss the theoretical and methodological hurdles facing GeoAI research—such as the "black box" nature of some models, data privacy concerns, and the need for spatially explicit, explainable models. We also identify future directions, including the development of more interpretable GeoAI tools, improved data integration, and stronger interdisciplinary collaborations. Overall, this chapter provides insights into current advancements, challenges, and the potential of GeoAI in health geography.

Keywords GeoAI · Health geography · Environmental exposures · Social determinants · Disease surveillance

C. Wang (✉)
Department of Geography and the Environment, The University of Alabama, Tuscaloosa, AL, USA
e-mail: cwang94@ua.edu

M. Zhang
Department of Health Systems and Implementation Science, Carilion School of Medicine, Virginia Tech, Roanoke, VA, USA
e-mail: mengxizhang@vt.edu

© The Author(s), under exclusive license to Springer Nature Switzerland AG 2025

X. Huang et al. (eds.), *GeoAI and Human Geography*, Springer Geography,
https://doi.org/10.1007/978-3-031-87421-5_14

14.1 Introduction

Health geography uses geographic data, methods, technologies, and perspectives to study geographic variations in health, disease, and health care within social, cultural, and environmental contexts (Moon, 2020). Research interest in health geography has surged rapidly in the wake of the COVID-19 pandemic due to its significant global impact and the growing awareness of geographic influence on disease transmission and health care. Numerous health care and health-related data including medical claims, electronic health records (EHRs), environmental data, mobility data, satellite and street view imagery, social media, and survey data become more accessible than ever. To explore their value and better address existing or potential key questions and research gaps in health geography, a series of novel spatial models has been developed (Georganos et al., 2021; Georganos & Kalogirou, 2022; Wang & Wang, 2022; Wang et al., 2021, 2024). Among them, geographic artificial intelligence (GeoAI), as the integration of geospatial data and analysis with machine learning (ML) and deep learning (ML) (Gao et al., 2023), has received more attentions. A growing number of research seek to harness the power of GeoAI to better understanding the spatial patterns and variations of health outcomes and provide better insights for informing more effective health policies.

As the rise of GeoAI shifts the research paradigm in health geography, this chapter aims to provide a critical overview of how and to what extent GeoAI has been used to address critical issues in health geography. Note that the term GeoAI used in this chapter refers to AI models used for spatial data or spatialized AI models. Based on existing literature, our discussion will focus on three aspects directly related to geography of health outcome or disease outbreak detection (see Fig. 14.1). The first aspect will explore how GeoAI are used to estimate and

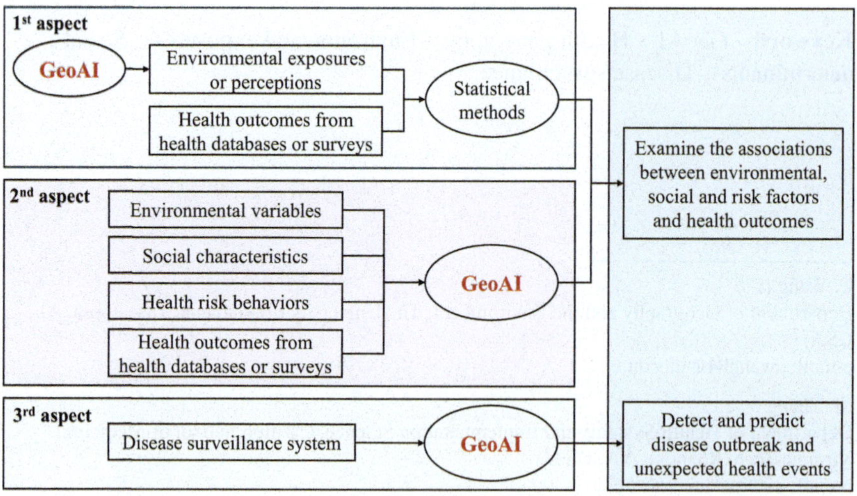

Fig. 14.1 GeoAI applications in health geography

predict environmental exposures or individual's perceptions of those exposures. The second aspect will examine using GeoAI as alternatives to traditional statistical models to directly investigate the relationships between social, risk behaviors, and environmental factors and health outcomes, hereafter refer to social determinants of health (SDOH). The third aspect will discuss how GeoAI are used to detect and forecast diseases. Due to limited space, we may not explain all GeoAI or statistical models or cover all aspects of this topic. However, we believe these comprehensive discussions capture the main applications of GeoAI in health geography. We hope that this chapter will inspire readers to better understand the potential of GeoAI in addressing health issues and the challenges or directions that may have been or would be raised in the near future.

14.2 Estimate Environmental Exposures and Perceptions

14.2.1 Motivation for Using GeoAI

Long-term environmental and neighborhood exposures, such as air pollutants, water contamination, heat, extreme weather, hazards, and built environments have disproportionately influenced human health and well-being, particularly for those living in historically and economically vulnerable communities. It was reported that around 24% of global deaths and disease burden is attributable to environmental causes (Prüss-Üstün et al., 2016). Identifying these modifiable environmental exposure factors is critical to develop better medicine and treatment plans and inform actionable policies to mitigate their potential health risks.

However, the assessment of environmental exposures or perceptions has historically been difficult and costly, often with insufficient coverage or generating limited amounts of data at a low level of granularity. For example, monitoring stations of air pollutants such as particulate matter (PM) and ozone are sparsely distributed in urban areas and only record data every several days. The limited data may not adequately and accurately capture the real spatial or temporal distributions of these pollutants (Chen et al., 2023; Reid et al., 2019). Field audits, interviews, surveys, and questionnaires are often used to measure human perceptions or individuals' feeling of the built environments. However, these approaches are time-consuming and laborious, and they only cover a small sample that may have (un)conscious biases. To address these challenges, many research seek to combine these small samples with GeoAI to predict different pollutants at finer resolutions, characterize micro-level built environments, or measure large-scale human perceptions, and link them with health data to examine their associations.

14.2.2 Estimate Pollutants

Existing studies have used various data, such as PM 2.5 monitoring data, satellite-based aerosol optical depth, meteorological and land use data, to train artificial neural network (ANN) (Di et al., 2017), geographically weighted regression (GWR) (Wu et al., 2020), generalized boosting model (Reid et al., 2016, 2019), or ensemble ML model (Chen et al., 2023) for predicting daily PM 2.5 and/or ozone concentrations at different finer resolutions across the U.S. By aggregating them to ZIP codes or counties, and combing with medical data, area-level covariates, or demographic data, these studies further used statistical methods such as Cox proportional-hazards model (Di et al., 2017), negative binominal mixed model (Wu et al., 2020), Poisson generalized estimating equations (Reid et al., 2016, 2019), and quasi-Poisson regression (Chen et al., 2023) to examine their associations with Medicare mortality (Di et al., 2017), Covid-19 mortality (Wu et al., 2020), respiratory hospitalization and emergency department visits (Chen et al., 2023; Reid et al., 2016, 2019), respectively. The four statistical models are developed to deal with survival or count data with well-established theories and assumptions. In contrast, the four ML models are more conducive to handling more complex and larger datasets that may have nonlinearities. All studies demonstrated that higher PM 2.5 exposure was positively associated with higher mortalities or more healthcare utilizations. These findings can help inform immediate and important policy actions such as prioritizing prevention resources to highly polluted areas or offering more scientific evidence to lower the national air quality standards (Di et al., 2017; Wu et al., 2020).

14.2.3 Characterize Built Environments

Most of the existing research has used various DL models to extract micro-level or eye-level features from thousands or millions of street view images and linked them to surveys or questionnaires to analyze their associations with various health outcomes. To list some, Kim et al. (2023) adopted the Deeplab V3+ to segment 261,290 Google's street view (GSV) images to identify visual environments including street greenery and sidewalk pavement around the home locations of 1000 respondents in Seoul, Korea. By integrating macro-level neighborhood walkable environments, health behaviors, and sociodemographic variables, the study further used structural equation models to examine their relationship with survey obesity. Although Maharana and Nsoesie (2018) conducted a similar research in four US urban areas, they used a different DL model, termed VGG-CNN-F network to extract 125 features of the built environments (e.g., parks and housing types) from 150,000 GSV images. After aggregating these features to census tracts, they applied Elastic Net, a regularized regression method designed to prevent overfitting of high-dimensional data, to quantify the associations between the built environments and obesity prevalence extracted from the behavioral risk factor surveillance system (BRFSS). Xiao et al. (2021) used the FCN-8s to segment green vegetations from 235,928 Tencent's street view (TSV) images. By linking their average proportions to the satellite-based NDVI, predicted air pollutants, and sociodemographic characteristics of 24,845 participants from a health survey, the study further applied generalized linear mixed regression models to examine their associations with obesity-related indicators such as body mass index (BMI). The FCN-8s model was also used by other researchers to segment streetscape green and blue spaces from TSV images (Helbich et al., 2019; Yu et al., 2021). In combination with questionnaires or surveys, air pollutants and others, they were further used to analyze their associations with geriatric depression (Helbich et al., 2019) and childhood asthma (Yu et al., 2021). All three DL models, Deeplab V3+, VGG-CNN-F, and FCN-8s, are developed on the framework of convolutional neural networks (CNNs) with different architectures and complexities for semantic segmentation. Regarding their findings, unlike the negative effects of air pollutants, these built or natural environments have positive effects on various health outcomes which may serve as important evidence for promoting outdoor activities in healthy, walkable, and green environments to reduce obesity, depression, and asthma and informing better urban design.

14.2.4 *Measure Human Perceptions of the Built Environment*

Other studies have harnessed GeoAI to measure human perceptions of the neighborhood built or natural environments where they live, work, or play. For instance, Wang et al. (2019a) applied the FCN-8s to segment 150 types of common features (e.g., trees and buildings) from 134,778 TSV images and used random forest (RF) model on sampled perceptual scores of 30 volunteers to predict six perceptual scores of TSV images (e.g., safety, wealth, and depression). By linking these perceptual scores and macro-level neighborhood covariates to the survey of 1231 seniors in Beijing, China, the study further used multilevel regression models to examine the associations between six perceptions of neighborhood environments and self-reported physical and mental health outcomes (i.e., health condition, depression, and anxiety). Only perceived safety was significantly associated with both physical and mental health outcomes. The authors conducted a similar study in Guangzhou, China, and confirmed that perceived neighborhood safety was positively associated with self-reported mental health (Wang et al., 2019b). Instead of using self-reported health outcomes from surveys, Xu et al. (2023) applied a different CNN model to estimate six neighborhood environmental perceptions in Wuhan, China, and extracted cardiovascular patients from medical records to examine their associations through negative binomial regression models. While perceived wealth was negatively associated with cardiovascular prevalence among senior populations, the perceived depression and vitality had a positive association.

14.3 Examine Social Determinants of Health Outcomes

14.3.1 Motivation for Using GeoAI

Social determinants of health (SDOH) are any non-medical factor that influences health outcomes (U.S. Department of Health and Human Services (HHS), 2024b). As key drivers of health inequities, SDOH have gained increased attention globally. The World Health Organization (WHO, 2024) has incorporated SDOH into sustainable development goals (SDG) to assess the achievement of health equity and improvement of global health. The United States has launched the Healthy People 2030 initiative which features SDOH with health equity and health literacy as three priority areas (HHS, 2024a). SDOH encompass five key areas: economic stability, education access and quality, health-care access and quality, social and community context, and neighborhood and built environments (HHS, 2024b). Because each of them has numerous spatial and non-spatial factors, examining their relationships with health outcomes may need to process large datasets with intensive computing and account for nonlinearity and spatial non-stationarity. Using traditional statistical methods may not always be effective. Therefore, many researchers have sought to apply more flexible GeoAI models as alternatives to examine the interplay of multiple determinants in shaping health outcomes. This section will focus on summarizing how GeoAI models are used to examine the associations between a wide range of environmental, social, and behavioral risk factors and health outcomes.

14.3.2 Examine the Associations Between Environmental Exposures and Health Outcomes

Research in this line has attempted to collect numerous variables and use or improve popular ML algorithms to investigate the combined synergetic effects of different environmental exposures on predicting health outcomes, such as cancer, COVID-19, and self-reported health. These studies, while considering fewer social and risk factors, have favored environmental exposures. The most common environmental exposure is air pollution, followed by chemicals, water pollution, temperature, and living conditions. Unlike most environmental exposures that are identified or interpolated from publicly available data sources, most health outcome variables are obtained from surveys or official health databases. For instance, Cazzolla Gatti et al. (2023) applied k-means clustering and RF models to explore possible links between 35 environmental pollutions (e.g., air pollutants and chemicals), socioeconomic factors, and cancer mortality at the regional and province levels in Italy. Here cancer mortality was extracted from the Italian Cancer Registers. The authors conducted a similar study examining the importance of air quality on SARS-CoV-2 mortality and infectivity in Italy (Cazzolla Gatti et al., 2020). Both found that long-term exposure to air pollutants was associated with mortality.

Although RF is an effective, generalizable, and non-spatial ML algorithm, it struggles with handling count or spatially heterogeneous data. To overcome these limitations, Lee et al. (2024) incorporated Poisson regression into RF to examine the relationship between county-level radiation and non-radiation exposures (e.g., indoor radon and air quality), health risk behaviors, demographic and socioeconomic factors, and lung cancer incidence rates extracted from the Surveillance, Epidemiology, and End Results (SEER) database in the U.S. Labib (2024) used RF and spatial RF models to explore the exposure effects of NO_2, greenness, and temperature on premature mortality and morbidity extracted from statistics in Greater Manchester. Gu et al. (2023) used geographically weighted random forest (GWRF) model to examine the spatial associations between 21 surface water pollutants and six digestive cancer incidences obtained from China Cancer Registry. Similar to the findings in Italy (Cazzolla Gatti et al., 2020, 2023), air pollution ranked the most influential factor in predicting lung cancer incidence (Lee et al., 2024) and premature mortality (Labib, 2024). The common water pollutants, fluoride and arsenic, were associated with most digestive cancer incidences such as stomach and colorectal cancer (Gu et al., 2023). In addition to the RF, XGBoost is another well-known ensemble ML algorithm that has been used to explore the joint effects of industrial pollutants such as mercury and lead on provincial lung cancer incidence in China (Wang et al., 2024) and the impact of living quality on self-reported health outcomes from surveys in the Netherlands (Viljanen et al., 2022).

14.3.3 Examine the Associations Between Social and Risk Factors and Health Outcomes

Research in this line tends to use well-known ML algorithms to examine the combined effects of various social and behavioral risk factors on health outcomes or costs using county or census tract as a unit of analysis. For example, Dong et al. (2022a) used classification and regression tree (CART) and RF to identify the geographic patterns of phenotypes, a combination of various socioeconomic and risk factors for late-stage breast cancer identified from the SEER database in 680 U.S. counties. Using the same database, the lead author collaborated with other researchers and applied RF to evaluate and compare the associations between county-level social and behavioral risk factors including race/ethnicity, healthcare access, health status, and socioeconomic status, and colorectal cancer incidence rates at different ages (Dong et al., 2023a). Later on, a series of similar studies were conducted to identify geographic variations and risk factors associated with county-level cardio-oncology mortality and premature cardiovascular mortality extracted from the Centers for Disease Control and Prevention (CDC) WONDER database (Dong et al., 2023b; Motairek et al., 2023). In addition to CART and RF, other studies used XGBoost to identify social determinants of county-level cardiovascular disease prevalence and care costs in the USA (Sun et al., 2023). Because using a single ML algorithm may not adequately show its performance relative to other ML or statistical models, some studies have applied and compared multiple ML algorithms, such as RF, gradient boosting machine (GBM), support vector machine (SVM), and regression tree with traditional statistical methods to examine social and risk factors associated with county-level obesity and lower respiratory infections and block-group level drug overdose, respectively (Bozorgi et al., 2021; Mollalo et al., 2020; Scheinker et al., 2019). In general, ML algorithms have some advantages in achieving higher accuracy and explaining more variations in health outcomes, but they do not have readily coefficients for variables, thus are less interpretable than traditional statistical methods.

Despite this concern, some recent studies have developed some spatialized ML algorithms, such as the geographic random forest (GRF) (Georganos et al., 2021) and previously mentioned GWRF (Georganos & Kalogirou, 2022). The GRF and GWRF are local and nonlinear GeoAI models. Both are built on the concept of the classic GWR (Fotheringham et al., 2009) to decompose the RF into several sub-models for spatial data while accounting for nearby observations in a predefined bandwidth. A difference is that GWF assigns equal weights to all data within the same bandwidth whereas the GWRF assigns different weights that allow observations with higher weights to be more likely selected in the sub-models. The early released GRF has been used to assess the associations between various socioeconomic and behavioral risk factors and the SEER-based cancer mortality across the US counties (Dong et al., 2022b), census-tract-level self-reported hypertension in Chicago (Lotfata et al., 2023a), and EHR-based, census tract level emergency and inpatient visits in North Carolina (Bhavsar et al., 2018), respectively. The more recently developed GWRF has also been used to examine socioeconomic and behavioral risk determinants of asthma and diabetes in the US counties (Lotfata et al., 2023b; Quiñones et al., 2021) and self-reported physical and mental health in Australia (Wang et al., 2023).

14.4 Detect and Forecast Diseases

14.4.1 Support Disease Surveillance Systems

Disease surveillance systems continuously and systematically collect, analyze, and interpret data for disease and injury prevention and control (Thacker & Berkelman, 1988). Traditional surveillance systems rely on passive routing reporting by people in healthcare facilities and diagnostic laboratories which is lengthy and may potentially delay the detection of disease outbreaks.

The advent of GeoAI addresses this issue as it is able to harness rich data from social media such as Twitter data, search engines such as Google, crowdsourcing, and other sources to predict disease outbreaks and provide early warnings prior to official reports (Aiello et al., 2020). For instance, Allen et al. (2016) used a SVM to distinguish between tweets reflecting cases of influenza and tweets irrelevant to actual illness. Thapen et al. (2016) introduced DEFENDER system that integrates social and news media data while incorporating GeoAI for outbreak detection, situational awareness, and forecasting. Zhang et al. (2022) used fine-tuning bidirectional encoder representation from transformers (BERT) to classify health-related Twitter data and developed an intelligent COVID-19 early warning system. Venkatramanan et al. (2021) created a ML-based anonymized mobility map that aggregated smartphone-based mobility data to evaluate its effectiveness in forecasting influenza epidemics in the USA and Australia. In addition, other studies applied long short-term memory (Ayyoubzadeh et al., 2020), vector error correction model (Turk et al., 2021), and RF (Peng et al., 2021), principal component analysis and time series modeling (Rabiolo et al., 2021) to explore the potential of Google Trends for COVID-19 surveillance and outbreak prediction. Most of those models performed well in predicting diseases in advance, demonstrating the effectiveness of using GeoAI to develop disease surveillance systems.

In addition, federal agencies and industries have led the efforts in developing disease surveillance systems. For example, CDC has supported a collaborative influenza forecasting initiative, called FluSight that integrated various statistical, mechanistic, AI or ML models to more accurately predict influenza hospital admissions in the USA (Mathis et al., 2024). AI companies like BlueDot (2024) and Metabiota (2024) also leverage various ML algorithms to monitor outbreaks of various infectious diseases such as coronavirus, HIV, and tuberculosis across different regions globally for early warning.

14.5 Potential Challenges and Future Directions

As GeoAI continues to contribute to the development of health geography, it is important to summarize potential challenges to guide future research directions.

The first noteworthy challenge is theoretical innovation. Currently, many relevant applications are data- or model-driven. They tend to explore geographic patterns in massive datasets or study the associations with health care and diseases to make predictions. While these applications are expected to engage or advance theories in geography or GIScience, like other geospatial research, related theories are missing or fall behind the development of methodology and technology. In addition, there are some concerns about the explainability and generalizability of GeoAI as the mechanistic processes of generating results from input variables via GeoAI models and the model's adaptability from one study area to another remain unclear (Hu et al., 2024). Given the unique insights generated by existing research, there is a need to distill and systematically summarize the findings to create new general principles or develop new conceptual frameworks to transform our understanding of the world.

The second noteworthy challenge is GeoAI models. While most research tends to use AI models to spatial data to address health geography, relatively fewer studies have developed spatially explicit GeoAI models. Interestingly, studies on estimating environmental exposures or human perceptions favor DL models while SDOH, health and disease monitoring, and forecasting studies prefer non-DL-based ML models. Despite the fact that GeoAI can better handle nonlinear relationships, spatially heterogeneity, high-dimensional data, or missing data with less restrictive assumptions, many GeoAI models are still considered "black boxes." The lack of transparency poses challenges for interpretability and may discourage individuals from using these models. Fortunately, there is a growing body of research focusing on developing explainable GeoAI, and we hope that these models can also benefit health geography research. In addition, it is not common to evaluate and compare the performance of different GeoAI models used in health geography. The fairness and transferability of these models may require further investigation.

The third challenge is data privacy and integration. Using GeoAI in health geography requires extensive data. Current research either combines GeoAI predicted big data with survey-based small data or feeds massive data to GeoAI models to study social or environmental impacts on health or disease outbreaks. Among these data, medical data can be particularly challenging to access due to privacy, confidentiality, or cost constraints. Most of them are only provided at the aggregated levels and lack detailed patient characteristics. Traditional surveys, although collected at the individual level, are usually not representative due to challenges to implement randomly sampling strategies. Additionally, data collected from different sources may be presented in diverse formats with different resolutions, structures, and accuracies, their integration may be complex and full of uncertainty. Interdisciplinary collaborations may be needed to address these issues.

14.6 Conclusion

GeoAI is revolutionizing the field of health geography by integrating spatial intelligence to enhance our understanding of health outcomes. By analyzing the relationship between environment and social determinants and human health, as well as exploring health events and disease monitoring and prediction, GeoAI provides a deeper understanding of how these factors influence health outcomes with a stronger analytical power. However, challenges remain in areas such as theoretical innovation, model development, and privacy protection, which must be addressed as we continue to explore this emerging field.

References

Aiello, A. E., Renson, A., & Zivich, P. (2020). Social media-and internet-based disease surveillance for public health. *Annual Review of Public Health, 41*, 101.

Allen, C., Tsou, M.-H., Aslam, A., Nagel, A., & Gawron, J.-M. (2016). Applying GIS and machine learning methods to Twitter data for multiscale surveillance of influenza. *PLoS One, 11*(7), e0157734. https://doi.org/10.1371/journal.pone.0157734

Ayyoubzadeh, S. M., Ayyoubzadeh, S. M., Zahedi, H., Ahmadi, M., & S, R. N. K. (2020). Predicting COVID-19 incidence through analysis of google trends data in Iran: Data mining and deep learning pilot study. *JMIR Public Health and Surveillance, 6*(2), e18828. https://doi.org/10.2196/18828

Bhavsar, N. A., Gao, A., Phelan, M., Pagidipati, N. J., & Goldstein, B. A. (2018). Value of neighborhood socioeconomic status in predicting risk of outcomes in studies that use electronic health record data. *JAMA Network Open, 1*(5), e182716–e182716. https://doi.org/10.1001/jamanetworkopen.2018.2716

BlueDot. (2024). BlueDot.. Retrieved November 24 from https://bluedot.global/

Bozorgi, P., Porter, D. E., Eberth, J. M., Eidson, J. P., & Karami, A. (2021). The leading neighborhood-level predictors of drug overdose: A mixed machine learning and spatial approach. *Drug and Alcohol Dependence, 229*, 109143. https://doi.org/10.1016/j.drugalcdep.2021.109143

Cazzolla Gatti, R., Velichevskaya, A., Tateo, A., Amoroso, N., & Monaco, A. (2020). Machine learning reveals that prolonged exposure to air pollution is associated with SARS-CoV-2 mortality and infectivity in Italy. *Environmental Pollution, 267*, 115471. https://doi.org/10.1016/j.envpol.2020.115471

Cazzolla Gatti, R., Di Paola, A., Monaco, A., Velichevskaya, A., Amoroso, N., & Bellotti, R. (2023). The spatial association between environmental pollution and long-term cancer mortality in Italy. *Science of the Total Environment, 855*, 158439. https://doi.org/10.1016/j.scitotenv.2022.158439

Chen, A. I., Ebisu, K., Benmarhnia, T., & Basu, R. (2023). Emergency department visits associated with wildfire smoke events in California, 2016–2019. *Environmental Research, 238*, 117154. https://doi.org/10.1016/j.envres.2023.117154

Di, Q., Wang, Y., Zanobetti, A., Wang, Y., Koutrakis, P., Choirat, C., Dominici, F., & Schwartz, J. D. (2017). Air pollution and mortality in the Medicare population. *New England Journal of Medicine, 376*(26), 2513–2522. https://doi.org/10.1056/NEJMoa1702747

Dong, W., Bensken, W. P., Kim, U., Rose, J., Berger, N. A., & Koroukian, S. M. (2022a). Phenotype discovery and geographic disparities of late-stage breast cancer diagnosis across U.S. counties: A machine learning approach. *Cancer Epidemiology, Biomarkers & Prevention, 31*(1), 66–76. https://doi.org/10.1158/1055-9965.Epi-21-0838

Dong, W., Bensken, W. P., Kim, U., Rose, J., Fan, Q., Schiltz, N. K., Berger, N. A., & Koroukian, S. M. (2022b). Variation in and factors associated with US county-level cancer mortality, 2008-2019. *JAMA Network Open, 5*(9), e2230925–e2230925. https://doi.org/10.1001/jamanetworkopen.2022.30925

Dong, W., Kim, U., Rose, J., Hoehn, R. S., Kucmanic, M., Eom, K., Li, S., Berger, N. A., & Koroukian, S. M. (2023a). Geographic variation and risk factor association of early versus late onset colorectal cancer. *Cancers, 15*(4), 1006. https://www.mdpi.com/2072-6694/15/4/1006

Dong, W., Motairek, I., Nasir, K., Chen, Z., Kim, U., Khalifa, Y., Freedman, D., Griggs, S., Rajagopalan, S., & Al-Kindi, S. G. (2023b). Risk factors and geographic disparities in premature cardiovascular mortality in US counties: A machine learning approach. *Scientific Reports, 13*(1), 2978. https://doi.org/10.1038/s41598-023-30188-9

Fotheringham, A. S., Brunsdon, C., & Charlton, M. (2009). Geographically weighted regression. In *The Sage handbook of spatial analysis* (Vol. 1, pp. 243–254). Sage.

Gao, S., Hu, Y., & LI, W. (2023). *Handbook of geospatial artificial intelligence*. CRC Press.

Georganos, S., Grippa, T., Niang Gadiaga, A., Linard, C., Lennert, M., Vanhuysse, S., Mboga, N., Wolff, E., & Kalogirou, S. (2021). Geographical random forests: A spatial extension of the random forest algorithm to address spatial heterogeneity in remote sensing and population modelling. *Geocarto International, 36*(2), 121–136. https://doi.org/10.1080/10106049.2019.1595177

Georganos, S., & Kalogirou, S. (2022). A forest of forests: A spatially weighted and computationally efficient formulation of geographical random forests. *ISPRS International Journal of Geo-Information, 11*(9), 471. https://www.mdpi.com/2220-9964/11/9/471

Gu, W., Xue, F., Han, W., Wang, Z., Zhao, J., Zhang, L., Yang, C., & Jiang, J. (2023). Assessment of the spatial association between multiple pollutants of surface water and digestive cancer incidence in China: A novel application of spatial machine learning. *Ecological Indicators, 154*, 110897. https://doi.org/10.1016/j.ecolind.2023.110897

Helbich, M., Yao, Y., Liu, Y., Zhang, J., Liu, P., & Wang, R. (2019). Using deep learning to examine street view green and blue spaces and their associations with geriatric depression in Beijing, China. *Environment International, 126*, 107–117. https://doi.org/10.1016/j.envint.2019.02.013

Hu, Y., Goodchild, M., Zhu, A.-X., Yuan, M., Aydin, O., Bhaduri, B., Gao, S., Li, W., Lunga, D., & Newsam, S. (2024). A five-year milestone: Reflections on advances and limitations in GeoAI research. *Annals of GIS, 30*(1), 1–14.

Kim, J.-I., Yu, C.-Y., & Woo, A. (2023). The impacts of visual street environments on obesity: The mediating role of walking behaviors. *Journal of Transport Geography, 109*, 103593. https://doi.org/10.1016/j.jtrangeo.2023.103593

Labib, S. M. (2024). Greenness, air pollution, and temperature exposure effects in predicting premature mortality and morbidity: A small-area study using spatial random forest model. *Science of the Total Environment, 928*, 172387. https://doi.org/10.1016/j.scitotenv.2024.172387

Lee, H., Hanson, H. A., Logan, J., Maguire, D., Kapadia, A., Dewji, S., & Agasthya, G. (2024). Evaluating county-level lung cancer incidence from environmental radiation exposure, PM2.5, and other exposures with regression and machine learning models. *Environmental Geochemistry and Health, 46*(3), 82. https://doi.org/10.1007/s10653-023-01820-4

Lotfata, A., Grekousis, G., & Wang, R. (2023a). Using geographical random forest models to explore spatial patterns in the neighborhood determinants of hypertension prevalence across Chicago, Illinois, USA. *Environment and Planning B: Urban Analytics and City Science, 50*(9), 2376–2393. https://doi.org/10.1177/23998083231153401

Lotfata, A., Moosazadeh, M., Helbich, M., & Hoseini, B. (2023b). Socioeconomic and environmental determinants of asthma prevalence: A cross-sectional study at the U.S. county level using geographically weighted random forests. *International Journal of Health Geographics, 22*(1), 18. https://doi.org/10.1186/s12942-023-00343-6

Maharana, A., & Nsoesie, E. O. (2018). Use of deep learning to examine the association of the built environment with prevalence of neighborhood adult obesity. *JAMA Network Open, 1*(4), e181535–e181535. https://doi.org/10.1001/jamanetworkopen.2018.1535

Mathis, S. M., Webber, A. E., León, T. M., Murray, E. L., Sun, M., White, L. A., Brooks, L. C., Green, A., Hu, A. J., Rosenfeld, R., Shemetov, D., Tibshirani, R. J., McDonald, D. J., Kandula, S., Pei, S., Yaari, R., Yamana, T. K., Shaman, J., Agarwal, P., et al. (2024). Evaluation of FluSight influenza forecasting in the 2021–22 and 2022–23 seasons with a new target laboratory-confirmed influenza hospitalizations. *Nature Communications, 15*(1), 6289. https://doi.org/10.1038/s41467-024-50601-9

Metabiota. (2024). Improving health and wellness outcomes.. Retrieved November 24 from https://metabiota.com/

Mollalo, A., Vahedi, B., Bhattarai, S., Hopkins, L. C., Banik, S., & Vahedi, B. (2020). Predicting the hotspots of age-adjusted mortality rates of lower respiratory infection across the continental United States: Integration of GIS, spatial statistics and machine learning algorithms. *International Journal of Medical Informatics, 142*, 104248. https://doi.org/10.1016/j.ijmedinf.2020.104248

Moon, G. (2020). Health geography. In A. Kobayashi (Ed.), *International encyclopedia of human geography* (2nd ed., pp. 315–321). Elsevier. https://doi.org/10.1016/B978-0-08-102295-5.10388-9

Motairek, I., Dong, W., Salerno, P. R. V. O., Janus, S. E., Ganatra, S., Chen, Z., Guha, A., Makhlouf, M. H., Hassani, N. S., Rajagopalan, S., & Al-Kindi, S. G. (2023). Geographical patterns and risk factor association of cardio-oncology mortality in the United States. *The American Journal of Cardiology, 201*, 150–157. https://doi.org/10.1016/j.amjcard.2023.06.037

Peng, Y., Li, C., Rong, Y., Pang, C. P., Chen, X., & Chen, H. (2021). Real-time prediction of the daily incidence of COVID-19 in 215 countries and territories using machine learning: Model development and validation. *Journal of Medical Internet Research, 23*(6), e24285. https://doi.org/10.2196/24285

Prüss-Üstün, A., Wolf, J., Corvalán, C., Bos, R., & Neira, M. (2016). Preventing disease through healthy environments: A global assessment of the burden of disease from environmental risks. *World Health Organization.*

Quiñones, S., Goyal, A., & Ahmed, Z. U. (2021). Geographically weighted machine learning model for untangling spatial heterogeneity of type 2 diabetes mellitus (T2D) prevalence in the USA. *Scientific Reports, 11*(1), 6955. https://doi.org/10.1038/s41598-021-85381-5

Rabiolo, A., Alladio, E., Morales, E., McNaught, A. I., Bandello, F., Afifi, A. A., & Marchese, A. (2021). Forecasting the COVID-19 epidemic by integrating symptom search behavior into predictive models: Infoveillance study. *Journal of Medical Internet Research, 23*(8), e28876. https://doi.org/10.2196/28876

Reid, C. E., Considine, E. M., Watson, G. L., Telesca, D., Pfister, G. G., & Jerrett, M. (2019). Associations between respiratory health and ozone and fine particulate matter during a wildfire event. *Environment International, 129*, 291–298. https://doi.org/10.1016/j.envint.2019.04.033

Reid, C. E., Jerrett, M., Tager, I. B., Petersen, M. L., Mann, J. K., & Balmes, J. R. (2016). Differential respiratory health effects from the 2008 northern California wildfires: A spatiotemporal approach. *Environmental Research, 150*, 227–235. https://doi.org/10.1016/j.envres.2016.06.012

Scheinker, D., Valencia, A., & Rodriguez, F. (2019). Identification of factors associated with variation in US county-level obesity prevalence rates using epidemiologic vs machine learning models. *JAMA Network Open, 2*(4), e192884–e192884. https://doi.org/10.1001/jamanetworkopen.2019.2884

Sun, F., Yao, J., Du, S., Qian, F., Appleton, A. A., Tao, C., Xu, H., Liu, L., Dai, Q., Joyce, B. T., Nannini, D. R., Hou, L., & Zhang, K. (2023). Social determinants, cardiovascular disease, and health care cost: A nationwide study in the United States using machine learning. *Journal of the American Heart Association, 12*(5), e027919. https://doi.org/10.1161/JAHA.122.027919

Thacker, S. B., & Berkelman, R. L. (1988). Public health surveillance in the United States. *Epidemiologic Reviews, 10*(1), 164–190.

Thapen, N., Simmie, D., Hankin, C., & Gillard, J. (2016). DEFENDER: Detecting and forecasting epidemics using novel data-analytics for enhanced rsponse. *PLoS One, 11*(5), e0155417. https://doi.org/10.1371/journal.pone.0155417

Turk, P. J., Tran, T. P., Rose, G. A., & McWilliams, A. (2021). A predictive internet-based model for COVID-19 hospitalization census. *Scientific Reports, 11*(1), 5106. https://doi.org/10.1038/s41598-021-84091-2

U.S. Department of Health and Human Services (HHS). (2024a). *Priority areas—Healthy People 2030.*. Retrieved July 10 from https://odphp.health.gov/healthypeople/priority-areas

U.S. Department of Health and Human Services (HHS). (2024b). Social determinants of health—Healthy People 2030.. Retrieved July 10 from https://health.gov/healthypeople/priority-areas/social-determinants-health

Venkatramanan, S., Sadilek, A., Fadikar, A., Barrett, C. L., Biggerstaff, M., Chen, J., Dotiwalla, X., Eastham, P., Gipson, B., & Higdon, D. (2021). Forecasting influenza activity using machine-learned mobility map. *Nature Communications, 12*(1), 726.

Viljanen, M., Meijerink, L., Zwakhals, L., & van de Kassteele, J. (2022). A machine learning approach to small area estimation: Predicting the health, housing and Well-being of the population of Netherlands. *International Journal of Health Geographics, 21*(1), 4. https://doi.org/10.1186/s12942-022-00304-5

Wang, R., Liu, Y., Lu, Y., Zhang, J., Liu, P., Yao, Y., & Grekousis, G. (2019a). Perceptions of built environment and health outcomes for older Chinese in Beijing: A big data approach with street view images and deep learning technique. *Computers, Environment and Urban Systems, 78*, 101386. https://doi.org/10.1016/j.compenvurbsys.2019.101386

Wang, R., Yuan, Y., Liu, Y., Zhang, J., Liu, P., Lu, Y., & Yao, Y. (2019b). Using street view data and machine learning to assess how perception of neighborhood safety influences urban residents' mental health. *Health & Place, 59*, 102186. https://doi.org/10.1016/j.healthplace.2019.102186

Wang, C., & Wang, F. (2022). GIS-automated delineation of hospital service areas in Florida: From Dartmouth method to network community detection methods. *Annals of GIS, 28*(2), 93–109.

Wang, C., Wang, F., & Onega, T. (2021). Network optimization approach to delineating health care service areas: Spatially constrained Louvain and Leiden algorithms. *Transactions in GIS, 25*(2), 1065–1081.

Wang, S., Cai, W., Tao, Y., Sun, Q. C., Wong, P. P. Y., Huang, X., & Liu, Y. (2023). Unpacking the inter- and intra-urban differences of the association between health and exposure to heat and air quality in Australia using global and local machine learning models. *Science of the Total Environment, 871*, 162005. https://doi.org/10.1016/j.scitotenv.2023.162005

Wang, C., Onega, T., & Wang, F. (2024). Overlapping cancer service areas: Delineation and implications. *Transactions in GIS, 28*(3), 604–622.

Wang, S., Zhang, T., Li, Z., & Hong, J. (2024). Exploring pollutant joint effects in disease through interpretable machine learning. *Journal of Hazardous Materials, 467*, 133707. https://doi.org/10.1016/j.jhazmat.2024.133707

World Health Organization (WHO). (2024). *Summary of achievement—Social determinants of health.*. Retrieved July 10 from https://open.who.int/2016-17/our-work/category/03/programme/03.004/eba

Wu, X., Nethery, R. C., Sabath, M. B., Braun, D., & Dominici, F. (2020). Air pollution and COVID-19 mortality in the United States: Strengths and limitations of an ecological regression analysis. *Science Advances, 6*(45), eabd4049. https://doi.org/10.1126/sciadv.abd4049

Xiao, X., Wang, R., Knibbs, L. D., Jalaludin, B., Heinrich, J., Markevych, I., Gao, M., Xu, S.-L., Wu, Q.-Z., Zeng, X.-W., Chen, G.-B., Hu, L.-W., Yang, B.-Y., Yu, Y., & Dong, G.-H. (2021). Street view greenness is associated with lower risk of obesity in adults: Findings from the 33 Chinese community health study. *Environmental Research, 200*, 111434. https://doi.org/10.1016/j.envres.2021.111434

Xu, J., Liu, Y., Liu, Y., An, R., & Tong, Z. (2023). Integrating street view images and deep learning to explore the association between human perceptions of the built environment and cardiovascular disease in older adults. *Social Science & Medicine, 338*, 116304. https://doi.org/10.1016/j.socscimed.2023.116304

Yu, H., Zhou, Y., Wang, R., Qian, Z., Knibbs, L. D., Jalaludin, B., Schootman, M., McMillin, S. E., Howard, S. W., Lin, L.-Z., Zhou, P., Hu, L.-W., Liu, R.-Q., Yang, B.-Y., Chen, G., Zeng, X.-W., Feng, W., Xiang, M., & Dong, G.-H. (2021). Associations between trees and grass presence with childhood asthma prevalence using deep learning image segmentation and a novel green view index. *Environmental Pollution, 286*, 117582. https://doi.org/10.1016/j.envpol.2021.117582

Zhang, Y., Chen, K., Weng, Y., Chen, Z., Zhang, J., & Hubbard, R. (2022). An intelligent early warning system of analyzing twitter data using machine learning on COVID-19 surveillance in the US. *Expert Systems with Applications, 198*, 116882. https://doi.org/10.1016/j.eswa.2022.116882

Chapter 15
A Review on the Application of Geospatial Intelligence in Tourism Geography

Wenjie Xiao, Yi Liu, and Xuehan Wu

Abstract This chapter provides a critical review of how geospatial intelligence (GeoAI) technologies have been integrated into tourism geography. As artificial intelligence (AI) and big data continue to reshape research methodologies, tourism geography has evolved from relying primarily on qualitative approaches to embracing advanced quantitative techniques and computational models. We highlight key GeoAI-driven applications, including the identification of tourist destinations, sensing of tourists' emotions through text and image analytics, discovery of behavioral patterns from mobility data, and advanced decision-making tools for destination assessment, recommendation, and scenario simulation. Alongside these applications, we discuss the methodological innovations—from machine learning to convolutional neural networks and graph convolutional neural networks—that have enhanced spatial perception, cognition, and planning. We also examine persistent challenges in data integration, noise reduction, privacy, and the creation of truly multimodal, standardized tourism datasets. In addressing these issues, this chapter suggests future directions, including the improved fusion of diverse data sources, the integration of large language models, and the development of novel hybrid frameworks. Ultimately, this review demonstrates how GeoAI is reshaping tourism geography research and practice, fostering more efficient, accurate, and context-aware spatial insights.

W. Xiao
School of Tourism Management, Sun Yat-sen University, Guangzhou, China

School of Tourism, Jishou University, Zhangjiajie, China

Y. Liu (✉)
School of Tourism Management, Sun Yat-sen University, Guangzhou, China

Key Laboratory of Sustainable Tourism Smart Assessment Technology, Ministry of Culture and Tourism of China, Zhuhai, China
e-mail: liuyi89@mail.sysu.edu.cn

X. Wu
School of Tourism Management, Sun Yat-sen University, Guangzhou, China

© The Author(s), under exclusive license to Springer Nature Switzerland AG 2025
X. Huang et al. (eds.), *GeoAI and Human Geography*, Springer Geography,
https://doi.org/10.1007/978-3-031-87421-5_15

Keywords GeoAI · Tourism geography · Spatial analytics · Tourist behavior · Destination recommendation

15.1 Introduction

In recent decades, the advancement of information technology has significantly influenced the development of tourism geography. The emergence of the "Quantitative Revolution" in the 1960s disrupted the prevailing trend of research methodologies rooted in qualitative approaches. Studies on tourism geography consistently integrate mathematical statistics and modeling analysis using quantitative research approaches. The presentation discussed the trend of abandoning one type of research in favor of another, specifically qualitative and quantitative research. The advancement of information technology in the 1980s, particularly in the areas of remote sensing (RS), global positioning systems (GPS), and geographic information systems (GIS), significantly enhanced the efficiency and quality of research in tourism geography. During the mid-2010s, artificial intelligence exhibited remarkable proficiency in areas such as computer vision and natural language processing. Geospatial artificial intelligence (GeoAI) is the result of the close integration of artificial intelligence and geospatial research. The academic community widely accepted GeoAI during 2017 SIGSPATIAL conference. Subsequently, GeoAI has been widely applied in various tourism geographical studies, encompassing significant topics such as tourist behavior patterns, tourist emotional experiences, tourist destination evaluations, and tourist route recommendations. This has yielded fruitful research outcomes.

GeoAI, a state-of-the-art and widely used information technology, holds significant practical value and has given rise to several application instances. In addition, the ongoing development of large language models (LLMs) and deep learning has contributed to their thriving fields. There is an urgent need to explore the numerous valuable applications that arise from this progress. It is crucial to conduct a comprehensive analysis of the current research progress in geoAI within the field of tourism geography. This will assist in clearly defining the paths and regulations for applying geoAI in tourism geography, as well as exploring new situations and directions for its use.

15.2 Research Progress

The current application technologies used in tourism geography for geoAI include geospatial intelligent sensing (geoAIS), geospatial intelligent cognition (geoAIC), and geospatial intelligent decision-making (geoAIDM). GeoAIS employs support vector machines and convolutional neural networks, which are machine learning algorithms, to extract spatial elements and objects from large-scale geographic data. It also identifies the cognitive boundaries of tourism's geographic space and perceives the social functions and emotions associated with space. GeoAIC employs geographical data mining to uncover the patterns and connections present in geospatial data. GeoAIDM employs advanced algorithms to create models and forecast spatial scenarios. This section will provide an overview of the research advancements in such technologies related to tourism geography.

15.2.1 GeoAIS: Identify Tourist Areas and Sense the Emotions of Tourists

GeoAIS's primary use is to identify tourist areas and sense tourists' emotions. Since 2007, scientists have utilized geotagged photographs from Flickr to identify prominent metropolitan sites by employing algorithms that analyze geotags, location, and visual data (Kennedy et al., 2007; Giglio et al., 2019). Between 2017 and 2022, scholars used social media and tourist trajectory to estimate the level of tourist activity. They have identified various tourist areas with different spatial scales, such as tourism interest areas (Shao et al., 2017), tourist activity areas (Nolasco-Cirugeda et al., 2022), and global tourist activity areas (Dietz et al., 2020).

A 1982 study discovered that the environmental features of a location can affect how tourists perceive their experiences, thereby influencing their emotions (Zube et al., 1982). Scholars have used geospatial intelligence to examine two types of data, namely geospatial text and geographic imagery, after doing a thorough study using traditional survey methods including questionnaires, talks, and video simulations. As a result, there has been a notable decrease in the amount of time and energy required for conventional research approaches. In the context of geographic text analysis, researchers have found that neural network techniques such as dictionary and rule matching algorithms, support vector machines (SVM), and Naive Bayes may effectively identify visitors' positive and negative emotions. This can be achieved by analyzing data from platforms like Weibo and comments, as demonstrated by Liu et al. in 2019. The algorithms have the ability to detect the subtle emotions of travelers in different microenvironments, such as theme parks, including feelings of regret and chagrin (Park et al., 2020b). However, the lack of globally applicable geospatial intelligence algorithms and cross-cultural biases in geographic text data sources hinder their widespread use. The introduction of geographical photo and convolutional neural networks, which utilize deep learning methods, has effectively resolved the previously mentioned problems. Reputable researchers use visual deep learning algorithms to analyze the emotions transmitted by individuals' facial expressions in images. Subsequently, they employ the geographical tags linked to these photographs to create a comprehensive emotional map of the world (Zhang et al., 2018; Kang et al., 2019). However, privacy and security concerns regarding the geographically significant data obtained from facial expression images pose a major obstacle to emotion perception research.

15.2.2 GeoAIC: Mining Tourist Behavior Patterns

GeoAIC utilizes GPS, Weibo, and other spatiotemporal data that possess large-scale, multi-dimensional, noisy, and nonlinear attributes. It uses spatiotemporal mining algorithms, like spatiotemporal pattern discovery, spatiotemporal clustering, and spatiotemporal anomaly detection, to look for patterns in the data that show how tourist groups behave. GeoAIC is extensively employed in analyzing the behavioral patterns of tourists (Lew & McKercher, 2006).

Researchers have specifically studied tourists' behavior patterns at picturesque spots. For instance, Zheng et al. (2012) employed photographs from Flickr and Panoramio. They applied spatiotemporal clustering algorithms to group the travel pathways of tourists in various scenic locations into many groups. They then classified these categories into distinct patterns based on topological properties. Subsequently, scholars have expanded the spatial scope of their research to include tourism destinations (Hu et al., 2019) and cross-regional areas (Vu et al., 2015; Park et al., 2020a). Moreover, the research places special emphasis on mobile patterns in unique situations. For instance, studies have examined a variety of transportation patterns such as walking, automobiles, and trains (Brunauer et al., 2013); the movements of visitors with an interest in archaeological heritage (Payntar et al., 2021); and the movements of tourists within a specific time period (Luo et al., 2013). In summary, the research findings demonstrate a broad spectrum of richness and diversity.

Compared to macro- and mesoscales, research at micro-spatial dimensions has received less attention. Due to the decreased distance between stopping spots in the microsector, tourists tend to have shorter stays. Consequently, research necessitates a greater level of precision in space-time trajectories. Geotagged social media might not fulfill the aforementioned criteria. GPS devices possess the capability to precisely record the location and time of a trip. However, they present challenges in terms of recycling and are not well-suited for extensive utilization. Future research will focus on analyzing the specific patterns of tourist mobility at a small-scale level.

The aforementioned applications require highly accurate and frequent spatiotemporal data. Unfortunately, the current data has a lot of noise and might include the information of people who aren't tourists. It may also have the wrong records because of GPS positioning errors and signal issues (Zheng et al., 2019). All of these factors have an impact on spatial intelligence, albeit to varying degrees. In order to achieve this objective, sophisticated algorithms prioritize ways for gathering data (Sivalingam, et al., 2024). Ohashi et al. (2016) and Van Dijk et al. (2018) employ machine learning techniques to distinguish between tourist and non-tourist information, as well as activity points and itinerary points. Furthermore, Zheng (2015) have developed algorithms specifically designed to remove noise. Nevertheless, it is challenging to entirely eradicate these noises. In order to address the issue of data noise interference, future research can explore novel approaches, such as utilizing trajectory similarity measurements to extract trajectories. This can lead to enhanced trajectory accuracy (Zheng et al., 2019).

15.2.3 GeoAIDM: Tourism Destination Evaluation, Recommendation, and Simulation

In the 1970s, Hunt introduced the idea of a tourist destination image (Hunt, 1971). However, determining how to quantify it has consistently posed a challenge in the academic sphere. In 2007, the academic community utilized text analysis technologies for the first time to portray the image of Macau through the content of travel blogs (Choi et al., 2007). Ten years later, researchers employed machine learning models to evaluate destination pictures using geographical image data sources (Deng & Li, 2018). Subsequently, researchers began using similar concepts to assess various aspects of tourist destinations. For instance, they utilized big data from social media platforms, including textual and visual information, to evaluate the appeal of urban tourism from multiple perspectives (Giglio et al., 2019). They employed Weibo data and Dianping data to gauge the level of social and economic activities, respectively, and used this information to comprehensively assess the vitality of tourist cities (Huang et al., 2020).

Travel recommendation is the practical application that is most closely associated with the evaluation of tourism destinations. Regarding this matter, previous studies have undergone three distinct stages: The initial phase primarily utilizes geographical big data to suggest popular attractions and destinations. This is done by considering widely acknowledged indicators, such as transportation convenience (Horozov et al., 2006). Alternatively, recommendations can be generated by analyzing big data and taking into account various factors like distance, price, and rating to identify landmark attractions. However, research in this state lacks customization for individual users and overlooks the unique characteristics of the attractions. During the second stage, recommendation services mostly rely on collaborative filtering algorithms to suggest attractions based on the similarity between different users and their shared tastes. This research methodology necessitates dependence on historical records and may encounter a cold start issue (Zheng and Xie, 2011). The next phase of the research focuses on using the graph convolutional neural network algorithm (Bao et al., 2021) to forecast the user's next location by leveraging other users' GPS or trajectory data. At this stage, research offers a notable benefit in that it does not rely on the user's past records and can make precise forecasts.

Geospatial decision-making intelligence can also be utilized for simulating future scenarios. Land utilization and the advancement of tourism are significant subjects in this context. In the 1930s, McMurry brought up the topic of the connection between the development of tourism and changes in land use (McMurry, 1930). In recent years, the academic community has started utilizing spatial intelligence algorithms to simulate tourism's land-use scenarios. This involves employing machine learning algorithms to simulate and assess the effects of changes in tourist numbers on the environment, as demonstrated by Furgała-Selezniow (2021) in their research.

15.3 Application Cases

Geospatial intelligence has two significant application scenarios: place semantic perception and intelligent site selection. This section will use the case of social function perception to demonstrate the use of geospatial intelligence perception and decision-making in tourism geography.

15.3.1 Spatially Intelligent Perception: Convolutional Neural Network (CNN) Perceives Social Functions

Spatial intelligence perception utilizes geographic big data, such as points of interest (POIs) and street view images, to identify specific characteristics of a location. Machine learning, spatial clustering, deep learning, and other algorithms determine these characteristics, which include emotions (Dubey et al., 2016), social functions (Zhu et al., 2020), and cognitive boundaries (Zhu et al., 2020). This approach avoids the challenges of manually defining the concept of place or constructing a knowledge base (Purves et al., 2019). In recent years, machine learning and convolutional neural network (CNN) methods have dominated the field of spatial intelligence perception, and the integration of graph convolutional neural networks (GCNN) into spatial intelligence perception applications has overcome CNN's limitation to structured input.

One notable study among them is Zhu and Zhang Fan's research, which focuses on using GCNN to understand social functions. The study conducted research in Beijing, utilizing 240,000 points of interest (POI) data, taxi trajectory data, and street view picture data. These datasets were utilized to train place graphs, which were then employed in predicting the social functions of geographic locations(Zhu et al., 2020). A location graph is a visual representation where places are depicted as nodes and the spatial connections between places are shown as edges. The study uses a kernel density approach to determine the borders of POI in locations like Zhongguancun and Wudaokou. It then calculates the spatial interaction intensity by analyzing taxi trajectories and creates a place network consisting of over 200 nodes and 40,000 edges.

This study utilizes a visual deep learning algorithm to identify street scene pictures as the independent variable and the social function of the place as the dependent variable. The obtained place features are then inputted into a place graph. By training a GCNN algorithm, a relationship model between the visual features of the place and the social function is established.

The GCNN incorporates the level of spatial interaction between geospatial areas, resulting in improved training variables for the spatial intelligent perception algorithm. This upgrade from first order to second order enhances perception accuracy. This approach overcomes the issue of under-representation of individual geographic big data.

15.3.2 Spatial Intelligent Decision-Making: Gym Location Selection

Spatial intelligent decision-making strives to attain integrated services through the utilization of big data and the management of spatial information. It concentrates on the realization of the spatial aspects of industrial terminals, as well as the process of layout and location selection. This includes monitoring and evaluating the current situation, as well as providing assistance in the decision-making process for spatial solutions. It depends on multiple modes of operation. Geographic big data employs a bottom-up approach to address the drawbacks of time-consuming and inadequately grounded top-down decision-making. It also fulfills the future-oriented requirements for selecting spatial locations.

The process of choosing commercial locations extensively applies spatial intelligence decision-making, which includes assessing the quality of hotel locations (Yang et al., 2015) and selecting restaurant locations with a focus on sales (Han et al., 2022). Gyms offer consumer groups a venue for recreation and social engagement, in contrast to commercial establishments like restaurants and hotels. They have the potential to evolve into a third space and serve as representations of the leisure sector. Scholars have expressed interest in gym locations. Xu et al.'s intelligent placement selection model for gyms initiated the intelligent selection of leisure facilities.

The researchers employed a sophisticated approach that involved utilizing a five-dimensional integrated multi-source heterogeneous data fusion algorithm, a spatial clustering algorithm, and an artificial intelligence neural network. Their objective was to develop an intelligent location decision-making model for gyms, with a focus on business districts. To validate the model's effectiveness, they conducted a case study in Shapingba, Chongqing City. The researchers utilized gym data, location information (including GDP level and night-light images within a 500-meter grid area surrounding the gym), population data, and industry-related data (such as density of commercial residential areas, catering services, and living facilities). They employed a deep learning four-layer neural network that incorporates multiple sources of heterogeneous spatiotemporal data streams. By setting user-defined conditions (such as buffers, shortest paths, and time intersectionality), they were able to determine the time and space characteristics of the gym grid. Multi-dimensional relationships efficiently achieve the integration and extraction of diverse data from several sources, as well as the establishment of a quality evaluation index system for gym site selection. Based on this approach, they utilized the Chongqing Shapingba gym terminal POI data obtained from Amap. They employed a spatial aggregation technique to identify the spatial symbiotic relationship between the gym and other commercial forms. To achieve this, they employed a feedforward error back-propagation three-layer perceptron (BP). The goal of TLP training is to anticipate the suitability of the grid layout gym and provide a scientific basis for selecting the gym's site. This will enable rational decision-making when it comes to choosing the gym location.

Spatial intelligent decision-making relies on diverse data sources to assess the effectiveness of layouts from various perspectives and establish an integrated decision-making system that includes indicator systems, evaluation methods, and future predictions. This system can provide more efficient and accurate practical solutions for commercial facility layout and location selection.

15.4 Commentary and Outlook

Upon reviewing the current body of research, it is evident that spatial intelligence applications predominantly utilize textual data, trajectory data, points of interest (POI), and picture data. The data exhibit significant disparities in terms of production techniques, storage formats, and information values. Therefore, how to integrate and utilize these diverse and varied data sources has become a crucial undertaking in geospatial intelligence research. Various types of modal data have distinct application scenarios. Active digital footprints, which are visual images, are more suitable for explaining travel behaviors (Fan et al., 2023). On the other hand, passive digital footprints, represented by mobile roaming data, provide a more objective measurement of tourists' spatio-temporal behaviors. Despite the similarity of the modal data, the collection of tourism data lacks standardized and regulated channels and methods. Different statistical methods and standards may be employed by various scenic spots, attractions, and tourism online platforms, resulting in noticeable differences in their data structures. Current research trends often involve integrating data from various sources or platforms to enhance the effectiveness of analysis. For instance, researchers have combined POI data with Sina Weibo data to identify the types of activities undertaken by tourists (Shao et al., 2017). Similarly, multiple platforms of Flick's User-Generated Content (UGC) image data have been combined to identify tourist activity areas by utilizing complementary data (Nolasco-Cirugeda et al., 2022). Existing research in the field of data application has yet to achieve true integration. One specific challenge that many current studies have been unable to effectively address is the problem of geographical segmentation. Geographical intelligence is primarily used to divide this data into geographical and emotional fundamental components, creating a varied and continuous sample area. For instance, trajectory data is divided into several lines, picture data is divided into multiple basic pixel units, and POI data is divided into latitude and longitude. These fundamental components just function as supplements to one another and do not accomplish genuine fusion. This impedes the development of intricate linkages within the multi-source data space and poses challenges in integrating these data into the fundamental architecture of the transformer-based large model. In order to achieve full integration, it is necessary to divide multi-source multimodal geographic data into benchmarks of other data. It is also important to address the challenges of data heterogeneity, spatiotemporal alignment, and semantic alignment in this process. Currently, there is a noticeable trend in tourism geography where AI technology is being heavily relied upon. Specifically, deep learning models, such as CNN and attentional mechanisms, are being extensively used to identify patterns and predict the movements and locations of individuals in the field of tourism (Bao et al., 2021; Zhao et al., 2019). Furthermore, deep learning algorithms continue to be widely used in research areas such as destination image (Díaz-Pacheco et al., 2024), sentiment prediction (Iqbal et al., 2024), and travel recommendation (Shrestha et al., 2024).

Additionally, there are emerging tendencies in contemporary study. For instance, the transfer of GCNN methods or huge language models to the geographical domain. GCNN methods excel at capturing unstructured data relationships by facilitating message passing between nodes and edges. This process can be likened to the temporal and spatial transfer of event processes (Huang et al., 2024). On the other hand, the BERT language model's ability to understand the relationship between words and sentences can be seen as the spatial structure and composition of geospatial elements based on specific relationships. Furthermore, there is a growing introduction of machine learning techniques that combine model-driven and data-driven approaches to sense physical information that is known in advance. This introduces a new research paradigm for geospatial intelligence research and enables detailed description and prediction of complex geographic phenomena. However, the use of intelligent technologies in geospatial research will continue to be expanded in the future. This is because our geographic environment and the widespread use of Internet technologies and AI are reconstruing, and new patterns and mechanisms are constantly emerging. The complexity of these emerging phenomena needs to embrace and advance novel technologies and methodologies to facilitate study in this area.

References

Bao, Y., Huang, Z., Li, L., Wang, Y., & Liu, Y. (2021). A BiLSTM-CNN model for predicting users' next locations based on geotagged social media. *International Journal of Geographical Information Science, 35*(4), 639–660.

Brunauer, R., Hufnagl, M., Rehrl, K., & Wagner, A. (2013, October). Motion pattern analysis enabling accurate travel mode detection from GPS data only. 16th International IEEE Conference on Intelligent Transportation Systems (ITSC 2013) (pp. 404–411). IEEE.

Choi, S., Lehto, X. Y., & Morrison, A. M. (2007). Destination image representation on the web: Content analysis of Macau travel related websites. *Tourism Management, 28*(1), 118–129.

Deng, N., & Li, X. R. (2018). Feeling a destination through the "right" photos: A machine learning model for DMOs' photo selection. *Tourism Management, 65*, 267–278.

Díaz-Pacheco, Á., Guerrero-Rodríguez, R., Álvarez-Carmona, M. Á., Rodríguez-González, A. Y., & Aranda, R. (2024). Quantifying differences between UGC and DMO's image content on Instagram using deep learning. *Information Technology & Tourism, 26*(2), 293–329.

Dietz, L. W., Sen, A., Roy, R., & Wörndl, W. (2020). Mining trips from location-based social networks for clustering travelers and destinations. *Information Technology & Tourism, 22*(1), 131–166.

Dubey, A., Naik, N., Parikh, D., Raskar, R., & Hidalgo, C. A. (2016). Deep learning the city: Quantifying urban perception at a global scale. In *Computer Vision–ECCV 2016: 14th*

European Conference, Amsterdam, The Netherlands, October 11–14, 2016, proceedings, part I 14 (pp. 196–212). Springer.

Fan, Z., Zhang, F., Loo, B. P., & Ratti, C. (2023). Urban visual intelligence: Uncovering hidden city profiles with street view images. *Proceedings of the National Academy of Sciences, 120*(27), e2220417120.

Furgała-Selezniow, G., Jankun-Woźnicka, M., Kruk, M., & Omelan, A. A. (2021). Land use and land cover pattern as a measure of tourism impact on a lakeshore zone. *Land, 10*(8), 787.

Giglio, S., Bertacchini, F., Bilotta, E., & Pantano, P. (2019). Using social media to identify tourism attractiveness in six Italian cities. *Tourism Management, 72*, 306–312.

Han, S., Jia, X., Chen, X., Gupta, S., Kumar, A., & Lin, Z. (2022). Search well and be wise: A machine learning approach to search for a profitable location. *Journal of Business Research, 144*, 416–427.

Horozov, T., Narasimhan, N., & Vasudevan, V. (2006, January). Using location for personalized POI recommendations in mobile environments. In *International Symposium on Applications and the Internet (SAINT'06)* (p. 6). IEEE.

Hu, F., Li, Z., Yang, C., & Jiang, Y. (2019). A graph-based approach to detecting tourist movement patterns using social media data. *Cartography and Geographic Information Science, 46*(4), 368–382.

Huang, B., Zhou, Y., Li, Z., Song, Y., Cai, J., & Tu, W. (2020). Evaluating and characterizing urban vibrancy using spatial big data: Shanghai as a case study. *Environment and Planning B: Urban Analytics and City Science, 47*(9), 1543–1559.

Huang, F., Lv, J., & Yue, Y. (2024). Jointly spatial-temporal representation learning for individual trajectories. *Computers, Environment and Urban Systems, 112*, 102144.

Hunt, J. D. (1971). *Image–A factor in tourism*. Colorado State University.

Iqbal, M., Renjie, L., & Xu, J. (2024). *Behavior analysis of photo-taking tourists at attraction-level using deep learning and latent Dirichlet allocation in conjunction with kernel density estimation* (Vol. 12, p. 92945). IEEE Access.

Kang, Y., Jia, Q., Gao, S., Zeng, X., Wang, Y., Angsuesser, S., et al. (2019). Extracting human emotions at different places based on facial expressions and spatial clustering analysis. *Transactions in GIS, 23*(3), 450–480.

Kennedy, L., Naaman, M., Ahern, S., et al. (2007). How flickr helps us make sense of the world: context and content in community-contributed media collections[C]. In *Proceedings of the 15th ACM international conference on multimedia* (pp. 631–640).

Lew, A., & McKercher, B. (2006). Modeling tourist movements: A local destination analysis. *Annals of Tourism Research, 33*(2), 403–423.

Liu, Y., Huang, K., Bao, J., & Chen, K. (2019). Listen to the voices from home: An analysis of Chinese tourists' sentiments regarding Australian destinations. *Tourism Management, 71*, 337–347.

Luo, W., Tan, H., Chen, L., & Ni, L. M. (2013, June). Finding time period-based most frequent path in big trajectory data. In *Proceedings of the 2013 ACM SIGMOD international conference on management of data* (pp. 713–724).

McMurry, K. C. (1930). The use of land for recreation. *Annals of the Association of American Geographers, 20*(1), 7–20.

Nolasco-Cirugeda, A., García-Mayor, C., Lupu, C., & Bernabeu-Bautista, A. (2022). Sco** out urban areas of tourist interest though geolocated social media data: Bucharest as a case study. *Information Technology & Tourism, 24*(3), 361–387.

Ohashi, H., Nguyen, P. X., Akiyama, T., Yamamoto, M., & Sato, A. (2016). Trip-extraction method based on characteristics of sensors and human-travel behavior for sensor-based travel survey. *Journal of Information Processing, 24*(1), 39–48.

Park, S., Xu, Y., Jiang, L., Chen, Z., & Huang, S. (2020a). Spatial structures of tourism destinations: A trajectory data mining approach leveraging mobile big data. *Annals of Tourism Research, 84*, 102973.

Park, S. B., Kim, J., Lee, Y. K., & Ok, C. M. (2020b). Visualizing theme park visitors' emotions using social media analytics and geospatial analytics. *Tourism Management, 80*, 104127.

Payntar, N. D., Hsiao, W. L., Covey, R. A., & Grauman, K. (2021). Learning patterns of tourist movement and photography from geotagged photos at archaeological heritage sites in Cuzco, Peru. *Tourism Management, 82*, 104165.

Purves, R. S., Winter, S., & Kuhn, W. (2019). Places in information science. *Journal of the Association for Information Science and Technology, 70*(11), 1173–1182.

Shao, H., Zhang, Y., & Li, W. (2017). Extraction and analysis of city's tourism districts based on social media data. *Computers, Environment and Urban Systems, 65*, 66–78.

Shrestha, D., Wenan, T., Shrestha, D., Rajkarnikar, N., & Jeong, S. R. (2024). Personalized tourist recommender system: A data-driven and machine-learning approach. *Computation, 12*(3), 59.

Sivalingam, P., Asirvatham, D., Marjani, M., Masood, J. A. I. S., Chakravarthy, N. K., Veerisetty, G., & Lestari, M. T. (2024). A review of travel behavioural pattern using GPS dataset: A systematic literature review. *Measurement: Sensors, 101031*.

Van Dijk, J. (2018). Identifying activity-travel points from GPS-data with multiple moving windows. *Computers, Environment and Urban Systems, 70*, 84–101.

Vu, H. Q., Li, G., Law, R., & Ye, B. H. (2015). Exploring the travel behaviors of inbound tourists to Hong Kong using geotagged photos. *Tourism Management, 46*, 222–232.

Yang, Y., Tang, J., Luo, H., & Law, R. (2015). Hotel location evaluation: A combination of machine learning tools and web GIS. *International Journal of Hospitality Management, 47*, 14–24.

Zhang, F., Zhou, B., Liu, L., Liu, Y., Fung, H. H., Lin, H., & Ratti, C. (2018). Measuring human perceptions of a large-scale urban region using machine learning. *Landscape and Urban Planning, 180*, 148–160.

Zhao, L., Song, Y., Zhang, C., Liu, Y., Wang, P., Lin, T., et al. (2019). T-GCN: A temporal graph convolutional network for traffic prediction. *IEEE Transactions on Intelligent Transportation Systems, 21*(9), 3848–3858.

Zheng, Y. (2015). Trajectory data mining: An overview. *ACM Transactions on Intelligent Systems and Technology, 6*(3), 1–41.

Zheng, Y., & Xie, X. (2011). Learning travel recommendations from user-generated GPS traces. *ACM Transactions on Intelligent Systems and Technology (TIST), 2*(1), 1–29.

Zheng, Y. T., Zha, Z. J., & Chua, T. S. (2012). Mining travel patterns from geotagged photos. *ACM Transactions on Intelligent Systems and Technology (TIST), 3*(3), 1–18.

Zheng, W., Zhou, R., Zhang, Z., Zhong, Y., Wang, S., Wei, Z., & Ji, H. (2019). Understanding the tourist mobility using GPS: How similar are the tourists? *Tourism Management, 71*, 54–66.

Zhu, D., Zhang, F., Wang, S., Wang, Y., Cheng, X., Huang, Z., & Liu, Y. (2020). Understanding place characteristics in geographic contexts through graph convolutional neural networks. *Annals of the American Association of Geographers, 110*(2), 408–420.

Zube, E. H., Sell, J. L., & Taylor, J. G. (1982). Landscape perception: Research, application and theory. *Landscape Planning, 9*(1), 1–33.

Chapter 16
Artificial Intelligence for Cartography and Maps

Yuhao Kang, Chenglong Wang, Yu Feng, Guillaume Touya, and Jina Kim

Abstract The integration of Artificial Intelligence (AI) into cartography represents a transformative opportunity for advancing mapmaking, geovisualization, and geospatial analysis. This chapter explores the applications of AI in cartography, focusing primarily on two major streams of AI methods: deep learning and generative AI. In particular, notable deep learning methods include Deep Convolutional Neural Networks (DCNNs), Graph Convolutional Neural Networks (GCNs), and Generative Adversarial Networks (GANs), while generative AI methods include Stable Diffusion-based models and Large Language Models (LLMs). These approaches have the potential not only to improve the performance of traditional cartographic design decisions but also to enhance human creativity. Through four example case studies, including map object detection, map generalization, map style transfer, and map evaluation, we illustrate how AI methods could be employed in cartographic studies. Beyond technological advancements, this chapter also addresses the ethical and social implications associated with the use of AI in

Y. Kang (✉)
GISense Lab, Department of Geography and the Environment, The University of Texas at Austin, Austin, TX, USA
e-mail: yuhao.kang@austin.utexas.edu

C. Wang
School of Urban Planning and Design, Peking University Shenzhen Graduate School, Shenzhen, China
e-mail: chenglongw@stu.pku.edu.cn

Y. Feng
Chair of Cartography and Visual Analytics, Technical University of Munich, Munich, Germany
e-mail: y.feng@tum.de

G. Touya
LASTIG, University of Gustave Eiffel, ENSG, IGN, Saint-Mande, France
e-mail: guillaume.touya@ign.fr

J. Kim
Department of Computer Science and Engineering, University of Minnesota, Minneapolis, MN, USA
e-mail: kim01479@umn.edu

© The Author(s), under exclusive license to Springer Nature Switzerland AG 2025

X. Huang et al. (eds.), *GeoAI and Human Geography*, Springer Geography,
https://doi.org/10.1007/978-3-031-87421-5_16

cartography. Issues such as bias, trustworthiness, commodification, geoprivacy, and transparency are discussed to ensure the responsible use of AI for cartography.

Keywords Cartography · Maps · Mapmaking · AI · Deep learning · Generative AI · Map style · Map generalization

16.1 Introduction

Cartography, the art and science of making maps, has a history spanning over a thousand years (Krygier, 1995). From prehistoric cave drawings, such as those found in the Cave of Altamira, Spain (Smith, 1982), to modern printed and digital maps, people have relied on visual representations to convey complex geographic information and facilitate spatial thinking. Over the centuries, cartography, the discipline of mapmaking, has experienced significant evolution, driven by societal demands and technological progress. Back to the Age of Discovery, the focus was making accurate paper maps using techniques from surveying and geodesy. With the advent of the printing press, cartographers shifted their attention to cartographic design, emphasizing clarity, aesthetics, and usability in printed maps. In the digital era, cartography advanced due to the emergence of technologies like Geographic Information Systems (GIS) (Kitchin & Dodge, 2007). The creation of digital maps using GIS has significantly improved the mapmaking process, enhancing accuracy, efficiency, and data integration. Later, the development of interactive maps has enabled users to explore and manipulate geovisualizations and maps in real time, thereby enhancing user engagement and understanding of geographic information (Robinson et al., 2017).

Today, a new and exciting opportunity is emerging: the integration of Artificial Intelligence (AI) into cartography (Kang et al., 2024; Touya et al., 2019). This convergence offers new possibilities for mapmaking, geovisualization, and geospatial analysis. Given that the concepts of AI have evolved with the progress of computer science techniques, we focus on two major streams of AI methods in this chapter: (1) deep learning; and (2) generative AI. Deep learning has been widely adopted since 2012, particularly through the use of Deep Convolutional Neural Networks (DCNNs) and other variants such as Graph Convolutional Neural Networks (GCNs) and Generative Adversarial Networks (GANs). These deep learning methods have benefited a broad range of applications across numerous fields, including cartography. Cartographers have successfully leveraged deep learning to support cartographic design decisions such as map generalization, map style transfer, and map readings, demonstrating the impact of these techniques in mapmaking. Generative AI, on the contrary, has attracted significant attention more recently, due to the success of models such as Stable Diffusion[1] and ChatGPT.[2] These models

[1] https://stability.ai/

[2] https://chatgpt.com/

have introduced new possibilities for creative mapmaking, ranging from improving the mapmaking efficiency to generating artistic styles in maps. In this chapter, we will showcase several applications of AI including map object detection, map generalization, map style transfer, and map evaluation, to illustrate how they could support cartographic design decisions.

The motivations of integrating of AI in cartography stand from the following three aspects. First, AI methods, including deep learning, machine learning, and generative AI methods, can significantly improve the efficiency and performance of certain cartographic design decisions compared to traditional approaches. These methods could enhance tasks such as map feature extraction and map generalization, enabling more precise and efficient decision-making in map production. Second, the advance in AI facilitates the creative and artistic dimensions of cartography. It is challenging for existing GIS tools to assist the design processes in mapmaking. Through generative AI, cartographers are empowered to design more personalized and creative cartographic representations. Third, AI helps preserve and advance cartographic knowledge by capturing traditional expertise and making geographic information more accessible. AI may store classic map knowledge by reading and understanding maps and could facilitate the translation of map knowledge into other formats, such as integrating with media and generating textual contents.

This chapter provides an overview of how AI methods could be utilized in cartography and mapmaking alongside four example applications. In Sect. 16.2, an introduction of map data including vector format and raster format and their corresponding AI methods are provided. Then, several examples of using AI in cartographic design decisions are demonstrated in Sect. 16.3, including map object detection, map generalization, map style transfer, and map evaluation. Section 16.4 discusses several ethical and social aspects that need attention when leveraging AI for cartography. Section 16.5 concludes the chapter.

16.2 Map Data

To leverage AI methods to process map data and facilitate the mapmaking process, it is essential to first understand the characteristics of map data. Map data is primarily represented in two formats: vector and raster. Vector data represents geographic features using points, lines, and polygons, making it ideal for capturing discrete objects such as roads, buildings, and administrative boundaries. Raster data, on the other hand, represents geographic information as a grid of cells or pixels, where each cell has a value that corresponds to a specific characteristic of the area, such as elevation, temperature, or land cover. These two data formats serve as the foundation for creating maps, each being suitable for different AI approaches depending on the nature of the data they represent.

16.2.1 Vector Map Data

Vector map data represents geographic features through discrete geometries, such as points (e.g., cities or landmarks), lines (e.g., roads and rivers), and polygons (e.g., lakes, parks). Vector data follows an object-based structure, storing not only the spatial geometry of each feature but also attribute information associated with it. This makes vector data particularly suitable for AI methods that model individual geographic objects and their spatial relationships.

The following notable AI methods have been applied to vector map data. Decision tree algorithms could model vector data based on various attributes and have been widely utilized for map generalizations (Karsznia & Weibel, 2018). GCNs are particularly well-suited for vector data as they can model spatial relationships and connectivity between features. GCNs convert points and polygons into graph structures (e.g., using Delaunay triangulation), while polylines are inherently represented as graphs, making GCNs ideal for tasks involving map object representation and map generalization (Yan et al., 2019).

16.2.2 Raster Map Data

Raster map data represents geographic information as a grid of cells or pixels, with each cell containing a value that corresponds to a specific attribute of the area it covers. Raster data is appropriate for representing gradual variations in geographic attributes, making it well-suited for image processing-based analysis.

Certain AI methods, particularly deep learning models, are highly effective for analyzing raster map data due to their ability to modeling and understanding high-dimensional visual features from images. Notable algorithms include DCNNs, GANs, and generative AI models. DCNNs have been leveraged for tasks such as map object detection and map generalization. GANs are another powerful AI methods for raster data, particularly in generating map images, and have been leveraged for map style transfer (Christophe et al., 2022; Kang et al., 2019). Additionally, the latest advancements in generative AI, such as DALL-E[3] and Midjourney,[4] can also be leveraged to generate maps. Despite that there are several ethical concerns, the usage of generative AI has brought new opportunities to support the mapmaking process.

[3] https://openai.com/index/dall-e-3/

[4] https://www.midjourney.com/

16.3 Using GeoAI for Cartographic Design Decisions

In this section, we introduce several common cartographic design decisions that are benefited from the use of AI. In particular, the following cartographic design decisions are illustrated including map object detection, map generalization, map style transfer, and map evaluation.

16.3.1 Map Object Detection

A map object refers to an element or feature in a map. Map objects could represent physical entities as points, lines, polygons, or certain map elements like labels, north arrows, and legends. Map objects convey geographic details, offering valuable insights into the environment and physical landscape of the real world (Chiang, 2010; Chiang et al., 2014, 2019). There have been numerous efforts to automatically detect various map objects, such as wetlands (Chiang et al., 2014, 2020; Duan et al., 2021b), buildings (Pérez Nava & Sánchez Berriel, 2024), and lines including railroads (Chiang et al., 2019; Duan et al., 2020, 2021a). Recent significant advancements in deep learning models have enabled time-efficient and accurate object detection by leveraging image representations learned from the DCNNs-based models.

Compared to objects in standard scene images (e.g., a photograph of a street), map object detection presents unique challenges. Map objects are often small, may overlap with other objects or text labels, can be affected by poor scanning conditions and the original quality of the map, and vary significantly across different authors and their cartographic styles (Chiang et al., 2009; Chiang & Knoblock, 2009; Pérez Nava & Sánchez Berriel, 2024). Additionally, there is a lack of annotated data for map objects (Chiang et al., 2019; Duan et al., 2021a, 2021b) that are unique from public image datasets used in computer vision (e.g., ImageNet (Deng et al., 2009)). Such distinct challenges in map object detection require careful decisions when employing object detection techniques in computer vision. In efforts to address the detection of small map objects, Chiang et al. (2019) introduced a modified version of the Pyramid Scene Parsing Network (PSPNet) (Zhao et al., 2017), incorporating skip connections from Very Deep Convolutional Networks (VGG) (Simonyan & Zisserman, 2014). Their proposed approach aims to overcome the limitations of DCNNs in preserving local spatial information (i.e., the locations of objects), assisting in distinguishing small map objects from complex backgrounds. Duan et al. (2021a) addressed the challenge of the limited availability of annotated map objects by introducing an algorithm to automatically generate labeled data and resolve the misalignment between external vector data and map objects on scanned maps. This misalignment arises due to differences in scale, coordinate projection systems, and publication years between maps and external vector data. To address this, the system's labeled data generation component uses the shape of vector data as prior knowledge and develops optimization objectives based on the Chan-Vese algorithm (Chan & Vese, 2001), which segments an image into foreground and background regions. This approach effectively aligns the vector data with the map content, improving the quality of labeled training data for cartographic tasks.

Here, we showcase a recent work for cartographic symbol detection using the Target-Guided Generative Model (TGGM) (Duan et al., 2021b). Maps often use clusters of cartographic symbols to represent area features, such as groups of wetland symbols indicating wetland regions. Traditional semi-supervised or supervised object detection methods require extensive annotations for both target and non-target symbols. However, these methods often cannot handle visual inconsistencies between symbols in the training set and those in new map regions. To address this challenge, TGGM utilizes the idea of regions of interest (ROIs) in maps where symbols (e.g., a group of wetland symbols) and background within an ROI (e.g., a wetland region) show visual similarities, enabling one-shot, iterative symbol detection using a single annotation of the target symbol.

TGGM begins by scanning an ROI with a sliding window and encoding each window image using a Conditional Variational Autoencoder (CVAE) (Sohn et al., 2015). Simultaneously, TGGM models a Mixture of Gaussians (MoG) with two components: one for the distribution of CVAE-encoded window images containing target symbols (i.e., target windows) and another for non-target windows. Initially, TGGM updates the CVAE and the target component in the MoG using the encoded image of the annotated target window and its visual augmentations. For other window images, TGGM uses the CVAE to guide the prediction process. Specifically, if a window image, when conditioned on being a target window, yields a lower reconstruction error than when conditioned as a non-target window, TGGM classifies it as a target window; otherwise, it is classified as a non-target window. In subsequent iterations, TGGM refines the distributions of the CVAE-encoded target and non-target windows in the MoG based on the prediction results from previous iterations, continuously improving its performance. TGGM's experiments on detecting wetlands in scanned United States Geological Survey (USGS) historical topographic maps demonstrate the model's effectiveness, using a single human annotation of a wetland symbol and a few labeled ROIs from USGS contemporary wetland vector data, which approximate wetland areas on historical maps. Figure 16.1 derived from Duan et al. (2021b) illustrates example results, with green boxes indicating correctly detected windows containing wetland symbols (true positive windows) and red boxes highlighting incorrect detections (false positive windows). This study demonstrates the potential of leveraging the unique spatial characteristics of geographic features and cartographic principles to significantly enhance object detection from maps, addressing specific challenges in map object detection.

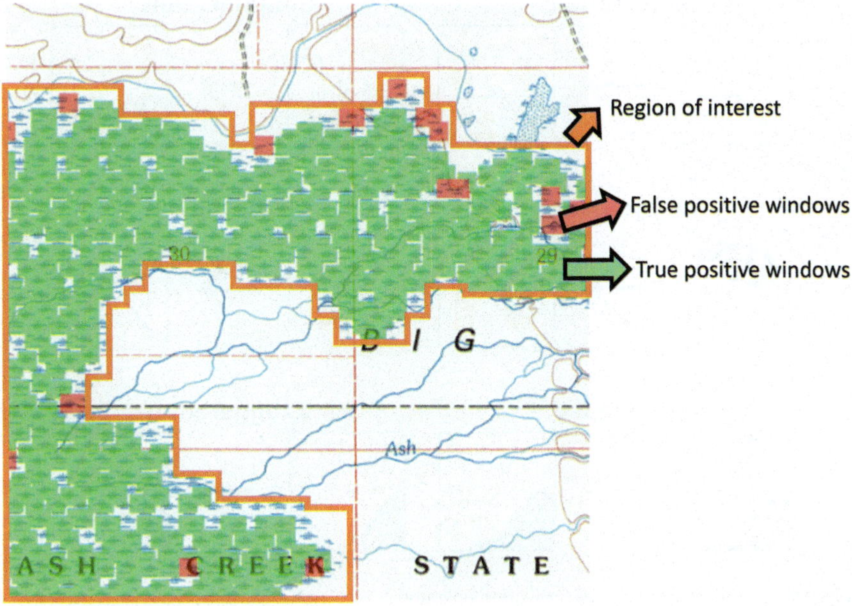

Fig. 16.1 Original figure from Duan et al. (2021b). Visualization of experimental results for wetland detection using the TGGM. The red and green boxes indicate false positive and true positive detection results, respectively

16.3.2 Map Generalization

Map generalization is a fundamental task in cartography, particularly in generating multi-scale map representations. Its main goal is to simplify the representation of geographic features on a map so that the map can maintain readability and information validity at different scales while avoiding information overload or confusion. Extensive research has focused on the specific generalization operators tailored to different map features. Each operator has a specific goal and function, such as selection, simplification, aggregation, collapse, exaggeration, and displacement. Examples of some of the key operators are demonstrated in Fig. 16.2.

However, with the combined application of multiple operators, especially in the representation of multi-scale maps, the challenge lies in the mutual constraints and conflicts between these operators. To address conflicts between operators, end-to-end solutions have gained increasing attention. Ideally, multiple operators can act simultaneously without needing to be handled independently. This is one of the reasons why deep learning techniques are well-suited for this task.

Since 2018, deep learning models have been increasingly applied to the field of map generalization, demonstrating significant potential in tasks such as building generalization (Feng et al., 2019; Fu et al., 2024a, 2024b; Kang et al., 2020; Sester

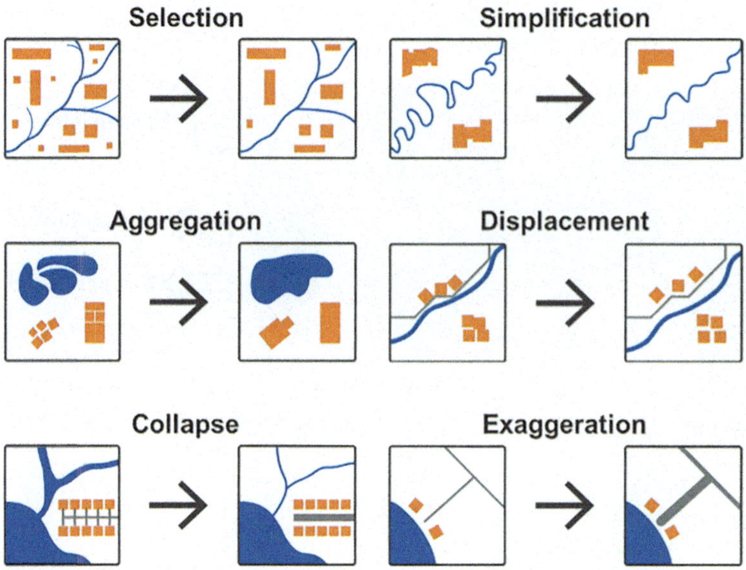

Fig. 16.2 Examples of key generalization operators[5]

et al., 2018; Touya et al., 2019; Yan & Yang, 2024; Zhou et al., 2023), road simplification (Courtial et al., 2020, 2023), coastline simplification (Du et al., 2022a, 2022b), contour line simplification (Yan & Yang, 2024; Yu & Chen, 2022), and point selection (Xiao et al., 2024). These methods can be categorized into two main types based on their input formats: (1) rasterized vector data as input and (2) vector data as input.

The former involves rasterizing the vector data, such as lines and polygons, which allows for easy integration with computer vision models like DCNNs or GANs. These end-to-end models allow for the simultaneous application of multiple operators. For example, in building generalization, this method can perform selection, simplification, and aggregation all at once (Feng et al., 2019; Fu et al., 2024a). Figure 16.3 provides several examples of generalization using deep learning methods. Through layered representation, generalization across different map features, such as lines and building polygons, can be handled in a unified workflow (Courtial et al., 2022b). This rasterized input also integrates well with building mapping tasks based on remote sensing imagery, demonstrating the potential for generating a series of multi-scale maps directly from remote sensing data (Feng et al., 2020). However, one of the major concerns is the loss of data details during the rasterization process (Harrie et al., 2024). Furthermore, the generated shapes, such as for buildings, may exhibit "wabbly-wall effect" (Fu et al., 2024a), while roads

[5] https://cartography-playground.gitlab.io/playgrounds/cartographic-generalization/.

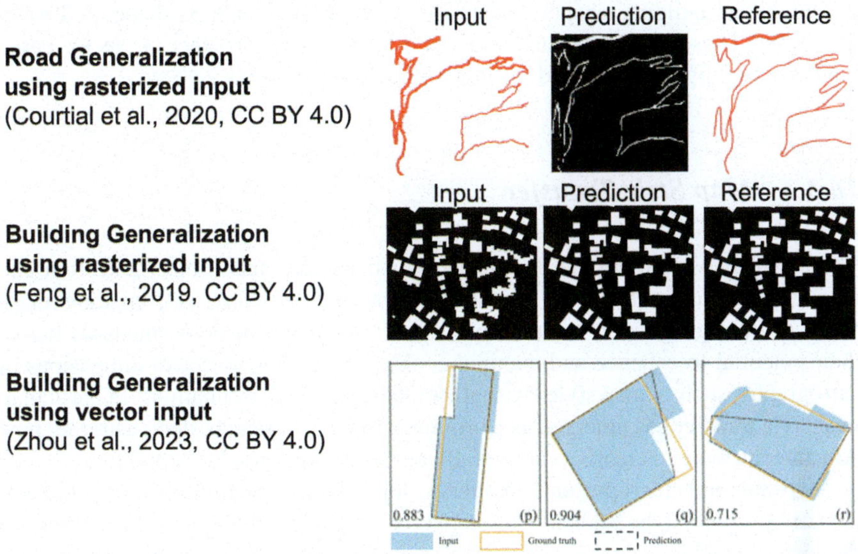

Fig. 16.3 Examples of generalization results for roads and buildings using rasterized vector data as input or vector data as input

may show issues like discontinuities in the road network topology (Courtial et al., 2023). Although several methods have been proposed to mitigate these challenges, there remains significant potential for further enhancing and refining the quality of generalization.

The latter approaches directly utilize vector data—points, lines, and polygons—as input. Inspired by successful studies in shape cognition, such as using graph representations (Yan et al., 2019, 2021; Yan & Yang, 2022) or sequence representations (Yan & Yang, 2022), line simplification in the field of map generalization has gained significant attention and yielded promising results. Du et al. (2022b) applied a multi-layer perceptron for node selection. Yan & Yang (2024) designed a Graph Autoencoder (GAE), using pooling operations to reduce nodes while predicting new node positions to simplify the shapes of lines and buildings. This method has also been applied to building shape simplification. Similarly, Zhou et al. (2023) utilized GCNs to predict node movements and removals for simplifying building shapes. Feng et al. (2023) adopted a sequence-based approach using a Transformer model to predict node movements to simplify shapes. However, vector-based building shape simplification still faces significant challenges, such as limitations on the number of nodes. Zhou et al. (2023) allow for 4–19 nodes, while Yan & Yang (2024) use a stride of 2, reducing approximately half the nodes, neither of which is ideal. In addition, most research has focused on simplifying individual lines or polygons without much consideration of surrounding entities. Methods for simplification or aggregation that effectively account for neighboring objects have not been widely explored. Meanwhile, incorrect node removal predictions may often lead to distorted or redundant building shapes, especially with more complex shapes (Zhou et al., 2023). These persistent limitations constrain the performance of vector-based methods, highlighting the need for more innovative solutions.

16.3.3 Map Style Transfer

The above two cartographic design decisions, namely, map object detection and map generalization, focus on the scientific aspect of maps, i.e., creating maps with accurate geographic information. A key advancement of AI methods lies in their potential to enhance and manipulate the artistic and subjective dimensions of cartography, such as map styles, visual variables, and map evaluations. In particular, map style transfer has emerged as a prominent area of research, demonstrating how AI can contribute to creative and visually engaging cartographic outputs.

Map style transfer represents an intersection between two broader research areas: (1) cartographic style studies (Christophe et al., 2016), which aim to manipulate the visual variables of map elements to enhance map aesthetics and functionality; and (2) style transfer (Gatys et al., 2016) that tries to transform an image by adopting the style of another image. While past attempts of style transfer required vector cartographic data as input (Christophe & Hoarau, 2012), the development of DCNNs and their derivatives like GANs enabled this crossover with style transfer.

Map style transfer research has primarily focused on two key directions. First, to transfer artistic styles from existing maps, paintings, and other visual artwork, to a map differently styled (Kang et al., 2019, 2024). The second focuses on generating styled maps from satellite imagery (Christophe et al., 2022). By leveraging abundant external sources such as online web maps, paintings, and high-temporal-frequency satellite images, map style transfer can help to update maps much quicker than with the traditional process, offering an innovative approach to map generation. In addition, map style transfer also supports stylistic diversity and personalization, offering users customized cartographic experiences.

Since the introduction of GANs such as Pix2Pix and CycleGAN (Isola et al., 2017; Zhu et al., 2017), significant strides have been made in map style transfer and map generation from satellite images. For instance, Kang et al. (2019) first generated a Google-styled map from OpenStreetMap data, and Christophe et al. (2022) created historical map styles from contemporary satellite images.

To further illustrate the recent and the remaining challenges posed by map style transfer, we focus here on the seminal use case: how to transfer the style of Google Maps to a satellite image? Figure 16.4 shows three satellite images, their counterparts in Google Maps (the version without text), and the images generated by a Pix2Pix model trained with image pairs from HuggingFace.[6]

[6] https://huggingface.co/

Fig. 16.4 Three examples of the sat2map use case that show the current challenges of this use case. The left column contains satellite images used as input in this use case (©Google). The middle column shows the reference Google Maps images of the same location that the model tries to predict (©Google). The right column shows the predictions of a Pix2Pix model (Isola et al., 2017) trained with a dataset of image pairs from HuggingFace

Overall, the maps produced by the Pix2Pix model show promising results, demonstrating a visual similarity to the true Google Maps images. However, several challenges remain in the process of achieving high-quality map style transfer. For instance, we can see topology problems (e.g., road connectivity in the middle row image). Specific loss functions or architectures were proposed to improve the topology preservation in the generated maps (Chen et al., 2020; Xu et al., 2023; Zhang et al., 2020). We can also see blurred features, e.g., in the top image with ramps of the interchange. The attempts to solve these problems are only partly successful (Ganguli et al., 2019; Li et al., 2020a; Tang et al., 2023), and the resulting images still appear as unrealistic because of this problem. Maps also contain complex patterns that might be abstracted and generalized (e.g., the highway interchange in the top image), and this abstraction problem (Kang et al., 2019) is quite different from other classical style transfer tasks such as the transformation from day to night with photographs. The final problem with these images is the lack of information in the output images. For instance, the traffic direction arrows, or the footpaths hidden by the canopy, are not visible on the bottom image, and similar problems occur when the output is supposed to include text. Recently, convincing propositions were made to solve this missing information problem (Song et al., 2023; Wu et al., 2024). Furthermore, researchers have explored the potential of editing map style sheets with generative AI. A CartoAgent framework is developed with multiple agents working collaboratively. By separating map data and styling files with different agents, it is possible to perform map style transfer by merely focusing on the artistic dimension of maps. This offers innovative solutions for map style transfer while ensuring geographic accuracy (Wang et al., forthcoming).

16.3.4 Map Evaluation

Map critique has long been an essential component of the mapmaking process, involving the systematic evaluation and enhancement of map quality (Courtial et al., 2022a). This evaluation includes assessing both geographic accuracy and visual design, ensuring that maps accurately express geographic information and are visually appealing (Kang et al., 2023). Such evaluative practices are crucial for producing maps that are not only informative but also aesthetically engaging. The integration of AI into map evaluation further enhances these processes by enabling automated quality assessments, ensuring that maps meet rigorous standards of accuracy and visual effectiveness.

Cartographers often perform map evaluation from both objective and subjective aspects. Regarding objective map evaluation methods, researchers often design quantifiable metrics to measure map quality. For example, Yang et al. (2022) assessed the effectiveness of cartographic generalization by calculating changes in position, size, orientation, and shape. Courtial et al. (2022a) evaluated the results of mountain road generalization using seven raster-based constraints, such as clutter and coalescence reduction, among others. Researchers have also deployed AI models to evaluate maps. For instance, Kang et al. (2019) trained a classifier to determine whether distinct map characteristics are retained after map style transfer. On the other hand, subjective map evaluation methods commonly involve user studies. Chen et al. (2020), for example, invited graduate students to participate in a cognitive experiment where they were asked to select the best-styled maps. Further, some researchers have collected data from participants to train models that simulate the decision-making process. This approach has led to the development of human-in-the-loop AI models for map evaluation (Kang et al., 2024). For example, Wang et al. (2024) curated a dataset containing thousands of maps along with quality ratings,

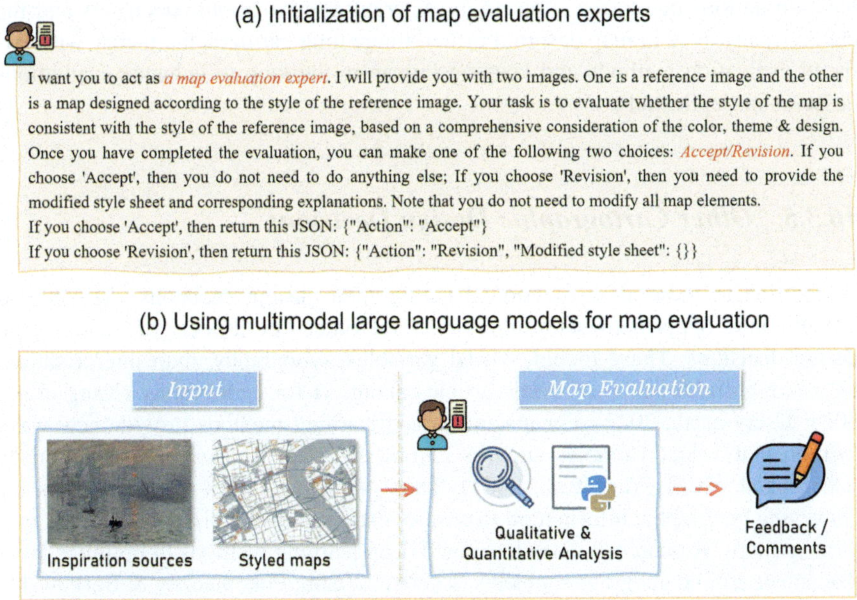

Fig. 16.5 Using a generative AI agent for map evaluation

and a Siamese neural network was subsequently trained to align with these expert ratings.

Recent advancements in generative AI have created new opportunities for map evaluation. The multimodal capabilities of generative AI models, supported by Large Language Models (LLMs), such as the groundbreaking GPT-4,[7] provide a promising solution for automating map evaluation. Generative AI models could analyze and understand data from multiple modalities, particularly visual data. Researchers have begun exploring the integration of these models into the map evaluation process. For example, as shown in Fig. 16.5, during map style transfer, researchers can utilize well-designed prompts to initialize a generative AI agent and assign it the role of a map evaluation expert. This generative AI agent can qualitatively analyze whether the style of the generated map is consistent with the reference image in terms of color, theme, and design. Additionally, it can quantitatively use tools to compare color histograms between the two images and provide constructive feedback. Through multiple rounds of map evaluation, the quality of the map product is significantly enhanced.

In summary, the advancements in AI, both deep learning-based and generative AI-based map evaluation methods, have enabled the automatic assessment of the quality of large-scale cartographic outputs. However, existing approaches still

[7] https://openai.com/index/gpt-4/

have limitations due to the subjectivity inherent in cartographic decisions and the creativity involved in map design. Further, integrating cartographic theory with AI-based evaluation methods and keeping human-in-the-loop to enhance evaluations deserves in-depth study.

16.3.5 Other Cartographic Design Decisions

In addition to these abovementioned cartographic design decisions, AI has also brought new opportunities to support several other key aspects of cartographic design decisions. These include visual variables, typography, map interpretation, etc., as highlighted in two review articles about AI for cartography (Kang et al., 2024; Usery et al., 2022). For instance, deep learning methods provide new ways to represent visual variables such as colors, size, and shapes of map elements (Chen et al., 2021; Yan et al., 2021). These visual variables are basic units for conveying geographic information to ensure readers understand the geographic patterns in maps. Researchers have leveraged deep learning methods to optimize label placement around geographic objects (Li et al., 2020b; Pokonieczny & Borkowska, 2019). AI could also help advance the understanding of relationships among map features. For instance, researchers have employed deep learning methods to identify the geographic location and extent of features depicted in maps (Evans et al., 2017). AI methods have also be leveraged to align map features across different maps to understand temporal changes of places (Shbita et al., 2020). In summary, the advent of AI has created new opportunities to enhance a wide range of cartographic design decisions. These advancements not only support cartographers in automating the mapmaking process, but also contribute to creating more visually appealing and informative maps for readers.

16.4 Ethical and Social Implications of GeoAI for Cartography

Despite the advancements in AI for cartography, the ethical and social implications of AI have not been fully addressed and discussed (Kang et al., 2024; Zhao et al., 2021). Questions such as “How to ensure maps created by AI adhere to ethical guidelines?” and “What ethical guidelines should we develop to guide AI for cartographic design decisions?” remain underexplored. To address these questions, researchers have explored several key aspects that require attention as AI continues to be leveraged into cartographic practices.

The commodification of maps as marketable products raises questions regarding intellectual property and ownership to AI-assisted cartographic products. For example, who holds the copyright for AI-generated maps? Given that AI algorithms are trained on map datasets that may include copyrighted materials, and AI developers may claim the ownership of the algorithms, more discussions around legality and ethics of ownership are essential.

Trustworthiness is another critical consideration, especially with the emergence of generative AI. Maps are always considered as a combination of scientific rigor and artistic visual communication. Given that the scientific aspect emphasizes the need for accurate representation of geographic locations, a fundamental question arises: "To what extent can we trust a machine-generated map?" (Griffin, 2020, p. 9). This question becomes even more important with the use of generative AI in mapmaking. The phenomenon of "hallucination" in generative AI—where models produce information that may not accurately reflect reality—poses a significant challenge for AI-generated maps. Such hallucinations can lead to inaccurate or misleading geographic representations, which may result in unethical maps and misinterpretation of geographic information (Kang et al., 2023). Thus, ensuring accuracy and reliability of AI-generated maps is crucial to build trust for their adoption and use in cartographic practices.

A significant factor that undermines the trustworthiness of AI in mapmaking refers to the presence of various forms of bias. When leveraging AI to support various cartographic design decisions, bias may arise at multiple stages, including dataset preparation, model and algorithm training, and the interpretations of maps. These biases often result from inaccurate and skewed representations of geographic information. More critically, AI models may amplify existing biases when learning from datasets that contain uncertainties or inherent biases. Thus, there is a special need to monitor, measure, and reduce bias throughout the process of integrating AI for cartography.

In addition to the key aspects of commodification, trustworthiness, and bias, geoprivacy and transparency are also crucial considerations for ensuring the ethical use of AI in cartography. Geoprivacy refers to protecting individuals' location information from unauthorized disclosure, suggesting the need to ensure that AI-assisted maps do not reveal personal geographic information, even if it is derived from personal data. Moreover, given the "black-box" nature of many AI models, it is essential to understand the underlying mechanism behind AI algorithms to foster trust with audience and ensure their responsible applications. To summarize, addressing the ethical and social implications of AI in cartography is necessary for ensuring its responsible use. We call for the cartographic and GIScience communities to develop comprehensive guidelines to maximize the benefits of AI in mapmaking and spatial understanding while minimizing associated risks and ethical challenges.

16.5 Conclusions

In this chapter, we provide an overview of AI in cartography, highlighting several applications to showcase how AI could benefit cartographic studies. There is no doubt that AI has brought new opportunities to the field by improving the efficiency and quality of mapmaking processes. More impressively, AI has the potential to understand and enhance creativity and preserve cartographic knowledge, pushing the boundaries of cartographic practices. In this chapter, we discussed several important topics that can be benefited from the usage of AI, including map object detection, map generalization, map style transfer, and map evaluation. In addition to these opportunities brought by AI, these advancements also bring ethical and social challenges that need more attention to ensure their responsible use. Issues such as commodification, trustworthiness, bias, geoprivacy, and transparency need to be acknowledged and actively mitigated. We call for establishing comprehensive guidelines and ethical frameworks to help navigate the evolving landscape of cartography in the era of AI and moving toward a human-centered Geospatial Data Science practice.

References

Chan, T. F., & Vese, L. A. (2001). Active contours without edges. *IEEE Transactions on Image Processing, 10*(2), 266–277.

Chen, X., Chen, S., Xu, T., Yin, B., Peng, J., Mei, X., & Li, H. (2020). SMAPGAN: Generative adversarial network-based semisupervised styled map tile generation method. *IEEE Transactions on Geoscience and Remote Sensing*, 1–19.

Chen, T., Chen, M., Zhu, A.-X., & Jiang, W. (2021). A learning-based approach to automatically evaluate the quality of sequential color schemes for maps. *Cartography and Geographic Information Science, 48*(5), 377–392.

Chiang, Y.-Y. (2010). *Harvesting geographic features from heterogeneous raster maps* [PhD Thesis]. University of Southern California.

Chiang, Y.-Y., & Knoblock, C. A. (2009). Classification of raster maps for automatic feature extraction. In *Proceedings of the 17th ACM SIGSPATIAL International Conference on Advances in Geographic Information Systems* (pp. 138–147).

Chiang, Y.-Y., Knoblock, C. A., Shahabi, C., & Chen, C.-C. (2009). Automatic and accurate extraction of road intersections from raster maps. *GeoInformatica, 13*, 121–157.

Chiang, Y.-Y., Leyk, S., & Knoblock, C. A. (2014). A survey of digital map processing techniques. *ACM Computing Surveys, 47*(1), 1–44.

Chiang, Y.-Y., Duan, W., Leyk, S., Uhl, J. H., & Knoblock, C. A. (2019). *Using historical maps in scientific studies: Applications, challenges, and best practices*. Springer.

Chiang, Y.-Y., Duan, W., Leyk, S., Uhl, J. H., & Knoblock, C. A. (2020). Creating structured, linked geographic data from historical maps: Challenges and trends. In Y.-Y. Chiang, W. Duan, S. Leyk, J. H. Uhl, & C. A. Knoblock (Eds.), *Using historical maps in scientific studies: Applications, challenges, and best practices* (pp. 37–63). Springer.

Christophe, S., & Hoarau, C. (2012). Expressive map design based on pop art: Revisit of semiology of graphics? *Cartographic Perspectives, 73*, Article 73.

Christophe, S., Duménieu, B., Turbet, J., Hoarau, C., Mellado, N., Ory, J., Loi, H., Masse, A., Arbelot, B., Vergne, R., Brédif, M., Hurtut, T., Thollot, J., & Vanderhaeghe, D. (2016). Map style formalization: Rendering techniques extension for cartography. In P. Bénard & H. Winnemöller (Eds.), *Expressive 2016 the joint symposium on computational aesthetics and sketch-based interfaces and modeling and non-photorealistic animation and rendering* (pp. 59–68). The Eurographics Association.

Christophe, S., Mermet, S., Laurent, M., & Touya, G. (2022). Neural map style transfer exploration with GANs. *International Journal of Cartography, 8*(1), 18–36.

Courtial, A., El Ayedi, A., Touya, G., & Zhang, X. (2020). Exploring the potential of deep learning segmentation for mountain roads generalisation. *ISPRS International Journal of Geo-Information, 9*(5), Article 5.

Courtial, A., Touya, G., & Zhang, X. (2022a). Constraint-based evaluation of map images generalized by deep learning. *Journal of Geovisualization and Spatial Analysis, 6*(1), 13.

Courtial, A., Touya, G., & Zhang, X. (2022b). Representing vector geographic information as a tensor for deep learning based map generalisation. *AGILE: GIScience Series, 3*, 1–8.

Courtial, A., Touya, G., & Zhang, X. (2023). Deriving map images of generalised mountain roads with generative adversarial networks. *International Journal of Geographical Information Science, 37*(3), 499–528.

Deng, J., Dong, W., Socher, R., Li, L.-J., Li, K., & Fei-Fei, L. (2009). Imagenet: A large-scale hierarchical image database. *2009 IEEE Conference on Computer Vision and Pattern Recognition*, 248–255.

Du, J., Wu, F., Xing, R., Gong, X., & Yu, L. (2022a). Segmentation and sampling method for complex polyline generalization based on a generative adversarial network. *Geocarto International, 37*(14), 4158–4180.

Du, J., Wu, F., Yin, J., Liu, C., & Gong, X. (2022b). Polyline simplification based on the artificial neural network with constraints of generalization knowledge. *Cartography and Geographic Information Science, 49*(4), 313–337.

Duan, W., Chiang, Y.-Y., Leyk, S., Uhl, J. H., & Knoblock, C. A. (2020). Automatic alignment of contemporary vector data and georeferenced historical maps using reinforcement learning. *International Journal of Geographical Information Science, 34*(4), 824–849.

Duan, W., Chiang, Y.-Y., Leyk, S., Uhl, J. H., & Knoblock, C. A. (2021a). A label correction algorithm using prior information for automatic and accurate geospatial object recognition. *2021 IEEE International Conference on Big Data (Big Data)*, 1604–1610.

Duan, W., Chiang, Y.-Y., Leyk, S., Uhl, J. H., & Knoblock, C. A. (2021b). Guided generative models using weak supervision for detecting object spatial arrangement in overhead images. *2021 IEEE International Conference on Big Data (Big Data)*, 725–734.

Evans, M. R., Mahmoody, A., Yankov, D., Teodorescu, F., Wu, W., & Berkhin, P. (2017). LiveMaps: Learning geo-intent from images of maps on a large scale. In *Proceedings of the 25th ACM SIGSPATIAL International Conference on Advances in Geographic Information Systems* (pp. 1–9).

Feng, Y., Thiemann, F., & Sester, M. (2019). Learning cartographic building generalization with deep convolutional neural networks. *ISPRS International Journal of Geo-Information, 8*(6), Article 6.

Feng, Y., Yang, C., & Sester, M. (2020). Multi-scale building maps from aerial imagery. *The International Archives of the Photogrammetry, Remote Sensing and Spatial Information Sciences, XLIII-B3-2020*, 41–47.

Feng, Y., Qian, N., Fu, C., Zhou, Z., & Sester, M. (2023). PolygonTranslator—Learning to simplify building footprints from one scale to another. *Abstracts of the ICA, 6*, 1–2.

Fu, C., Zhou, Z., Feng, Y., & Weibel, R. (2024a). Keeping walls straight: Data model and training set size matter for deep learning in building generalization. *Cartography and Geographic Information Science, 51*(1), 130–145.

Fu, C., Zhou, Z., Xin, Y., & Weibel, R. (2024b). Reasoning cartographic knowledge in deep learning-based map generalization with explainable AI. *International Journal of Geographical Information Science, 38*(10), 2061–2082.

Ganguli, S., Garzon, P., & Glaser, N. (2019). GeoGAN: A conditional GAN with reconstruction and style loss to generate standard layer of maps from satellite images. *arXiv Preprint arXiv: 1902.05611.*

Gatys, L. A., Ecker, A. S., & Bethge, M. (2016). Image style transfer using convolutional neural networks. *Proceedings of the IEEE Conference on Computer Vision and Pattern Recognition*, 2414–2423.

Griffin, A. L. (2020). Trustworthy maps. *Journal of Spatial Information Science, 2020*(20), 5–19.
Harrie, L., Touya, G., Oucheikh, R., Ai, T., Courtial, A., & Richter, K.-F. (2024). Machine learning in cartography. *Cartography and Geographic Information Science, 51*(1), 1–19.
Isola, P., Zhu, J.-Y., Zhou, T., & Efros, A. A. (2017). Image-to-image translation with conditional adversarial networks. *Proceedings of the IEEE Conference on Computer Vision and Pattern Recognition*, 1125–1134.
Kang, Y., Gao, S., & Roth, R. E. (2019). Transferring multiscale map styles using generative adversarial networks. *International Journal of Cartography, 5*(2–3), 115–141.
Kang, Y., Rao, J., Wang, W., Peng, B., Gao, S., & Zhang, F. (2020). Towards cartographic knowledge encoding with deep learning: A case study of building generalization. *Proceedings of the AutoCarto*, 1–6.
Kang, Y., Zhang, Q., & Roth, R. (2023). The ethics of AI-generated maps: A study of DALLE 2 and implications for cartography. *arXiv Preprint arXiv: 2304.10743*.
Kang, Y., Gao, S., & Roth, R. E. (2024). Artificial intelligence studies in cartography: A review and synthesis of methods, applications, and ethics. *Cartography and Geographic Information Science, 51*(4), 599–630.
Karsznia, I., & Weibel, R. (2018). Improving settlement selection for small-scale maps using data enrichment and machine learning. *Cartography and Geographic Information Science, 45*(2), 111–127.
Kitchin, R., & Dodge, M. (2007). Rethinking maps. *Progress in Human Geography, 31*(3), 331–344.
Krygier, J. B. (1995). Cartography as an art and a science? *The Cartographic Journal, 32*(1), 3–10.
Li, J., Chen, Z., Zhao, X., & Shao, L. (2020a). MapGAN: An intelligent generation model for network tile maps. *Sensors, 20*(11), Article 11.
Li, Y., Sakamoto, M., Shinohara, T., & Satoh, T. (2020b). Automatic label placement of area-features using deep learning. *The International Archives of the Photogrammetry, Remote Sensing and Spatial Information Sciences*, *XLIII-B4-2020*, 117–122.
Pérez Nava, F., & Sánchez Berriel, I. (2024). Automatic detection and 3D reconstruction of buildings from historical maps. *Applied Artificial Intelligence, 38*(1), 2378274.
Pokonieczny, K., & Borkowska, S. (2019). Using artificial neural network for labelling polygon features in topographic maps. *GeoScape, 13*(2), 125–131.
Robinson, A. C., Demšar, U., Moore, A. B., Buckley, A., Jiang, B., Field, K., Kraak, M.-J., Camboim, S. P., & Sluter, C. R. (2017). Geospatial big data and cartography: Research challenges and opportunities for making maps that matter. *International Journal of Cartography, 3*(sup1), 32–60.
Sester, M., Feng, Y., & Thiemann, F. (2018). Building generalization using deep learning. *The International Archives of the Photogrammetry, Remote Sensing and Spatial Information Sciences, XLII–4*, 565–572.
Shbita, B., Knoblock, C. A., Duan, W., Chiang, Y.-Y., Uhl, J. H., & Leyk, S. (2020). Building linked spatio-temporal data from vectorized historical maps. In I. A. Harth, S. Kirrane, A.-C. Ngonga Ngomo, H. Paulheim, A. Rula, A. L. Gentile, P. Haase, & M. Cochez (Eds.), *The semantic web* (pp. 409–426). Springer.
Simonyan, K., & Zisserman, A. (2014). Very deep convolutional networks for large-scale image recognition. *arXiv Preprint arXiv:1409.1556*.
Smith, C. D. (1982). The emergence of "Maps" in European Rock Art: A prehistoric preoccupation with place. *Imago Mundi, 34*, 9–25.
Sohn, K., Lee, H., & Yan, X. (2015). Learning structured output representation using deep conditional generative models. *Advances in Neural Information Processing Systems*, 28.
Song, J., Chen, H., Du, C., & Li, J. (2023). Semi-MapGen: Translation of remote sensing image into map via semisupervised adversarial learning. *IEEE Transactions on Geoscience and Remote Sensing, 61*, 1–19.
Tang, H., Liu, H., Xu, D., Torr, P. H. S., & Sebe, N. (2023). AttentionGAN: Unpaired image-to-image translation using attention-guided generative adversarial networks. *IEEE Transactions on Neural Networks and Learning Systems, 34*(4), 1972–1987.

Touya, G., Zhang, X., & Lokhat, I. (2019). Is deep learning the new agent for map generalization? *International Journal of Cartography, 5*(2–3), 142–157.
Usery, E. L., Arundel, S. T., Shavers, E., Stanislawski, L., Thiem, P., & Varanka, D. (2022). GeoAI in the US Geological Survey for topographic mapping. *Transactions in GIS, 26*(1), 25–40.
Wang, C., Chen, T., Liu, Y., Kang, M., Su, S., & Li, B. (2024). TransMI: A transfer-learning method for generalized map information evaluation. *Cartography and Geographic Information Science*, 1–17.
Wang, C., Kang, Y., Gong, Z., Pengjun, Z., Feng, Y., Zhang, W., & Li, G. (forthcoming). CartoAgent: A multimodal large language model-powered multi-agent cartographic framework for map style transfer and evaluation.
Wu, S., Chen, Y., Mermet, S., Hurni, L., Schindler, K., Gonthier, N., & Landrieu, L. (2024). SteGoGAN: Leveraging steganography for non-bijective image-to-image translation. *Proceedings of the IEEE/CVF Conference on Computer Vision and Pattern Recognition*, 7922–7931.
Xiao, T., Ai, T., Yu, H., Yang, M., & Liu, P. (2024). A point selection method in map generalization using graph convolutional network model. *Cartography and Geographic Information Science, 51*(1), 20–40.
Xu, J., Zhou, X., Han, C., Dong, B., & Li, H. (2023). SAM-GAN: Supervised learning-Based aerial image-to-map translation via generative adversarial networks. *ISPRS International Journal of Geo-Information, 12*(4), Article 4.
Yan, X., & Yang, M. (2022). A comparative study of various deep learning approaches to shape encoding of planar geospatial objects. *ISPRS International Journal of Geo-Information, 11*(10), Article 10.
Yan, X., & Yang, M. (2024). A deep learning approach for polyline and building simplification based on graph autoencoder with flexible constraints. *Cartography and Geographic Information Science, 51*(1), 79–96.
Yan, X., Ai, T., Yang, M., & Yin, H. (2019). A graph convolutional neural network for classification of building patterns using spatial vector data. *ISPRS Journal of Photogrammetry and Remote Sensing, 150*, 259–273.
Yan, X., Ai, T., Yang, M., & Tong, X. (2021). Graph convolutional autoencoder model for the shape coding and cognition of buildings in maps. *International Journal of Geographical Information Science, 35*(3), 490–512.
Yang, M., Yuan, T., Yan, X., Ai, T., & Jiang, C. (2022). A hybrid approach to building simplification with an evaluator from a backpropagation neural network. *International Journal of Geographical Information Science, 36*(2), 280–309.
Yu, W., & Chen, Y. (2022). Data-driven polyline simplification using a stacked autoencoder-based deep neural network. *Transactions in GIS, 26*(5), 2302–2325.
Zhang, Y., Yin, Y., Zimmermann, R., Wang, G., Varadarajan, J., & Ng, S.-K. (2020). An enhanced GAN model for automatic satellite-to-map image conversion. *IEEE Access, 8*, 176704–176716.
Zhao, H., Shi, J., Qi, X., Wang, X., & Jia, J. (2017). Pyramid scene parsing network. *Proceedings of the IEEE Conference on Computer Vision and Pattern Recognition*, 2881–2890.
Zhao, B., Zhang, S., Xu, C., Sun, Y., & Deng, C. (2021). Deep fake geography? When geospatial data encounter Artificial Intelligence. *Cartography and Geographic Information Science, 48*(4), 338–352.
Zhou, Z., Fu, C., & Weibel, R. (2023). Move and remove: Multi-task learning for building simplification in vector maps with a graph convolutional neural network. *ISPRS Journal of Photogrammetry and Remote Sensing, 202*, 205–218.
Zhu, J.-Y., Park, T., Isola, P., & Efros, A. A. (2017). Unpaired image-to-image translation using cycle-consistent adversarial networks. *Proceedings of the IEEE international conference on computer vision*, 2223–2232.

Chapter 17
GeoAI and Transportation

Xiao Li and Si Qiao

Abstract This chapter examines the transformative potential of geospatial artificial intelligence (GeoAI) in reshaping transportation systems, offering a concise overview of its innovative applications. Centered on four key components of transportation—modes, infrastructures, networks, and flows—we explore how emerging geospatial datasets and representative artificial intelligence (AI) solutions are being applied to address complex transportation challenges. Rather than providing an exhaustive review, we highlight representative applications of GeoAI, including transportation mode identification, road asset management, connectivity and accessibility assessment, and traffic demand prediction. By showcasing these innovations, this chapter serves as a starting point for further exploration and discussion on how GeoAI can revolutionize transportation systems, paving the way for more efficient, accessible, and sustainable mobility futures.

Keywords GeoAI · Transportation modes · Road asset management · Network connectivity · Traffic demand forecasting

X. Li (✉)
Transport Studies Unit, School of Geography and the Environment, University of Oxford, Oxford, UK

Linacre College, University of Oxford, Oxford, UK
e-mail: xiao.li@ouce.ox.ac.uk

S. Qiao
Department of Urban Planning and Design, The University of Hong Kong, Hong Kong, SAR, China

Center on Poverty and Inequality, Stanford University, Stanford, CA, USA
e-mail: qiaosi@hku.hk

© The Author(s), under exclusive license to Springer Nature Switzerland AG 2025

X. Huang et al. (eds.), *GeoAI and Human Geography*, Springer Geography,
https://doi.org/10.1007/978-3-031-87421-5_17

17.1 Introduction

> Transportation is the center of the world! It is the glue of our daily lives. When it goes well, we don't see it. When it goes wrong, it negatively colors our day, makes us feel angry and impotent, curtails our possibilities. (Robin Chase)

Transportation is a cornerstone of modern society, significantly influencing daily life across various domains, including the economy, technology, environment, and public health, among others (Giuliano, 2001; Rodrigue, 2024).

From an economic perspective, transportation is closely linked to economic growth. For example, over 10 million people are employed in the transport industry, contributing to 5% of the European Union's gross domestic product (GDP). The transportation sector is also at the forefront of groundbreaking technological advancements. Powered by renewable energy and artificial intelligence (AI), emerging transportation technologies such as connected and autonomous vehicles (CAVs), delivery drones, micromobility vehicles, and high- or super-sonic speed vehicles are becoming a reality, revolutionizing how people live and travel. However, it is worth noting that the transport sector remains a major contributor to energy-related emissions, accounting for approximately one-quarter of global greenhouse gas emissions. Approximately 95% of the world's transportation energy is still derived from fossil fuels. From a public health perspective, transportation significantly influences human lives in multiple ways. It enables access to healthcare services, facilitates social interactions and inclusion, and promotes physical and mental health through activities such as walking and cycling. However, road traffic injuries persist as a leading cause of preventable deaths, with more than a million people losing their lives worldwide each year.

As Elijah Cummings said: "*Our transportation decisions determine much more than where roads or bridges or tunnels or rail lines will be built. They determine the connections and barriers that people will encounter in their daily lives, and thus how hard or easy it will be for people to get where they need and want to go.*" With the emergence of geospatial artificial intelligence (GeoAI), it is worthwhile for transportation researchers and practitioners to consider how best to embrace this transformative technology to continue strengthening mobility capabilities while navigating and addressing the challenges posed by our transportation systems.

17.1.1 Key Components of Transportation

The study of transportation is closely related to understanding the mobility of people, goods, and information across space. From the perspective of transportation geography, mobility is shaped by three key concepts: nodes, networks, and demand. Nodes function as the origin, transit, and destination points for movements. Networks represent the infrastructures and means that support these movements. Demand reflects the underlying "forces" that drive the generation of different types of movements.

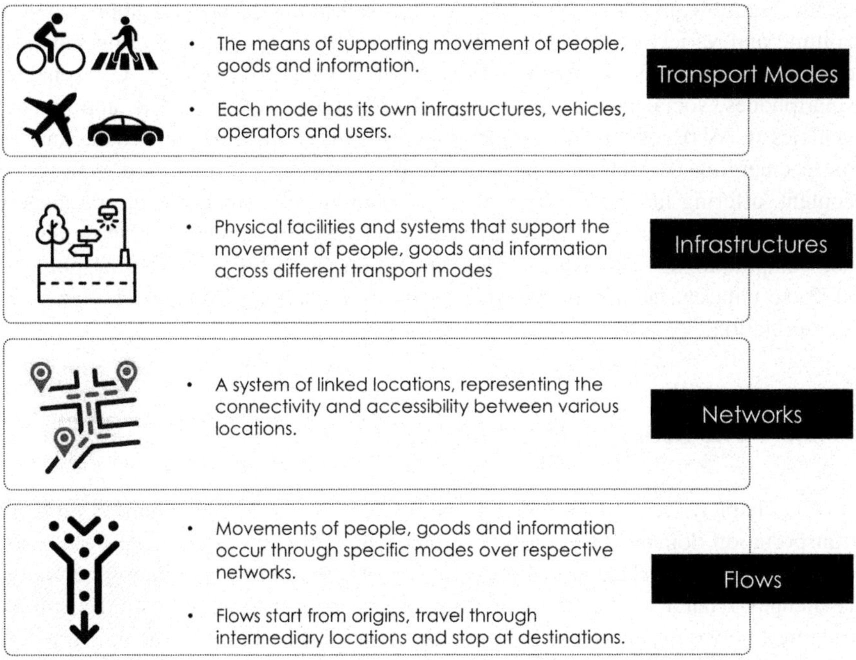

Fig. 17.1 Core components of transportation. (Adapted from Rodrigue (2024))

Similarly, Dr. Jean-Paul Rodrigue identified four interconnected core components of transportation: modes, infrastructures, networks, and flows, as illustrated in Fig. 17.1 (Rodrigue, 2024). Modes and infrastructures explain "how" people, goods, and information are transported, focusing on the means and the physical facilities that enable transportation. Networks focus on the linkages between locations, illustrating how places are connected and served via transportation systems. Flows emphasize the processes of movements, examining where "flows" originate, travel through, and end, as well as who participates in and is influenced by these processes.

17.1.2 Key Components of AI

While the idea of combining AI with geographical problem-solving dates back to the 1980s, the term GeoAI was officially coined at an Association for Computing Machinery (ACM) workshop in 2017 (Li, 2022). Since then, GeoAI has rapidly emerged as one of the most exciting and disruptive technologies in the field of geospatial analytics. Its growing popularity is fueled by advancements in three key areas: geospatial Big Data, AI, and high-performance computing (HPC).

We are now in the era of "Big Data," characterized by the unprecedented volume and variety of geotagged data continually generated and collected from diverse sources such as satellites, Internet of Things (IoT) sensors, GPS-enabled smartphones, social media, street view images (SVI), connected & autonomous vehicles (CAVs), among others. Compared to traditional geospatial data sources, these emerging datasets are massive in size, diverse in format, and granular in content, offering far richer information for studying the environment and society. Coupled with recent breakthroughs in AI and HPC, researchers are now able to use computationally intensive models and solutions to fully harness the potential of these datasets, addressing a wide range of complex geospatial challenges in transportation.

17.1.3 The Scope of This Chapter

In this chapter, we will provide an overview of GeoAI applications within the transportation domain, offering a "sampler plate" to showcase the innovative use of "new" geospatial data and "representative" AI solutions in addressing geospatial challenges related to the four key components of transportation: modes, infrastructures, networks, and flows. While transportation encompasses the movement of people, goods, and information, this chapter primarily focuses on people-centered transportation.

The purpose of this chapter is not to provide an exhaustive collection of GeoAI technologies and applications, but rather to serve as a starting point for exploring how GeoAI can transform transportation systems and shape their future.

17.2 Example Applications of GeoAI in Transportation

GeoAI has demonstrated transformative potential in addressing various challenges and improving efficiency in transportation systems. In this section, we present some representative examples of how GeoAI is applied across different components of transportation, including transportation mode identification (TMI), road asset management (RAM), connectivity and accessibility, and traffic demand forecasting.

17.2.1 GeoAI in Transportation Modes—Supporting Transportation Mode Identification

When exploring foundational applications of GeoAI in transportation, transportation mode identification (TMI) stands out as one of the most extensively studied and applied areas. TMI underpins the understanding of mobility patterns and informs planning interventions across transportation systems. By determining whether individuals travel by walking, cycling, driving, or using public transit, accurate mode detection becomes essential for studying travel behavior, emissions modeling, transportation planning, and infrastructure optimization.

Advancements in GeoAI have greatly enhanced the accuracy and scalability of TMI. Leveraging multi-source geospatial data—including GPS trajectories, accelerometer readings, smartphone gyroscopes, and external datasets like weather information and road networks—GeoAI employs advanced machine learning and deep learning algorithms to achieve high precision in distinguishing transportation modes (e.g., Qiao & Yeh, 2023b; Reddy et al., 2008; Semanjski et al., 2017; Shin et al., 2015; Stenneth et al., 2011; Zhou et al., 2016). These models utilize features such as speed, acceleration, stop frequency, and route patterns to classify modes of transport effectively.

Moreover, GeoAI introduces a transformative layer to TMI by integrating spatial context data into movement pattern analysis. This approach goes beyond traditional indicators like speed and acceleration, incorporating characteristics of the surrounding environment—such as proximity to roads, cycleways, transit stops, and railways—for a nuanced understanding of mobility. For example, Semanjski et al. (2017) demonstrated the effectiveness of combining spatial context mining with mobile-sensed data to infer transportation modes. Using OpenStreetMap (OSM) data to define spatial features and GPS data from 8000 participants, their Support Vector Machine (SVM) model achieved a 94% success rate in identifying modes such as walking, cycling, driving, and public transit. Similarly, Hong et al. (2023) assessed the impact of various geospatial context features on travel mode detection using GNSS trajectory data. Their study employed a random forest model combined with SHapley Additive exPlanation (SHAP) values to determine feature importance. Graells-Garrido et al. (2023) proposed a method that combines mobile phone application usage data with official datasets to infer transportation modes, including biking and ride-hailing services. This data fusion approach enables the updating of transportation mode distributions, providing insights into evolving urban mobility patterns and supporting sustainable transportation planning. The TSANET model exemplifies the application of deep learning in TMI (Zhu et al., 2024). By extracting and integrating global and local spatiotemporal features from GPS trajectory data, TSANET effectively distinguishes between similar transportation modes, such as taxis and private cars. This approach addresses challenges like traffic congestion, where traditional indicators may fall short, by leveraging the spatial context of transportation networks.

The value of accurate TMI extends beyond classification. It serves as a critical input for applications like urban planning, emissions monitoring, and personalized mobility services. A prominent example is its application in autonomous vehicles (AVs), which rely on TMI to make informed, context-aware decisions. By discerning the modes of surrounding entities—such as pedestrians, cyclists, and other vehicles—AVs can navigate urban environments more effectively, improving reaction times and reducing collision risks. Advanced models like MapsTP integrate high-definition (HD) maps with multimodal trajectory prediction to provide AVs with granular environmental details, including lane boundaries, road geometry, and traffic signs, alongside real-time sensor data like speed, acceleration, and yaw rate (Sharma et al., 2024). This integration enables AVs to predict the behavior of surrounding road users and assign probabilities to multiple potential trajectories, facilitating safer navigation. Similarly, models like Trajectron++ utilize heterogeneous data sources, such as vehicle positions and pedestrian dynamics, to forecast movement patterns. These models enhance AV perception by distinguishing transportation modes, allowing vehicles to adapt to diverse scenarios (Salzmann et al., 2020). Mode-aware systems like GoalNet further refine trajectory prediction by incorporating spatial contexts such as traffic direction and road classifications, directly modeling goal-directed behaviors of road users, and ensuring safe and efficient navigation. These advancements demonstrate how mode identification not only improves AV decision-making but also supports cooperative transportation systems by integrating with Vehicle-to-Everything (V2X) communication.

17.2.2 GeoAI in Transportation Infrastructures—Facilitating Road Asset Management

Transportation infrastructures form the backbone of mobility, enabling the movement of people and goods. Effectively monitoring, managing, and maintaining road assets, commonly referred to as road asset management (RAM), is crucial for supporting transportation agencies and authorities in planning, constructing, maintaining, and replacing road assets (Kargah-ostadi et al., 2020; Li et al., 2023a). Key RAM tasks include road asset digitization and inventory, condition monitoring and assessment, and semantic information detection and evaluation.

Digitizing and inventorying road assets are foundational tasks in RAM. Traditionally, road assets were manually digitized, which was both labor-intensive and time-consuming. Recent advancements in geospatial data technologies, such as street view images (SVIs), LiDAR, and high-resolution aerial imagery, have revolutionized this process. SVIs, in particular, have increasingly been used for road asset detection due to their ability to capture detailed streetscapes similar to human vision. A growing body of research has leveraged SVIs in combination with AI algorithms to achieve road asset digitization. For example, X. Li et al. (2021a) used road network data and SVIs to digitize traffic signal-controlled road intersections. Their method involved generating intersections from road network shapefiles, retrieving SVIs for these locations, and applying a deep learning-based segmentation model—Seamless Scene Segmentation (SeamSeg)—to identify traffic signals in the imagery. M. Li et al. (2023b) systematically reviewed the application of SVIs and LiDAR, combined with various AI methods, including machine and deep learning, to recognize and map traffic control infrastructures, such as traffic signs, signals, and pavement markings. Zhou et al. (2022) innovatively fused SVIs with point cloud data obtained through mobile laser scanning and implemented a multi-task neural network to recognize and inventory multiple street infrastructures and furniture (e.g., trash bins, benches) simultaneously.

In addition to identifying the location of road assets, assessing and monitoring their quality is equally important. Many studies have used GeoAI solutions to assess and monitor road surface conditions, such as slippery/roughness assessment, pavement distress evaluation, and pothole detection, among others (Li et al., 2014; Ranyal et al., 2022; Sattar et al., 2018). Various sensors (e.g., accelerometer, gyroscope, GPS) from smartphones, telematics devices, and vehicle onboard systems have been used to capture abnormal vehicle vibrations when driving on bumpy roads or hitting potholes (Li et al., 2019; Ranyal et al., 2022). Combined with AI algorithms, these road surface anomalies can be effectively identified and mapped. In addition to vibration data, vision-based data such as RGB or multispectral images obtained through drones, videos from dashboard cameras, and point clouds from LiDAR are also being used, achieving good performance in monitoring road surface conditions (Gopalakrishnan et al., 2017; Wu et al., 2019). Ranyal et al. (2022) systematically reviewed the application of various emerging spatial data obtained via smart sensing techniques combined with AI to perform road condition monitoring, further cementing the potential and growing popularity of applying GeoAI in this field.

Beyond the physical characteristics, such as location, type, and condition, the semantic information of road assets, such as speed limits, lane-turning indications, as well as alerts and warnings, plays an important role in traffic control. However, the inventorying and assessment of semantic information consistency across different facilities along roadways have largely been overlooked. For example, you may have encountered driving situations where a pavement marking indicates that a U-turn is allowed, while a traffic sign indicates differently. Such semantic conflicts can confuse both human-drivers and self-driving cars. To address this challenge, Shi et al. (2024) introduced the concept of semantic space—structured semantic information conveyed by all pavement markings and traffic signs along a road segment—to assess semantic consistency. They used SVIs combined with deep learning algorithms to extract semantic information from different road facilities, which was then geolocated and structured using IMU and GNSS data to assess consistency. This research highlights a new application of GeoAI in RAM, emphasizing the inventorying and quality assessment of semantic information.

17.2.3 GeoAI in Transportation Networks—Revolutionizing Connectivity and Accessibility

Transportation networks represent linkages between places, comprising interconnected nodes (e.g., stations, intersections) and links (e.g., roads, rail lines, airways). GeoAI offers a transformative toolkit to analyze, model, and enhance transportation networks, key to improving efficiency, accessibility, and sustainability for future cities.

First, in terms of analyzing transportation networks, GeoAI provides a comprehensive framework by using graph-based models and machine learning techniques. These methods allow researchers to identify bottlenecks, evaluate resilience, and uncover hidden inefficiencies in complex systems. GeoAI-enabled network analyses have been used to identify critical nodes in urban road networks, such as major intersections or transit hubs (Zhang et al., 2019; Zhao et al., 2024; Zhou et al., 2015). These nodes often represent high-flow areas that, if disrupted, could significantly impact overall network performance. By incorporating real-time traffic data, Wang et al. (2022) proposed a general framework that integrates statistical learning with traffic flow models, using low-penetration-rate vehicle trajectories to estimate parameters in signalized arterial corridors.

GeoAI also improves network connectivity by integrating multimodal datasets and simulating traveler behaviors. It facilitates the identification of last-mile connectivity gaps, guiding infrastructure improvements like pedestrian walkways or bike-sharing stations (Matthes et al., 2023). Similarly, machine learning models have been deployed to optimize bus routes by predicting passenger demand and adapting schedules to align with peak travel times (Wu et al., 2020), optimizing service in low travel density areas (Shu & Li, 2022), and integrating multimodal travel modes (Qiao & Yeh, 2023a).

GeoAI's integration with other emerging technologies and real-time data streams enables the creation of smart transportation networks. These networks can adapt to changing conditions by dynamically rerouting flows, optimizing traffic signals, and managing demand for public transit. The Dynamic Traffic Optimization in Smart Cities (DTOS) framework utilizes real-time data from IoT devices to continuously update traffic network models, allowing for proactive adjustments that minimize travel time and reduce congestion (Le et al., 2024). Similarly, IoT-enabled real-time traffic monitoring systems have been developed to enhance intelligent transportation systems by processing traffic flow data and enabling automated control measures (Dui et al., 2024). Studies using deep reinforcement learning models have also been applied to real-time traffic management, enabling systems to automatically adapt to changing traffic conditions and optimize flow (Singh, 2023). These advancements demonstrate how the convergence of GeoAI with other emerging technologies facilitates the development of responsive and efficient transportation networks.

17.2.4 GeoAI in Transportation Flows—Advancing Traffic Demand Prediction

Understanding traffic demand is critically important for various stakeholders. Accurate demand forecasting, for example, enables transport authorities to design, optimize, and implement infrastructure and policies that effectively meet current and future transportation needs. For providers of shared mobility services, such as ride-hailing, bike-sharing, and shared e-scooters, accurately estimating service demand is essential to optimize operations, improve user satisfaction, and maximize profitability. Ultimately, such optimizations improve transportation experiences for all users.

To achieve accurate traffic prediction, acquiring high-quality traffic flow data is essential. Traditionally, traffic volume was collected manually by data officers or through fixed-location sensors, such as inductive loop detectors, cameras, or LiDAR, which typically only capture data from spatially limited areas. With the prevalence of IoT technologies, the trajectories of travel activities can now be effectively recorded by various means, including mobile phones, telematics devices, fleet vehicles, connected and autonomous vehicles (CAVs), shared bikes, and e-scooters, among others. Additionally, trip record data, such as origin and destination (OD) pairs with associated timestamps for departure and arrival times, distance, and speed, are aggregated by various providers of public and shared transport services and have now become publicly accessible (Jalilifar et al., 2024; Li et al., 2021b; Shaygan et al., 2022).

Based on these data sources, various AI algorithms have been employed to learn the spatial and temporal patterns of traffic volumes. Convolutional Neural Networks (CNNs) and Recurrent Neural Networks (RNNs), along with their variants, are commonly used due to their excellent performance in capturing spatial variations with regular convolutional kernels and in extracting temporal dynamics from sequential data, respectively (Ma et al., 2017; Shaygan et al., 2022).

Some studies have innovatively combined CNN and RNN to propose hybrid AI frameworks that better capture the spatiotemporal dynamics of traffic demand. For example, Ren et al. (2020) combined the Long Short-Term Memory (LSTM) neural network—a specific type of RNN—with CNNs to comprehensively capture both dynamic temporal dependencies and spatial dependencies from historical traffic flows. This hybrid model was trained on GPS-based trajectory data collected from city-wide taxis and demonstrated superior prediction accuracy compared to several standard deep learning models.

On the other hand, some researchers have challenged the baseline assumption of CNNs—that neighboring regions (adjacent pixels) are highly correlated— and argue that this assumption limits CNN performance in certain applications. For example, X. Li et al. (2023b) examined the traffic demand of bike-sharing services and argued that bike usage in neighboring urban regions can vary significantly due to differences in the built environment and spatial variations in travel behavior. Conversely, some distant regions may exhibit similar usage patterns due to shared land use or commuting behavior, challenging the assumption that spatial proximity equates to behavioral similarity, they proposed the concept of "semantic neighbors"—the top eight regions (cells) with the highest similarity in temporal bike usage patterns to the central cell—and developed an Irregular Convolutional Long Short-Term Memory Network (IrConv+LSTM) model. The model was applied to bike-sharing datasets, including departure and arrival times and station locations, from five cities. Results showed that this new model outperformed traditional CNN models, suggesting that GeoAI should think beyond spatial neighbors.

17.3 Concluding Remarks

The integration of GeoAI into transportation systems presents transformative opportunities to enhance mobility, but several significant challenges must be addressed to ensure its successful and equitable application.

First, GeoAI heavily relies on high-quality data. However, deprived communities and underdeveloped countries often lack access to such data, rendering many GeoAI applications infeasible in these areas. This disparity in data availability also limits the ability of GeoAI algorithms to "learn" from these regions, resulting in biased performance in data-deficient environments.

Second, GeoAI solutions may inadvertently perpetuate or exacerbate inequities if they fail to account for underrepresented groups or regions. Current transportation datasets often lack information on the mobility needs and challenges of vulnerable populations, such as children, older adults, and people with disabilities. Without incorporating this critical information, the implementation of GeoAI may unintentionally reinforce existing disparities, leading to inequitable outcomes in transportation design and management.

Finally, collecting and analyzing sensitive geospatial and transportation data (e.g., GPS traces or traffic camera feeds) raises significant privacy concerns. Navigating legal and policy requirements related to data use, AI deployment, and infrastructure modification poses a substantial challenge to the adoption of GeoAI in transportation systems.

References

Dui, H., Zhang, S., Liu, M., Dong, X., & Bai, G. (2024). IoT-enabled real-time traffic monitoring and control management for intelligent transportation systems. *IEEE Internet of Things Journal., 11*, 15842.

Giuliano, G. (2001). Transportation geography. *In International Encyclopedia of the Social & Behavioral Sciences., 10*(4), 15866–15873.

Gopalakrishnan, K., Khaitan, S. K., Choudhary, A., & Agrawal, A. (2017). Deep convolutional neural networks with transfer learning for computer vision-based data-driven pavement distress detection. *Construction and Building Materials, 157*, 322.

Graells-Garrido, E., Opitz, D., Rowe, F., & Arriagada, J. (2023). A data fusion approach with mobile phone data for updating travel survey-based mode split estimates. *Transportation Research Part C: Emerging Technologies, 155*, 104285.

Hong, Y., Stüdeli, E., & Raubal, M. (2023). Evaluating geospatial context information for travel mode detection. *Journal of Transport Geography, 113*, 103736.

Jalilifar, E., Li, X., Martin, M., & Huang, X. (2024). Exploring the potential of market-available connected vehicle data in border crossing time estimation. Transactions in urban data, science, and technology.

Kargah-ostadi, N., Waqar, A., & Hanif, A. (2020). Automated real-time roadway asset inventory using artificial intelligence. *Transportation Research Record, 2674*(11), 220–234.

Le, T. V., Tran, H. T., & Le, D. L. (2024). Dynamic Traffic Optimization in Smart Cities (DTOS): Integrating OpenStreetMap, IoT, and Fog Computing. *SN Computer Science, 5*(7), 823.

Li, W. (2022). GeoAI in social science. In *Handbook of spatial analysis in the social sciences*. https://doi.org/10.4337/9781789903942.00025

Li, X., Chen, R., & Chu, T. (2014). A crowdsourcing solution for road surface roughness detection using smartphones. 27th International Technical Meeting of the Satellite Division of the Institute of Navigation, ION GNSS 2014.

Li, X., Huo, D., Goldberg, D. W., Chu, T., Yin, Z., & Hammond, T. (2019). Embracing crowdsensing: An enhanced mobile sensing solution for road anomaly detection. *ISPRS International Journal of Geo-Information, 8*(9), 412.

Li, X., Ning, H., Huang, X., Kang, Y., Dadashova, B., & Ma, A. (2021a). Urban infrastructure audit: An effective protocol to digitize signalized intersections by mining street view images. *Cartography and Geographic Information Science, 49*(1), 32–49.

Li, X., Xu, H., Huang, X., Guo, C., Kang, Y., & Ye, X. (2021b). *Emerging geo-data sources to reveal human mobility dynamics during COVID-19 pandemic: Opportunities and challenges* (Vol. 1, p. 22). Computional Urban Science.

Li, M., Li, X., Wu, D., Zou, L., & Huang, X. (2023a). Digitizing and inventorying traffic control infrastructures: A review of practices. *Transportation Research Interdisciplinary Perspectives, 21*, 100879.

Li, X., Xu, Y., Zhang, X., Shi, W., Yue, Y., & Li, Q. (2023b). Improving short-term bike sharing demand forecast through an irregular convolutional neural network. *Transportation Research Part C: Emerging Technologies, 147*, 103984.

Ma, X., Dai, Z., He, Z., Ma, J., Wang, Y., & Wang, Y. (2017). Learning traffic as images: A deep convolutional neural network for large-scale transportation network speed prediction. *Sensors (Switzerland), 17*(4).

Matthes, P., Jeschor, D., & Springer, T. (2023). GeoAI-powered lane matching for bike routes in GLOSA Apps. Proceedings of the 31st ACM International Conference on Advances in Geographic Information Systems.

Qiao, S., & Yeh, A. G.-O. (2023a). Mobility-on-demand public transport toward spatial justice: Shared mobility or mobility as a service. *Transportation Research Part D: Transport and Environment, 123*, 103916.

Qiao, S., & Yeh, A. G.-O. (2023b). Understanding the effects of environmental perceptions on walking behavior by integrating big data with small data. *Landscape and Urban Planning, 240*, 104879.

Ranyal, E., Sadhu, A., & Jain, K. (2022). Road condition monitoring using smart sensing and artificial intelligence: A review. *Sensors, 22*(8), 3044.

Reddy, S., Burke, J., Estrin, D., Hansen, M., & Srivastava, M. (2008). Determining transportation mode on mobile phones. 2008 12th IEEE international symposium on wearable computers.

Ren, Y., Chen, H., Han, Y., Cheng, T., Zhang, Y., & Chen, G. (2020). A hybrid integrated deep learning model for the prediction of citywide spatio-temporal flow volumes. *International Journal of Geographical Information Science, 34*(4), 802.

Rodrigue, J.-P. (2024). The geography of transport systems. In *The geography of transport systems* (6th ed.). Routledge.

Salzmann, T., Ivanovic, B., Chakravarty, P., & Pavone, M. (2020). Trajectron++: Dynamically-feasible trajectory forecasting with heterogeneous data. Computer Vision–ECCV 2020: 16th European Conference, Glasgow, UK, August 23–28, 2020, Proceedings, Part XVIII 16.

Sattar, S., Li, S., & Chapman, M. (2018). Road surface monitoring using smartphone sensors: A review. *Sensors (Switzerland), 18*(11) MDPI AG.

Semanjski, I., Gautama, S., Ahas, R., & Witlox, F. (2017). Spatial context mining approach for transport mode recognition from mobile sensed big data. *Computers, Environment and Urban Systems, 66*, 38–52.

Sharma, S., Das, A., Sistu, G., Halton, M., & Eising, C. (2024). MapsTP: HD map images based multimodal trajectory prediction for automated vehicles. IET Conference Proceedings CP887.

Shaygan, M., Meese, C., Li, W., Zhao, X., & (George), & Nejad, M. (2022). Traffic prediction using artificial intelligence: Review of recent advances and emerging opportunities. *Transportation Research Part C: Emerging Technologies, 145*.

Shi, J., Li, X., Lin, B., Zhou, L., & Lv, G. (2024). Assessing semantic consistency of pavement markings and signs using street view images—A case study on lane-turning information. *International Journal of Digital Earth, 17*(1).

Shin, D., Aliaga, D., Tunçer, B., Arisona, S. M., Kim, S., Zünd, D., & Schmitt, G. (2015). Urban sensing: Using smartphones for transportation mode classification. *Computers, Environment and Urban Systems, 53*, 76–86.

Shu, W., & Li, Y. (2022). A novel demand-responsive customized bus based on improved ant colony optimization and clustering algorithms. *IEEE Transactions on Intelligent Transportation Systems, 24*(8), 8492–8506.

Singh, D. (2023). Deep Reinforcement Learning (DRL) for Real-Time Traffic Management in Smart Cities. 2023 international conference on communication, Security and Artificial Intelligence (ICCSAI).

Stenneth, L., Wolfson, O., Yu, P. S., & Xu, B. (2011). Transportation mode detection using mobile phones and GIS information. Proceedings of the 19th ACM SIGSPATIAL international conference on advances in geographic information systems.

Wang, S., Patwary, A. U. Z., & Huang, W. (2022). A general framework for combining traffic flow models and Bayesian network for traffic parameters estimation. *Transportation research part C: emerging technologies, 139*, 103664.

Wu, H., Yao, L., Xu, Z., Li, Y., Ao, X., Chen, Q., Li, Z., & Meng, B. (2019). Road pothole extraction and safety evaluation by integration of point cloud and images derived from mobile mapping sensors. *Advanced Engineering Informatics, 42*, 100936.

Wu, W., Li, P., Liu, R., Jin, W., Yao, B., Xie, Y., & Ma, C. (2020). Predicting peak load of bus routes with supply optimization and scaled Shepard interpolation: A newsvendor model. *Transportation Research Part E: Logistics and Transportation Review, 142*, 102041.

Zhang, Y., Marshall, S., & Manley, E. (2019). Network criticality and the node-place-design model: Classifying metro station areas in Greater London. *Journal of Transport Geography, 79*, 102485.

Zhao, Z., Tang, L., Yang, X., Zhang, H., Li, G., & Li, Q. (2024). Identifying critical urban intersections from a fine-grained spatio-temporal perspective. *Travel Behaviour and Society, 34*, 100649.

Zhou, Y., Fang, Z., Thill, J.-C., Li, Q., & Li, Y. (2015). Functionally critical locations in an urban transportation network: Identification and space–time analysis using taxi trajectories. *Computers, Environment and Urban Systems, 52*, 34–47.

Zhou, X., Yu, W., & Sullivan, W. C. (2016). Making pervasive sensing possible: Effective travel mode sensing based on smartphones. *Computers, Environment and Urban Systems, 58*, 52–59.

Zhou, Y., Han, X., Peng, M., Li, H., Yang, B., Dong, Z., & Yang, B. (2022). Street-view imagery guided street furniture inventory from mobile laser scanning point clouds. *ISPRS Journal of Photogrammetry and Remote Sensing, 189*, 63.

Zhu, F., Xu, W., Liu, D., & Shi, H. (2024). TSANET: Transportation mode recognition model with global and local spatiotemporal features. *The Journal of Supercomputing, 80*(7), 9194–9219.

Chapter 18
GeoAI and Urban Geography

Pengyuan Liu, Yujun Hou, Binyu Lei, Xiucheng Liang, and Filip Biljecki

Abstract Recognising GeoAI as an emerging and rapidly evolving field that has been increasingly adopted in urban geography, this chapter provides an overarching overview of the GeoAI methods for urban analytics. It begins by revisiting the theoretical underpinnings of urban theory and mapping the evolution of urban spatial analytics, tracing the journey from traditional statistical methods to the cutting-edge AI-driven approaches reshaping the discipline today. Beyond examining the current state of GeoAI, the chapter also identifies current trending topics and investigates future directions for developing human-centric methodologies that prioritise the needs and experiences of urban residents. By emphasising the human dimension of urban analytics, the chapter seeks to contribute to the ongoing discourse on how GeoAI can be harnessed to enhance city governance, urban planning, and the overall quality of urban life.

Keywords GeoAI · Urban geography · Spatial analytics · Human-centric approaches · City governance

18.1 Introduction

Cities are inherently complex systems, characterised by entangled networks of people, infrastructure, and natural environments (Batty, 2009). Such urban systems are in a state of constant flux, continually evolving as a result of the

P. Liu
Future Cities Lab Global, Singapore-ETH Centre, Singapore, Singapore

Y. Hou · B. Lei · X. Liang
Department of Architecture, National University of Singapore, Singapore, Singapore

F. Biljecki (✉)
Department of Architecture, National University of Singapore, Singapore, Singapore

Department of Real Estate, National University of Singapore, Singapore, Singapore
e-mail: filip@nus.edu.sg

© The Author(s), under exclusive license to Springer Nature Switzerland AG 2025

X. Huang et al. (eds.), *GeoAI and Human Geography*, Springer Geography,
https://doi.org/10.1007/978-3-031-87421-5_18

ongoing interactions between social, economic, and environmental forces (Brenner & Schmid, 2015; Pred, 2017). Urban geography, a vibrant and dynamic sub-field of geography, seeks to understand the spatial structures, patterns, and processes that shape urban environments and people's lives within them (Hall & Barrett, 2012). It offers a spatial perspective on how human activities, infrastructure, and natural landscapes interact and shape in tandem with the multifaceted urban systems that define contemporary society (Pacione, 2009). As cities continue to expand and transform, they face a growing array of challenges, ranging from sustainability and resilience (Folke, 2006) to issues of equity and liveability (Ruth & Franklin, 2014), which are becoming increasingly complex. Addressing these challenges necessitates innovative approaches that capture, analyse, and interpret the vast quantities of information generated within urban environments.

In the era of big data (Kitchin, 2014), thanks to the increasingly accessible GPS-enabled devices and equipment, up to 80% of data produced every day is geo-referenced (Leszczynski & Crampton, 2016). Traditional approaches to studying urban geography have been profoundly transformed by the advent of new technologies, with GeoAI (Geospatial Artificial Intelligence) emerging as one of the most pioneering advancements. GeoAI harnesses the capabilities of artificial intelligence (AI), high-performance computing (HPC), and spatial big data (SBD), offering cutting-edge tools that allow for the enhanced and automated analysis of urban spaces with unprecedented precision and enhanced accuracy (Li, 2020; Janowicz et al., 2020; Liu & Biljecki, 2022). Through advanced AI-enabled methodologies, GeoAI enables novel use cases and provides invaluable insights into urban environments' complexities, facilitating more informed and effective urban planning and management strategies and, eventually, enhancing the quality of life for urban inhabitants.

This chapter offers a comprehensive overview of GeoAI as a cutting-edge collection of techniques within the field of urban geography. It begins by revisiting the theoretical foundations of urban theory and tracing the evolution of urban spatial analytics, from traditional statistical methods to the advanced AI-assisted approaches that are redefining the discipline today. The chapter also critically examines the uncertainties inherent in GeoAI analysis, highlighting the potential biases, limitations, and challenges that must be navigated when applying these technologies to urban studies. In addition to addressing the current state of GeoAI, this chapter explores future directions for developing human-centric GeoAI methodologies that prioritise the needs and experiences of urban residents. By focusing on the human dimension of urban analytics, the chapter aims to contribute to the ongoing discourse on how GeoAI can be leveraged to improve city governance, planning, and the overall quality of urban life.

18.2 Quantitative Evolution of Urban Geography

Urban geography has evolved significantly over the past century, with key milestones marking the development and transformation of this field, influencing both what and how geographic knowledge is produced for cities and the people living in them.

In the mostly descriptive and fieldwork-driven works in the early days (Reclus et al., [1876–94]; Sauer, 1941, 1952), geographers sought to catalogue instances of phenomena, ranging from physical landscape traits, to socioeconomic and cultural practices, and associate them with their locations and maps to narrate the distinct or similar features of different regions (Cope, 2010). Later, descriptive theories and models were developed in the early twentieth century to understand the spatial organisation, function, and processes of cities, such as the Concentric Zone Model (Burgess, 1925), the Sector Model (Hoyt, 1939), and the Multiple Nuclei Model (Harris & Ullman, 1945).

Meanwhile, much progress had been made in economic geography, spatial analysis, and quantitative social science, paving the way for what was later called the 'Quantitative Revolution' in geography in the 1950s and 1960s (Adams, 2001; Barnes, 2001). Influenced by works from other disciplines, such as the *Theory of Games and Economic Behavior* by von Neumann & Morgenstern (1944) and Human Behaviour and the Principle of Least Effort by Zipf (1949), quantitative techniques such as mathematical and statistical modelling began to gain interest among urban geographers to study the spatial arrangement of human activities (Garrison, 1959a,b, 1960; Isard, 1960; Burton, 1963).

The movement saw the re-introduction of location theory, which originated from the seminal work of Johann Heinrich von Thünen, 'The Isolated State' (1826), and laid the foundations for systematically understanding the principles of land use (Hoover & Giarratani, 1999). Benefiting from the development of location theory, the Central Place Theory developed by Christaller (1933) explained the spatial distribution and size hierarchy of human settlements, which were postulated to function as 'central places' to supply economic services to surrounding areas. Such work put forth new research agendas and methods and demonstrated the values of quantification to urban geographers—reproducible results, robust foundation for theory-testing, understanding and policy-making, and knowledge production that was cumulative instead of additive (Adams, 2001).

Another significant work that profoundly re-shaped the research methods in this field was Locational Analysis in Human Geography by Haggett (1965), which provided a thorough framework for understanding spatial patterns through quantitative methods. Such an exploration marked a milestone that transformed urban geography from a primarily descriptive field to one that is analytical, theoretical, and data-driven, and set the stage for modern spatial analysis and Geographic Information Systems (GIS), which have become important methods and tools to analyse urban phenomena.

Since the Quantitative Revolution, a range of statistical and mathematical techniques have been incorporated into the study of urban geography, including regression analysis (Barnes, 1998), factor analysis (Clark et al., 1974), and clustering methods (Webber & Craig, 1976). These methods enabled urban geographers to explore the relationships between different economic, social, and environmental variables, uncover their underlying patterns, and investigate their similarities and differences among spatial clusters of communities. Yet, even though the phenomena being studied are geographical, these modelling methods often do not take into account spatial information (e.g. coordinates, distances, topological relationships, interactions, and flows among spatial objects) and could overlook the spatial variability underlying the data (De Sabbata & Liu, 2023).

In 1970, Tobler's First Law of Geography was introduced. It states that 'everything is related to everything else, but near things are more related than distant things' (Tobler, 1970). This pivotal theory gave rise to the now fundamental concepts of spatial dependence and spatial autocorrelation, which underpin all spatial analysis. A range of spatial modelling and geostatistics techniques were developed, such as Inverse Distance Weighting and Kriging. These interpolation methods predict the values of unmeasured locations based on known data points by considering their distances and relative positions and have been useful in generating, from merely a set of sampled locations, detailed maps of certain urban variables such as population density (Liu et al., 2008; Wu & Murray, 2005), microclimate variables (Han et al., 2024), and air quality variables (Gardner-Frolick et al., 2022; Shukla et al., 2020; Jerrett et al., 2001).

As computers evolved rapidly, more advanced techniques became possible, such as agent-based modelling (ABM), cellular automata (CA), and spatial regression models. ABM and CA have been used to simulate urban growth, land use changes, and transportation systems by modelling the interactions of individual agents and cells, offering insights into urban sprawl and traffic dynamics (Batty, 1997; Clarke et al., 1997; Batty, 2005; Torrens, 2006; Crooks et al., 2008). Spatial regression models help analyse spatial dependencies in housing markets, crime patterns, and health outcomes, providing a deeper understanding of urban socio-economic and environmental dynamics (Fotheringham et al., 2003; Anselin, 1988; LeSage & Pace, 2009). Another analytical-computational technique developed in the late twentieth century is space syntax (Hillier et al., 1976; Hillier & Hanson, 1984), which is used to study the spatial configurations of built environments (e.g. street networks) and how these configurations influence societal behaviours within urban areas, such as human movement, accessibility, connectivity, and social interactions (Penn et al., 1998). Collectively, these techniques enable urban geographers to better model, manage, and plan urban environments.

Other than methodological advancements, technological progress has also brought forth a multitude of new data sources, ranging from satellite imagery to street view imagery, social media, and crowdsourced data. Long gone are the days when urban geographers were confined to fieldwork measurements or the often limited census survey data. With the advent of the 'digital turn' in geography (Ash et al., 2018a,b), researchers now tap into a burgeoning volume and variety of urban data that provides countless new ways and dimensions to measure human activities, urban characteristics, and social interactions. Yet, at the same time, such data is increasingly rapid, complex, noisy, and multi-sourced, and traditional modelling methods could face challenges in adapting to this era of spatial big data (Evans et al., 2014; Lee & Kang, 2015). GeoAI has emerged as a crucial and transformative element in urban geography research and applications. By integrating AI with spatial (and temporal) aspects, GeoAI offers promising solutions for mining, interpreting, and predicting urban and human dynamics (Gao, 2021; Gao et al., 2023), revolutionising ways to study complex urban systems.

18.3 Geospatial Artificial Intelligence

GeoAI is an interdisciplinary field representing the intersection of geography and AI (Gao, 2021). It empowers the research to investigate geospatial phenomena and enhance the understanding of human habitation (Liu & Biljecki, 2022; Mai et al., 2022).

At its core foundation, GeoAI builds on the principles and theories of GIS, automating the processing and analysis of vast and intricate geospatial datasets. Machine learning is integral to the functionality of GeoAI, providing the analytical power needed to handle the complexity and scale of modern geospatial data. However, conventional machine learning methods face limitations in interpreting the ever-growing and increasingly complex urban data, which often presents high-dimensionality and non-linear relationships that characterise the dynamic and complex nature of urban-human interactions. The rise of deep learning and neural networks has marked a turning point in addressing these challenges (Grekousis, 2019). Deep learning algorithms, particularly convolutional neural networks (CNNs) and recurrent neural networks (RNNs), have the ability to automatically learn hierarchical features from raw data, rendering them particularly well-suited for tasks involving large-scale and heterogeneous geospatial datasets. For instance, CNNs are highly effective in image recognition tasks, such as detecting urban scenes and objects from street view images (Biljecki & Ito, 2021; Ito et al., 2024), while RNNs excel at capturing temporal patterns in time-series data, such as predicting traffic flows (Medina-Salgado et al., 2022) or urban growth (Quadri et al., 2024).

Yet, GeoAI is more than merely applying deep learning methods to geographical problems. The emergence of graph neural networks (GNNs) has introduced a new dimension to GeoAI by integrating geographic theories directly into the models, thereby making them spatially explicit (Liu & Biljecki, 2022). While traditional deep learning techniques, such as CNNs and RNNs, have been effective in handling grid-based and sequential data, they often struggle to represent and analyse the complex, non-Euclidean structures inherent in certain types of geographic data. For example, urban networks, transportation systems, and social interactions are more accurately represented as graphs, where nodes correspond to entities like locations or individuals, and edges denote the spatial relationships or interactions between them. GNNs, which are able to process graph data, effectively address these limitations by providing a framework specifically designed to accommodate the irregular and interconnected nature of geospatial data (Mai et al., 2022). By incorporating geographic theories, such as Tobler's First Law of Geography, GNNs allow GeoAI models to account for essential spatial concepts like spatial autocorrelation, proximity effects, and network dynamics, allowing for a more robust understanding of geographic phenomena and deeper insights into a variety of complex issues (Liu et al., 2024; Lei et al., 2024).

The field of GeoAI is rapidly evolving, with ongoing advancements and continuous expansion of its capabilities. Emerging technologies, such as Large Language Models (LLMs) and other innovative approaches, provide geographers with unprecedented tools for analysing urban complexity. In the following sections, we offer a scoping review of how GeoAI has been integrated into urban geography, highlighting the opportunities that modern technologies present for an in-depth, in-situ understanding of urban environments, as well as examining their potential limitations.

18.4 GeoAI in Urban Geography: Integration and Applications

Figure 18.1 provides an overview of the key types of urban data, GeoAI approaches, and emerging trends and opportunities in this field. In the following sections, we will delve deeper into each of these components and their applications in urban geography.

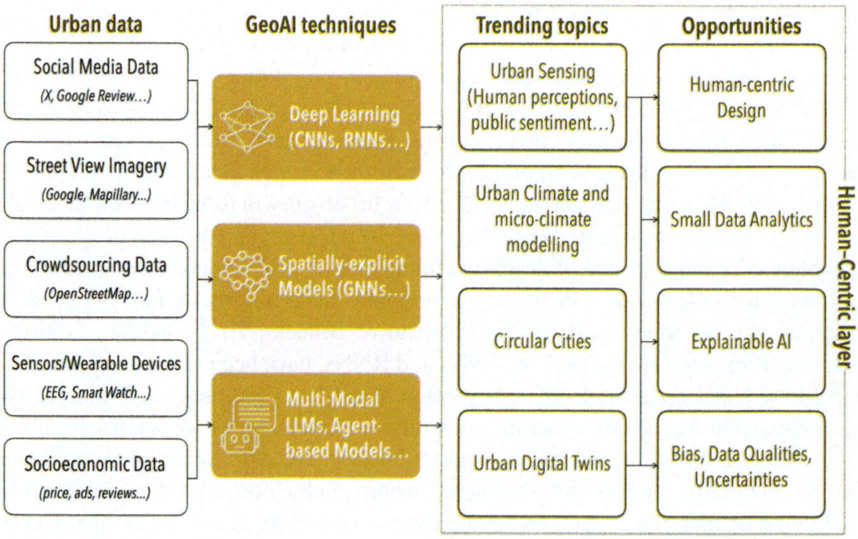

Fig. 18.1 Overview of key types of urban data, relevant GeoAI approaches, and emerging trends and opportunities

18.4.1 Emerging Urban Data

The evolvement of GeoAI techniques provides powerful tools to analyse complex urban spatial data, facilitating insights into urban dynamics and built environments that were previously unavailable and improving existing understandings by increasing spatial and temporal resolution and reliability. This section includes very recent examples of various developments, with a focus on novel sources or forms of urban data that have been leveraged with AI-assisted spatial analytics to support urban geography research.

Social media data has been spotlighted (Liu et al., 2015). For example, it has been used to understand urban park visitation (Wei et al., 2024) and park perception (Zhao et al., 2024), infer the design of riverscapes (Yang et al., 2022), predict traffic pollution (Zhang et al., 2024), and uncover semantic footprints (Berragan et al., 2024), among many examples in the past years.

Another prominent medium for urban studies in the past several years has been street view imagery (SVI) (Liang et al., 2024a; Liu & Sevtsuk, 2024). It has been used to map greenery and the urban form (Biljecki et al., 2023b), conduct virtual audits of dwellings (Yan et al., 2024), and assess walkability (Chen et al., 2024; Li et al., 2024). Sensing information on humans, facilitated by SVI, has also been the focus of many studies recently. For example, perception studies (Ito et al., 2024), which will be introduced more in the next section, have been commonly conducted in the past years thanks to the ease of reaching users around the world and providing the required data. The latest research efforts appear to be focused on change detection (Liang et al., 2023; Stalder et al., 2024), contributing to sensing the dynamics of cities and understanding the impacts of urbanisation.

Apart from the continued use of such data sources, there is a continuous trend of investigating the usability of latent urban data for urban sensing. For example, Liang et al. (2024b) demonstrated that scraping information on the website of a multi-purpose indoor arena in New York City can enhance human mobility prediction under public events. Such research has been facilitated by LLMs, demonstrating how emerging sources of data, coupled with new means to process information, can unlock new applications or enhance existing ones. Another relatively underutilised data source is real estate ads posted in online marketplaces. For example, they have been used to detect new buildings and amenities (Chen & Biljecki, 2022) and to understand decoration patterns across neighbourhoods and cities (Liu et al., 2019). In that space, Wang et al. (2023) have investigated the usability of reviews of homestays (Airbnb) to reveal the perception of neighbourhoods across multiple dimensions, for example, greenery, noise, and crime.

We outline a few trends that can be observed recently. First, urban analytics is becoming multi-modal, e.g. social media data is being fused with SVI to reveal new insights (Wang, 2024). Second, while in the past, a lot of such data was made available by governments and companies, now it is being increasingly crowdsourced. For example, weather data is progressively available from personal weather stations (Brousse et al., 2024), and crowdsourced data on buildings is becoming growingly available thanks to initiatives such as OpenStreetMap (Biljecki, Chow, & Lee, 2023a). Third, thanks to the wide availability of urban data and their harmonisation, research is becoming increasingly global and includes multiple cities (Hou et al., 2024). Finally, some of these lines of research have started being supported with open data, e.g. social media (Poorthuis et al., 2024) and real estate (Rey-Blanco et al., 2024), lowering the entry barriers for technical research and simplifying the process of obtaining relevant datasets.

18.4.2 Current Trends and Applications

The integration of GeoAI techniques has benefitted an array of urban geography studies. In particular, it has facilitated notable advancements in applications of rising importance, such as urban climate mitigation, circular cities, and urban perception, owing to the increasing focus on sustainable and inclusive urban development.

In the realm of urban climate, deep learning models are being leveraged to analyse complex environmental data and support decision-making processes (Patel et al., 2023). A notable development in this area is the use of Multimodal LLMs, which offer a comprehensive method for integrating diverse data types, including satellite imagery, SVI, and text descriptions, to measure urban environments holistically. For example, Fujiwara et al. (2024) developed a multimodal model that combines microclimate data with street-level and satellite imagery to predict microclimate variables (e.g. air temperature, relative humidity, wind speed, and global horizontal irradiance) at high spatial and temporal resolutions. These models allow researchers to consider more comprehensive features and depict cities as inherently complex systems, enhancing our understanding and management of urban spaces.

Benefiting from the recent trends of the digital twin and circular economy, GeoAI is poised to play a transformative role in advancing the circular digital built environment to enhance sustainability, efficiency, and resilience in urban spaces. One of the primary ways GeoAI contributes to this is through resource optimisation and lifecycle management (Mortaheb & Jankowski, 2023). For instance, it can model the potential effects of different urban layouts on energy consumption, waste generation, and resource use, enabling planners to design cities that are both efficient and resilient (Mortaheb & Jankowski, 2023). Additionally, GeoAI can aid in designing flood-resistant infrastructure that not only protects urban areas but also incorporates materials and designs that can be easily maintained, repurposed, or recycled (Zhou et al., 2023), thus contributing to a more liveable urban environment for the commonwealth of its residents.

While urban living conditions are impacted by external causes such as climate change and the increasing focus on circular economy development, human perception consistently shapes our understanding of the urban environment. Here, we feature the urban perception studies that apply perception-based labelling to images, providing a quantitative and extensive evaluation of human responses to built environments. These studies typically collect subjective impressions from participants using images, categorise the images with perceptual labels, and transform these labels into measurable attributes of urban environments (Ito et al., 2024). A foundational contribution to this field is the crowdsourced dataset known as Place Pulse 2.0, developed by Dubey et al. (2016). This dataset encompasses six attributes—depressing, boring, beautiful, safe, lively, and wealthy—and covers data from 56 cities, significantly expanding the scope of urban perception research. Not only has this dataset enabled extensive subsequent research on human perceptions across various regions (Zhang et al., 2018; Liang et al., 2023; Hou et al., 2024), but it has also broadened the urban perception framework to encompass other sectors of the built environment, such as soundscapes (Zhao et al., 2023), waterscapes (Luo et al., 2022), and building facades (Liang et al., 2024a).

Facing the growing need for comprehensive urban understanding, GeoAI, with its powerful capabilities to integrate and analyse multi-source data, is demonstrating significant potential in this evolving field. Meanwhile, due to the increasing attention on the concept of 'Technology for Social Good', which urges the shift from traditional top-down technological development that often focuses on big data analysis to individual-level 'small data' analytics, there is a growing need to bridge the technology on the urban-scale analysis to better support individuals' living experiences, thus enabling the GeoAI methods to be human-centric.

18.4.3 Towards Human-Centric Integrated GeoAI

The integration of GeoAI with human-centric approaches in urban geography is emerging as a crucial direction for future research. This convergence aims to bridge the gap between advanced technological capabilities and the nuanced understanding of urban systems that geographers and urban planners possess, to bring adequate attention to not only the *environment*, but also more importantly, the *people*.

Psychological-integrated urban sensing studies represent an evolving interdisciplinary field that synergises traditional psychological insights with cutting-edge urban sensing technologies. The psychological patterns embedded within urban environments offer a deeper understanding of how these patterns influence and shape the experiences and well-being of residents. Unlike conventional methods that primarily rely on SVI for urban observation, as discussed in the previous section, there is a noticeable shift towards incorporating psychological data, which underscores the growing recognition of the importance of human perception and mental states in shaping urban landscapes, leading to more nuanced analyses and interventions that consider both the physical and psychological dimensions of urban life (Helbich, 2018).

Wearable devices like smartwatches with sensors for tracking heart rate variability and skin conductance have become essential in urban sensing studies (Tartarini et al., 2023). These tools monitor stress levels as people move through the city, capturing real-time physiological responses alongside contextual data like temperature, noise, and air quality, allowing researchers to understand how urban environments impact well-being and identify stress patterns over time. Beyond stress monitoring, these devices evaluate urban design interventions aimed at improving comfort and reducing stress (Liu et al., 2023).

Additionally, sophisticated physiological devices like EEGs are increasingly used in urban research (Mavros et al., 2016). EEGs measure brain activity, offering insights into cognitive and emotional states in response to urban environments. Their portability and ease of use now allow data collection in real-world settings, enhancing our understanding of the relationship between the built environment and psychological well-being (Aspinall et al., 2015; Bolouki, 2023).

Through integrating physiological and machine insights, human-centric GeoAI approaches prioritise urban citizens' needs, behaviours, and experiences, ensuring that AI-driven analyses and solutions are technically sophisticated and socially relevant. A recent example proposed by Liu et al. (2023) constructs the pedestrians and their interactions with surrounding environments as human-centric dynamic graphs to study outdoor comfort. These graphs account for spatio-temporal variations observed in human walking patterns, including changes in sound levels, solar intensity, and visual perceptions. This research enables a two-way interaction between pedestrians and the GeoAI model, hence, providing individual tailored comfort prediction to navigate human outdoor activities.

18.4.4 Uncertainties

While the integration of AI with urban geography offers significant potential, it also introduces several uncertainties that researchers and practitioners must carefully consider. These uncertainties stem from various sources, including data quality, model limitations, and the inherent complexity of urban systems, necessitating a cautious and informed approach.

A primary source of uncertainty in GeoAI applications stems from the quality and representativeness of the data used (Crampton et al., 2013; Shelton et al., 2015; Poorthuis et al., 2023). Urban data, particularly when sourced from user-generated content or areas with limited digital infrastructure, can be incomplete, biased, inconsistent, or unrepresentative of the population. For instance, social media users in London are predominantly wealthy, young, and educated (Ballatore & De Sabbata, 2018). These limitations not only pose challenges in constructing accurate GeoAI models but also risk introducing biases in the interpretation of results (Graham et al., 2014). Furthermore, a study of OpenStreetMap's building data completeness reveals significant regional variability, with larger metropolitan areas typically having more complete datasets (Herfort et al., 2023). SVI also suffers quality and availability issues that are heterogeneous both spatially and temporally (Hou & Biljecki, 2022). Such variability in urban data raises crucial questions about the generalisability of AI models trained on data from one urban context to other cities with differing socio-economic, cultural, or physical characteristics, potentially limiting the broad applicability of GeoAI solutions.

Furthermore, the inherent complexity and unpredictability of urban systems add another layer of uncertainty. Cities are dynamic, adaptive systems influenced by countless interacting factors, many of which are difficult to quantify or predict. These factors include economic shifts, population movements, policy changes, and environmental conditions, all of which evolve in non-linear and sometimes unexpected ways (Jacobi et al., 2010). AI models, despite their sophistication, may struggle to capture the full complexity of urban phenomena, potentially overlooking important nuances or emergent patterns. For instance, social dynamics such as neighbourhood gentrification are subtle and multi-faceted processes that can be inadequately represented in datasets typically used for training AI systems. These intricacies challenge accurate modelling, often requiring advanced, context-specific adaptations that can complicate the development and scaling of general GeoAI solutions. A critical need remains for ongoing refinement of these data and modelling capabilities to better understand and capture a richer contextual representation of urban environments (Yap et al., 2023).

Addressing these uncertainties requires interdisciplinary collaboration, robust validation techniques, and a commitment to ethical and transparent AI practices. A notable effort towards this direction is the development of explainable AI (XAI) techniques specifically tailored for urban geographic applications enhancing the transparency and accountability of AI-driven urban research (Liu et al., 2024). As the field of GeoAI in urban geography continues to evolve, developing strategies to navigate these uncertainties will ensure that AI-driven approaches contribute positively to urban understanding and development.

18.5 Conclusion

The integration of GeoAI into urban geography marks a significant leap forward. It reforms the way we study how cities are constructed, governed, and experienced. By incorporating advanced computing techniques with domain knowledge, it became possible to analyse complex urban data with unprecedented precision, efficiency, and scale. Such an advancement facilitates more timely, granular, and data-driven insights that can fundamentally improve urban planning and decision-making, and, at the same time, enable a more humanistic focus. However, integrating GeoAI comes with challenges, particularly in managing the uncertainties inherent in data quality, model transparency, and the dynamic nature of urban environments. To fully leverage the potential of GeoAI, it is crucial to adopt strategies that enhance data reliability, adequately quantify and communicate uncertainty, and ensure that AI models are both interpretable and contextually relevant. This will help GeoAI provide more accurate and actionable insights, ultimately leading to more informed and effective urban planning and policy-making.

References

Adams, J. S. (2001). The quantitative revolution in urban geography. *Urban Geography, 22*(6), 530–539. https://doi.org/10.2747/0272-3638.22.6.530

Anselin, L. (1988). *Spatial econometrics: Methods and models*. Kluwer Academic Publishers.

Ash, J., Kitchin, R., & Leszczynski, A. (2018a). *Digital geographies*. SAGE Publications.

Ash, J., Kitchin, R., & Leszczynski, A. (2018b). Digital turn, digital geographies? *Progress in Human Geography, 42*(1), 25–43.

Aspinall, P., Mavros, P., Coyne, R., & Roe, J. (2015). The urban brain: Analysing outdoor physical activity with mobile EEG. *British Journal of Sports Medicine, 49*(4), 272–276.

Ballatore, A., & De Sabbata, S. (2018). Charting the geographies of crowdsourced information in greater london. In *Geospatial technologies for all: Selected papers of the 21st agile conference on geographic information science*, 21 (pp. 149–168).

Barnes, T. J. (1998). A history of regression: Actors, networks, machines, and numbers. *Environment and Planning A, 2*, 203–223.

Barnes, T. J. (2001). Lives lived and lives told: Biographies of geography's quantitative revolution. *Environment and Planning D: Society and Space, 19*(4), 409–429.

Batty, M. (1997). Cellular automata and urban form: A primer. *Journal of the American Planning Association, 63*(2), 266–274.

Batty, M. (2005). *Cities and complexity: Understanding cities with cellular automata, agent-based models, and fractals*. MIT Press.

Batty, M. (2009). Cities as complex systems: Scaling, interaction, networks, dynamics and urban morphologies. In R. A. Meyers (Ed.), *Encyclopedia of complexity and systems science* (pp. 1041–1071). Springer.

Berragan, C., Singleton, A., Calafiore, A., & Morley, J. (2024). Mapping Great Britain's semantic footprints through a large language model analysis of reddit comments. *Computers, Environment and Urban Systems, 110*, 102121.

Biljecki, F., & Ito, K. (2021). Street view imagery in urban analytics and gis: A review. *Landscape and Urban Planning, 215*, 104217.

Biljecki, F., Chow, Y. S., & Lee, K. (2023a). Quality of crowdsourced geospatial building information: A global assessment of OpenStreetMap attributes. *Building and Environment, 237*, 110295.

Biljecki, F., Zhao, T., Liang, X., & Hou, Y. (2023b). Sensitivity of measuring the urban form and greenery using street-level imagery: A comparative study of approaches and visual perspectives. *International Journal of Applied Earth Observation and Geoinformation, 122*, 103385.

Bolouki, A. (2023). Neurobiological effects of urban built and natural environment on mental health: Systematic review. *Reviews on Environmental Health, 38*(1), 169–179.

Brenner, N., & Schmid, C. (2015). Towards a new epistemology of the urban? *City, 19*(2–3), 151–182.

Brousse, O., Simpson, C. H., Poorthuis, A., & Heaviside, C. (2024). Unequal distributions of crowdsourced weather data in England and Wales. *Nature Communications, 15*(1), 4828.

Burgess, E. W. (1925). The growth of the City: An introduction to a research project. In R. E. Park, E. W. Burgers, & R. D. McKenzie (Eds.), *The city*. University of Chicago Press.

Burton, I. (1963). The quantitative revolution and theoretical geography. *Canadian Geographies/Ge'ographies canadiennes, 7*(4), 151–162.

Chen, X., & Biljecki, F. (2022). Mining real estate ads and property transactions for building and amenity data acquisition. *Urban Informatics, 1*, 12.

Chen, Y., Huang, X., & White, M. (2024). A study on street walkability for older adults with different mobility abilities combining street view image recognition and deep learning—The case of chengxianjie community in Nanjing (China). *Computers, Environment and Urban Systems, 112*, 102151.

Christaller, W. (1933). *Die zentralen orte in süddeutschland: eine ökonomisch-geographische untersuchung über die gesotzmässigkeit der verbreitung und entwicklung der siedlungen mit städtischen funktionen*. Gustav Fischer.

Clark, D., Davies, W. K., & Johnston, R. J. (1974). The application of factor analysis in human geography. *Journal of the Royal Statistical Society Series D: The Statistician, 3-4*, 259–281.

Clarke, K. C., Hoppen, S., & Gaydos, L. J. (1997). A self-modifying cellular automaton model of historical urbanization in the San Francisco bay area. *Environment and Planning B: Planning and Design, 24*(2), 247–261.

Cope, M. (2010). The SAGE handbook of qualitative geography. .

Crampton, J. W., Graham, M., Poorthuis, A., Shelton, T., Stephens, M., Wilson, M. W., & Zook, M. (2013). Beyond the geotag: Situating 'big data'and leveraging the potential of the geoweb. *Cartography and Geographic Information Science, 40*(2), 130–139.

Crooks, A. T., Castle, C. J. E., & Batty, M. (2008). Key challenges in agent-based modelling for geo-spatial simulation. *Computers, Environment and Urban Systems, 6*, 417–430.

De Sabbata, S., & Liu, P. (2023). A graph neural network framework for spatial geodemographic classification. *International Journal of Geographical Information Science, 37*, 1–23.

Dubey, A., Naik, N., Parikh, D., Raskar, R., & Hidalgo, C. A. (2016). Deep learning the city: Quantifying urban perception at a global scale. In *European conference on computer vision* (pp. 196–212).

Evans, M. R., Oliver, D., Zhou, X., & Shekhar, S. (2014). Spatial big data. In H. A. Karimi (Ed.), *Big data: Techniques and technologies in geoinformatics*. CRC Press.

Folke, C. (2006). Resilience: The emergence of a perspective for social-ecological systems analyses. *Global Environmental Change, 16*(3), 253–267.

Fotheringham, A. S., Brunsdon, C., & Charlton, M. (2003). *Geographically weighted regression: The analysis of spatially varying relationships*. Wiley.

Fujiwara, K., Khomiakov, M., Yap, W., Ignatius, M., & Biljecki, F. (2024). Micro-climate vision: Multimodal prediction of climatic parameters using street-level and satellite imagery. *Sustainable Cities and Society, 114*, 105733.

Gao, S. (2021). *Geospatial artificial intelligence (GeoAI)*. Oxford University Press.

Gao, S., Hu, Y., & Li, W. (Eds.). (2023). *Handbook of geospatial artificial intelligence* (1st ed.). CRC Press.

Gardner-Frolick, R., Boyd, D., & Giang, A. (2022). Selecting data analytic and modeling methods to support air pollution and environmental justice investigations: A critical review and guidance framework. *Environmental Science & Technology, 5*, 2843–2860.

Garrison, W. L. (1959a). Spatial structure of the economy: I. *Annals of the Association of American Geographers, 49*(2), 232–239.

Garrison, W. L. (1959b). Spatial structure of the economy: II. *Annals of the Association of American Geographers, 49*(4), 471–482.

Garrison, W. L. (1960). Spatial structure of the economy: III. *Annals of the Association of American Geographers, 50*(3), 357–373.

Graham, M., Hogan, B., Straumann, R. K., & Medhat, A. (2014). Uneven geogra phies of user-generated information: Patterns of increasing informational poverty. *Annals of the Association of American Geographers, 104*(4), 746–764.

Grekousis, G. (2019). Artificial neural networks and deep learning in urban geography: A systematic review and meta-analysis. *Computers, Environment and Urban Systems, 74*, 244–256.

Haggett, P. (1965). *Locational analysis in human geography*. Edward Arnold.

Hall, T., & Barrett, H. (2012). *Urban geography*. Routledge.

Han, J., Chong, A., Lim, J., Ramasamy, S., Wong, N. H., & Biljecki, F. (2024). Microclimate spatio-temporal prediction using deep learning and land use data. *Building and Environment, 253*, 111358.

Harris, C. D., & Ullman, E. L. (1945). The nature of cities. *The Annals of the American Academy of Political and Social Science, 242*, 7–17.

Helbich, M. (2018). Toward dynamic urban environmental exposure assessments in mental health research. *Environmental Research, 161*, 129–135.

Herfort, B., Lautenbach, S., Porto de Albuquerque, J., Anderson, J., & Zipf, A. (2023). A spatio-temporal analysis investigating completeness and inequalities of global urban building data in openstreetmap. *Nature Communications, 14*(1), 3985.

Hillier, B., & Hanson, J. (1984). *The social logic of space*. Cambridge University Press.

Hillier, B., Leaman, A., Stansall, P., & Bedford, M. (1976). Space syntax. *Environment and Planning B: Planning and Design, 3*(2), 147–185.

Hoover, E. M., & Giarratani, F. (1999). *An introduction to regional economics*.

Hou, Y., & Biljecki, F. (2022). A comprehensive framework for evaluating the quality of street view imagery. *International Journal of Applied Earth Observation and Geoinformation, 115*, 103094.

Hou, Y., Quintana, M., Khomiakov, M., Yap, W., Ouyang, J., Ito, K., et al. (2024). Global streetscapes—A comprehensive dataset of 10 million street-level images across 688 cities for urban science and analytics. *ISPRS Journal of Photogrammetry and Remote Sensing, 215*, 216–238.

Hoyt, H. (1939). *The structure and growth of residential neighborhoods in American cities*. Federal Housing Administration. Gleerup.

Isard, W. (1960). *Methods of regional analysis*. MIT Press.

Ito, K., Kang, Y., Zhang, Y., Zhang, F., & Biljecki, F. (2024). Understanding urban perception with visual data: A systematic review. *Cities, 152*, 105169.

Jacobi, P., Kjellen, M., McGranahan, G., Songsore, J., & Surjadi, C. (2010). *The citizens at risk: From urban sanitation to sustainable cities*. Routledge.

Janowicz, K., Gao, S., McKenzie, G., Hu, Y., & Bhaduri, B. (2020). Geoai: Spatially explicit artificial intelligence techniques for geographic knowledge discovery and beyond. *International Journal of Geographical Information Science*. Taylor & Francis, *34*(4), 625.

Jerrett, M., Burnett, R. T., Kanaroglou, P., Eyles, J., Finkelstein, N., Giovis, C., & Brook, J. R. (2001). A gis–environmental justice analysis of particulate air pollution in Hamilton, Canada. *Environment and Planning A, 6*, 955–973.

Kitchin, R. (2014). Big data, new epistemologies and paradigm shifts. *Big Data & Society, 1*(1), 2053951714528481.

Lee, J.-G., & Kang, M. (2015). Geospatial big data: Challenges and opportunities. *Big Data Research, 2*(2), 74–81. https://doi.org/10.1016/j.bdr.2015.01.003

Lei, B., Liu, P., Milojevic-Dupont, N., & Biljecki, F. (2024). Predicting building characteristics at urban scale using graph neural networks and street-level context. *Computers, Environment and Urban Systems, 111*, 102129.

LeSage, J., & Pace, R. K. (2009). *Introduction to spatial econometrics*. CRC Press.

Leszczynski, A., & Crampton, J. (2016). Introduction: Spatial big data and everyday life. *Big Data & Society, 3*(2), 2053951716661366.

Li, W. (2020). Geoai: Where machine learning and big data converge in giscience. *Journal of Spatial Information Science, 20*, 71–77.

Li, J., Zhang, X., Li, L., Wang, X., Cheng, J., Gao, C., & Ling, J. (2024). An estimation method for multidimensional urban street walkability based on panoramic semantic segmentation and domain adaptation. *Engineering Applications of Artificial Intelligence, 136*, 108905.

Liang, X., Zhao, T., & Biljecki, F. (2023). Revealing spatio-temporal evolution of urban visual environments with street view imagery. *Landscape and Urban Planning, 237*, 104802.

Liang, X., Chang, J. H., Gao, S., Zhao, T., & Biljecki, F. (2024a). Evaluating human perception of building exteriors using street view imagery. *Building and Environment, 111875*, 111875.

Liang, Y., Liu, Y., Wang, X., & Zhao, Z. (2024b). Exploring large language models for human mobility prediction under public events. *Computers, Environment and Urban Systems, 112*, 102153.

Liu, P., & Biljecki, F. (2022). A review of spatially-explicit geoai applications in urban geography. *International Journal of Applied Earth Observation and Geoinformation, 112*, 102936.

Liu, L., & Sevtsuk, A. (2024). Clarity or confusion: A review of computer vision street attributes in urban studies and planning. *Cities, 150*, 105022.
Liu, X., Kyriakidis, P. C., & Goodchild, M. F. (2008). Population-density estimation using regression and area-to-point residual kriging. *International Journal of Geographical Information Science, 4*, 431–447.
Liu, Y., Liu, X., Gao, S., Gong, L., Kang, C., Zhi, Y., et al. (2015). Social sensing: A new approach to understanding our socioeconomic environments. *Annals of the Association of American Geographers, 105*(3), 512–530.
Liu, X., Andris, C., Huang, Z., & Rahimi, S. (2019). Inside 50,000 living rooms: An assessment of global residential ornamentation using transfer learning. *EPJ Data Science, 8*(1).
Liu, P., Zhao, T., Luo, J., Lei, B., Frei, M., Miller, C., & Biljecki, F. (2023). Towards human-centric digital twins: Leveraging computer vision and graph models to predict outdoor comfort. *Sustainable Cities and Society, 93*, 104480.
Liu, P., Zhang, Y., & Biljecki, F. (2024). Explainable spatially explicit geospatial artificial intelligence in urban analytics. *Environment and Planning B: Urban Analytics and City Science, 51*(5), 1104–1123.
Luo, J., Zhao, T., Cao, L., & Biljecki, F. (2022). Semantic Riverscapes: Perception and evaluation of linear landscapes from oblique imagery using computer vision. *Landscape and Urban Planning, 228*, 104569.
Mai, G., Janowicz, K., Hu, Y., Gao, S., Yan, B., Zhu, R., et al. (2022). A review of location encoding for geoai: Methods and applications. *International Journal of Geographical Information Science, 36*(4), 639–673.
Mavros, P., Austwick, M. Z., & Smith, A. H. (2016). Geo-eeg: Towards the use of eeg in the study of urban behaviour. *Applied Spatial Analysis and Policy, 9*, 191–212.
Medina-Salgado, B., Sa'nchez-DelaCruz, E., Pozos-Parra, P., & Sierra, J. E. (2022). Urban traffic flow prediction techniques: A review. *Sustainable Computing: Informatics and Systems, 35*(100), 739.
Mortaheb, R., & Jankowski, P. (2023). Smart city re-imagined: City planning and geoai in the age of big data. *Journal of Urban Management, 12*(1), 4–15.
Pacione, M. (2009). *Urban geography: A global perspective*. Routledge.
Patel, P., Kalyanam, R., He, L., Aliaga, D., & Niyogi, D. (2023). Deep learning-based urban morphology for city-scale environmental modeling. *PNAS Nexus, 2*(3), pgad027.
Penn, A., Hillier, B., Banister, D., & Xu, J. (1998). Configurational modelling of urban movement networks. *Environment and Planning B: Planning and Design, 25*(1), 59–84.
Poorthuis, A., Zook, M., Shelton, T., Graham, M., & Stephens, M. (2023). 17 using digital social data in geographical research. *Key Methods in Geography, 278*.
Poorthuis, A., Chen, Q., & Zook, M. (2024, July). A nationwide dataset of de-identified activity spaces derived from geotagged social media data. *Environment and Planning B: Urban Analytics and City Science*. https://doi.org/10.1177/23998083241264051
Pred, A. (2017). *City-systems in advanced economies: Past growth, present processes and future development options*. Routledge.
Quadri, R., et al. (2024). A survey and analysis of intelligent forecasting and decision-making evaluation of urban growth using artificial intelligence. *International Journal of Advanced Research in Computer Science, 15*(2).
Reclus, E., Ravenstein, E. G., & Keane, A. H. A. H. [1876–94] *The universal geography: Earth and its inhabitants*. J.S. Virtue & Co., Limited.
Rey-Blanco, D., Arbues, P., Lopez, F., & Paez, A. (2024). A geo-referenced micro-data set of real estate listings for spain's three largest cities. *Environment and Planning B: Urban Analytics and City Science*.
Ruth, M., & Franklin, R. S. (2014). Livability for all? conceptual limits and practical implications. *Applied Geography, 49*, 18–23.
Sauer, C. O. (1941). Foreword to historical geography. *Annals of the Association of American Geographers, 31*(1), 1–24.
Sauer, C. O. (1952). *Agricultural origins and dispersals*. The American Geographical Society.

Shelton, T., Poorthuis, A., & Zook, M. (2015). Social media and the city: Rethinking urban socio-spatial inequality using user-generated geographic information. *Landscape and Urban Planning, 142*, 198–211.

Shukla, K., Kumar, P., Mann, G. S., & Khare, M. (2020). Mapping spatial distribution of particulate matter using kriging and inverse distance weighting at supersites of megacity delhi. *Sustainable Cities and Society, 101*, 997.

Stalder, S., Volpi, M., Büttner, N., Law, S., Harttgen, K., & Suel, E. (2024). Self-supervised learning unveils urban change from street-level images. *Computers, Environment and Urban Systems, 112*(102), 156.

Tartarini, F., Frei, M., Schiavon, S., Chua, Y. X., & Miller, C. (2023). Cozie apple: An ios mobile and smartwatch application for environmental quality satisfaction and physiological data collection. *Journal of Physics: Conference Series, 2600*, 142003.

Tobler, W. R. (1970). A computer movie simulating urban growth in the detroit region. *Economic Geography, sup1*, 234–240.

Torrens, P. M. (2006). Simulating sprawl. *Annals of the Association of American Geographers, 2*, 248–275.

von Neumann, J., & Morgenstern, O. (1944). *Theory of games and economic behavior*. Princeton University Press.

Wang, Y. (2024). Fusing multi-source social media data and street view imagery to inform urban space quality: a study of user perceptions at kampong glam and haji lane. *Urban Informatics, 3*(1).

Wang, J., Chow, Y. S., & Biljecki, F. (2023). Insights in a city through the eyes of Airbnb reviews: Sensing urban characteristics from homestay guest experiences. *Cities, 140*(104), 399.

Webber, R., & Craig, J. (1976). Which local authorities are alike. *Population Trends, 5*, 13–19.

Wei, D., Wang, Y., Jiang, Y., Guan, X., & Lu, Y. (2024). Deciphering the effect of user-generated content on park visitation: A comparative study of nine chinese cities in the pearl river delta. *Land Use Policy, 144*(107), 259.

Wu, C., & Murray, A. T. (2005). A cokriging method for estimating population density in urban areas. *Computers, Environment and Urban Systems, 5*, 558–579.

Yan, T., Yang, X. R., Sun, H., & Cantor, D. (2024). Using google street view for virtual observations of neighborhoods and dwelling units: A feasibility study. *PLOS ONE, 19*(8), e0307272.

Yang, C., Liu, T., & Zhang, S. (2022). Using flickr data to understand image of urban public spaces with a deep learning model: A case study of the Haihe River in Tianjin. *ISPRS International Journal of Geo-Information, 11*(10), 497.

Yap, W., Stouffs, R., & Biljecki, F. (2023). Urbanity: Automated modelling and analysis of multidimensional networks in cities. *NPJ Urban Sustainability, 3*(1), 45.

Zhang, F., Zhou, B., Liu, L., Liu, Y., Fung, H. H., Lin, H., & Ratti, C. (2018). Measuring human perceptions of a large-scale urban region using machine learning. *Landscape and Urban Planning, 180*, 148–160.

Zhang, Y., Sun, T., Wang, L., Huang, B., Pan, X., Song, W., et al. (2024). Portraying on-road co2 concentrations using street view panoramas and ensemble learning. *Science of The Total Environment, 946*, 174326.

Zhao, T., Liang, X., Tu, W., Huang, Z., & Biljecki, F. (2023). Sensing urban soundscapes from street view imagery. *Computers, Environment and Urban Systems, 99*(101), 915.

Zhao, X., Lu, Y., Huang, W., & Lin, G. (2024). Assessing and interpreting perceived park accessibility, usability and attractiveness through texts and images from social media. *Sustainable Cities and Society, 112*(105), 619.

Zhou, S., Wang, Y., Jia, W., Wang, M., Wu, Y., Qiao, R., & Wu, Z. (2023). Automatic responsive-generation of 3d urban morphology coupled with local climate zones using generative adversarial network. *Building and Environment, 245*(110), 855.

Zipf, G. K. (1949). *Human behavior and the principle of least effort*. Addison-Wesley Press.

Part IV
Emerging Trends Future Perspectives

Chapter 19
Identifying Gentrification with GeoAI: Synthesizing Physical and Socio-economic Dimensions

Feicui Gou and Zhigang Li

Abstract Gentrification is a multidimensional and context-specific process involving transformations in neighborhoods' physical and socio-economic fabric. However, due to limitations in data availability as well as existing modeling techniques, the measurement of the physical dimensions has long been underexplored. In recent years, the rapid growth of big data and advancements in GeoAI technologies have provided new paths to facilitate large-scale, automated, and precise analyses of gentrification. To fill the gap, this chapter explores the application of deep mapping based on computer vision algorithms to integrate physical dimensions into gentrification identification models, highlighting the advantages of GeoAI in gentrification studies. Taking Wuhan City, China, as a study case, we construct a multidimensional framework combining a deep convolutional neural network (CNN) based on Baidu Street View (BSV) with a threshold model informed by socioeconomic indicators. The integration captures both physical and social upgrade processes, mapping the spatial distribution and identifying clusters of gentrification in Wuhan between 2016 and 2019, thereby providing a nuanced understanding of gentrification across the city. The chapter concludes by discussing the challenges and future directions for incorporating deep learning approaches into gentrification research, offering new insights into integrating GeoAI technologies into gentrification studies.

Keywords GeoAI · Gentrification · Built environment · Socioeconomic indicators · Deep learning

19.1 Introduction

As a classic field of urban geography studies, gentrification was originally coined by Ruth Glass in 1964 to describe the social and spatial transformations that

F. Gou · Z. Li (✉)
School of Urban Design, Wuhan University, Wuhan, China
e-mail: zhigangli@whu.edu.cn

© The Author(s), under exclusive license to Springer Nature Switzerland AG 2025
X. Huang et al. (eds.), *GeoAI and Human Geography*, Springer Geography,
https://doi.org/10.1007/978-3-031-87421-5_19

occurred when middle-class households moved into deteriorated working-class neighborhoods in the inner city of London (Glass, 1964). Over the last half-century, however, the scope of the concept has expanded considerably, driven by temporal and social diversity (e.g., sub-types of gentrification) and spatial heterogeneity (e.g., gentrification in the Global South) (Brown-Saracino, 2017; Clark, 2005; Maloutas, 2012; Silver & Clark, 2016). Nowadays, gentrification is recognized as a complicated and far-reaching process of urban economic, social, and spatial transformation, featured by two key components: the displacement of lower socioeconomic classes and the reinvestment of capital (Smith, 2013; Zuk et al., 2018). From an operational perspective, gentrification involves three processes of neighborhood changes: (1) reinvestment of capital, (2) landscape changes, and (3) the influx of high(er) socioeconomic classes accompanied by the displacement of low(er) socioeconomic classes (Davidson & Lees, 2005).

Along with planetary gentrification, a substantial body of research has empirically explored gentrification, both qualitatively and quantitatively. Qualitative studies adopt a micro-level perspective, focusing on the dynamic changes within specific communities or neighborhoods and examining gentrification's profound social and cultural impacts, such as lifestyle shifts and the replacement of cultural spaces. These studies also delve into the mechanisms and institutional forces driving gentrification, along with the strategies for resisting it. In contrast, quantitative research employs data-driven models to depict gentrification's geographic distribution, expansion rates, and socioeconomic dynamics, aiming to describe and explain its patterns and effects (Brown-Saracino, 2017). Basically, identifying the locations of gentrification remains a central focus of empirical research and a critical foundation for formulating effective policy responses (Cohen & Pettit, 2019).

Despite significant advancements in gentrification research, due to the ambiguity of conceptualization and operational uncertainty, gentrification identification and mapping remain fraught with challenges and controversies (Liu et al., 2019). For instance, qualitative studies may offer in-depth ethnographic insights, employing detailed narratives to uncover the complexities of gentrification (Maya et al., 2024). However, they are often highly localized and time-intensive, making them unsuitable for cross-regional model construction. Moreover, qualitative gentrification studies frequently focus on specific "signifying locations" (Reades et al., 2019), potentially overlooking other districts experiencing similar social transformations, thereby limiting the generalizability of the findings. Quantitative methods, by comparison, take a macro perspective, aiming to identify gentrification through socioeconomic indicators, such as changes in income levels (Freeman, 2005) or rental prices (Zapatka & Beck, 2021). Nevertheless, these approaches are heavily dependent on high spatial resolution and temporally consistent data, particularly census data (Finio, 2022), which can be inconsistent or unavailable in certain regions, thus restricting their applicability (Raya-Tapia et al., 2024). Furthermore, the reliance on socioeconomic changes may result in the risk of misclassification, such as conflating aging populations (Wang & Durst, 2023) or immigration influxes (Hwang & Sampson, 2014) with gentrification (Thackway et al., 2023b).

Given that gentrification encompasses multifaceted and complex socioeconomic phenomena and processes, there is a growing consensus in the literature that measuring gentrification across multiple dimensions can effectively reduce the risk of misclassification. However, extant studies mainly operationalize gentrification based on socioeconomic changes, primarily due to the relative ease of access and data availability. In contrast, studies focusing on the physical upgrading as a critical dimension are relatively rare, and research integrating both socioeconomic and physical dimensions remains even scarcer (Podagrosi et al., 2011). A few studies addressing the physical dimension of gentrification largely relied on direct audits of neighborhood architectural features or related visually observable changes (Chapple, 2009; Hammel & Wyly, 1996; Wyly & Hammel, 1999). While such site-based observation approaches provide some spatial information, it is labor-intensive and technologically constrained, limiting their application to small geographic scales. Additionally, they are difficult to replicate. The substantial labor required to collect and analyze built environment data has thus become a significant barrier to identifying gentrification from the perspective of built environment transformation (Ilic et al., 2019).

In this vein, rapid growth of big data and advancements in artificial intelligence (AI) may unlock new opportunities for identifying gentrification, particularly in the following three aspects: (1) the widespread application of geospatial big data provides abundant information for analyzing socio-spatial characteristics, such as remote sensing imagery and street-level panoramic services (e.g., Google Street View, GSV) (Kwon et al., 2014); (2) recent breakthroughs in AI and machine learning, especially deep learning, enable AI to effectively imitate human perception in social-spatial sensing and reasoning (Liu & Biljecki, 2022; Wang et al., 2024a); and (3) advancements in algorithms and computational technologies have made it feasible to process large-scale, computationally intensive geospatial data (Li, 2020). The GeoAI framework holds significant potential to enhance our understanding of gentrification, notably by expanding the analytical dimensions, spatial scales, and precision of gentrification identification.

19.2 GeoAI for Gentrification

19.2.1 The Advantages of GeoAI Techniques for Gentrification Studies

The essence of GeoAI lies in leveraging advanced AI technologies to analyze and emulate human perception and reasoning processes, enabling the exploration and visualization of complex geographic data (Li et al., 2024; Liu & Biljecki, 2022; Wang et al., 2024a). This approach uncovers spatial patterns that traditional methods often struggle to detect, providing deeper insights into geospatial phenomena. Based on different learning paradigms, the GeoAI method can be broadly classified into top-down and bottom-up approaches (Mortaheb & Jankowski, 2023). In a top-down approach, knowledge is explicitly encoded in machine-readable languages, facilitating formal and logical reasoning. In contrast, bottom-up approaches operate directly on data to identify, predict, and model geospatial features and phenomena without requiring predefined knowledge, and the data-driven approaches can be classified into clustering, classification, and regression (Li, 2022).

Machine learning (ML) and deep learning (DL) serve as the methodological foundation for integrating GeoAI into gentrification research (Liu & Song, 2024). Machine learning algorithms are trained on labeled data, where "labels" represent descriptor values for data samples (e.g., the binary yes/no indicator of whether an area is gentrified). It can be applied to predictive classification tasks (e.g., determining whether a specific location is undergoing gentrification) or to regression tasks that estimate continuous outputs (e.g., measuring the degree of socioeconomic change in an area). Deep learning, a branch of machine learning that relies on neural network architectures, has been increasingly employed in a wide range of applications, such as black-box models that imitate perceptual judgements of a scene (e.g., categorizing street view imagery as "boring" or "vivid"), and extracts semantic information from images (e.g., buildings, roads, facilities, trees). The capabilities of DL are valuable for scene classification tasks (e.g., categorizing areas as "gentrified" or "non-gentrified") in gentrification studies (Maya et al., 2024).

Advancements in GeoAI have provided researchers with innovative tools to address the complex challenge of modeling gentrification. GeoAI stands out for its ability to analyze multidimensional data and uncover meaningful patterns within high-volume, stochastic datasets. Although scholars have acknowledged the multifaceted nature of gentrification (Anguelovski et al., 2018), addressing multiple variables simultaneously in operationalization has remained a challenge. Traditional approaches, including threshold-based methods—which compare various indicators (e.g., income, housing prices) against predefined benchmarks (median or mean) values to determine whether gentrification has occurred (Choi et al., 2018)—and dimensionality reduction techniques like Principal Component Analysis (PCA) (Walks & Maaranen, 2008), both face significant limitations, particularly with respect to low dimensionality and the interpretability of results (Liu et al., 2019). In contrast, GeoAI excels in processing large-scale, high-dimensional, and heterogeneous datasets (Li & Hsu, 2022). Whether through supervised or unsupervised learning, machine learning methods can leverage automatic feature learning to extract gentrification patterns from datasets without being constrained by the number of indicators or dimensions analyzed (Maya et al., 2024).

Importantly, recent advancements in computer vision within GeoAI have allowed researchers to incorporate previously overlooked indicators into gentrification modeling, particularly building development and investment, which are critical dimensions of gentrification often neglected in traditional studies (Ilic et al., 2019; Thackway et al., 2023b). Visual data sources like Google Street View (GSV) offer valuable insights into the urban built environment at lower costs and with higher data integrity (Zhang et al., 2023), yet earlier approaches relying on manual annotation to detect image changes severely limited scalability and practical applicability. Integration of deep learning techniques has facilitated large-scale automated analysis of visual data (LeCun et al., 2015). For example, Convolutional Neural Networks (CNNs) effectively integrate features across spatial scales—from low-level positional details to high-level semantic information—to derive comprehensive conclusions about environmental changes (Bai et al., 2023). By integrating the scene parsing capabilities of advanced computer vision with geographical knowledge, GeoAI techniques can potentially address the limitations of earlier approaches (Liu & Biljecki, 2022; Reichstein et al., 2019), providing a strong rationale for their application in gentrification research.

19.2.2 GeoAI in Gentrification Studies

Supported by multi-source geospatial big data—describing both the physical and social environments—including raster-based imagery (e.g., remote sensing images, street view images, and geotagged photos), vector-based spatial data (e.g., points of interest, areas of interest, and buildings), and statistic-based data (e.g., census, economic data), GeoAI techniques have been increasingly integrated into gentrification studies (Wang et al., 2024a). As illustrated in Fig. 19.1, the critical applications include: (1) predicting gentrification through feature learning, (2) identifying gentrification using classification models, and (3) incorporating physical environment dimension into gentrification models with computer vision techniques.

The pioneering application of GeoAI techniques in gentrification analysis can be traced back to the work of (Reades et al., 2019). They introduced the Random Forest (RF) algorithm to analyze and learn patterns and processes of community change between 2001 and 2011 using census data, identifying areas likely to experience "uplift" or "decline" by 2021, and thereby forecasting the trajectory of gentrification. Similarly, Gray et al. (2023) utilized three ensemble models (Boosting Machine, Extreme Gradient Boost, and Bootstrap Aggregation) to predict gentrification's presence, type, and temporal properties across the rest of England. More recently, Thackway et al. (2023a) employed three decision tree-based

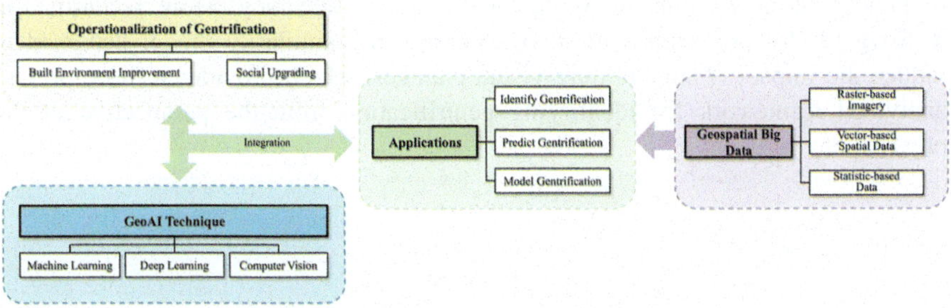

Fig. 19.1 Conceptual diagram of the integration of GeoAI and gentrification

algorithms (Random Forest, Gradient Boosting Machine, and Extreme Gradient Boosting) to predict future gentrification in Sydney by analyzing changes in the socioeconomic index of advantage and disadvantage, as well as reporting on the performance characteristics of different machine learning models in gentrification prediction.

Another growing body of literature has applied machine learning classification models to identify gentrification. Unsupervised learning algorithms were the earliest to be employed (Ye et al., 2015). For instance, Liu et al. (2019) utilized K-means clustering with seven variables (including measures of capital reinvestment, property tenure, social upgrading, and ethnicity) to classify gentrification in Auckland (Liu et al., 2019). Similarly, Best et al. (2023) employed K-means clustering to highlight vulnerabilities across environmental, socioeconomic, and housing indicators, identifying new patterns of climate gentrification. Following the developments, supervised machine learning algorithms have also been adopted for gentrification identification. For example, Zeng et al. (2022) used XGBoost to dynamically track gentrification by analyzing localized services represented by points of interest (POI), a big data category providing detailed business classifications. More recently, this approach has been further extended to identify gentrification sub-types. For instance, Assaad & Jezzini (2024) proposed an integrated framework combining unsupervised and supervised machine learning algorithms to classify green gentrification sensitivity in New York City into six levels using 35 relevant indicators.

A significant advancement in GeoAI applications for gentrification studies is the use of image data, particularly street view imagery, which has created new opportunities for integrating architectural forms into gentrification models and incorporating physical environment changes as a critical dimension for identifying gentrification. Ilic et al. (2019) pioneered in leveraging deep mapping techniques with Google Street View (GSV) to identify gentrification. Their research introduced the concept of "visual proxies" to represent gentrification. The study employed a Siamese Convolutional Neural Network (SCNN) to detect physical improvements of individual properties over time in Ottawa's Greenbelt area. Following this modeling approach, Thackway et al. (2023b) further integrated the analysis of changes in the built environment with socioeconomic indicators. Their study developed an SCNN to process GSV images, identify physical improvements to properties, and cross-validate these findings with socioeconomic trends, which enabled the mapping of gentrification hotspots in Sydney with greater accuracy. More recently, Liu & Song (2024) proposed a masked deep pyramid similarity model designed to reduce the impact of nonbuilding visual changes, which provides a more precise analytical framework for identifying gentrification from the perspective of the physical environment.

19.3 Case Study: Mapping Gentrification in Wuhan Through the Integration of Physical and Socioeconomic Dimensions

This section examines the application of GeoAI techniques for identifying gentrification and their potential to bridge the gap between its social and physical dimensions. By leveraging a comprehensive Street View image (SVI) database and a Siamese Convolutional Neural Network (SCNN) deep learning model, it presents a method to evaluate the physical dimension of gentrification on a urban-scale. Furthermore, applying a threshold model based on a geo-economic database evaluates the socioeconomic dimension of gentrification, offering a more integrated and holistic perspective.

19.3.1 Methodology

To identify gentrification, a two-part model was developed (Fig. 19.2). The first component is a Siamese Convolutional Neural Network (SCNN) model, which extracts features from scene understanding and the spatiotemporal rules of urban physical changes to determine transformations in the built environment (Liu & Song, 2024; Thackway et al., 2023b). The second component is a threshold model, which intuitively identifies socioeconomic changes based on statistical and geospatial data. The results from both dimensions are then overlaid to determine the

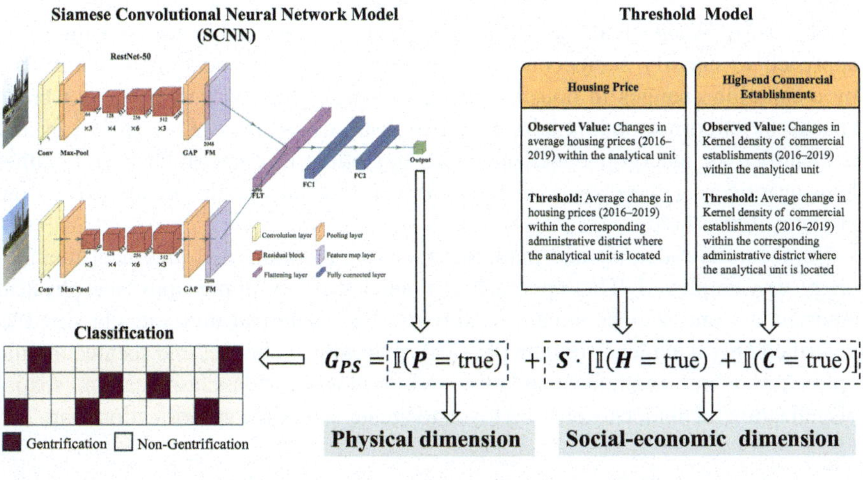

Fig. 19.2 Framework for identifying gentrification through the integration of physical and social dimensions

spatial distribution of gentrification, which is subsequently aggregated at standard analytical units for visualization.

A Siamese Convolutional Neural Network (SCNN) is a specialized neural network architecture designed to assess the similarity between two inputs. The core principle involves leveraging convolutional neural networks with shared weights to extract features from both inputs, followed by the computation of their similarity using a designated metric, such as Euclidean distance or cosine similarity (Chicco, 2021). Convolutional deep learning techniques have been extensively applied across various change detection scenarios, including land use transformations and road network modifications (Lei et al., 2020). In this study, we utilize an SCNN as a change detection framework, employing time-series street view image pairs—captured at the same location but at two different time points—as inputs. By extracting image features through convolutional layers, conducting feature vector interaction computations, and applying a carefully designed loss function, the model generates a binary classification outcome. This output generates a binary classification of whether specific changes, such as physical infrastructure upgrades, have occurred between the two input street view images.

The threshold method involves comparing changes in specific indicators against predefined thresholds (e.g., median or mean) to determine whether socioeconomic upgrading has occurred. This study uses housing prices and the density of high-end commercial establishments as key socioeconomic indicators. Among these, housing sale prices stand out as the only available socioeconomic data offering both a long time series and high spatial resolution. Consequently, many scholars have regarded housing prices as a primary, or even sole, indicator for observing gentrification (Anguelovski et al., 2017; Wu & Rowe, 2022). Meanwhile, high-end commercial places, which reflect middle-class lifestyles and tastes, are increasingly becoming an important dimension for representing socioeconomic status (Huang et al., 2024; Wang et al., 2024b). This study employs a mean-based threshold approach to classify socioeconomic upgrading. The classification is determined by comparing changes in housing prices within an analytical unit to the average change in its corresponding administrative district. Specifically, an analytical unit is identified as undergoing socioeconomic upgrading if temporal changes in either housing prices or the density of high-end commercial establishments exceed the predefined threshold.

To accurately identify and map gentrification, a conditional judgment methodology was employed. This approach evaluated two key dimensions: the physical environment and socioeconomic conditions. The analytical unit was classified as gentrified only if evidence of both physical environment changes and socioeconomic upgrading were simultaneously observed. Following this classification, spatial visualization techniques were employed to create a detailed gentrification map.

19.3.2 Study Area and Data Source

This study takes Wuhan as a case study. As a major city in central China, Wuhan has undergone rapid economic growth and urban transformation in recent years. This economic dynamism has driven the swift expansion of the middle class and attracted a substantial influx of migrants. The interplay between population mobility and social stratification creates a critical context for the socioeconomic transitions linked to gentrification. Furthermore, Wuhan has implemented extensive urban renewal programs involving large-scale redevelopment of aging urban areas and shantytowns, significantly reshaping specific neighborhoods' built environments. These transformations highlight profound socioeconomic changes and provide a viable context for examining the material upgrading characteristic of gentrification. This study focuses on the urban development area rather than the entire city (Fig. 19.3), as the area accounts for approximately 85% of the total population and hosts the majority of the city's urban functions and activities.

The core datasets used in this study include Street View Imagery (SVI), housing transaction records, and POI data (points of interests). The input for the Siamese Convolutional Neural Network (SCNN) model consists of paired street view images, comprising current and historical imagery, which were downloaded through internal Baidu API calls (Fig. 19.4). To ensure comprehensive spatial coverage, a total of 85,515 sampling points were generated at 100-meter intervals along the road network provided by Open Street Map (OSM). Considering SVI's spatial distribution and temporal availability, the study focuses on two specific time points: 2016 and 2019. After filtering, 14,760 sampling points with complete image coverage for both years were identified. For each sampling point, four images were captured from different angles (0°, 90°, 180°, and 270°). To make the built environment analysis

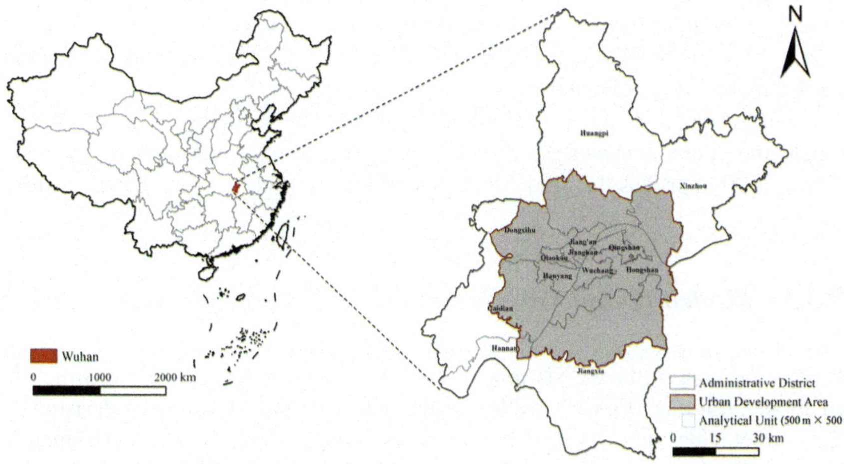

Fig. 19.3 The study area

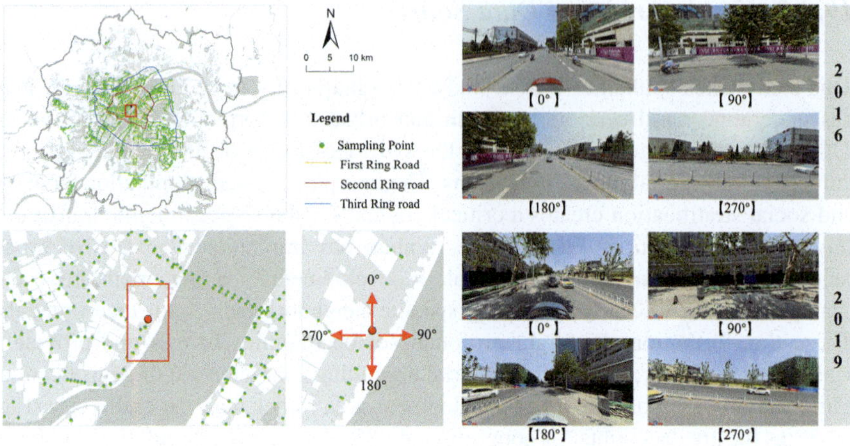

Fig. 19.4 Baidu street view data collection and sample data

more targeted, this study included only street-facing images (90° and 270°). As a result, the final SVI database comprises 59,040 images, organized into 29,520 time-series pairs.

The input for the threshold model consists of housing transaction records and points of high-end commercial POI data. Housing transaction data were sourced from Lianjia (https://wh.lianjia.com/), a leading real estate platform in China. We considered the transaction records of 2016 and 2019 for each neighborhood, collecting a total of 20,751 transactions. The POI dataset was obtained from Baidu Maps, which provides the geographic locations of various commercial establishments in Wuhan. Five categories of high-end commercial establishments closely related to gentrification were selected: upscale dining, coffee and tea shops, beauty and hair services, yoga and fitness studios, and creative stores (Huang et al., 2024; Wang et al., 2024b). Across 2016 and 2019, 19,733 POI records for high-end commercial establishments were collected.

To ensure detailed spatial analysis and enhance the accuracy of interpretations by reducing potential biases associated with aggregated data, this study employed a 500 m × 500 m grid as the fundamental analytical unit for mapping gentrification.

19.3.3 Modeling and Training

The development of the SCNN model involves data cleaning, dataset construction, and model training. First, considering the visual composition characteristics of urban street view imagery in Chinese cities, image cropping was performed to remove the bottom one-third of the images (approximately 170 pixels in height), which primarily depicted roads. To minimize potential interference from elements such as trees, vehicles, and roads in the deep learning process for the built environment, a semantic segmentation algorithm (SegNet) was employed to classify images into basic object types (e.g., buildings, trees, sky, vehicles). This process identified and excluded unsuitable image pairs for classification. After filtering, a final dataset of 23,946 image pairs was obtained.

To train the SCNN model, a subset comprising 30% of the entire SVI dataset was selected and further divided into training, validation, and test sets in a 60:20:20 ratio. A total of 7184 time-series SVI pairs (14,368 street view images) were randomly selected from image pairs with less than 30% noise elements to construct the dataset for deep learning. For the training dataset, five experts from the case city (Wuhan) were invited to label the data after receiving a detailed briefing on the characteristics of built environment upgrades. The labeling process classified each SVI pair based on whether physical improvements were observed: pairs in which the latest image exhibited changes indicating physical improvements, were labeled as "1," while pairs with no discernible changes were labeled as "0." To address class imbalance, a balanced training dataset with a 50:50 split between positive and negative samples was created. The final dataset consisted of 4310 time-series SVI pairs for training, 1252 pairs for validation, and 1252 pairs for testing.

Following the methodology outlined by Thackway et al. (2023b), the SCNN was trained for image similarity analysis using the Keras library in Python 3.10. The core architecture of the model is ResNet-50, with weights pre-trained on the ImageNet database. Input images were resized to a uniform size of 224 $\times$ 224 pixels and normalized based on statistical parameters from the ImageNet database. To enhance data diversity and reduce the risk of overfitting, various random augmentation strategies, including random cropping, scaling, and normalization, were applied during training. To further prevent overfitting, the model comprises two dropout layers and four fully connected layers. The model was trained using a Sigmoid activation function in the output layer to generate similarity scores. During training, binary cross-entropy was used as the loss function, with Adadelta as the optimizer. The batch size of 64 and the 14 training epochs ensured adequate training and strong generalization performance.

The SCNN model exhibited strong performance, achieving a test accuracy of 80.88% and an AUC of 80.43%. It demonstrated effective identification of similar image pairs with a recall of 68.63%, while maintaining a high precision of 90.91%, highlighting its robustness in minimizing false positives. All subsequent analyses presented in this study are based on this fully trained model, referred to as the SCNN_BSV.

Figure 19.5 illustrates examples of change detection results using the SCNN_BSV model. A similarity score below 0.5 signifies a predicted change between two images. Figure 19.5a-d display true positives, where the model accurately identified changes. Conversely, Fig. 19.5i-j depict false negatives, where the model failed to detect actual changes. Additionally, Fig. 19.5e-f present false positives, where the model incorrectly identified changes in unchanged images,

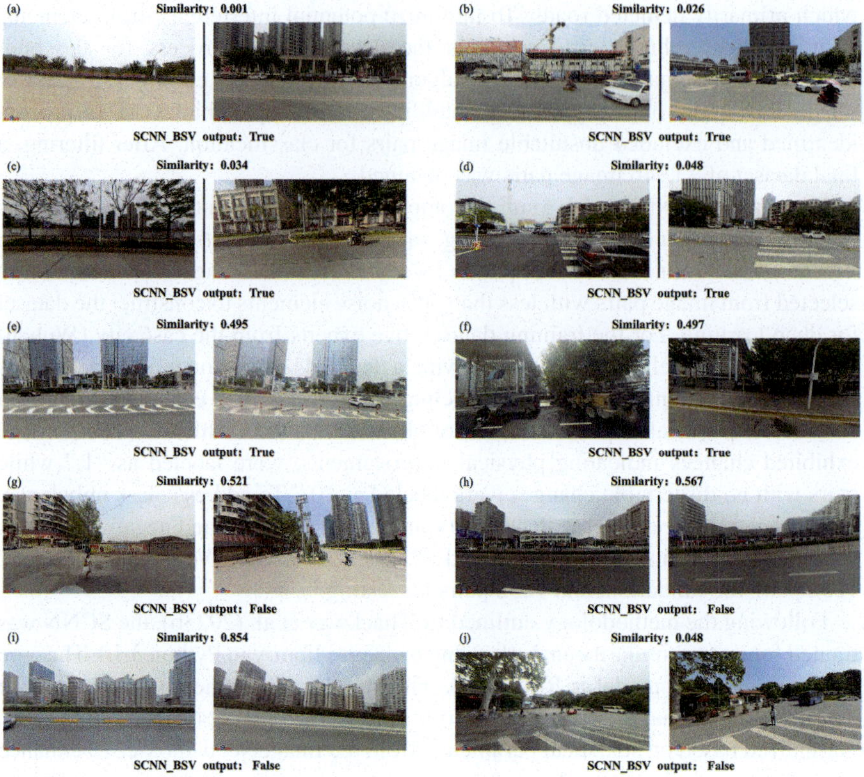

Fig. 19.5 Examples of SCNN_BSV model detections Note: the vertical symbol (|) captures actual change between the time-series SVI pairs

while Fig. 19.5g-h show false negatives, where the model did not detect actual changes.

19.3.4 Results and Findings

Using SCNN_BSV, we detected time-series SVI pairs at 14,760 unique locations within the study area. The results are illustrated in Fig. 19.6. Among the image pairs captured at a 90° street-facing angle, 1846 pairs were identified as positive for physical improvements (Fig. 19.6a). Similarly, among the image pairs captured at a 270° angle, 2235 pairs were identified as positive (Fig. 19.6b). These positive detections were aggregated into 500 × 500-meter spatial grids, as shown in Fig. 19.6c–d. Of the 2572 grids with time-series SVI pairs coverage, 437 grids exhibited a high density of positive points (>0.3), ultimately identifying them as areas that underwent visual changes indicating physical improvements between 2016 and 2019.

To assess the reliability of the deep mapping approach in detecting physical improvements, we compared the mapped results on 500 × 500-m spatial grids from SCNN_BSV to the spatial distribution of building rooftop changes in Wuhan from 2016 to 2019 (Fig. 19.6e–f). The physical improvement maps of the SCNN_BSV detected results, and the building rooftop changes exhibit very similar patterns, with an overlap rate of 95.42%, underscoring the SCNN_BSV's effectiveness in capturing physical improvements.

Using the threshold model, we analyzed housing price changes in 3106 spatial grids containing residential communities. As shown in Fig. 19.7, 998 grids exhibited housing price growth levels exceeding the average growth of their respective administrative districts and were classified as having significant housing price increases. Additionally, an analysis of changes in Kernel Density Estimation (KDE) results for high-end POI distributions across the 2 years identified 492 grids with density increases above the administrative district average, classifying them as having experienced significant consumer environment upgrades. Through spatial overlay analysis, 271 grids were ultimately identified as having undergone socioeconomic upgrading between 2016 and 2019 (Fig. 19.7d).

By overlaying the physical improvement map generated by the SCNN model with the socioeconomic upgrading map derived from the threshold model, we identified 234 grids where physical improvements aligned with socioeconomic upgrades. These grids were ultimately identified as having undergone gentrification. Figure 19.8 visually illustrates the detected gentrification grids between 2016 and 2019.

Through integrating physical and socioeconomic dimensions, this study developed a more comprehensive view of gentrification, and six distinct gentrification clusters have been identified within Wuhan's urban development zone (Fig. 19.9). Among these, the Sixin area in Hanyang District exhibited the most pronounced gentrification characteristics. As an emerging urban sub-center and the core development hub of Wuhan's New District, the Sixin area has undergone rapid advancements in infrastructure, transportation networks, and industries. These developments have attracted significant middle-class migration, serving as the region's primary driver of gentrification. In addition, areas such as Wansong Jiedao (subdistrict) in Jianghan District and Hongwei Road Jiedao in Qingshan District have emerged as prominent gentrification hotspots due to the continued concentration of high-end commercial, financial, and service industries. Resource-rich locales like Heping Jiedao and Guandong Jiedao in Hongshan District have similarly become focal points of gentrification, leveraging their superior residential environments, natural landscapes, and the proliferation of high-end housing developments. Urban renewal zones, including the northern section of Gutian Jiedao in Qiaokou District and Zhangjiawan Jiedao in Hongshan District, exhibit strong gentrification characteristics driven by significant infrastructure improvements and the construction of new residential developments. These transformations have facilitated the upward

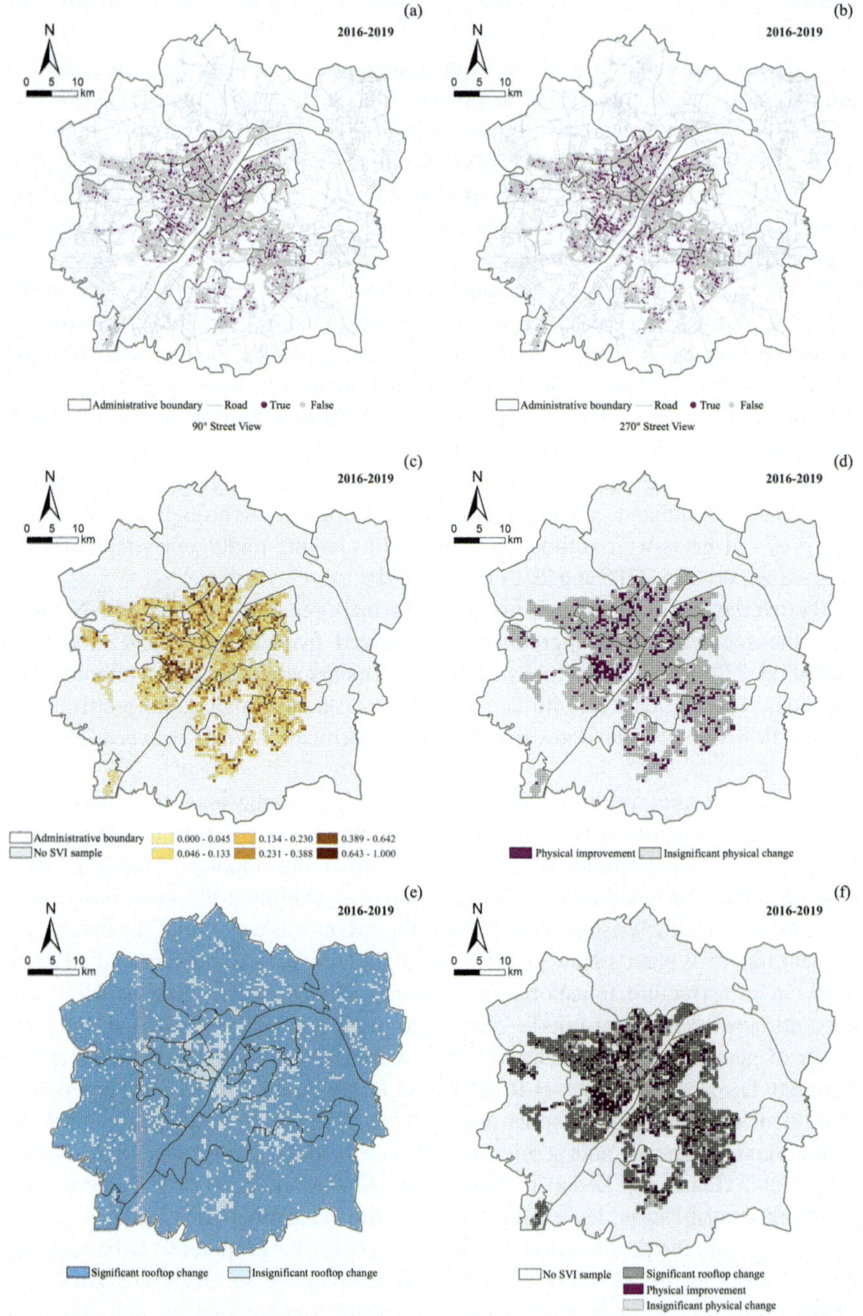

Fig. 19.6 The physical improvement detected using the SCNN_BSV model, in Wuhan between 2016 and 2019

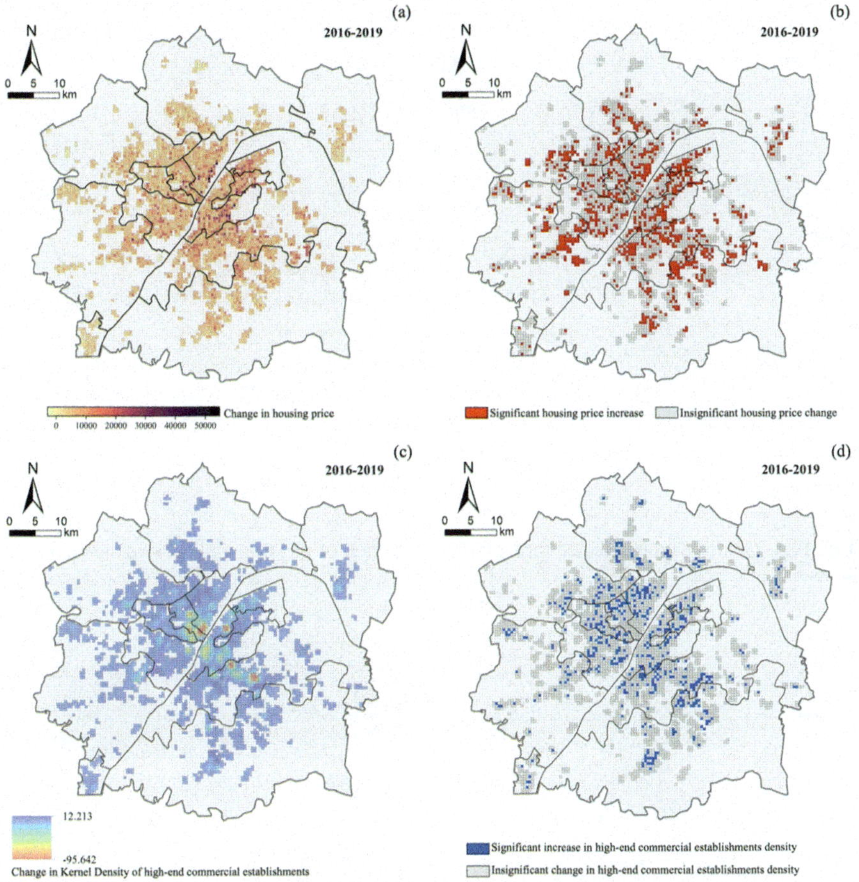

Fig. 19.7 The socioeconomic upgrade detected using the threshold model in Wuhan between 2016 and 2019

socioeconomic restructuring of local communities. Moreover, urban expansion and enhanced transportation infrastructure have spurred gentrification trends in sub-center districts on the outskirts of Wuhan's core city. Notable examples include Zhifang Jiedao in Jiangxia District and Jiangjun Road Jiedao in Dongxihu District, where these factors are contributing to a gradual shift in socioeconomic dynamics.

19.4 Discussions

GeoAI technologies, particularly recent advancements in image-based deep learning, have enabled researchers to incorporate previously unavailable built environment indicators into gentrification models, providing new capabilities for detection

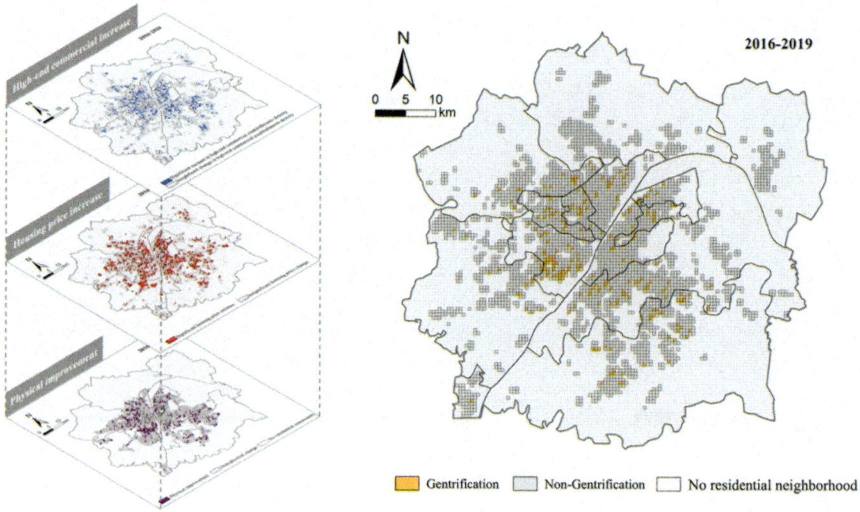

Fig. 19.8 Gentrification detected through the integration of physical and socioeconomic dimensions

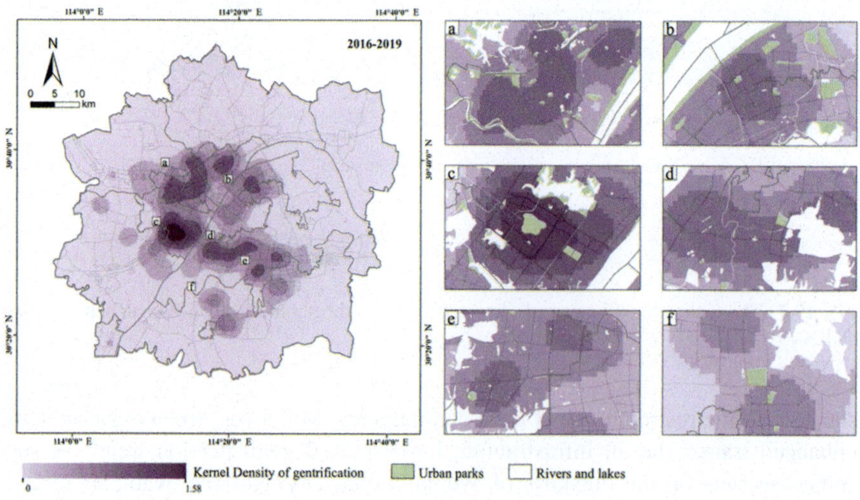

Fig. 19.9 The distribution of gentrification clusters in Wuhan between 2016 and 2019

and analysis. However, the application of GeoAI in gentrification studies remains in an early stage, facing several challenges.

Data Limitations While SVI data supports the detection of physical improvements and enables gentrification identification with high spatial and temporal resolution, it presents certain limitations compared to more structured datasets like remote sensing imagery. Spatially, SVI cannot capture data for gated or closed communities where roads are not publicly accessible, potentially under-detecting upgrades in areas that often serve as gentrification hotspots due to their spatial isolation. Temporally, SVI availability is constrained by infrequent updates, with multi-year gaps and limited historical coverage before 2014. These constraints restrict the analytical time span and hinder the ability to infer the precise timing of physical improvements, limiting insights into the lifecycle of gentrification.

Performance Challenges in DL Models The performance of DL models is susceptible to the accuracy of image segmentation. Poor visual cues, such as temporary objects like vehicles, introduce noise and result in false positives in built environment change detection. Although techniques like masking can mitigate these effects, balancing data cleaning with maintaining a sufficient dataset remains a challenge. Additionally, slight shifts in camera positions can alter structural elements in images, causing misjudgments in detecting changes.

Transferability Challenges in DL Models While DL models offer automated workflows and can adapt to new urban environments, their implementation requires context-specific adjustments, including deployment strategies, training data preparation, and validation methods. Manual labeling of training data remains labor-intensive but critical for ensuring model accuracy and reliability.

Interpretability Challenges in DL Models The interpretability of DL models remains a significant issue. While effective, architectures like ResNet-50 do not explicitly identify semantic features (e.g., windows, porches). Instead, they mimic human judgments of pixel-level changes without extracting interpretable features. This "black-box" nature makes it difficult to ascertain whether the model targets relevant pixels (e.g., those forming a new window) or to define thresholds for structural changes. These limitations impede discussions on the role of specific architectural attributes in driving gentrification.

Future Directions Future research should address these challenges to advance GeoAI in gentrification studies. First, expanding data sources to include satellite imagery could mitigate camera angle issues and offer better temporal granularity for tracking physical changes. Second, enhancing DL models' expressive and predictive capabilities through advanced architectures and innovative techniques could improve their ability to capture complex patterns and generate accurate predictions. Third, developing explainable DL models using frameworks like Explainable Artificial Intelligence (XAI) is crucial. XAI can provide insights into model predictions, facilitate auditing, and identify architectural attributes, offering a deeper understanding of how built forms influence community change. By addressing these areas, future studies can enhance the robustness, precision, and interpretability of GeoAI applications in gentrification research.

19.5 Conclusion

This chapter explores the role of advanced GeoAI technologies in gentrification research. It begins by outlining the advantages of GeoAI and summarizing three critical applications: (1) predicting gentrification through feature learning, (2) identifying gentrification using classification models, and (3) integrating physical environment dimensions into gentrification models with computer vision techniques. A case study in Wuhan illustrates how image-based deep learning technologies bridge the gap between the physical and social dimensions of gentrification analysis. By integrating a Siamese Neural Network (SCNN) based on Baidu Street View (BSV) with a traditional threshold model, the study develops a framework to capture the multidimensional nature of gentrification. The spatial distribution and hotspots of gentrification in Wuhan from 2016 to 2019 are identified and mapped based on physical improvements detected by the SCNN_BSV model and socioeconomic changes identified through the threshold model. GeoAI techniques significantly enhance the efficiency and scalability of urban-scale gentrification automated detection while enabling a more comprehensive and nuanced analytical perspective. Finally, the chapter addresses challenges in integrating deep learning approaches with gentrification research, including data limitations, model performance, transferability, and interpretability, and highlights critical directions for future investigation.

References

Anguelovski, I., Connolly, J. J. T., Masip, L., & Pearsall, H. (2017). Assessing green gentrification in historically disenfranchised neighborhoods: a longitudinal and spatial analysis of Barcelona. In I. Anguelovski, J. J. T. Connolly, L. Masip, & H. Pearsall (Eds.), *Urban geography* (Vol. 39, pp. 458–491).

Anguelovski, I., Connolly, J. J., Masip, L., & Pearsall, H. (2018). Assessing green gentrification in historically disenfranchised neighborhoods: A longitudinal and spatial analysis of Barcelona. *Urban Geography, 39*(3), 458–491.

Assaad, R. H., & Jezzini, Y. (2024). Green gentrification vulnerability index (GGVI): A novel approach for identifying at-risk communities and promoting environmental justice at the census-tract level. *Cities, 148*, 104858.

Bai, T., Wang, L., Yin, D., Sun, K., Chen, Y., Li, W., & Li, D. (2023). Deep learning for change detection in remote sensing: A review. *Geo-spatial Information Science, 26*(3), 262–288.

Best, K. B., Jouzi, Z., Islam, M. S., Kirby, T., Nixon, R., Hossan, A., & Nyiawung, R. A. (2023). Typologies of multiple vulnerabilities and climate gentrification across the East Coast of the United States. *Urban Climate, 48*, 101430.

Brown-Saracino, J. (2017). Explicating divided approaches to gentrification and growing income inequality. *Annual Review of Sociology, 43*(1), 515–539.

Chapple, K. (2009). *Mapping susceptibility to gentrification: The early warning toolkit*. The Center for Community Innovation/University of California.

Chicco, D. (2021). Siamese neural networks: An overview. *Artificial neural networks*, 73–94.

Choi, M., Van Zandt, S., & Matarrita-Cascante, D. (2018). Can community land trusts slow gentrification? *Journal of Urban Affairs, 40*(3), 394–411.

Clark, E. (2005). The order and simplicity of gentrification ication—A political challenge. In E. Clark (Ed.), *Gentrification in a global context* (Vol. 261). Policy Press.
Cohen, M., & Pettit, K. L. (2019). *Guide to measuring neighborhood change to understand and prevent displacement*. National Neighborhood Indicators Partnership (NNIP). https://www.urban.org/research/publication/guide-measuring-neighborhoodchange-understand-and-prevent-displacement
Davidson, M., & Lees, L. (2005). New-build 'gentrification' and London's riverside renaissance. *Environment and Planning a-Economy and Space, 37*(7), 1165–1190. https://doi.org/10.1068/a3739
Finio, N. (2022). Measurement and definition of gentrification in urban studies and planning. *Journal of Planning Literature, 37*(2), 249–264.
Freeman, L. (2005). Displacement or succession? Residential mobility in gentrifying neighborhoods. *Urban Affairs Review, 40*(4), 463–491.
Glass, R. (1964). *Aspects of change*. London, MacGibbon & Kee.
Gray, J., Buckner, L., & Comber, A. (2023). Predicting gentrification in England: A data primitive approach. *Urban Science, 7*(2), Article 64. https://doi.org/10.3390/urbansci7020064
Hammel, D. J., & Wyly, E. K. (1996). A model for identifying gentrified areas with census data. In D. J. Hammel & E. K. Wyly (Eds.), *Urban geography* (Vol. 17, pp. 248–268).
Huang, Y., Hong, X., Yao, X., & Yin, M. (2024). Which characteristics represent the gentrification affected by parks? A study case in Nanjing, China. *Ecological Indicators, 160*, 111862.
Hwang, J., & Sampson, R. J. (2014). Divergent pathways of gentrification: Racial inequality and the social order of renewal in Chicago neighborhoods. *American Sociological Review, 79*(4), 726–751.
Ilic, L., Sawada, M., & Zarzelli, A. (2019). Deep mapping gentrification in a large Canadian city using deep learning and Google street view. *PLoS One, 14*(3), e0212814.
Kwon, O., Lee, N., & Shin, B. (2014). Data quality management, data usage experience and acquisition intention of big data analytics. *International Journal of Information Management, 34*(3), 387–394.
LeCun, Y., Bengio, Y., & Hinton, G. (2015). Deep learning. *Nature, 521*(7553), 436–444.
Lei, Y., Peng, D., Zhang, P., Ke, Q., & Li, H. (2020). Hierarchical paired channel fusion network for street scene change detection. *IEEE Transactions on Image Processing, 30*, 55–67.
Li, W. (2020). GeoAI: Where machine learning and big data converge in GIScience. *Journal of Spatial Information Science, 20*, 71–77.
Li, W. (2022). GeoAI in social science. In *Handbook of spatial analysis in the social sciences* (pp. 291–304). Edward Elgar Publishing.
Li, W., & Hsu, C.-Y. (2022). GeoAI for large-scale image analysis and machine vision: Recent progress of artificial intelligence in geography. *ISPRS International Journal of Geo-Information, 11*(7), 385.
Li, W., Hsu, C.-Y., Wang, S., & Kedron, P. (2024). GeoAI reproducibility and replicability: A computational and spatial perspective. *Annals of the American Association of Geographers, 114*(9), 2085–2103.
Liu, P., & Biljecki, F. (2022). A review of spatially-explicit GeoAI applications in urban geography. *International Journal of Applied Earth Observation and Geoinformation, 112*, 102936.
Liu, C., & Song, W. X. (2024). Mapping property redevelopment via GeoAI: Integrating computer vision and socioenvironmental patterns and processes. *Cities, 144*, Article 104644. https://doi.org/10.1016/j.cities.2023.104644
Liu, C., Deng, Y., Song, W. X., Wu, Q. Y., & Gong, J. (2019). A comparison of the approaches for gentrification identification. Cities, 95, Article 102482. doi:https://doi.org/10.1016/j.cities.2019.102482, 102482.
Maloutas, T. (2012). Contextual diversity in gentrification research. *Critical Sociology, 38*(1), 33–48.
Maya, M., Simi, H., & Pearsall, H. (2024). Machine learning to model gentrification: A synthesis of emerging forms. *Computers, Environment and Urban Systems, 111*, 102119.

Mortaheb, R., & Jankowski, P. (2023). Smart city re-imagined: City planning and GeoAI in the age of big data. *Journal of Urban Management, 12*(1), 4–15.
Podagrosi, A., Vojnovic, I., & Pigozzi, B. (2011). The diversity of gentrification in Houston's urban renaissance: From cleansing the urban poor to supergentrification. *Environment and Planning A, 43*(8), 1910–1929.
Raya-Tapia, A. Y., Ramírez-Márquez, C., & Ponce-Ortega, J. M. (2024). Clustering of zones according to the level of gentrification by using an unsupervised learning algorithm. *Cities, 151*, 105133.
Reades, J., De Souza, J., & Hubbard, P. (2019). Understanding urban gentrification through machine learning. *Urban Studies, 56*(5), 922–942.
Reichstein, M., Camps-Valls, G., Stevens, B., Jung, M., Denzler, J., Carvalhais, N., & Prabhat, F. (2019). Deep learning and process understanding for data-driven earth system science. *Nature, 566*(7743), 195–204.
Silver, D. A., & Clark, T. N. (2016). *Scenescapes: How qualities of place shape social life*. University of Chicago Press.
Smith, N. (2013). Gentrification, the frontier, and the restructuring of urban space. In *Gentrification of the City* (pp. 15–34). Routledge.
Thackway, W., Ng, M., Lee, C. L., & Pettit, C. (2023a). Building a predictive machine learning model of gentrification in Sydney. In W. Thackway, M. Ng, C. L. Lee, & C. Pettit (Eds.), *Cities* (Vol. 134).
Thackway, W., Ng, M., Lee, C. L., & Pettit, C. (2023b). Implementing a deep-learning model using Google street view to combine social and physical indicators of gentrification. Computers Environment and Urban Systems, 102, Article 101970. doi:https://doi.org/10.1016/j.compenvurbsys.2023.101970.
Walks, R. A., & Maaranen, R. (2008). Gentrification, social mix, and social polarization: Testing the linkages in large Canadian cities. *Urban Geography, 29*(4), 293–326.
Wang, W., & Durst, N. J. (2023). Planning for active aging: Exploring housing preferences of elderly populations in the United States. *Journal of Housing and the Built Environment, 38*(2), 795–809.
Wang, S., Huang, X., Liu, P., Zhang, M., Biljecki, F., Hu, T., Fu, X., Liu, L., Liu, X., & Wang, R. (2024a). Mapping the landscape and roadmap of geospatial artificial intelligence (GeoAI) in quantitative human geography: An extensive systematic review. *International Journal of Applied Earth Observation and Geoinformation, 128*, 103734.
Wang, L., Zhou, J., Wang, P., Pan, C., & Li, H. (2024b). Green gentrification and urban parks: Exploring the complex dynamics of commercial transformation. *Journal of the Knowledge Economy, 15*, 1–33.
Wu, L., & Rowe, P. G. (2022). Green space progress or paradox: Identifying green space associated gentrification in Beijing. In L. Wu & P. G. Rowe (Eds.), *landscape and urban planning* (Vol. 219, pp. 104321).
Wyly, E. K., & Hammel, D. J. (1999). *Islands of decay in seas of renewal: Housing policy and the resurgence of gentrification* (Vol. 10, p. 711).
Ye, M., Vojnovic, I., & Chen, G. (2015). The landscape of gentrification: Exploring the diversity of "upgrading" processes in Hong Kong, 1986–2006. *Urban Geography, 36*(4), 471–503.
Zapatka, K., & Beck, B. (2021). Does demand lead supply? Gentrifiers and developers in the sequence of gentrification, new York City 2009–2016. *Urban Studies, 58*(11), 2348–2368.
Zeng, J., Yue, Y., Gao, Q., Gu, Y., & Ma, C. (2022). Identifying localized amenities for gentrification using a machine learning-based framework. In J. Zeng, Y. Yue, Q. Gao, Y. Gu, & C. Ma (Eds.), *Applied geography* (Vol. 145, p. 102748).
Zhang, Y., Liu, P., & Biljecki, F. (2023). Knowledge and topology: A two layer spatially dependent graph neural networks to identify urban functions with time-series street view image. *ISPRS Journal of Photogrammetry and Remote Sensing, 198*, 153–168.
Zuk, M., Bierbaum, A. H., Chapple, K., Gorska, K., & Loukaitou-Sideris, A. (2018). Gentrification, displacement, and the role of public investment. *Journal of Planning Literature, 33*(1), 31–44.

Chapter 20
Geographic Limitations of Generative AI Models

Junghwan Kim

Abstract Generative AI tools, such as ChatGPT, have rapidly influenced diverse fields, including human geography. They offer opportunities for advanced computational analyses, qualitative data handling, and the generation of synthetic data for sensitive topics. Their capabilities, such as text-to-image modeling, can potentially democratize quantitative methods and assist in auditing built environments, reducing the need for extensive coding or high-performance computing infrastructure. However, generative AI tools also pose challenges. Hallucination effects—where AI confidently fabricates misinformation—can mislead users, while political, social, and geographic biases risk uneven outcomes and interpretations. Geographic biases manifest as reduced accuracy or specificity for certain areas, particularly rural or low-population-density regions. Similarly, environmental justice data queries for less populous U.S. counties reveal ChatGPT's limited local knowledge. Although methods like retrieval-augmented generation may address some issues, the root lies in disparities in training data quality and coverage. Human geographers must carefully weigh the potential of generative AI against these risks, adopt strategies for responsible use, and emphasize improved, equitable data collection practices to ensure positive and reliable impacts on human geography research.

Keywords Generative AI · Geographic bias · Hallucination · Human geography · Multimodal data

20.1 Introduction to Generative AI in Human Geography

Since the launch of OpenAI's ChatGPT in November of 2022, various generative AI tools have substantially impacted society, government, industry, and research sectors. Aligning with the growing interest in geospatial AI (GeoAI) among human

J. Kim (✉)
Department of Geography, Virginia Polytechnic Institute and State University, Blacksburg, VA, USA
e-mail: junghwankim@vt.edu

© The Author(s), under exclusive license to Springer Nature Switzerland AG 2025

X. Huang et al. (eds.), *GeoAI and Human Geography*, Springer Geography,
https://doi.org/10.1007/978-3-031-87421-5_20

geographers in recent years (Wang et al., 2024a), generative AI has attracted human geographers' attention (Wang et al., 2024b). Researchers have explored how these emerging generative AI tools can be used for various topics in human geography research, such as environmental studies (Atkins et al., 2024; Kim et al., 2024; Nguyen et al., 2024), transportation studies (Devunuri et al., 2024; Kim & Lee, 2023; Voß, 2023a, 2023b), health studies (Mbakwe et al., 2023; Yiannakoulias, 2024), course material development assistance (Day & Perkins, 2024), and so on. Since generative AI tools provide a convenient means of communicating data and information with users, some researchers are focusing on generative AI tools to retrieve geographic data and generate information, which may have the potential to replace conventional web search engines and GIS tools (Jiang & Yang, 2024; Li & Ning, 2023; Tao & Xu, 2023; Wang et al., 2024; Xu & Tao, 2024; Yiannakoulias, 2024).

Additionally, researchers have explored the potential of generative AI to democratize use cases of advanced computational geospatial analyses because of the powerful capabilities of generative AI tools' creating programming codes or debugging errors of codes (Bucaioni et al., 2024; Chen et al., 2023; Guo et al., 2024). It would further broaden the participation in quantitative human geography research since generative AI tools can lower barriers to adopting computational approaches, such as coding (Jayachandran et al., 2024). On the other hand, generative AI tools have been explored as a new auxiliary tool for supporting qualitative research (Carvalho et al., 2024; Katz et al., 2024). Despite the limitations of generative AI tools' capabilities in conducting in-depth analysis of emotions and meaning that can be dissected through immersive conservations with human subjects, generative AI tools can efficiently assist qualitative researchers. For instance, speech-to-text (STT) technologies play an essential role in transcribing interviews with human subjects, which typically costs a lot of money and time when humans transcribe manually. Additionally, as natural language processing (NLP) techniques lie at the core of generative AI tools, human geographers can utilize generative AI tools to conduct thematic analysis, content analysis, or sentiment analysis (Carvalho et al., 2024; Katz et al., 2024). In practical settings, customized chatbots powered by generative AI assist local planners in communicating with citizens (Androutsopoulou et al., 2019; Senadheera et al., 2024).

Another crucial domain of generative AI applications in human geography is synthetic data (e.g., Bisbee et al., 2023; Haydari et al., 2024; Wang et al., 2023). Human geography research often involves utilizing confidential or sensitive information, such as questionnaire survey data or individuals' daily activity-travel patterns captured by GPS trajectories (Huang et al., 2021; Kim et al., 2021; Wang et al., 2022). While beneficial for enriching our fundamental understanding of topics in human geography, collecting those data is often limited because of difficulties in recruiting survey participants of large populations or geoprivacy issues (Kwan et al., 2004; Richardson et al., 2015). To overcome this issue, generative AI technologies can simulate and create synthetic data that is not real but precisely reflects real-world human behavior and perceptions (Haydari et al., 2024; Wang et al., 2023). Because synthetic data are not real data, concerns about privacy and confidentiality can be alleviated. Another advantage is that human geographers can generate different synthetic datasets by controlling settings and model parameters, so researchers can test multiple scenarios, which can hardly be achieved in a conventional survey-based research setting (Konduri, 2016).

Last but not least, generative AI tools embedded with multimodal large language models (LLMs) have been explored actively (Akinboyewa et al., 2024; Jang et al., 2024; Jang & Kim, 2025; Xu & Tao, 2024). Unlike unimodal LLMs (e.g., GPT-3) that only handle text-type data, multimodal LLMs can handle diverse data types, including texts, images, audio, and videos. For instance, OpenAI's Sora and Synthesia are cutting-edge text-to-video tools. However, those text-video generative AI technologies are still immature at the time of writing (September 2024). Perhaps text-to-image (or image-to-text) types of generative AI (e.g., OpenAI's Dall-E, Adobe's Firefly) are more widely used and discussed their applicability in practice (Akinboyewa et al. 2024; Jang et al., 2024; Jang & Kim, 2025; Xu & Tao, 2024). This text-image generative AI shows great potential, particularly in a subfield of human geography addressing the built environment or urban design.

For instance, a recent study empirically investigated the potential capabilities of generative AI tools in visually auditing built environment elements from street-view images (Jang & Kim, 2025). The study found that OpenAI's GPT-4o and Google's Gemini-1.5-pro performed at approximately 90% of the agreement rate when compared to a widely used conventional deep learning-based computer vision method. This is a fairly impressive result because it implies human geographers can readily utilize generative AI tools to examine built environment features visually. Using generative AI tools is much easier than adopting traditional deep learning-based methods that ask researchers to be equipped with coding skills and high-performance computing resources, such as GPU.

20.2 Limitations: Hallucination and Biases

However, despite the potential application of generative AI models to human geography research, one caveat that should be acknowledged is the potential limitations of generative AI tools, such as hallucination effects, biases, responsible use, and so on. A hallucination effect refers to a phenomenon where generative AI "confidently" provides inaccurate or false answers to prompts while pretending that it provides true, accurate, and precise answers to humans' requests Day (2023). Day (2023) investigated how ChatGPT (GPT-3) produced counterfeit references of papers when he asked ChatGPT to refer to geography-related questions. Although this weakness is being overcome by advanced GPT models that adopt a retrieval-augmented generation (RAG) method to access information databases (Guu et al., 2020; Lewis et al., 2020), the paper calls for researchers' attention to the hallucination effects of generative AI tools, which can further be problematic in use cases. For instance, imagine a scenario where laypeople utilize generative AI tools to learn about local urban planning-related issues (e.g., environmental pollution). If users are not fully aware of the limitations of generative AI tools and those AI tools provide fake or fabricated information, it would be problematic as public debates on local topics might be "contaminated" by counterfeit information.

Another critical limitation of generative AI tools is related to biases. For instance, recent studies have demonstrated that generative AI tools are prone to political biases (Baum & Villasenor, 2023; Faisal & Anastasopoulos, 2022; Fujimoto & Takemoto, 2023; Kim & Lee, 2023; Rozado, 2023). This is an interesting finding because there is no need to have a premise that generative AI models are toward specific political orientations, but the testing results report that those AI models are indeed biased. Beyond academic curiosity about the political biases of generative AI tools, considering the potential application of generative AI to law enforcement and police domains Halford and Webster (2024), biases in generative AI tools could lead to favoring specific groups regarding their political orientations.

20.3 Geographic Biases in Generative AI Tools

In addition to political biases, geographic biases should be discussed (Fulman et al., 2024; Manvi et al., 2024; Kim et al., 2024). As one of earlier studies, Kim et al. (2024) utilized OpenAI's ChatGPT (GPT-4) API to investigate how ChatGPT provides local-specific answers to prompts that ask about environmental justice issues of 3108 counties in the United States. If ChatGPT is "omniscient," it should provide local-specific answers to all counties because each county in the U.S. has its own environmental justice issues. For instance, counties in the West (e.g., Nevada, Arizona, California) would have heat or water-related environmental justice issues, while counties in the Appalachian region (e.g., Kentucky, West Virginia) would have coal mining-related issues. However, the results reported that ChatGPT is limited in providing local-specific information to some counties, and those weaknesses in capabilities show a geographic pattern. Counties with lower population densities and lower median income levels have a higher chance (statistically significant) of not being able to get local-specific responses from the ChatGPT. This is indeed an interesting finding because of the presence of systematic geographic biases in generative AI tools. Despite the great potential of using generative AI tools for human geography research, the study's finding calls for more careful attention considering ChatGPT's geographic limitations.

Recently, many studies have investigated the geographic dimensions of generative AI tools. Jang and Kim (2025) focused on multimodal large language models. They investigated the spatial pattern of generative AI models' performance in visually detecting built environment features from selected street-view images from Columbus, Ohio. They found a significantly different performance of the Gemini-1.5-Pro model between high-density urban and low-density rural areas. The generative AI model performs better in urban than rural areas, suggesting the geographic biases in multimodal LLMs' capabilities in detecting key built environments from images. Renshaw et al. (2025) also investigated ChatGPT and Gemini's competencies in spatial querying tasks, such as listing queen-type contiguity spatial units to a given spatial unit. Focusing on counties in the United States, they found that ChatGPT and Gemini demonstrated poorer performance (measured by accuracy) for "boundary" counties that share boundaries with other states than their counterparts. The study also found that poorer spatial querying capabilities were reported in rural counties than in urban counties. Similar to the studies above, results from Renshaw et al. (2025) also reveal geographic differences in generative AI tools' capabilities and limitations.

Another interesting observation is reported by Atkins et al. (2024), where the authors examined how ChatGPT reported country-specific responses about climate change-related hazard issues and compared their responses with the IPCC report, which was considered to be a ground truth dataset. Their findings do not report strong systematic patterns regarding the accuracy of ChatGPT, which was different from other studies. One possible explanation is that there are more available training data at a country level (i.e., low spatial resolution) than at a county level (i.e., high spatial resolution). It is not surprising that more training data for generative AI tools exist at a global level. Country-level climate change-related natural hazard information is more general and easily found than county-level environmental justice issues.

20.4 Concluding Remarks

In concluding remarks, this chapter explored generative AI tools' competencies and limitations for human geography research. This chapter demonstrates the "double-edged sword" nature of generative AI tools for human geography research. There are advantages and strengths, such as assisting in advanced computational human geography studies (e.g., coding) and efficiently analyzing qualitative data. However, generative AI tools should be utilized carefully because of the presence of hallucination effects and biases in those tools. AI developers, GIScientists, and human geographers are actively working on overcoming the limitations of generative AI tools. The retrieval-augmented generation (RAG), an advanced method that retrieves information from external sources (Guu et al., 2020; Lewis et al., 2020), is a feasible solution to overcome the effects of hallucination. More importantly, the fundamental behind this limitation is poor training data—quantity and quality—for specific geographic regions (Graham et al., 2014; Liu et al., 2022; Mai et al., 2023), which has been one of the crucial research topics related to AI ethics in recent years. A rudimentary solution would be reducing gaps in training data, as differences in quantity and quality of training data are attributed to geographic biases in the capabilities of generative AI tools. Thus, it is imperative for human geographers to be equipped with ethical awareness to ensure the responsible use of generative AI tools, not only to mitigate potential geographic biases but also to ensure that these tools have a positive influence on the progress of human geography research.

References

Akinboyewa, T., Ning, H., Lessani, M. N., & Li, Z. (2024). Automated floodwater depth estimation using large multimodal model for rapid flood mapping. *Computational Urban Science, 4*(1), 12.

Androutsopoulou, A., Karacapilidis, N., Loukis, E., & Charalabidis, Y. (2019). Transforming the communication between citizens and government through AI-guided chatbots. *Government Information Quarterly, 36*(2), 358–367.

Atkins, C., Girgente, G., Shirzaei, M., & Kim, J. (2024). Generative AI tools can enhance climate literacy but must be checked for biases and inaccuracies. *Communications Earth & Environment, 5*(1), 226.

Baum, J., & Villasenor, J. (2023). *The politics of AI: ChatGPT and political bias*. Retrieved from https://www.brookings.edu/articles/the-politics-of-ai-chatgpt-and-political-bias/

Bisbee, J., Clinton, J. D., Dorff, C., Kenkel, B., & Larson, J. M. (2023). Synthetic replacements for human survey data? The perils of large language models. *Political Analysis, 32*, 1–16.

Bucaioni, A., Ekedahl, H., Helander, V., & Nguyen, P. T. (2024). Programming with ChatGPT: How far can we go? *Machine Learning with Applications, 15*, 100526.

Carvalho, T., Negm, H., & El-Geneidy, A. (2024). A comparison of the results from artificial intelligence-based and human-based transport-related thematic analysis. *Findings*.

Chen, E., Huang, R., Chen, H. S., Tseng, Y. H., & Li, L. Y. (2023). GPTutor: A ChatGPT-powered programming tool for code explanation. In *International conference on artificial intelligence in education* (pp. 321–327). Springer Nature.

Day, T. (2023). A preliminary investigation of fake peer-reviewed citations and references generated by ChatGPT. *The Professional Geographer, 75*(6), 1024–1027.

Day, T., & Perkins, A. J. (2024). Empowering Canadian geography students and faculty: New approaches to curriculum and course material development using generative AI as a writing assistant. In M. Solem, R. G. Boehm, & J. Zadrozny (Eds.), *Powerful geography*. Cham: Springer. https://doi.org/10.1007/978-3-031-54845-1_5

Devunuri, S., Qiam, S., & Lehe, L. J. (2024). ChatGPT for GTFS: Benchmarking LLMs on GTFS semantics... and retrieval. *Public Transport, 16*, 1–25.

Faisal, F., & Anastasopoulos, A. (2022). *Geographic and geopolitical biases of language models*. arXiv preprint arXiv:2212.10408.

Fujimoto, S., & Takemoto, K. (2023). Revisiting the political biases of ChatGPT. *Frontiers in Artificial Intelligence, 6*, 1232003.

Fulman, N., Memduhoğlu, A., & Zipf, A. (2024). Distortions in judged spatial relations in large language models. *The Professional Geographer, 76*, 1–9.

Graham, M., Hogan, B., Straumann, R. K., & Medhat, A. (2014). Uneven geographies of user-generated information: Patterns of increasing informational poverty. *Annals of the Association of American Geographers, 104*, 746–764.

Guo, Q., Cao, J., Xie, X., Liu, S., Li, X., Chen, B., & Peng, X. (2024, February). Exploring the potential of ChatGPT in automated code refinement: An empirical study. In *Proceedings of the 46th IEEE/ACM International Conference on Software Engineering* (pp. 1–13).

Guu, K., Lee, K., Tung, Z., Pasupat, P., & Chang, M. (2020). Retrieval augmented language model pre-training. In *International conference on machine learning* (pp. 3929–3938). PMLR.

Halford, E., & Webster, A. (2024). Using chat GPT to evaluate police threats, risk and harm. *International Journal of Law, Crime and Justice, 78*, 100686. https://doi.org/10.1016/j.ijlcj.2024.100686

Haydari, A., Chen, D., Lai, Z., & Chuah, C. N. (2024). *Mobilitygpt: Enhanced human mobility modeling with a GPT model*. arXiv preprint arXiv:2402.03264.

Huang, J., Kwan, M. P., & Kim, J. (2021). How culture and sociopolitical tensions might influence people's acceptance of COVID-19 control measures that use individual-level georeferenced data. *ISPRS International Journal of Geo-Information, 10*(7), 490.

Jang, K. M., & Kim, J. (2025). Multimodal large language models as built environment auditing tools. *The Professional Geographer, 77*(1), 84–90. https://doi.org/10.1080/00330124.2024.2404894

Jang, K. M., Chen, J., Kang, Y., et al. (2024). Place identity: A generative AI's perspective. *Humanities and Social Sciences Communications, 11*, 1156. https://doi.org/10.1057/s41599-024-03645-7

Jayachandran, D., Maldikar, P., Love, T. S., & Blum, J. J. (2024). Leveraging generative artificial intelligence to broaden participation in computer science. *Proceedings of the AAAI Symposium Series, 3*(1), 486–492.

Jiang, Y., & Yang, C. (2024). Is ChatGPT a good geospatial data analyst? Exploring the integration of natural language into structured query language within a spatial database. *ISPRS International Journal of Geo-Information, 13*(1), 26.

Katz, A., Gerhardt, M., & Soledad, M. (2024). Using generative text models to create qualitative codebooks for student evaluations of teaching. *International Journal of Qualitative Methods, 23*, 16094069241293283. https://doi.org/10.1177/16094069241293283

Kim, J., & Lee, J. (2023). How does ChatGPT introduce transport problems and solutions in North America? *Findings*.

Kim, J., Lee, J., Jang, K. M., & Lourentzou, I. (2024). Exploring the limitations in how ChatGPT introduces environmental justice issues in the United States: A case study of 3,108 counties. *Telematics and Informatics, 86*, 102085. https://doi.org/10.1016/j.tele.2023.102085

Kim, J., Kwan, M. P., Levenstein, M. C., & Richardson, D. B. (2021). How do people perceive the disclosure risk of maps? Examining the perceived disclosure risk of maps and its implications for geoprivacy protection. *Cartography and Geographic Information Science, 48*(1), 2–20.

Konduri, K. C., You, D., Garikapati, V. M., & Pendyala, R. M. (2016). Enhanced synthetic population generator that accommodates control variables at multiple geographic resolutions. *Transportation Research Record, 2563*(1), 40–50.

Kwan, M. P., Casas, I., & Schmitz, B. (2004). Protection of geoprivacy and accuracy of spatial information: How effective are geographical masks? *Cartographica: The International Journal for Geographic Information and Geovisualization, 39*(2), 15–28.

Lewis, P., Perez, E., Piktus, A., Petroni, F., Karpukhin, V., Goyal, N., Küttler, H., Lewis, M., Yih, W. T., Rocktäschel, T., & Riedel, S. (2020). Retrieval-augmented generation for knowledge-intensive nlp tasks. *Advances in Neural Information Processing Systems, 33*, 9459–9474.

Li, Z., & Ning, H. (2023). Autonomous GIS: The next-generation AI-powered GIS. *International Journal of Digital Earth, 16*(2), 4668–4686.

Liu, Z., Janowicz, K., Cai, L., Zhu, R., Mai, G., & Shi, M. (2022). Geoparsing: Solved or biased? An evaluation of geographic biases in geoparsing. *AGILE: GIScience Series, 3*, 9.

Mai, G., Huang, W., Sun, J., Song, S., Mishra, D., Liu, N., & Lao, N. (2023). *On the opportunities and challenges of foundation models for geospatial artificial intelligence*. arXiv preprint arXiv:2304.06798.

Manvi, R., Khanna, S., Burke, M., Lobell, D., & Ermon, S. (2024). *Large language models are geographically biased*. arXiv preprint arXiv:2402.02680.

Mbakwe, A. B., Lourentzou, I., Celi, L. A., Mechanic, O. J., & Dagan, A. (2023). ChatGPT passing USMLE shines a spotlight on the flaws of medical education. *PLOS Digital Health, 2*(2), e0000205.

Nguyen, H., Nguyen, V., López-Fierro, S., Ludovise, S., & Santagata, R. (2024). Simulating climate change discussion with Large Language Models: Considerations for science communication at scale. In *Proceedings of the Eleventh ACM Conference on Learning@ Scale* (pp. 28–38).

Renshaw, A., Lourentzou, I., Lee, J., Crawford, T., & Kim, J. (2025). Comparing the spatial querying capacity of large language models: OpenAI's ChatGPT and Google's Gemini Pro. *The Professional Geographer, 77*(2), 186–198. https://doi.org/10.1080/00330124.2024.2434455

Richardson, D. B., Kwan, M. P., Alter, G., & McKendry, J. E. (2015). Replication of scientific research: Addressing geoprivacy, confidentiality, and data sharing challenges in geospatial research. *Annals of GIS, 21*(2), 101–110.

Rozado, D. (2023). The political biases of ChatGPT. *Social Sciences, 12*(3), 148.
Senadheera, S., Yigitcanlar, T., Desouza, K. C., Mossberger, K., Corchado, J., Mehmood, R., & Cheong, P. H. (2024). Understanding Chatbot adoption in local governments: A review and framework. *Journal of Urban Technology*, 1–35.
Tao, R., & Xu, J. (2023). Mapping with ChatGPT. *ISPRS International Journal of Geo-Information, 12*(7), 284.
Voß, S. (2023a). Bus bunching and bus bridging: What can we learn from generative AI tools like ChatGPT? *Sustainability, 15*(12), 9625.
Voß, S. (2023b). Successfully using ChatGPT in logistics: Are we there yet? In *International Conference on computational logistics* (pp. 3–17). Springer Nature.
Wang, J., Kim, J., & Kwan, M. P. (2022). An exploratory assessment of the effectiveness of geomasking methods on privacy protection and analytical accuracy for individual-level geospatial data. *Cartography and Geographic Information Science, 49*(5), 385–406.
Wang, X., Fang, M., Zeng, Z., & Cheng, T. (2023). *Where would I go next? Large language models as human mobility predictors*. arXiv preprint arXiv:2308.15197.
Wang, S., Zhuang, S., & Zuccon, G. (2024). Large language models based stemming for information retrieval: Promises, pitfalls and failures. In *Proceedings of the 47th International ACM SIGIR Conference on Research and Development in Information Retrieval* (pp. 2492–2496).
Wang, S., Huang, X., Liu, P., Zhang, M., Biljecki, F., Hu, T., & Bao, S. (2024a). Mapping the landscape and roadmap of geospatial artificial intelligence (GeoAI) in quantitative human geography: An extensive systematic review. *International Journal of Applied Earth Observation and Geoinformation, 128*, 103734.
Wang, S., Hu, T., Xiao, H., Li, Y., Zhang, C., Ning, H., & Ye, X. (2024b). GPT, large language models (LLMs) and generative artificial intelligence (GAI) models in geospatial science: A systematic review. *International Journal of Digital Earth, 17*(1), 2353122.
Xu, J., & Tao, R. (2024). Map Reading and analysis with GPT-4V (ision). *ISPRS International Journal of Geo-Information, 13*(4), 127.
Yiannakoulias, N. (2024). Spatial intelligence and contextual relevance in AI-driven health information retrieval. *Applied Geography, 171*, 103392.

Chapter 21
Generative AI and Geography Education

Jongwon Lee

Abstract This chapter explores the transformative impact of generative Artificial Intelligence (AI) on geography education through the lenses of the SAMR model, Bloom's Taxonomy, and the AI-TPACK framework. It examines how AI is reshaping teaching methods, learning experiences, and teacher competencies in geography education. The SAMR model demonstrates AI's potential to redefine geographical learning experiences, moving beyond mere substitution to enable novel educational approaches. Bloom's Taxonomy analysis reveals how AI alters students' cognitive experiences, enhancing higher order thinking skills while potentially risking over-reliance on AI-generated content. The AI-TPACK framework highlights the evolving competencies required for geography teachers in the AI era, emphasizing the need for integrated technical, pedagogical, and content knowledge.

Keywords Generative AI · Geography education · SAMR model

21.1 Introduction

> Technology will never replace great teachers, but technology in the hands of great teachers can be transformational.—George Couros

Artificial Intelligence (AI) has rapidly established itself as a transformative force across various industries, with a particularly profound impact in education. This evolution is notably marked by the swift development of generative AI technologies—systems capable of creating new content such as text, images, audio, and video. These advancements are expanding educational possibilities and challenging traditional pedagogical approaches (Kamalov et al., 2023; Luckin et al., 2022; Hanover Research, 2024; Felix & Webb, 2024).

J. Lee (✉)
Department of Social Studies Education, Ewha Womans University, Seoul, Republic of Korea
e-mail: jongwonlee@ewha.ac.kr

© The Author(s), under exclusive license to Springer Nature Switzerland AG 2025

X. Huang et al. (eds.), *GeoAI and Human Geography*, Springer Geography,
https://doi.org/10.1007/978-3-031-87421-5_21

Generative AI holds significant potential to revolutionize education by offering personalized and engaging learning experiences, which could allow teachers to focus on more meaningful tasks and alter both the content and methods of instruction in schools (Celik, 2023). Additionally, it could democratize access to quality education, particularly benefiting students for whom the traditional "one-size-fits-all" model of schooling is ineffective (OECD, 2023). However, the rise of sophisticated generative AI tools also introduces new challenges. As Bower et al. (2024) highlight, "the ability of ChatGPT and similar generative AI tools to provide students and teachers with often high-quality responses to a wide range of common educational tasks raised fundamental questions about what educators worldwide should be teaching and how students should be assessed."

In the context of geography education, the integration of generative AI presents unique opportunities and challenges. As a discipline that has historically embraced digital tools like Geographic Information Systems (GIS), remote sensing, and virtual reality, geography is well-positioned to leverage AI (Peter & Sprenger, 2022). The proliferation of geospatial big data and public datasets further amplifies this potential, facilitating innovative inquiry and analysis (Lee, 2023).

Nevertheless, it is crucial to note that the integration of AI technologies alone does not guarantee improved educational outcomes. It requires a structured approach to ensure that AI tools are effectively utilized to enhance learning experiences. As Ouyang and Jiao (2021) emphasize, "good educational outcomes typically do not occur by virtue of merely using advanced AI computing technologies" (p. 1). Mishra et al. (2023) further this argument, asserting that knowledge for the successful integration of technology extends beyond understanding specific content, pedagogical approaches, or technology in isolation. Instead, it demands a dynamic amalgam of all three elements, creating a synergistic framework that optimizes the potential of AI in educational settings.

In response to these challenges, Chang and Kidman (2023) argue for a robust framework to guide the use of AI tools in supporting meaningful geographical and environmental education. This chapter aims to address this need by critically analyzing the transformative impact of generative AI on geography education while simultaneously developing a comprehensive framework for its effective integration.

To address the complex landscape of generative AI in geography education, this chapter is structured around three key questions:

1. How is generative AI transforming geography teaching and learning, and what changes are desirable? Through the SAMR model, this question will explore the types of changes generative AI integration can bring to geography education and propose desirable transformations.
2. How is generative AI altering geography learning, particularly in terms of cognitive experiences? While the first question addresses overall changes in curriculum, teaching, and learning within geography education, this second question focuses specifically on how generative AI affects learners' cognitive experiences. The benefits and challenges associated with these changes will be examined through the lens of Bloom's Taxonomy.

3. What knowledge and competencies are required for geography teachers to effectively integrate generative AI into their teaching? This question will investigate the necessary expertise for geography teachers to effectively utilize generative AI and how this expertise differs from the skills and knowledge required for integrating existing technologies. The Technological Pedagogical Content Knowledge (TPACK) framework will be used to analyze this aspect.

By integrating the SAMR model, Bloom's Taxonomy, and the TPACK framework, this chapter aims to provide a comprehensive understanding of the transformative impact of generative AI on geography education, offering a strategic roadmap for its effective incorporation into the geography classroom and curriculum.

21.2 Generative AI in Education

Defining AI is a complex task, primarily due to the various ways in which intelligence itself can be understood (Wang, 2019). Baker et al. (2019) define AI as "computers which perform cognitive tasks, usually associated with human minds, particularly learning and problem-solving" (p. 10). For the purposes of this paper, this chapter adopts the definition by Popenici and Kerr (2017), who describe AI as "computing systems that are able to engage in human-like processes such as learning, adapting, synthesizing, self-correction, and use of data for complex processing tasks" (p. 2).

AI primarily relies on two key technologies: "Machine Learning" and "Deep Learning." Machine Learning involves algorithms that learn from data and make predictions or decisions, improving their performance over time without explicit programming for each task. Deep Learning, a more advanced subset, utilizes artificial neural networks inspired by the human brain to recognize patterns in complex, unstructured data, enabling applications like image and speech recognition, natural language processing, and voice assistants (UNESCO, 2021, 2023).

In the context of education, a specific type of AI, known as generative AI, is particularly transformative due to its focus on natural language processing. Generative AI is a machine learning system trained on an extensive dataset comprising text from various internet sources, such as books, articles, and websites (Bessette, 2023). Unlike traditional AI models that primarily focus on identifying and classifying content, generative AI models can generate new content. This capability is especially powerful in educational settings where the ability to produce realistic images, short movies, or well-structured essays can offer unprecedented creative opportunities (OECD, 2023). A prominent example of generative AI is ChatGPT, which generates human-like text based on algorithms designed for deep learning.

Understanding the key capabilities of generative AI is crucial to fully grasping its potential contributions and transformative roles in education. These core abilities—Language Processing, Pattern Recognition, Content Generation, and Data Analysis—serve as the foundation for how generative AI can enhance and personalize learning experiences, automate feedback, and foster deeper inquiry (Lee, 2024; UNESCO, 2021, 2023; U.S. Department of Education, 2023). Leveraging these capabilities, generative AI is reshaping key aspects of education, including Personalized Learning, Automated Feedback, and Enhanced Inquiry.

Personalized Learning is an educational approach where learning experiences are tailored to meet the unique needs, preferences, and learning paces of individual students (Celik et al., 2022; Su & Yang, 2023). Generative AI plays a crucial role in enhancing personalized learning by utilizing both Language Processing and Pattern Recognition. AI tools like ChatGPT enable multi-turn interactions, remembering previous conversations and responding in context, which allows for a dialogue that is closer to human interaction (Chatterjee & Dethlefs, 2023; Lee, 2024). This capability enables learners to progress at their own pace and level of understanding, with interactions that are tailored to individual learning styles and motivations (OECD, 2023; Su & Yang, 2023).

Automated Feedback refers to the process by which students receive instant and precise feedback on their learning activities (Zhai et al., 2021). This benefit is heavily reliant on Language Processing and Pattern Recognition, which allow AI to provide immediate, contextually relevant feedback during interactions. This quick and personalized feedback helps students correct mistakes and reinforce their understanding in real time, making the learning process more effective and engaging. A practical example of this can be seen in Intelligent Tutoring Systems (ITS), which adjust to individual student needs by modifying the difficulty level and offering step-by-step guidance based on the student's progress (Holmes et al., 2019). By responding dynamically to each student's performance, these systems enhance the learning experience, ensuring that feedback is not only timely but also tailored to the student's current capabilities and learning trajectory.

Enhanced Inquiry in geography education is significantly empowered by generative AI, which leverages its core functionalities of Data Analysis, Pattern Recognition, and Content Generation. Data Analysis enables students to work with complex geographical data more effectively. When investigating environmental patterns, AI tools can assist students in processing large datasets, identifying relevant trends, and highlighting key insights that might be challenging to discern manually. This capability helps deepen students' understanding of geographical concepts and enhances their ability to formulate sophisticated questions. Pattern Recognition contributes to geographical inquiry by uncovering hidden patterns within data. For instance, AI can help students identify correlations between geographic variables, such as the relationship between deforestation rates and local climate changes. Content Generation allows AI to produce dynamic, customized resources that enhance the inquiry process. When students pose questions like "What are the impacts of rising sea levels on coastal regions?", AI can generate interactive maps or visualizations highlighting at-risk areas.

21.3 Transforming Geography Education with Generative AI: A SAMR Perspective

The integration of generative AI in geography education represents a significant shift in how geographical concepts are taught, learned, and applied. To understand and effectively implement this transformation, we can turn to the SAMR model developed by Puentedura (2006, 2014). This model serves as a valuable framework for educators to assess and enhance the integration of technology in teaching and learning. It evaluates the impact of technology through four progressive levels: Substitution, Augmentation, Modification, and Redefinition.

At the substitution level, technology directly replaces traditional tools without fundamentally changing the task. For instance, using Google Maps instead of a physical atlas to locate a city exemplifies this level. The Augmentation level involves using technology to perform the same task but with added functionalities that enhance learning outcomes. An example of this would be utilizing the measure function in Google Maps to calculate distances between cities. Modification describes the stage where technology allows for significant task redesign. For example, using Google Maps' Street View to conduct a virtual field trip and analyze urban landscapes demonstrates this level of integration. Here, the technology enables a new approach to geographical exploration that wasn't possible with traditional methods, modifying how students engage with and understand urban environments. At the Redefinition level, technology enables tasks that were previously inconceivable. An example of this would be creating a collaborative project where students use Google Maps to map local historical landmarks, adding their own photos, videos, and historical information (Fig. 21.1).

This transformative approach aligns with recent research on human-AI collaboration in education. Siemens et al. (2022) advocate for a strategy that focuses on complementing AI capabilities rather than competing with them. This perspective resonates with Erik Brynjolfsson's concept of the "Turing Trap" (2022), which argues that AI's true value lies in augmenting human abilities, not in automating human tasks. In the educational context, the goal should be to use AI to enhance human performance and enable novel tasks. Rather than merely automating existing tasks, AI can expand human capabilities by allowing individuals to perform new tasks that would have been impossible without machine assistance. This approach aligns with the higher levels of the SAMR model (Modification and Redefinition), which encourages educators to leverage AI in ways that transform traditional classroom tasks and enable entirely new educational opportunities.

To illustrate the practical applications of generative AI across the SAMR levels in geography education, we can consider a series of activities related to population analysis and climate change data analysis (Table 21.1). In the climate change lesson, we see a clear evolution from substitution (uploading data to ChatGPT instead of manual entry) to Augmentation (using AI to analyze temperature changes), and finally to Modification or even Redefinition (predicting future temperature trends over a century). This final task exemplifies the kind of novel, previously

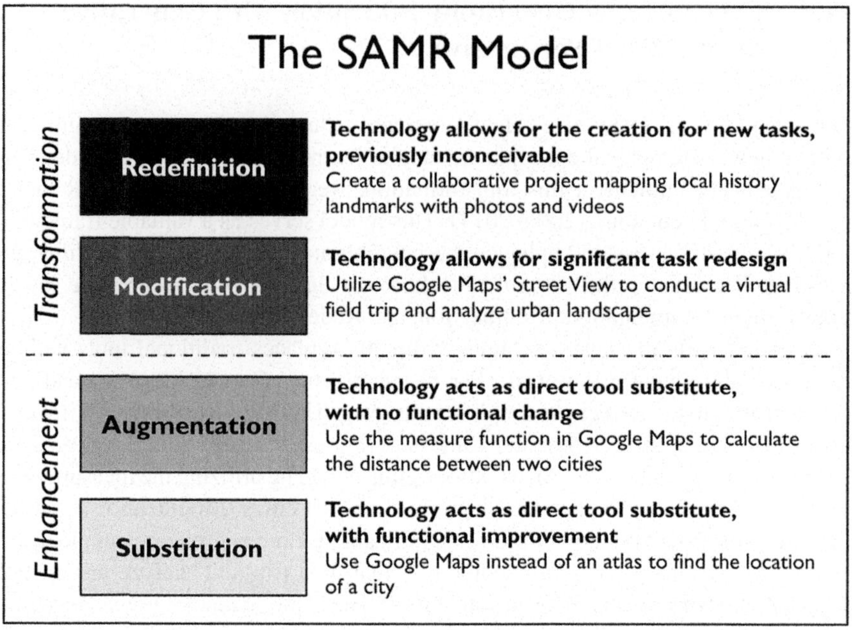

Fig. 21.1 The SAMR model (Puentedura, 2014). (Drawn with Google Maps examples by the author)

inconceivable activity that Brynjolfsson argues should be the goal of AI integration. It allows students to engage with complex predictive modeling that would be impossible without AI, fundamentally transforming the learning experience and expanding the boundaries of what's possible in the classroom.

In contrast, the population pyramid lesson, while benefiting from AI integration, remains largely within the lower levels of the SAMR model. The use of ChatGPT to generate population pyramids represents Augmentation, enhancing efficiency without fundamentally changing the task. However, the subsequent analysis of population characteristics and proposal of public services using ChatGPT falls under substitution, merely replacing manual processes without significantly altering the nature of the tasks. This limited application of AI, while improving efficiency, may not fully leverage its transformative potential in geography education and could potentially lead to passive learning experiences (Lee, 2024).

To overcome these limitations and move toward the higher levels of the SAMR model, educators have various options to extend activities beyond mere AI-driven analysis. For instance, after using ChatGPT to identify potential population issues in a given area, teachers might design activities that encourage students to engage in real-world investigations and problem-solving. Students could be invited to research actual solutions implemented by local governments or organizations, exploring public data such as city council reports or demographic studies. This approach allows

Table 21.1 Applications of Generative AI in Geography Education across the SAMR model

Traditional activity	Activity using Generative AI	SAMR level
Geography lesson: population pyramid		
1-1. Draw a population pyramid based on population data of Area A	1-1a. Upload population data of Area A to ChatGPT and request to generate a population pyramid	Augmentation
1-2. Analyze the population characteristics of Area A based on the population pyramid	1-2a. Request ChatGPT to analyze the population characteristics based on the population pyramid and review the results	Substitution
1-3. Propose necessary public services based on the population characteristics (population issues) of Area A	1-3a. Request ChatGPT to propose necessary public services based on the population characteristics (population issues) of Area A and review the results	Substitution
Geography lesson: climate change data analysis		
2-1. Download historical monthly average temperature data for Area A over the past 100 years	2-1a. Download historical monthly average temperature data for Area A over the past 100 years	
2-2. Students manually input the data into Excel	2-2a. Upload the data to ChatGPT for processing	Substitution
2-3. Analyze the changes in the number of days per year with an average monthly temperature above 30 °C	2-3a. Request ChatGPT to analyze changes in the number of days per year with an average monthly temperature above 30 °C, and review the results	Augmentation
	2-3b. Request ChatGPT to predict the number of days with an average monthly temperature above 30 °C over the next 100 years, and review the predictions	Modification/Redefinition

students to compare AI-generated solutions with real-world initiatives, potentially fostering critical thinking and active engagement. Additionally, educators could consider incorporating role-playing exercises where students take on the roles of urban planners or policymakers, using AI-generated data as a starting point for developing and defending comprehensive urban development proposals.

21.4 Bloom's Taxonomy: Benefits and Concerns of AI in Geography Learning

The integration of generative AI in geography education is profoundly altering the landscape of learning, particularly in terms of students' cognitive experiences. To understand these changes, we can turn to Bloom's Taxonomy, a framework that

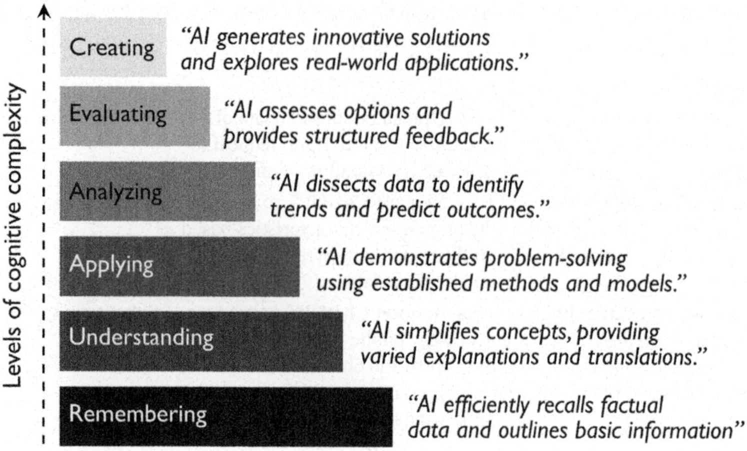

Fig. 21.2 Bloom's Digital Taxonomy and AI Integration

classifies educational learning objectives according to different levels of complexity within the cognitive domain. Originally created by Benjamin Bloom in 1956 and later revised by Anderson and Krathwohl in 2001, this taxonomy comprises six levels: Remembering, Understanding, Applying, Analyzing, Evaluating, and Creating.

However, as technology has evolved, so too has our understanding of how digital tools impact learning. Figure 21.2 illustrates how AI intersects with each level of Bloom's Taxonomy, showcasing the potential of AI to support and enhance cognitive processes across the spectrum. From efficiently recalling factual data at the "Remembering" level to generating innovative solutions at the "Creating" level, AI demonstrates its capacity to augment human cognition at every stage.

With the advent of generative AI, the impact on this taxonomy is profound. AI tools, capable of assisting across all levels of the cognitive domain, challenge the traditional hierarchical structure by potentially flattening it. As AI can efficiently provide information and perform tasks that traditionally required human cognitive effort, students may shift focus away from the lower cognitive levels of "Remember" and "Understand" and engage more readily with higher order tasks such as "Apply," "Analyze," "Evaluate," and even "Create."

Lee's (2024) research provides concrete examples of how generative AI transforms traditional educational approaches in geography studies. By developing AI-enhanced inquiry modules, Lee illustrates the profound differences in cognitive experiences when traditional methodologies are augmented with AI. For instance, in a Renewable Energy Inquiry, students used generative AI to map wind turbine locations, engaging in geographical data visualization and analysis of strategic resource placement. Lee concludes that the integration of generative AI empowers learners to perform authentic and sophisticated visualizations and analyses. From the perspective of Bloom's Taxonomy, generative AI not only emphasizes the understanding and analyzing stages, but also significantly enhances the applying and creating stages. This new process of visualization, analysis, and prediction is qualitatively different from textbook-based exploration and fundamentally aligns with the tasks performed by experts, marking a significant evolution in educational practices.

However, this transformation is not without challenges. While AI integration in geography education offers benefits in efficiency and advanced analysis, it also poses risks to student engagement and learning outcomes. As AI takes on significant aspects of geographical inquiry, there's a risk that students might become overly reliant on AI-generated analyses, potentially hindering the development of their critical thinking skills and personal insights (Kim, 2022; Lee, 2024).

To address these challenges and maximize the benefits of AI in education, it is crucial to consider how AI can be effectively integrated into the learning process while maintaining student engagement and fostering critical thinking. This consideration requires a nuanced understanding of the various ways AI can interact with learners in educational settings. In this context, the work of Ouyang & Jiao (2021) is particularly relevant. They propose three paradigms for the educational use of generative AI, each defining a different relationship between AI and learners:

1. *AI-directed, Learner-as-Recipient*: In this model, AI leads the learning process, with students primarily receiving content and instructions from the AI. This approach aligns with behaviorist learning theories, emphasizing structured sequences and immediate feedback.
2. *AI-supported, Learner-as-Collaborator*: Here, AI serves as a supportive tool, with learners actively collaborating with the technology. This paradigm draws from cognitive and social constructivist theories, promoting learning through interaction in socially situated contexts.
3. *AI-empowered, Learner-as-Leader*: In this paradigm, learners maintain control of their learning process, using AI as an augmentative tool. This approach prioritizes learner agency and views learning as a synergistic collaboration between learners, educators, and technology.

Among these, Ouyang & Jiao argue that the AI-empowered, Learner-as-Leader paradigm is the most desirable. This model encourages fully leveraging the capabilities of generative AI to tackle problems that were previously unsolvable in traditional learning or inquiry settings. Crucially, it emphasizes the importance of learners retaining agency in the inquiry process. By maintaining this balance, students can harness the power of AI while still developing critical thinking skills and personal insights through active, self-directed inquiry of geographical concepts and data.

21.5 TPACK Framework: Teacher Expertise for AI Integration in Geography

For teachers to fully leverage the opportunities presented by generative AI in education, it is essential that they not only understand the technical aspects of AI but also possess a deep, flexible understanding of how AI-based tools can contribute to pedagogical goals. This requires a systematic and adaptive knowledge of the diverse features and scenarios that generative AI can offer, as well as the ability to align these capabilities with the content being taught and the individual needs, interests, and experiences of students. Simply knowing how to use AI tools is not enough; teachers must recognize the unique contributions that generative AI can make to the learning process. This profound and nuanced understanding will enable educators to effectively integrate AI into their teaching strategies, creating AI-enhanced instructional methods that are truly aligned with both the content and the learners' needs.

The Technological, Pedagogical, and Content Knowledge (TPACK) framework, developed by Mishra and Koehler in 2006, provides a robust structure for understanding the knowledge teachers need to effectively integrate technology, including AI-based tools, into their instruction. This framework builds upon Lee Shulman's foundational concept of Pedagogical Content Knowledge (PCK), which emphasizes the importance of combining content knowledge (CK) and pedagogical knowledge (PK) to optimize student learning (Shulman, 1986). TPACK expands this idea by introducing Technological Knowledge (TK), highlighting the complex interplay between content, pedagogy, and technology in the digital age (Koehler & Mishra, 2009). It argues that effective teaching requires not just mastery of these three domains independently, but an integrated understanding of how they influence and inform one another. The TPACK represents a dynamic intersection where CK, PK, and TK coalesce, enabling teachers to harness technology to enhance their teaching strategies and improve student outcomes (Koehler et al., 2013).

However, the rapid advancement of AI technologies necessitates further evolution of the TPACK framework to address the unique challenges and opportunities that AI presents in education. This evolution has led to the development of the AI-TPACK framework, which places a new emphasis on AI literacy and the effective integration of AI tools into teaching practices (Lorenz & Romeike, 2023; Ning et al., 2024). In the AI-TPACK framework, each component of TPACK is re-envisioned through the lens of AI, offering educators a roadmap for incorporating AI technologies in ways that enrich the learning experience (Fig. 21.3). The following AI-related components of TPACK are based on the work of Celik (2023), Lorenz and Romeike (2023), and Ning et al. (2024):

- *AI-related Knowledge (AI-K)*: AI-K encompasses the understanding and skills needed to interact with, utilize, and critically evaluate AI technologies. This competence enables effective engagement with AI tools, understanding their functionalities, limitations, and societal impacts. Example question: "Describe

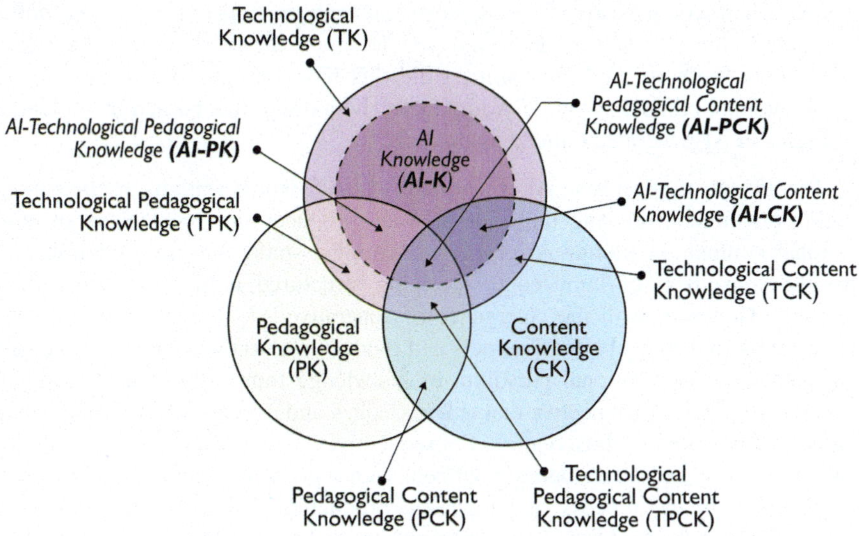

Fig. 21.3 The AI-TPACK Framework. (Adapted Ning et al., 2024, modified by the author)

a scenario where you used an AI-based tool to simplify a daily task. How did the tool improve the efficiency of the task?"

- *AI-related Pedagogical Knowledge (AI-PK)*: AI-PK refers to the ability to leverage AI technologies in educational contexts. It involves recognizing AI's potential to enhance teaching and learning, identifying its limitations, and understanding associated risks. Example question: "Can you describe a situation where you used AI-generated feedback in your teaching? How did you assess its impact on student learning and engagement?"
- *AI-related Content Knowledge (AI-CK)*: AI-CK represents the specialized understanding of how AI tools and methodologies are applied within a specific discipline. This knowledge enables critical analysis of AI's integration and impact in one's field, enhancing both professional practices and teaching strategies. For geography, this might encompass understanding how AI is used in professional geographical practices. Example question: "Could you provide an example of how you integrate AI-powered satellite image analysis tools into your teaching of environmental geography? How does this tool enhance your students' learning experience?"
- *AI-related Pedagogical Content Knowledge (AI-PCK)*: AI-PCK involves the strategic integration of AI technologies within a specific subject area to create more engaging and effective learning environments. It includes leveraging AI's capabilities for adaptive and personalized learning experiences while considering ethical implications. In geography education, this could involve designing learning experiences that integrate AI tools with geographical content in pedagogically sound ways. Example question: "Can you describe a lesson

plan where you used AI-powered geospatial analysis tools to teach about global climate change patterns? How did you integrate these tools with traditional teaching methods, and what impact did this have on students' understanding of complex climate systems? Additionally, how did you approach the potential biases in AI-generated climate predictions?"

The advent of generative AI is facilitating a shift where learners are becoming digital prosumers, blurring the lines between producers and consumers of educational content. In traditional classroom settings, students were predominantly passive consumers of knowledge, receiving structured information from their teachers. However, with the emergence of generative AI, students now have the opportunity to become both consumers and producers of knowledge. This dual role challenges the conventional paradigm of knowledge transmission and reception, necessitating a new perspective that acknowledges and encourages students' active participation in knowledge creation. Consequently, the clear-cut roles of knowledge delivery and reception are being redefined, with a growing emphasis on students' involvement in the process of knowledge production. This shift necessitates a reevaluation of the traditional teacher-centric model of education service delivery (Sung & Park, 2024). To illustrate this evolution, we can consider three stages of geography education (Fig. 21.4):

- *Traditional Geography Lesson*: In this stage, the focus was primarily on students' ability to absorb and reproduce knowledge transmitted by the teacher. Assessment largely measured how well students could replicate the input they received. Geography teachers in this era were primarily valued for their expertise in content knowledge and their ability to effectively transmit this information to students.
- *Geography Lesson in the Age of Internet Search*: As digitalization progressed, the emphasis shifted toward output. With vast information readily available, the focus moved to assessing students' problem-solving skills, critical thinking, and logical reasoning through their output. This shift led to the popularity of

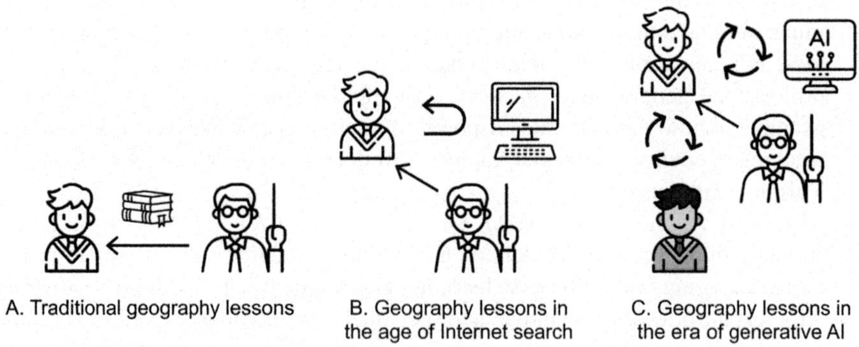

Fig. 21.4 Stages of Geography Education: Traditional, Internet Search, and Generative AI. (Adapted from Sung & Park, 2024, modified by the author)

performance-based and process-oriented assessments. In this period, geography teachers evolved into facilitators of learning, guiding students in navigating and critically evaluating the wealth of available information, and fostering skills in research and analysis.

- *Geography Lesson in the Era of Generative AI*: In this emerging paradigm, the key competency lies in students' ability to interact effectively with AI, reflect on their queries and the responses they receive, and adjust their thinking accordingly. Students need to develop skills in: Generating relevant queries for AI, evaluating AI-generated responses, refining questions based on AI feedback, synthesizing AI-assisted insights with their own knowledge, and developing metacognitive skills to navigate this human-AI collaborative learning environment. Geography teachers in this era are required to be supporters of student-AI interactions and metacognitive skill development. They need to understand AI capabilities, guide students in effective AI collaboration, and help develop students' critical thinking and metacognitive skills in an AI-enhanced learning environment.

21.6 Conclusion

This chapter has explored the transformative impact of generative AI on geography education through the lenses of the SAMR model, Bloom's Taxonomy, and the AI-TPACK framework. The analysis reveals several key findings:

1. *Transformation of Teaching and Learning*: The SAMR model demonstrates that generative AI has the potential to redefine geography education, moving beyond mere substitution to enable novel learning experiences. However, realizing this potential requires thoughtful integration and a focus on higher order learning activities.
2. *Cognitive Experiences*: Bloom's Taxonomy analysis shows that generative AI can significantly alter students' cognitive experiences in geography learning. While AI can enhance higher order thinking skills, there's a risk of over-reliance on AI-generated content, potentially hindering critical thinking development. The AI-empowered, Learner-as-Leader paradigm offers a promising approach to balance AI assistance with student agency.
3. *Teacher Competencies*: The AI-TPACK framework highlights the evolving competencies required for geography teachers in the AI era. Teachers need to develop not only technical skills but also the ability to integrate AI effectively into their pedagogical practices and content knowledge. The evolution toward AI-enhanced geography education demands a reimagining of teaching methods, assessments, and roles, emphasizing student-AI collaboration and metacognitive skill development in a dynamic learning ecosystem.

These findings underscore both the potential and challenges of integrating generative AI into geography education. Moving forward, it is crucial to consider how we can effectively harness these technologies while addressing the associated risks and maintaining the core values of education.

The integration of generative AI into geography education requires a thoughtful approach that goes beyond mere technological adoption. While AI offers powerful tools for personalized learning and data analysis, educators must ensure these technologies complement rather than replace critical human pedagogical elements. The goal should be to foster student agency and engagement, encouraging independent thought while harnessing AI's capabilities for enhanced learning experiences.

As geography education evolves, teacher competencies must expand to include both AI literacy and its ethical implications. This shift necessitates robust training programs that address technical skills, ethical considerations, and the potential risks of AI integration, including over-reliance and biases in AI-generated data.

Generative AI has the potential to redefine geography education through novel interactions with geographical data and dynamic learning experiences. However, successful integration depends on balancing AI's computational power with irreplaceable human teaching elements. By prioritizing student-centered, inquiry-driven learning and using AI as a supportive tool, geography education can leverage AI's transformative potential while preserving the essential qualities of human education.

References

Anderson, L. W., & Krathwohl, D. R. (2001). *A taxonomy for learning, teaching and assessing: A revision of Bloom's taxonomy of educational objectives*. Longman.

Baker, T., Smith, L., & Anissa, N. (2019). *Educ-AI-tion rebooted? Exploring the future of artificial intelligence in schools and colleges*. Retrieved from https://www.nesta.org.uk/report/education-rebooted/

Bessette, L. S. (2023). This isn't another piece on ChatGPT. *The National Teaching and Learning Forum, 32*(2), 11–12.

Bower, M., Torrington, J., Lai, J. W. M., Petocz, P., & Alfano, M. (2024). How should we change teaching and assessment in response to increasingly powerful generative artificial intelligence? Outcomes of the ChatGPT teacher survey. *Education and Information Technologies, 29*(12), 15403–15439. https://doi.org/10.1007/s10639-023-12405-0

Brynjolfsson, E. (2022). *The Turing Trap: The Promise & Peril of Human-Like Artificial Intelligence*. The Stanford Digital Economy Lab. Retrieved from https://digitaleconomy.stanford.edu/news/the-turing-trap-the-promise-peril-of-human-like-artificial-intelligence/

Celik, I. (2023). Towards intelligent-TPACK: An empirical study on teachers' professional knowledge to ethically integrate artificial intelligence (AI)-based tools into education. *Computers in Human Behavior, 138*, 107468. https://doi.org/10.1016/j.chb.2022.107468

Celik, I., Dindar, M., Muukkonen, H., & Järvelä, S. (2022). The promises and challenges of artificial intelligence for teachers: A systematic review of research. *TechTrends, 66*(4), 616–630. https://doi.org/10.1007/s11528-022-00715-y

Chang, C.-H., & Kidman, G. (2023). The rise of generative artificial intelligence (AI) language models—Challenges and opportunities for geographical and environmental education. *International Research in Geographical and Environmental Education, 32*(2), 85–89. https://doi.org/10.1080/10382046.2023.2194036

Chatterjee, T., & Dethlefs, N. (2023). This new conversational AI model can be your friend, philosopher, and guide ... And even your worst enemy. *Patterns, 4*(1), 100676. https://doi.org/10.1016/j.patter.2022.100676

Felix, J., & Webb, L. (2024). Use of artificial intelligence in education delivery and assessment. *UK Parliament Post*. Retrieved from https://post.parliament.uk/research-briefings/post-pn-0712/

Hanover Research. (2024). *Hanover report: Assessment and generative artificial intelligence in higher education*. University of Illinois Chicago. Retrieved from https://cms.ahs.uic.edu/inside-ahs/wp-content/uploads/sites/12/2024/04/Assessment-and-Generative-Artificial-Intelligence-in-Higher-Education.pdf

Holmes, W., Bialik, M., & Fadel, C. (2019). *Artificial intelligence in education: Promises and implications for teaching and learning*. The Center for Curriculum Redesign. Retrieved from https://curriculumredesign.org/wp-content/uploads/AIED-Book-Excerpt-CCR.pdf

Kamalov, F., Calonge, D. S., & Gurrib, I. (2023). New era of artificial intelligence in education: Towards a sustainable multifaceted revolution. *Sustainability, 15*, 12451. https://doi.org/10.3390/su151612451

Kim, M. (2022). Developing artificial intelligence-based geography teaching and learning strategies using goal-based scenario. *Journal of the Association of Korean Geographers, 11*(3), 271–286.

Koehler, M. J., & Mishra, P. (2009). What is technological pedagogical content knowledge? *Contemporary Issues in Technology and Teacher Education, 9*(1), 60–70.

Koehler, M. J., Mishra, P., & Cain, W. (2013). What is technological pedagogical content knowledge (TPACK)? *The Journal of Education, 193*(3), 13–19.

Lee, J. (2023). Beyond geospatial inquiry—How can we integrate the latest technological advances into geography education? *Education Sciences, 13*(11), 1128. https://doi.org/10.3390/educsci13111128

Lee, J. (2024). How can AI change geography teaching and learning? Focusing on geographical inquiry. *Journal of Korean Association of Geographical and Environmental Education, 32*(1), 95–112.

Lorenz, U., & Romeike, R. (2023). What is AI-PACK?—Outline of AI competencies for teaching with DPACK. In J. P. Pellet & G. Parriaux (Eds.), *Informatics in Schools: Beyond Bits and Bytes: Nurturing Informatics Intelligence in Education. 16th International Conference on Informatics in Schools: Situation, Evolution, and Perspectives, ISSEP 2023 Lausanne, Switzerland, October 23–25, 2023, Proceedings* (pp. 13–25).

Luckin, R., George, K., & Cukurova, M. (2022). *AI for school teachers*. CRC.

Mishra, P., & Koehler, M. J. (2006). Technological pedagogical content knowledge: A framework for integrating technology in teacher knowledge. *Teachers College Record, 108*(6), 1017–1054.

Mishra, P., Warr, M., & Islam, R. (2023). TPACK in the age of ChatGPT and generative AI. *Journal of Digital Learning in Teacher Education, 39*(4), 235–251. https://doi.org/10.1080/21532974.2023.2247480

Ning, Y., Zhang, C., Xu, B., Zhou, Y., & Wijaya, T. T. (2024). Teachers' AI-TPACK: Exploring the relationship between knowledge elements. *Sustainability, 16*, 978. https://doi.org/10.3390/su16030978

OECD. (2023). *Generative AI in the Classroom: From Hype to Reality?* Organization for Economic Co-operation and Development. Retrieved from https://one.oecd.org/document/EDU/EDPC(2023)11/en/pdf

Ouyang, F., & Jiao, P. (2021). Artificial intelligence in education: The three paradigms. *Computers and Education: Artificial Intelligence, 2*, 100020. https://doi.org/10.1016/j.caeai.2021.100020

Peter, C., & Sprenger, S. (2022). Digitalization and geography education—a curriculum analysis. *Erdkunde, 76*(1), 3–19.

Popenici, S. A. D., & Kerr, S. (2017). Exploring the impact of artificial intelligence on teaching and learning in higher education. *Research and Practice in Technology Enhanced Learning, 12*, 22. https://doi.org/10.1186/s41039-017-0062-8

Puentedura, R. (2006). *Transformation, technology, and education*. Retrieved from http://hippasus.com/resources/tte/

Puentedura, R. (2014). *Building transformation: An introduction to the SAMR model.* Retrieved from http://www.hippasus.com/rrpweblog/archives/2014/08/22/BuildingTransformation_AnIntroductionToSAMR.pdf

Shulman, L. (1986). Those who understand: Knowledge growth in teaching. *Educational Researcher, 15*(2), 4–14.

Siemens, G., Marmolejo-Ramos, F., Gabriel, F., Medeiros, K., Marrone, R., Joksimovic, S., & de Laat, M. (2022). Human and artificial cognition. *Computers and Education: Artificial Intelligence, 3*, 100107. https://doi.org/10.1016/j.caeai.2022.100107

Su, J., & Yang, W. (2023). Unlocking the power of ChatGPT: A framework for applying generative AI in education. *ECNU Review of Education, 6*(3), 355–366. https://doi.org/10.1177/20965311231168423

Sung, K., & Park, E. (2024). A preliminary study on the challenges of social studies education in the era of generative AI: Six key questions surrounding the use of ChatGPT. *Theory and Research in Citizenship Education, 56*(2), 247–288.

U.S. Department of Education. (2023). *Office of educational technology, artificial intelligence and future of teaching and learning: Insights and recommendations*. Retrieved from https://www2.ed.gov/documents/ai-report/ai-report.pdf

UNESCO. (2021). *AI and Education–Guidance for Policy-Makers*. United Nations Education, Scientific and Cultural Organization. Retrieved from https://gcedclearinghouse.org/resources/ai-and-education-guidance-policy-makers

UNESCO. (2023). *Guidance for Generative AI in Education and Research*, United Nations Education, Scientific and Cultural Organization. Retrieved from https://www.unesco.org/en/articles/guidance-generative-ai-education-and-research

Wang, P. (2019). On defining artificial intelligence. *Journal of Artificial General Intelligence, 10*(2), 1–37. https://doi.org/10.2478/jagi-2019-0002

Zhai, X., Chu, X., Chai, C. S., Jong, M. S. Y., Istenic, A., Spector, M., Liu, J.-B., Yuan, J., Li, Y., & Cai, N. (2021). A review of artificial intelligence (AI) in education from 2010 to 2020. *Complexity, 2021*, 1–18.

Chapter 22
Towards Ethical Spatial Decision-Making in Geospatial Artificial Intelligence

Yuhang Xie, Jikun Liu, Nanzhou Hu, Diya Li, Zhenlei Song, and Zhe Zhang

Abstract This chapter discusses the ethical considerations in spatial decision-making, focusing on the applications of Geospatial Artificial Intelligence (GeoAI). Spatial decision-making uses geographic information systems (GIS), spatial data, and analytical tools to support decisions in areas such as urban planning, environmental management, and public health. While widely applied, the ethical implications of these processes—particularly regarding data use, algorithm design, and human perception—are underexplored. Geoethics provides a framework for addressing these concerns by integrating ethical, social, and cultural dimensions into geospatial technologies. This chapter first explores the use of Volunteered Geographic Information (VGI) and social sensing, highlighting privacy, informed consent, and data security challenges. It then discusses the importance of transparency and explainability in GeoAI to ensure ethical decision-making in contexts like disaster management and public health. The chapter also emphasizes the significance of scale, including spatial, temporal, and demographic factors, in shaping ethical decisions. It examines how different scales can lead to varied ethical considerations and outcomes, particularly in complex decision-making scenarios. Additionally, the role of spatial uncertainty is analyzed, focusing on how errors in spatial data can affect decision accuracy and accountability. By integrating geoethical principles into GeoAI, this chapter promotes responsible, transparent, and equitable decision-making.

Keywords GeoAI · Geoethics · Transparency · Accountability

Y. Xie · J. Liu · N. Hu · D. Li · Z. Song
Department of Geography, Texas A&M University, College Station, TX, USA
e-mail: xieyuhang1997@tamu.edu; jikun@tamu.edu; nz972019@tamu.edu; diya.li@tamu.edu; songzl8_950606@tamu.edu

Z. Zhang (✉)
Department of Geography, Texas A&M University, College Station, TX, USA

Department of Electrical and Computer Engineering, Texas A&M University, College Station, TX, USA
e-mail: zhezhang@tamu.edu

© The Author(s), under exclusive license to Springer Nature Switzerland AG 2025

X. Huang et al. (eds.), *GeoAI and Human Geography*, Springer Geography,
https://doi.org/10.1007/978-3-031-87421-5_22

22.1 Introduction

Spatial decision-making is an interdisciplinary field that leverages spatial data, geographic information systems (GIS), and various analytical tools to support the analysis, visualization, and interpretation of spatial relationships and patterns for informed decision-making. This field is widely applied in areas such as urban planning, environmental management, transportation, public health, and resource management (Hillin et al., 2024; Li et al., 2024a; Densham & Goodchild, 1989; Crossland et al., 1995; Song et al., 2024). Despite extensive literature on these applications, there has been limited exploration of the geoethical issues inherent in the spatial decision-making process, particularly concerning the ethical implications of data usage, algorithm design, and the influence of human perception.

Geoethics explores the ethical, social, and cultural implications of earth sciences and geospatial technologies (Peppoloni et al., 2017). The term 'geoethics' combines the prefix 'geo', referring to Earth, with 'ethics', which encompasses both the social dimension of life and the individual sphere. Together, these elements form the fabric of society, highlighting the interconnection of our ethical responsibilities to both the planet and each other. Geoethics in spatial decision-making includes the application of ethical principles to the use of geographic information and technologies in planning and decision processes that affect environments and societies. Here, we need to ensure that decisions made with spatial data consider not only technical and scientific factors but also the broader implications for people and the planet. As Geospatial Artificial Intelligence (GeoAI) rapidly advances, its powerful capabilities greatly enhance spatial decision-making (Li et al., 2024a; Wang et al., 2024). GeoAI has become an essential tool for tackling complex spatial challenges due to its capability to process vast datasets, identify patterns, and simulate various scenarios. However, its use in spatial decision-making raises several ethical concerns, including potential biases in data and algorithms, issues related to privacy, and the need for transparency, accountability, and safeguards against misuse.

This book chapter offers an overview of ethical issues related to spatial decision-making, with a particular focus on GeoAI applications. The first section introduces several examples of ethical concerns in constructing social spaces using Volunteered Geographic Information (VGI). The second section centres on the transparency and explainability of GeoAI across various application domains. Following this, we delve into an in-depth discussion on spatial scale, exploring how it can influence ethical spatial decision-making. Finally, the chapter addresses the issue of spatial uncertainty in relation to data and GeoAI applications.

22.2 Social Construction of Spaces Using Volunteered Geographic Information

From a social perspective, geoethics offers insights into our societal interactions and responsibilities. It emphasizes responsible conduct and societal impact of earth sciences and geospatial technologies, advocating for principles such as respect for human rights, sustainability, and informed consent. Geoethics is deeply connected to the social implications of geospatial science, as illustrated by the modeling of VGI. VGI, which appeared in the literature in 2007 (Goodchild, 2007), refers to the use of tools to create, assemble, and disseminate geographic data voluntarily provided by individuals, a form of user-generated content. VGI often reflects the local knowledge and experiences of contributors, which can provide a rich understanding of places that might be missed by traditional mapping methods. An example of VGI is social sensing, where vast amounts of data, often including personal information, are gathered from various sources. For instance, Twitter posts frequently contain geolocation data that can be accessed by web scrapers and used to build a user's trajectory, tracing their real-world activities. This raises significant privacy concerns, as individuals may be unaware of how extensively their data is used or the potential implications. For example, Li et al. (2020a, 2020b) used social media to analyze depressive symptoms caused by COVID-19 to support decision-making in public health management. The analysis of mental health-related content from social media posts could potentially reveal sensitive information about a user's psychological state, medical conditions, or personal struggles without their explicit permission. This scenario highlights the ethical conflicts between technology and ethical concerns in social sensing.

Social media also has emerged as valuable sources for understanding human geography and spatial interactions. These data sources, such as geotagged tweets, can provide insights into patterns of human mobility and social connectivity that were previously difficult to capture at a global scale. For example, Li et al. (2021) introduced a place connectivity index that leverages geotagged social media data to measure spatial interactions between locations around the world. This work highlights how social sensing data can reveal multi-scale connectivity that may not be apparent from traditional geographic data sources alone. Additionally, research by Li et al. (2024a, 2024b) has examined the potential biases inherent in mobile location data, an increasingly common source for human geography studies. By understanding the demographic and spatial biases in these data, researchers can better analyze and mitigate such issues when using social sensing data to study human activity and mobility patterns.

GeoAI-based social sensing algorithms can inadvertently incorporate biases present in the training data, leading to socially unfair outcomes. These biases may arise from historical inequalities, sampling errors, flawed data collection methods, or platform-specific dynamics. For example, female users often post more content on some social media platforms than male users (e.g., Snapchat) (Villaronga et al., 2021). Furthermore, the data collected through social sensing could be used for surveillance purposes or to create detailed profiles of individuals, which can be exploited by governments and cause the risk of data misuse. Ensuring fairness in social sensing requires actively identifying and mitigating these biases through critical examination of data sources, algorithmic transparency, and the inclusion of diverse perspectives in developing and deploying GeoAI systems.

Another example of applying ethical frameworks in social sensing is the I-GUIDE (Institute for Geospatial Understanding through an Integrative Discovery Environment) project (I-GUIDE, 2024). I-GUIDE promotes geoethics by fostering responsible and sustainable relationships between human activities and Earth systems. The project illustrates the importance of ethical considerations in using geospatial data for AI applications. I-GUIDE collects and analyzes large volumes of geospatial data generated from various sources, including satellites, aircraft, scientific instruments, geolocation-based services, social media, and government surveys. The project emphasizes transparency, accountability, privacy, bias prevention, and data and model quality in handling this data. For instance, during humanitarian disaster responses or in risk mitigation for aging infrastructure like dams, I-GUIDE ensures that data collection and usage are conducted ethically, with appropriate safeguards to protect individual privacy and prevent bias. Despite these safeguards, the project acknowledges the possibility of unintended ethical harm and strives to mitigate such risks. By adhering to these ethical principles, I-GUIDE not only enhances the reliability and integrity of its findings but also builds public trust and promotes responsible data use in addressing major societal issues.

22.3 Transparency and Explainability in Geospatial Explainable Artificial Intelligence for Ethical Decision-Making

The importance of Geospatial Explainable AI (XAI) in geospatial decision-making lies in its ability to address the limitations of traditional 'black-box' AI models, which often lack transparency in their decision-making processes (Ali et al., 2023; Rudin, 2019; Alizadeh et al., 2022). In geospatial contexts, decisions driven by AI models can significantly impact application areas such as urban planning, disaster management, environmental monitoring, and public policy. However, the 'black-box' nature of many AI algorithms, where the internal workings and logic behind their predictions or classifications are not easily understood by humans, poses several risks and ethical concerns (Rudin, 2019). Geospatial XAI helps to mitigate these concerns by making AI models more transparent and interpretable. Beyond statistical metrics, the use of appropriate visualizations is critical for quick and accurately interpreting both geospatial data and the outputs of machine learning models. As spatial and temporal data grow increasingly complex, there is a pressing need for more research focused on Geospatial XAI. This area requires diverse techniques and methods to ensure transparency and explainability, including the use of visual tools like heat maps, plots, and natural language explanations to clarify the reasoning behind the system's decisions. By offering clear and understandable outputs, Geospatial XAI can help guide stakeholders in making better-informed decisions and facilitate more effective use of geospatial data (Roussel & Böhm, 2023).

However, the development of Geospatial XAI systems also raises significant ethical concerns that require careful attention. First, achieving the right balance between providing sufficient technical detail and ensuring understandability for non-experts is challenging. Wang et al. (2024) explained GeoAI systems often rely on complex algorithms and vast amounts of multi-dimensional spatial data, where the systems need to provide accurate explanations of their outputs to build trust and facilitate informed decision-making. Too much technical detail can overwhelm users, particularly those without a background in geospatial science or data analytics.

Misinterpretation of explanations is another significant concern in Geospatial XAI (Xing et al., 2023; Liu et al., 2024). In the context of geospatial data, explanations are often provided to help users understand the reasoning behind model predictions or decisions, particularly in complex systems. Geospatial data often involve multi-dimensional and multi-scale information. Explaining predictions from models that use such complex data can be challenging, leading to explanations that are difficult to interpret correctly. Furthermore, explanations provided by AI models may include technical terms and concepts that are not easily understood by non-experts. Visual explanations, such as heat maps or saliency maps, are commonly used in Geospatial XAI (Roussel and Böhm, 2023). However, these visualizations can be ambiguous or misleading if not properly contextualized. For instance, a heat map showing areas of high importance might be misinterpreted if users do not understand the underlying data or the model's decision-making process.

Additionally, privacy and data protection present significant ethical challenges in Geospatial XAI. Many applications rely on sensitive geospatial data that can reveal personal details, posing substantial privacy risks (Kounadi & Resch, 2018) highlight the difficulty of maintaining geoprivacy, especially with location-based data that can easily be linked to individuals. Even anonymized datasets can be re-identified by cross-referencing with other data sources, potentially leading to privacy breaches. Moreover, there is a risk that such data could be misused for purposes beyond its original intent, including surveillance or discriminatory practices. These ethical concerns underscore the importance of addressing bias, ensuring transparency, and protecting privacy to support the responsible and equitable use of Geospatial XAI in decision-making contexts.

The critical importance of ethical considerations in spatial decision-making becomes glaringly evident when reflecting on instances where these principles were neglected. An example is the response to Hurricane Katrina in 2005. The insufficient investment in the levee system, coupled with the disregard for the needs of vulnerable populations, resulted in devastating flooding and loss of life (Brinkley & Brown, 2017). Geospatial XAI could help model and communicate risks more effectively, ensuring the needs of all populations, especially for those vulnerable communities. For instance, if an AI model identifies areas at high risk of flooding due to inadequate levee systems or poor infrastructure, Geospatial XAI can offer a transparent explanation of the factors contributing to these risks to improve the trust of the AI models. This includes analyzing historical data on storm surge patterns, current conditions of levee structures, topographical vulnerabilities, and the density of populations in at-risk zones. By making these risk assessments more understandable and accessible, Geospatial XAI can empower decision-makers to prioritize investments in critical infrastructure, such as reinforcing levees or upgrading drainage systems.

Similarly, the Flint Water Crisis between 2014 and 2016 shows the consequences of ethical lapses in decision-making (Hanna-Attisha et al., 2016). During the Flint Water Crisis, decisions about water treatment and distribution were made without clear communication or understanding of the potential consequences. In this case, Geospatial XAI has the potential to address multiple dimensions of a complex issue like the Flint Water Crisis by making AI tools more accessible, transparent, and trustworthy. For example, if an AI model predicts a high risk of lead contamination in certain neighbourhoods, Geospatial XAI can provide detailed explanations of the factors contributing to this prediction, such as the age and material of the water pipes, historical data on water quality, and patterns of infrastructure wear and tear. By revealing how these variables interact to influence contamination risks, Geospatial XAI enables public officials, community leaders, and residents to understand the rationale behind the AI's predictions. As a result, decision-makers will be better equipped to implement targeted interventions, such as prioritizing pipe replacements in high-risk areas or adjusting water treatment processes to mitigate lead exposure. Geospatial XAI can provide understandable evidence of how decisions were made, potentially helping to identify where mistakes occurred and who might be held accountable. Explainable models can show, for instance, how certain decisions directly led to increased lead levels in the water.

22.4 Understanding Scale in Ethical Decision-Making

Scale is arguably one of the most critical factors in ethical decision-making, and it may be the most influential determinant in the process. The varying scales of entities involved in a decision-making framework significantly influence the final outcome, and the scalability of predetermined solutions often requires thorough evaluation before implementation. Scale refers to the magnitude, range, and depth of a decision-making logic chain, and its significance is evident from nearly every perspective. As such, scale is an irreplaceable consideration that must be carefully and holistically assessed before, during, and after a decision is made. In the context of ethics, it is essential to examine the geographical, demographic, and cultural implications for both decision-makers and those affected by the decision to fully understand its ethical impact (Healy, 2003).

A clear example of the effect of scale can be seen in the anticipated impact on populations. A spatial decision, even if made with similar purposes or under comparable circumstances, can vary significantly depending on the scale of the target audience. For instance, consider the allocation of a factory that may cause air pollution in a neighbouring region. The level of community resistance to the factory's establishment could differ drastically between a neighbourhood of

100 residents and one of 10,000 residents. This disparity in community response would, in turn, affect the ethical weight of concerns about air pollution and the responsibilities for compensation or productivity concessions, potentially leading to different decisions about the factory's allocation. Environmental impact assessments also tend to vary in stringency by spatial scale: county-level reviews can obscure pollution hotspots that only emerge when block or neighborhood scales are examined. As a result, mitigation measures that look adequate on higher grounds may actually leave the most vulnerable groups insufficiently protected. Moreover, scale can be understood on a more granular level. For example, differences in the proportions of elderly and pregnant populations between two equally populous communities could influence decisions about the factory's location, highlighting the impact of varying scales on what decision-makers prioritize. In some cases, allocating such polluting factories to communities more vulnerable to health risks might be deemed immoral or unethical. Population is just one of many factors where scale plays a crucial role, with others including the number of decision-makers, regions covered, natural resources impacted, and financial considerations. Therefore, the ethical evaluation of a spatial decision is inevitably tied to the scales of the involved entities, and any ethical justification must take these scales into account (Goodchild, 2008).

An important aspect of scale in decision ethics is the time scale. The time scale in decision-making refers to the consideration of both short-term and long-term effects of an action, with longer timelines often carrying greater ethical weight. Sustainable development serves as a fitting example. Sustainability has gained significant attention in recent years, driven not only by rising environmental awareness—fuelled by better education and the increasing frequency of severe weather events—but also by the recognition of past industrial and urban development practices that were short-sighted and ignored the long-term impact of spatial decisions. While some harmful policies may have been unavoidable to achieve overall development and survival, and while past decisions were constrained by the ethical and technological perspectives of their time, these decisions now serve as obstacles for future decision-makers and as reminders of what not to pursue. Maximizing the prospective view of decisions enhances their ethical merit and builds stronger confidence not only among the target audience but also among the decision-makers themselves (Goodchild, 2008).

The complexity and scalability of decisions are essential considerations. Decisions that aim to serve larger populations or broader areas often delve into more sophisticated decision-making structures and robust justification systems in pursuit of better scalability. Ethical values are complex in part because they continuously evolve and adapt to new challenges. Constructing adaptable and generalizable solution frameworks that are attentive to ethical considerations is challenging yet indispensable. AI decision support tools and quantitative technologies have encountered significant difficulties in this area since AI learns from existing social constructs and historical biases (Dhillon et al., 2024; Belenguer, 2022; Roselli et al., 2019). It has been repeatedly reported that AI decision support systems fail to recognize the specific traits of target audiences, resulting in severely biased outcomes (Norori et al., 2021). Some medical AI algorithms expressively require minority patients to be more severely ill to receive the same diagnosis and treatment as their white counterparts, label white patients more health-risky since they have previously paid more medical bills, or are found to exert biases through gender and races, although it has been found there are more significant variation within racial groups than between them (Backman, 2023; Ganz et al., 2021; Seyyed-Kalantari et al., 2021; Vyas et al., 2020). To avoid these ethical and sometimes dangerous issues, it is crucial to develop more complex, layered, and inclusive decision-making structures. Doing so will ultimately lead to more scalable, ethical, and trustworthy decisions (Healy, 2003).

22.5 Spatial Uncertainty in Ethical Decision-Making

In today's social and environmental decision-making landscape, spatial information plays a pivotal role. While it provides valuable support for decision-making, spatial information also brings a set of challenges due to the uncertainty inherent in such data (Hunter & Goodchild, 1996, 1997). The need for reliable and accurate information amid spatial uncertainty raises significant ethical concerns (Cornélis & Brunet, 2002; Healy, 2003). When decisions are made under such uncertainties, the potential for incorrect outcomes and serious consequences arises.

Spatial uncertainty is defined as the discrepancy between the contents of a spatial database and the corresponding phenomena in the real world (Goodchild, 2008; Hunter & Goodchild, 1996, 1997). Differences in understanding spatial uncertainty, opinions on issues, and the ambiguity of spatial information can all affect decision-making outcomes and processes (Denby et al., 2008). For instance, spatial data often suffers from collection errors, time lags, and other issues, making it difficult to achieve complete accuracy (Goodchild, 2008). Errors in spatial data collection and use can lead to mistakes in decision-making. For example, in geographic information systems (GIS), errors in digital elevation models (DEMs) can lead to incorrect viewshed analysis, where the visibility of certain areas from a specific point is assessed (Fisher, 1999). These inaccuracies may cause significant ethical issues in planning and decision-making. In urban planning or infrastructure development, viewshed analysis might be used to determine the placement of structures like cell towers, wind turbines, or surveillance systems (Bello & Aisabokhale, 2014). Errors in DEMs could result in these structures being placed in locations that unfairly disadvantage certain communities, either by exposing them to undesirable visual impacts or by depriving them of services. Viewshed analysis is also often used in environmental impact assessments, particularly for projects near protected landscapes or heritage sites (Fry et al., 2004). Errors in DEMs could lead to decisions that incorrectly assess the visual impact on these sensitive areas, potentially leading to the approval of projects that harm the environment.

Furthermore, spatial uncertainty is compounded by the discrepancies in opinion that often arise among decision-makers (McKelvey, 1980). Different decision-makers can interpret spatial data or predict future scenarios differently, and this inconsistency can increase decision-making uncertainty. Even among experts, different conclusions may be drawn from the same data due to varying professional backgrounds, further complicating decision-making and increasing uncertainty. For example, MacEacheren (1992) proposed a case study on population census: two neighbouring census blocks might have different census survey participation rates with variations in data quality. Even though one block's data might be considered more accurate, the inherent spatial and attribute aggregation introduces uncertainty. This means that different decision-makers could interpret the data in different ways, leading to diverse outcomes and increasing the overall uncertainty in the decision-making process. Such divergence is particularly problematic when handling certain concepts (such as 'vulnerability' and 'sustainability') that are inherently vague and difficult to define clearly (Zhang et al., 2014, 2018). For example, the extent to which a community can be considered 'vulnerable' may be subject to varying interpretations among scholars, further increasing uncertainty and the risk of incorrect decisions. In a study conducted in the Helsinki region, researchers applied a fuzzy multi-attribute decision-making model to assess vulnerability (Zhang et al., 2014). Due to the inherent uncertainty of the concepts involved, they gathered and integrated the opinions of multiple experts into the model to minimize the impact of uncertainty on decision-making.

Errors are inevitable in ethical decision-making, especially when faced with complex spatial uncertainties. In the context of GeoAI, these discrepancies can be amplified, as AI systems may incorporate biased data or differing interpretations of spatial data, resulting in inconsistent or even biased outcomes (Pierdicca & Paolanti, 2022). Although the consequences of wrong decisions may lead to severe economic losses, environmental damage, or social injustice, the more crucial issue is determining who should be held accountable for these errors. Should it be the model developer, the expert, or the user? Due to the spatial uncertainty involved in the spatial data and models' output, researchers must verify the model's accuracy through multiple steps to reduce uncertainty caused by the vagueness of the spatial information. This approach could serve as a dual verification method to establish an accountability mechanism for potentially erroneous decisions in the future. In other words, if rescue operations fail during actual disasters, the responsible party can demonstrate through this verification process that their decisions were based on reliable models and expert knowledge, effectively managing and assigning accountability. Solís and his team (Solís et al., 2017) pointed out that decision-makers often face challenges due to inconsistencies between the data spatial units on which their decisions are based and the actual impact range of their choices. This spatial inconsistency may prevent decision-makers from fully considering the consequences of their decisions at different scales, leading to erroneous outcomes. For example, decision-makers may make decisions within an administrative region, but the impact of those decisions may span multiple spatial units, complicating the implementation of accountability mechanisms.

The contributors to decision-making now include not only human actors but also policies, guidelines, and technologies (Goodwin, 2014). This means that responsibility for decisions extends beyond the decision-makers to include data providers and model developers, especially in scenarios where GeoAI systems are used to inform decisions. In disaster management, for example, the low spatial resolution of population data may prevent models from accurately identifying the most vulnerable areas. The propagation of errors in spatial data and how these errors can influence the outcome of spatial models, emphasizing the shared responsibility of data providers in maintaining data quality (Heuvelink & Burrough, 1993). Data providers are partially responsible for erroneous decisions resulting from data resolution or spatial mismatch, as GeoAI systems should be designed to incorporate high-quality, accurate data to function effectively. In scenarios like disaster management, where highly accurate data is essential, data providers must ensure that the spatial scale of their data aligns with decision-making needs. Failure to do so may exacerbate spatial inconsistency and undermine the accuracy of decisions and further undermine the confidence in the GeoAI-driven decision-making process.

22.6 Conclusion and Discussion

This book chapter explores the ethical considerations in constructing geospatial decision-making frameworks, with a particular focus on GeoAI. The integration of geoethics into spatial decision-making and GeoAI presents significant potential for future advancements. As technology continues to evolve, maintaining a strong ephasis on ethical frameworks is essential for addressing new chanllenges. Promising developments, such as improved data anonymization techniques, fairness and bias mitigation strategies, and enhanced approaches to accountability and explainability, are essential for ensuring that geospatial technologies and AI applications benefit society while minimizing potential harm.

Our discussion begins with the implementation of GeoAI in modelling VGI, such as social sensing, where we emphasize the importance of establishing guidelines that address privacy, bias, and accountability issues. Given the role of interpretability in ensuring transparency in GeoAI-based spatial decision-making, we present examples of geospatial XAI in decision-making contexts, highlighting how ethical concerns can be addressed.

In addition to the technical aspects of implementation, we stress the importance of considering "scale" when evaluating the social and ethical impacts of decisions. Through various examples, we demonstrate how neglecting temporal and spatial scales in decision-making can introduce bias, leading to ethical consequences. Finally, we discuss the spatial uncertainties that arise when collaborating with multiple stakeholder groups and examine the necessary accountability mechanisms. Developing a research agenda in ethical spatial decision-making involves examining how decisions concerning the use of space and resources are made with ethical considerations at the forefront. In this context, we propose the following research questions and outline future directions for formulating ethical decision-making frameworks using GeoAI:

1. How to identify and address biases in GeoAI models, especially those that arise from biased data, algorithms, or human inputs to produce more accurate interpretations and decision support?
2. What strategies can be employed to improve the transparency of GeoAI models, enabling users to understand the decision-making process and detect potential sources of bias?
3. In what ways can GeoAI be used to augment rather than replace human decision-making, ensuring that human judgement and ethical considerations are integrated into AI-driven spatial decisions?
4. How can existing legal frameworks be adapted or expanded to govern the ethical use of GeoAI in spatial decision-making?
5. Is it possible to incorporate ethical reasoning into GeoAI systems to ensure that explanations account for ethical dimensions, such as the impact on vulnerable populations or environmental sustainability?

This research agenda aims to integrate ethical considerations into spatial decision-making, ensuring that decisions are made with a focus on equity, transparency, accountability, sustainability, and justice. By addressing these research questions, scholars and practitioners can develop better tools, frameworks, and policies to guide ethical spatial decision-making, ultimately leading to more fair and just outcomes for all stakeholders involved.

References

Ali, S., Abuhmed, T., El-Sappagh, S., Muhammad, K., Alonso-Moral, J. M., Confalonieri, R., & Herrera, F. (2023). Explainable artificial intelligence (XAI): What we know and what is left to attain trustworthy artificial intelligence. *Information Fusion, 99*, 101805.

Alizadeh, B., Li, D., Hillin, J., Meyer, M. A., Thompson, C. M., Zhang, Z., & Behzadan, A. H. (2022). Human-centered flood mapping and intelligent routing through augmenting flood gauge data with crowdsourced street photos. *Advanced Engineering Informatics, 54*, 101730.

Backman, I. (2023). *Eliminating racial bias in health care AI: Expert panel offers guidelines*. Yale School of Medicine.

Belenguer, L. (2022). AI bias: Exploring discriminatory algorithmic decision-making models and the application of possible machine-centric solutions adapted from the pharmaceutical industry. *AI and Ethics, 2*(4), 771–787.

Bello, I. E., & Aisabokhale, I. (2014). 3D cartographic modeling and GIS visibility analysis of built environment for security management. *The Nigerian Journal of Cartography & GIS, 9*(1), 70–87.

Brinkley, J., & Brown, C. (2017). Levee failure and hurricane Katrina: A New Orleans Metropolis revisited. *Natural Hazards Review, 18*(1).

Cornélis, B., & Brunet, S. (2002). A policy-maker point of view on uncertainties in spatial decisions. In W. Shi, P. F. Fisher, & M. F. Goodchild (Eds.), *Spatial data quality* (pp. 168–185). London: Taylor & Francis.

Crossland, M. D., Wynne, B. E., & Perkins, W. C. (1995). Spatial decision support systems: An overview of technology and a test of efficacy. *Decision Support Systems, 14*(3), 219–235.

Denby, B., De Leeuw, F., De Smet, P., & Horálek, J. (2008). Sources of uncertainty and their assessment in spatial mapping. In *ETC/ACC Technical Paper* (p. 20).

Densham, P. J., & Goodchild, M. F. (1989). Spatial decision support systems: A research agenda. In *Scanning Electron Microsc Meet At* (pp. 707–716).

Dhillon, G., Grewal, H., Buddhavarapu, V. S., & Kashyap, R. (2024). Humanity at the crossroads: Ethical implications of AI in medicine. *Journal of SHM, 28*(5), 12–13.

Fisher, P. F. (1999). Models of uncertainty in spatial data. *Geographical Information Systems, 1*, 191–205.

Fosch-Villaronga, E., Poulsen, A., Søraa, R. A., & Custers, B. H. M. (2021). A little bird told me your gender: Gender inferences in social media. *Information Processing & Management, 58*(3), 102541.

Fry, G. L., Skar, B., Jerpåsen, G., Bakkestuen, V., & Erikstad, L. (2004). Locating archaeological sites in the landscape: A hierarchical approach based on landscape indicators. *Landscape and Urban Planning, 67*(1–4), 97–107.

Ganz, M., Holm, S. H., & Feragen, A. (2021). Assessing bias in medical AI. In *Workshop on Interpretable ML in Healthcare at International Conference on Machine Learning (ICML)*.

Goodchild, M. F. (2007). Citizens as sensors: The world of volunteered geography. *GeoJournal, 69*, 211–221.

Goodchild, M. (2008). Imprecision and spatial uncertainty. In S. Shekhar & H. Xiong (Eds.), *Encyclopedia of GIS*. Springer. https://doi.org/10.1007/978-0-387-35973-1_592

Goodwin, D. (2014). Decision-making and accountability: Differences of distribution. *Sociology of Health & Illness, 36*(1), 44–59.

Hanna-Attisha, M., LaChance, J., Sadler, R. C., & Champney Schnepp, A. (2016). Elevated blood lead levels in children associated with the Flint drinking water crisis: A spatial analysis of risk and public health response. *American Journal of Public Health, 106*(2), 283–290.

Healy, T. C. (2003). Ethical decision making: Pressure and uncertainty as complicating factors. *Health & Social Work, 28*(4), 293–301.

Heuvelink, G. B., & Burrough, P. A. (1993). Error propagation in cartographic modelling using Boolean logic and continuous classification. *International Journal of Geographical Information Systems, 7*(3), 231–246.

Hillin, J., Alizadeh, B., Li, D., Courtney, T., Meyer, M., Zhang, Z., & Behzadan, A. (2024). Designing user-centered decision support Systems for Climate Disasters: What information do communities and rescue responders need during floods? *Journal of Emergency Management, 22*(7), 71–85.

Hunter, G. J., & Goodchild, M. F. (1996). Communicating uncertainty in spatial databases. *Transactions in GIS, 1*(1), 13–24.

Hunter, G. J., & Goodchild, M. F. (1997). Modeling the uncertainty of slope and aspect estimates derived from spatial databases. *Geographical Analysis, 29*(1), 35–49.

I-GUIDE. (2024). *Institute for geospatial understanding through an integrative discovery environment*. Retrieved August 28, 2024, from https://i-guide.io/

Kounadi, O., & Resch, B. (2018). A Geoprivacy by Design Guideline for Research Campaigns That Utilize Location-Based Data Collection. *Journal of Empirical Research on Human Research Ethics, 13*, 203.

Li, D., Chaudhary, H., & Zhang, Z. (2020a). Modeling spatiotemporal patterns of depressive symptoms caused by COVID-19 using social media data mining. *International Journal of Environmental Research and Public Health, 17*(14), 4988.

Li, W., Zhang, C., & Chen, C. (2020b). GeoAI: Artificial intelligence for geographic knowledge discovery. *International Journal of Geographical Information Science, 34*(4), 625–650.

Li, Z., Huang, X., Ye, X., Jiang, Y., Martin, Y., Ning, H., & Li, X. (2021). Measuring global multi-scale place connectivity using geotagged social media data. *Scientific Reports, 11*(1), 14694.

Li, D., Zhang, Z., Alizadeh, B., Zhang, Z., Duffield, N., Meyer, M. A., Thompson, C. M., Gao, H., & Behzadan, A. H. (2024). A reinforcement learning-based routing algorithm for large street networks. *International Journal of Geographical Information Science, 38*(2), 183–215.

Li, Z., Ning, H., Jing, F., & Lessani, M. N. (2024b). Understanding the bias of mobile location data across spatial scales and over time: A comprehensive analysis of SafeGraph data in the United States. *PLoS One, 19*(1), e0294430.

Liu, P., Zhang, Y., & Biljecki, F. (2024). Explainable spatially explicit geospatial artificial intelligence in urban analytics. *Environment and Planning B: Urban Analytics and City Science, 51*(5), 1104–1123.

MacEachren, A. M. (1992). Visualizing uncertain information. *Cartographic Perspectives, 13*, 10–19.

McKelvey, R. D. (1980). Ambiguity in spatial models of policy formation. *Public Choice*, 385–402.

Norori, N., Hu, Q., Aellen, F. M., Faraci, F. D., & Tzovara, A. (2021). Addressing bias in big data and AI for health care: A call for open science. *Patterns, 2*(10), 100347.

Peppoloni, S., & Di Capua, G. (2017). Geoethics: Ethical, social and cultural implications in geosciences. *Annals of Geophysics, 60*(7).

Pierdicca, R., & Paolanti, M. (2022). GeoAI: A review of artificial intelligence approaches for the interpretation of complex geomatics data. *Geoscientific Instrumentation, Methods and Data Systems Discussions, 2022*, 1–35.

Roselli, D., Matthews, J., & Talagala, N. (2019). Managing Bias in AI. In *WWW '19: Companion Proceedings of the 2019 World Wide Web Conference.*

Roussel, C., & Böhm, K. (2023). Geospatial xai: A review. *ISPRS International Journal of Geo-Information, 12*(9), 355.

Rudin, C. (2019). Stop explaining black box machine learning models for high stakes decisions and use interpretable models instead. *Nature Machine Intelligence, 1*(5), 206–215.

Seyyed-Kalantari, L., Zhang, H., McDermott, M. B. A., Chen, I. Y., & Ghassemi, M. (2021). Underdiagnosis bias of artificial intelligence algorithms applied to chest radiographs in under-served patient populations. *Nature Medicine, 27*(12), 2176–2182.

Solís, P., Vanos, J. K., & Forbis, R. E. (2017). The decision-making/accountability spatial incongruence problem for research linking environmental science and policy. *Geographical Review, 107*(4), 680–704.

Song, Z., Chapman, P., Tao, J., Chang, P., Gao, H., Liu, H., Brannstrom, C., & Zhang, Z. (2024). Mapping the unheard: Analyzing tradeoffs between fisheries and offshore wind farms using multicriteria decision analysis. *Annals of the American Association of Geographers, 114*(3), 536–554.

Vyas, D. A., Eisenstein, L. G., & Jones, D. S. (2020). Hidden in plain sight — Reconsidering the use of race correction in clinical algorithms. *New England Journal of Medicine, 383*(9), 874–882.

Wang, S., Huang, X., Liu, P., Zhang, M., Biljecki, F., Hu, T., & Bao, S. (2024). Mapping the landscape and roadmap of geospatial artificial intelligence (GeoAI) in quantitative human geography: An extensive systematic review. *International Journal of Applied Earth Observation and Geoinformation, 128*, 103734.

Xing, J., & Sieber, R. (2023). The challenges of integrating explainable artificial intelligence into GeoAI. *Transactions in GIS, 27*(3), 626–645.

Zhang, Z., Demšar, U., Rantala, J., & Virrantaus, K. (2014). A fuzzy multiple-attribute decision-making modelling for vulnerability analysis on the basis of population information for disaster management. *International Journal of Geographical Information Science, 28*(9), 1922–1939.

Zhang, Z., Demšar, U., Wang, S., & Virrantaus, K. (2018). A spatial fuzzy influence diagram for modelling spatial objects' dependencies: A case study on tree-related electric outages. *International Journal of Geographical Information Science, 32*(2), 349–366.

Chapter 23
Sustainable GeoAI in Human Geography: Reproducible, Replicable, and Expandable

Lingbo Liu, Xiao Huang, Siqin Wang, and Xiaokang Fu

Abstract Sustainable Geospatial Artificial Intelligence (GeoAI) in human geography focuses on enhancing reproducibility, replicability, and expandability (RRE) to ensure reliability, scalability, and impact in geospatial research and applications. Achieving RRE is essential for verifying, extending, and applying geospatial findings across various contexts and scales. This chapter discusses the significance of RRE in GeoAI, highlighting how these principles contribute to sustainable outcomes in urban planning, agriculture, and public health. It explores the current advancements and challenges in implementing RRE within GeoAI, such as the complexity of workflows, scalability issues, and ethical considerations. By addressing these challenges, we aim to establish a framework for sustainable GeoAI that supports reliable, scalable, and equitable development across diverse domains.

Keywords GeoAI · Human geography · Reproducibility · Replicability and expandability (RRE) · Sustainability

23.1 Introduction

Sustainable Geospatial Artificial Intelligence (GeoAI) in human geography becomes centered on the core principles of reproducibility, replicability, and expandability (RRE). These principles are essential for advancing the applicability of GeoAI in sustainable research, aiming to produce outcomes that demonstrate

L. Liu · X. Fu
Center for Geographic Analysis, Harvard University, Cambridge, MA, USA
e-mail: lingboliu@fas.harvard.edu; xiaokang_fu@fas.harvard.edu

X. Huang (✉)
Department of Environmental Sciences, Emory University, Atlanta, GA, USA
e-mail: xiao.huang2@emory.edu

S. Wang
Spatial Sciences Institute, University of Southern California, Los Angeles, CA, USA
e-mail: siqinwan@usc.edu

© The Author(s), under exclusive license to Springer Nature Switzerland AG 2025
X. Huang et al. (eds.), *GeoAI and Human Geography*, Springer Geography,
https://doi.org/10.1007/978-3-031-87421-5_23

robustness, reliability, and scalability across various spatial and temporal contexts. Within human geography, integrating RRE principles enhances the sustainability and equity of GeoAI-driven research, equipping it to address complex societal challenges in urban planning, agricultural management, and public health improvement (Gao et al., 2024; Jayant et al., 2023; Li, 2020; Mohammed et al., 2023; VoPham et al., 2018).

In scientific research, reproducibility and replicability form the backbone of credible findings. Reproducibility is the capacity to achieve consistent results using the same dataset and methodology, thus ensuring the accuracy and reliability of findings. Replicability goes a step further by testing the robustness of findings across different datasets or under varied conditions. Expandability focuses on adapting existing models and methodologies to novel contexts or applications, enhancing their broader relevance and impact (Hsu & Li, 2023; Hu et al., 2019a; Nüst et al., 2018). Together, these principles establish a solid framework for sustainable GeoAI that upholds scientific rigor and practical utility.

In the domain of human geography, reproducibility is especially critical due to the inherent spatial heterogeneity and complexity of geographic data. Reproducible results build trust in spatial analyses, which is essential for evidence-based decision-making in areas such as urban development and resource management (Fisher, 2012; Wilson et al., 2020). Replicability further solidifies the credibility of GeoAI studies by demonstrating that findings are not overly reliant on specific datasets but hold true across different spatial and temporal scenarios (Liu & Biljecki, 2022). This is crucial in policymaking, where reliable insights inform interventions with significant societal impact (Campagna, 2005). Expandability also plays a crucial role in GeoAI's potential. By ensuring that GeoAI models and tools can adapt to new data, technological advancements, and emerging societal needs, researchers can address evolving challenges like climate change adaptation and urban expansion. This adaptability makes GeoAI a dynamic, versatile tool for supporting sustainable development (Hu et al., 2019b; Li & Hsu, 2022).

However, achieving RRE in GeoAI presents unique challenges. The complexity of geospatial data, characterized by its heterogeneity, high dimensionality, and diverse data sources, often complicates the establishment of reproducible workflows. Additionally, proprietary software and computationally intensive models create obstacles that limit other researchers' ability to replicate or build on existing studies (Sahana et al., 2023; VoPham et al., 2018). The increasing volume and complexity of geospatial data demand scalable computational solutions to handle this growth effectively (Nüst et al., 2018; Yang et al., 2022). Ethical considerations, such as algorithmic bias and data privacy, must also be addressed to ensure that GeoAI solutions are fair and do not inadvertently exacerbate social disparities (Janowicz, 2024; Musgrave et al., 2020).

This chapter explores the advancements, challenges, and potential solutions for enhancing RRE in GeoAI within human geography. By examining these facets, we aim to elucidate how GeoAI can reach its full potential, fostering sustainable and equitable development across key sectors such as urban planning, agriculture, and public health. Ultimately, this chapter seeks to contribute to a robust and sustainable GeoAI framework prioritizing reliability, scalability, and inclusivity, thereby positioning GeoAI as a powerful tool to address pressing societal challenges (Gonzales-Inca et al., 2022; Sk & Sudhir Kumar, 2021).

23.2 Conceptual Framework of RRE in GeoAI

23.2.1 Definitions of Reproducibility, Replicability, and Expandability

In the realm of GeoAI, three essential principles—reproducibility, replicability, and expandability—are foundational to ensure that scientific research yields reliable and broadly applicable outcomes. Reproducibility refers to the ability of different researchers to achieve the same results using the same dataset and analytical methods. It is a critical quality control mechanism that helps validate findings and ensures that the results are not influenced by arbitrary factors like errors in the analytical process (Gundersen, 2019; Haryo & Pulungan, 2022).

There exist different levels of replicability in various types of geospatial research or weak replicability due to the inherent spatiotemporal heterogeneity (Ostermann & Granell, 2016; Wainwright, 2020). The pursuit of generalizable conclusions or the laws in geographic sciences through new data utilization and model expansion is a crucial aspect of scientific research (Dangermond & Goodchild, 2019; Kedron & Holler, 2022), necessitating broader interdisciplinary collaboration and expansion. Figure 23.1 summarizes the framework for literature analysis on the RRE framework in geospatial analysis, which comprises a comparison of concepts like reproduction, replication, reanalysis, generalization, and expansion, and an overview of the key stages in geospatial analysis and the evolution of RRE solutions (Liu et al., 2024b).

Replicability, while closely related to reproducibility, entails applying the original methods to different datasets or in different environments to verify that the conclusions are robust and generalizable (Goodchild et al., 2020; Nishant et al., 2020). In human geography, replicability ensures that findings based on a specific location or dataset are not simply unique to that context but instead have broader applicability to other areas with similar conditions.

Expandability is the third critical component of this framework, and it involves extending existing research models or methods to address new data types or application scenarios. It is crucial in the context of rapidly evolving geospatial technologies and datasets, as expandability ensures that research conducted today can remain relevant by adapting to new technologies and research needs (Fan et al., 2023; Larsson, 2019; Marasinghe et al., 2024). For example, a GeoAI model developed to analyze urban growth in a specific region can be expanded with additional features to study other urban environments, integrating additional datasets such as environmental or demographic data to enhance model robustness and applicability.

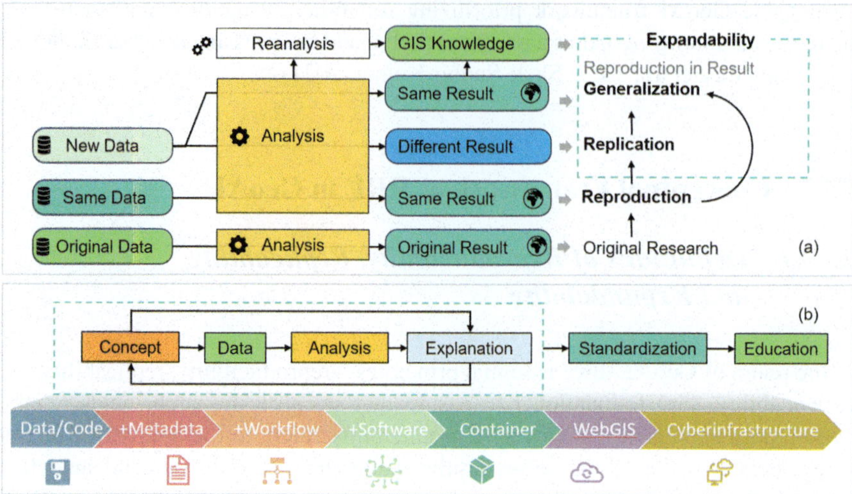

Fig. 23.1 Literature review summary on the Reproducible, Replicable, and Extensibility (RRE) framework for geospatial analysis. (**a**) Concepts comparison on reproduction, replication, reanalysis, generalization, and expansion, (**b**) key stages for geospatial analysis and the evolution of RRE solutions (Liu et al., 2024b)

These three principles—reproducibility, replicability, and expandability—form the basis of a sustainable GeoAI framework, ensuring that research is not only valid and reliable but also capable of adapting to emerging scientific and practical challenges. Achieving these objectives requires careful consideration of data standards, methodological transparency, and tools that support the seamless integration of new techniques and data sources.

23.2.2 *The Role of RRE in Human Geography*

The application of RRE principles in human geography is fundamental to advancing the reliability and impact of geospatial research. In human geography, where spatial heterogeneity, temporal dynamics, and socio-environmental interactions add layers of complexity, ensuring that research results are reproducible and replicable is of paramount importance (Wynsberghe, 2021). Such consistency is critical for urban planning, disaster management, public health interventions, and other areas that directly impact people's lives.

The reproducibility of geospatial analyses enables stakeholders, such as urban planners and policymakers, to make decisions based on tested and reliable findings. A reproducible GeoAI framework allows researchers to transparently share their workflows, data, and analytical methods, which is essential in building trust among scientific communities and decision-makers (Gao et al., 2020; Persello et al., 2022). For example, in public health research, ensuring that predictive models for disease spread are reproducible means that other researchers can validate the predictions and apply similar approaches in different geographical contexts to achieve reliable outcomes (Sagar et al., 2022).

Replicability in GeoAI ensures that models are adaptable to different datasets, thus extending their utility to new contexts. For example, a replicable urban accessibility model designed for one city should be adaptable for use in another city with minimal modifications (Tullis & Kar, 2020). This adaptability is particularly useful for comparative studies that aim to understand differences and similarities in urban dynamics across different regions. By demonstrating replicability, researchers can support the generalization of findings, thereby contributing to a broader understanding of human-environment interactions.

Expandability is critical for ensuring that GeoAI models can grow with technological advancements and emerging data types. For instance, a land-use change model may initially be built using remote sensing data but should be expandable to incorporate other data types, such as social media feeds or sensor network data, as these become more widely available (Janowicz et al., 2022; Waters, 2020). This expandability enhances the capacity of GeoAI to remain current and effective in the face of rapidly changing environmental and societal conditions, such as urbanization, climate change, and migration patterns (Bolón-Canedo et al., 2024).

The integration of RRE principles in human geography also plays a key role in fostering interdisciplinary collaboration. By providing reproducible and replicable frameworks, GeoAI facilitates collaboration among geographers, data scientists, urban planners, and other stakeholders, thereby enhancing the collective ability to address complex societal challenges. Such interdisciplinary efforts are especially crucial for tackling problems that span multiple domains, such as climate resilience and urban health, which require the expertise of diverse fields.

23.3 Current Progress and Approaches in GeoAI for Human Geography

The integration of AI into human geography has driven significant advancements in how spatial phenomena are analyzed and understood. Progress in reproducibility, replicability, and expandability has been supported by the adoption of open-source tools, transparent workflows, and scalable cloud computing solutions. Ethical considerations have also gained prominence, with increased emphasis on fairness, transparency, and community engagement.

23.3.1 Reproducibility and Replicability in GeoAI

Reproducibility and replicability are essential in GeoAI research, ensuring findings remain consistent and reliable across different contexts. However, the complexity of geospatial data and the variety of tools available make reproducibility challenging (Janowicz et al., 2019; Wang et al., 2024). Recent efforts to standardize data formats and workflows have notably enhanced reproducibility in the field. One key advancement is the use of open-source geospatial tools, such as QGIS and the Python-based Geospatial Data Abstraction Library (GDAL). These tools allow researchers to create and share workflows without proprietary software restrictions (Christopher et al., 2024). Open-source platforms also enable transparent versioning and documentation, ensuring that data, code, and models are accessible and verifiable by others (Shi et al., 2023).

Standardized geospatial formats like GeoJSON and Open Geospatial Consortium (OGC) standards have further supported reproducibility. These formats allow data from diverse sources to be integrated into common workflows, reducing discrepancies from heterogeneous formats and enabling collaborative analysis (Yang et al., 2020). Additionally, cloud-based platforms like Google Earth Engine facilitate access to large-scale geospatial datasets, enabling studies that require significant computational power to be replicated more easily (Philipe & Dalton, 2022).

To enhance replicability, researchers increasingly emphasize transparent, well-documented methodologies. "Reproducible research packages," including code, datasets, and detailed documentation, allow others to replicate analyses with different datasets, broadening insights into geographic phenomena across contexts (Fu, 2021; Persello et al., 2022).

23.3.2 Expandability of GeoAI Models

Expandability in GeoAI refers to adapting models and frameworks to new geographic contexts, datasets, or applications. This capability is crucial in human geography, where diverse, evolving challenges arise across varying spatial and socio-economic environments. Progress in expandability includes the use of modular machine learning frameworks such as TensorFlow and PyTorch, which enable researchers to modify or create models tailored to specific geographic. These frameworks facilitate the integration of diverse data types, such as satellite imagery, demographics, and climate information, broadening the scope of research questions GeoAI can address.

Cloud computing advancements also boost expandability. Platforms such as Amazon Web Services (AWS) and Microsoft Azure provide scalable environments that support the deployment and modification of GeoAI models across different geographic areas. By leveraging cloud infrastructure, researchers can expand their models to accommodate larger datasets or incorporate additional variables, enhancing their capacity to generate insights at multiple scales (Howe & Tullis, 2022). Transfer learning further supports expandability by reusing pre-trained models for new tasks. For example, a model trained on urban land cover in one area can be adapted to another, reducing retraining time and computational (Waters, 2020). This technique also addresses data scarcity in regions lacking high-quality training data (Zhang et al., 2022). Furthermore, interoperable tools and cross-platform workflows, integrating R, Python, and GIS, enhance expandability. Such workflows allow researchers to incorporate new datasets, explore diverse modeling techniques, and foster cross-disciplinary collaboration. For instance, combining Python-based machine learning with R-based spatial analysis can extend GeoAI to include real-time sensor data analytics, expanding the depth of geospatial research.

23.3.3 Ethical Considerations and Recent Progress in GeoAI

Ethical considerations are a critical aspect of the current progress in GeoAI, particularly concerning data privacy, algorithmic transparency, and equitable application. The use of GeoAI raises concerns about the potential for algorithmic bias, especially when models are trained on historical data that may contain inherent biases. Such biases can lead to outcomes that disproportionately impact marginalized communities, thereby reinforcing existing inequalities.

Recent progress in ethical GeoAI has focused on developing fairness-aware machine learning algorithms, which attempt to identify and mitigate biases during model development (Nüst & Pebesma, 2020). Furthermore, transparent model documentation and ethical guidelines have been established to ensure that GeoAI technologies are used responsibly and that their impacts are considered from multiple perspectives (DiSalvo et al., 2010). For example, the "model cards" framework provides standardized documentation that outlines the intended use, limitations, and potential biases of GeoAI models, promoting transparency and accountability (Marasinghe et al., 2024).

Community engagement has also emerged as a key strategy in ensuring the ethical application of GeoAI. By involving local communities in the development and implementation of GeoAI solutions, researchers can better align their work with community needs and priorities, thereby fostering greater trust and acceptance (Tullis & Kar, 2020). This participatory approach has been particularly effective in projects involving urban planning and environmental monitoring, where the direct involvement of affected communities can lead to more equitable and sustainable outcomes (Casali et al., 2022).

23.4 Challenges and Barriers in Achieving RRE in GeoAI

23.4.1 Data Complexity and Integration

Achieving repeatable, reproducible, and expandable (RRE) research in GeoAI faces significant challenges due to the complexity and diversity of geospatial data. Geospatial data exists in various forms, such as satellite imagery, GPS tracks, socio-economic datasets, and environmental data, each collected at differing spatial and temporal scales. This heterogeneity complicates integration into a unified GeoAI framework (Agustín García et al., 2020). Additionally, inconsistencies in spatial resolution, data quality, and temporal coverage can bias model predictions. For instance, combining high-resolution satellite images with low-resolution demographic data may yield misleading insights into population dynamics or land use. Overcoming these issues demands rigorous preprocessing, including data harmonization, normalization, and cleaning, to ensure compatibility across sources. Integrating real-time data, such as sensor networks and social media feeds, introduces further challenges due to the need for frequent updates and dynamic processing. These data types are essential for applications like disaster management, where quick decision-making is crucial, yet a lack of standardized formats and protocols hinders effective integration. Developing standardized formats is thus essential to support seamless integration and RRE in GeoAI research.

23.4.2 Scalability and Computational Challenges

Scalability is a key concern in GeoAI, particularly given the increasing volume, variety, and velocity of geospatial data being generated today. As geospatial datasets grow larger and more complex, the computational resources required to process, analyze, and store these data also increase substantially. This presents a significant challenge for researchers who may not have access to high-performance computing (HPC) facilities or cloud-based platforms. Training and deploying machine learning models, particularly deep learning models, on large-scale geospatial data can be computationally intensive. The need for substantial computational power often limits the ability of researchers to perform scalable analyses, especially when working with high-resolution satellite imagery or large spatial-temporal datasets. Additionally, the high cost of accessing commercial cloud computing platforms further exacerbates this issue, particularly for research teams with limited funding. To address these scalability challenges, researchers have begun to explore more efficient algorithms that can achieve similar levels of accuracy with reduced computational demands. Techniques such as model pruning, where redundant parameters in neural networks are removed, and quantization, where model precision is reduced to decrease computational requirements, have shown promise in making GeoAI models more scalable. Furthermore, the use of edge computing, which processes data closer to its source rather than relying solely on centralized servers, has been proposed as a solution to alleviate computational and data transfer bottlenecks.

23.4.3 *Ethical and Social Challenges*

GeoAI often involves sensitive data, raising ethical concerns around privacy, algorithmic bias, and equitable benefit distribution. Privacy is a key issue with detailed geospatial data, which can reveal personal locations and behaviors. The risk of re-identification in anonymized data is high, necessitating techniques like differential privacy to obscure individual details while preserving aggregate patterns. Balancing privacy with data accuracy remains a complex challenge. Algorithmic bias presents another ethical issue. GeoAI models trained on biased historical data may perpetuate or amplify existing biases, potentially disadvantaging marginalized communities. For example, resource allocation models can be biased if they underrepresent certain groups. Addressing this requires selecting fair training data and developing fairness-aware algorithms to reduce bias. Additionally, GeoAI benefits are not equally distributed, often favoring urban over rural areas or high-income over low-income regions. Rural communities, for instance, may lack the infrastructure or data needed for GeoAI, worsening the digital divide. Involving local stakeholders in GeoAI design and deployment helps ensure that all communities' needs are considered and addressed.

23.4.4 *Reproducibility Barriers*

Reproducibility is a cornerstone of scientific research, yet it remains a significant challenge in GeoAI. The complexity of geospatial workflows, combined with the use of proprietary tools and software, often makes it difficult for other researchers to replicate studies and verify findings. Many GeoAI projects rely on specialized software or custom-built scripts that are not documented in sufficient detail, thereby hindering reproducibility. A major barrier to reproducibility is the lack of standardized workflows and documentation practices. In many cases, researchers do not fully document their data preprocessing steps, parameter settings, or model configurations, making it challenging for others to replicate their work. This issue is exacerbated by the use of proprietary geospatial software, which may include built-in functions that are not openly accessible or well-documented, preventing others from reproducing the exact results. To enhance reproducibility, the use of open-source tools and standardized workflows has been advocated. Open-source platforms such as R and Python, coupled with detailed documentation and version-controlled code repositories, can help overcome many of the barriers associated with reproducibility in GeoAI research. Additionally, the development of containerized environments, such as those provided by Docker, allows researchers to encapsulate all dependencies needed for a given analysis, thus making it easier for others to reproduce their work regardless of differences in computing environments.

23.4.5 Stakeholder Engagement and Trustworthiness

The successful implementation of GeoAI requires engagement from various stakeholders, including local communities, policymakers, and private sector partners. Building trust in GeoAI is crucial, especially for applications impacting areas like urban planning or public health. One major trust challenge is the complexity and opacity of many GeoAI models, particularly deep learning-based ones, which often function as "black boxes" without clear insight into their decision-making processes. This lack of transparency can cause mistrust among stakeholders who struggle to understand or validate the outcomes. Enhancing model interpretability through explainable AI (XAI) methods is essential for trust-building, as XAI provides understandable explanations for model predictions, helping stakeholders grasp the factors that influence results. Another key challenge is aligning GeoAI solutions with the needs and priorities of target communities. Involving stakeholders throughout the project—from design to implementation and evaluation—ensures GeoAI applications are relevant, effective, and responsive to local contexts. Such participatory approaches increase the relevance and impact of GeoAI solutions and foster community ownership and trust in these technologies.

23.4.6 Lack of Skills and Training

A significant barrier to achieving RRE in GeoAI is the lack of skills and training among researchers and practitioners. GeoAI is an inherently interdisciplinary field that requires expertise in geography, data science, machine learning, and programming. Many researchers in traditional geography fields may not have the computational skills needed to develop or utilize GeoAI models effectively. The steep learning curve associated with GeoAI tools, particularly those involving coding and machine learning frameworks, can deter researchers from adopting these methods. To address this issue, more emphasis is needed on providing training and resources that are accessible to a wider range of users, including those without formal backgrounds in computer science. The development of visual programming platforms, such as KNIME, which allow users to build GeoAI workflows without extensive coding, represents a promising approach to lowering the barriers to entry and promoting broader adoption of GeoAI.

23.5 Solutions and Best Practices for RRE in GeoAI

23.5.1 Standardization of Data and Tools

One of the most critical solutions for enhancing RRE in GeoAI is the standardization of data, methods, and tools. Geospatial data comes from diverse sources with different formats, quality levels, and spatial resolutions, making it challenging to integrate into a unified analysis framework. To address this issue, researchers have emphasized the importance of adopting standardized data formats and metadata protocols to facilitate data sharing and integration. The Open Geospatial Consortium (OGC) standards, such as GeoJSON, Web Map Service (WMS), and Web Feature Service (WFS), have been instrumental in providing consistent formats for geospatial data exchange, thereby ensuring data compatibility across different platforms and software. By adopting these standards, researchers can ensure that data used in GeoAI analyses is accessible, well-documented, and interoperable, which is crucial for facilitating reproducibility and collaborative work.

Open-source geospatial tools like QGIS, GDAL, and Python-based libraries such as GeoPandas and Rasterio also play a crucial role in standardization. These tools offer transparent, reusable, and adaptable workflows, supported by extensive documentation and active communities. This availability allows researchers to replicate and expand on previous work, promoting a culture of open and reproducible science in GeoAI. Furthermore, using workflow management systems, like KNIME, to standardize analytical workflow enhances reproducibility. These systems enable researchers to create end-to-end pipelines for data preprocessing, analysis, and visualization, documenting each step for easy replication and broader use.

23.5.2 Visual Programming for Lowering Barriers

Visual programming tools play a significant role in reducing the barriers to entry for researchers and practitioners who may not have extensive programming expertise. Platforms like KNIME, Orange, and RapidMiner offer user-friendly, drag-and-drop interfaces that allow researchers to construct complex workflows without writing code, making GeoAI more accessible to a wider audience (Liu et al., 2024a; Liu & Wang, 2023). By using visual programming, researchers can ensure that their workflows are easily understood and replicated by others. This is particularly beneficial in interdisciplinary teams where not all members may have a strong background in data science or programming. Visual workflows provide an intuitive representation of the analysis process, making it easier to share and replicate across different teams and contexts. Additionally, visual programming platforms support the integration of external data science libraries and tools, allowing users to incorporate Python, R, or other programming languages as needed to extend the capabilities of their workflows. This combination of flexibility and ease of use helps promote expandability, as workflows can be adapted to incorporate new data types, models, or analytical methods with minimal effort.

23.5.3 Community Engagement and Interdisciplinary Collaboration

Engaging communities and fostering interdisciplinary collaboration are essential for ensuring GeoAI solutions are sustainable, ethical, and impactful. Integrating GeoAI into human geography requires diverse stakeholder input—including geographers, data scientists, policymakers, and local communities—to align GeoAI tools with the needs of those affected by them. Community engagement can be achieved through participatory approaches that involve stakeholders in all stages of GeoAI development, from problem definition and data collection to model evaluation and result interpretation. Participatory GIS (PGIS), for instance, allows local communities to contribute to data collection, ensuring GeoAI analyses reflect the realities of studied areas. This engagement also fosters trust, as community members directly shape research and experience its tangible benefits. Interdisciplinary collaboration is equally vital for promoting RRE practices in GeoAI. Addressing complex geospatial issues often requires combining expertise from geography, computer science, environmental science, and social sciences. Shared data repositories, collaborative coding platforms, and interdisciplinary research networks support this collaboration, bringing together technical and contextual knowledge to apply GeoAI results meaningfully.

23.5.4 Open Data and Cloud-Based Solutions

The availability of open data and cloud-based platforms has been a game-changer in making GeoAI research more reproducible, scalable, and accessible. Open data initiatives, such as those led by governments and international organizations, provide researchers with access to high-quality geospatial datasets that can be used to validate and replicate findings. By leveraging open data, researchers can reduce the cost and effort associated with data collection and ensure that their analyses are based on widely available, standardized datasets. Cloud-based platforms, such as Google Earth Engine and Microsoft Azure, offer scalable computing resources that can process large geospatial datasets and run computationally intensive models without requiring local hardware infrastructure. These platforms also provide collaborative features that allow researchers to share their code, data, and analysis results with others, making it easier to replicate studies and build upon previous work. The integration of cloud-based tools with open-source libraries allows for the development of reproducible and expandable workflows that are accessible to a broader audience, regardless of their computing capabilities. For example, Google Earth Engine's integration with Python and JavaScript APIs enables researchers to write scripts that can be shared and reused, providing a transparent and standardized way to perform complex geospatial analyses.

23.5.5 Documentation and Transparent Methodology

Comprehensive documentation and transparency in methodology are critical for ensuring reproducibility in GeoAI research. Researchers should document each step of their analytical process, including data preprocessing, model selection, parameter tuning, and validation procedures, in sufficient detail to allow others to follow and replicate their work. The adoption of best practices, such as creating reproducible research packages that include code, data, and documentation, is essential for promoting transparency. Version control systems, such as Git, are effective tools for tracking changes in code and workflows, ensuring that the development process is transparent and that previous versions of analyses can be revisited if needed. Additionally, the use of containerization technologies like Docker helps encapsulate the computing environment, including dependencies and software configurations, allowing others to recreate the exact environment in which the original analysis was conducted.

23.6 Future Directions for Sustainable GeoAI in Human Geography

23.6.1 Scalable and Privacy-Preserving Solutions

As the volume and variety of geospatial data continue to grow, the need for scalable GeoAI solutions that can efficiently process and analyze these data becomes increasingly important. Future GeoAI research should prioritize the development of more efficient machine learning algorithms that can handle large geospatial datasets with reduced computational costs. Techniques such as federated learning, which enables machine learning models to be trained across multiple decentralized devices or servers while keeping data local, show promise in this regard. Federated learning not only improves scalability but also enhances privacy by ensuring that sensitive geospatial data remains within its original context. Privacy-preserving techniques such as differential privacy and homomorphic encryption should also be further developed and integrated into GeoAI workflows. These techniques are crucial for ensuring that GeoAI applications do not compromise individual privacy while analyzing sensitive geospatial data. Implementing these methods at scale will be essential for gaining public trust and ensuring the ethical use of GeoAI in applications such as urban surveillance, health monitoring, and disaster response.

23.6.2 Ethical Frameworks and Explainable GeoAI

Ethical considerations will remain a major area of focus for future GeoAI research. As GeoAI is increasingly used to support decision-making in areas such as urban planning, environmental management, and social services, it is essential to develop ethical frameworks that guide the use of these technologies in an equitable and transparent manner. Ethical frameworks must address issues of algorithmic bias, fairness, and the potential for unintended consequences, particularly in contexts involving marginalized communities. Explainable AI (XAI) is a promising direction that seeks to enhance the transparency and interpretability of complex machine learning models. Future work should focus on developing explainable GeoAI models that provide meaningful insights into how decisions are made, helping stakeholders understand the factors driving model predictions. This is particularly important for applications in urban planning and public health, where transparency and accountability are key to gaining stakeholder support and ensuring equitable outcomes.

23.6.3 Community-Centered and Participatory GeoAI

To ensure that GeoAI contributes to sustainable and equitable development, future research must prioritize community-centered approaches that involve stakeholders throughout the research and implementation processes. Participatory GeoAI, which actively involves community members in data collection, analysis, and decision-making, can help ensure that GeoAI solutions are tailored to local needs and that their benefits are equitably distributed. This approach can also help mitigate the risks of algorithmic bias by incorporating local knowledge and context-specific insights that might otherwise be overlooked. Community engagement is particularly important for applications in rural and underserved areas, where the lack of adequate infrastructure and data availability can hinder the adoption of GeoAI. Future efforts should focus on developing participatory platforms that empower local communities to contribute to data collection and decision-making processes, ensuring that GeoAI technologies are inclusive and address the needs of all populations. Engaging local stakeholders can also help build trust in GeoAI technologies, increasing the likelihood of successful adoption and positive impact.

23.6.4 *Integration of Diverse Data Sources and Interdisciplinary Collaboration*

The integration of diverse data sources—such as remote sensing, social media, sensor networks, and crowd-sourced information—will be a key focus for future GeoAI research. Combining these data sources will enable more comprehensive analyses of human-environment interactions, providing richer insights into complex geographic phenomena such as migration, urban dynamics, and climate change adaptation. However, the successful integration of these diverse data types requires the development of advanced data fusion techniques that can handle varying data formats, spatial resolutions, and temporal scales. Interdisciplinary collaboration will continue to play a crucial role in advancing GeoAI research. Addressing complex challenges such as climate change, urban sustainability, and public health requires the combined expertise of geographers, data scientists, urban planners, social scientists, and policymakers. Future research should focus on building platforms and frameworks that facilitate such collaborations, allowing experts from different disciplines to work together seamlessly on shared projects and contribute their unique perspectives to GeoAI analyses.

23.6.5 *Automation and Real-Time GeoAI Analytics*

The future of GeoAI will increasingly involve automation and the ability to provide real-time insights, especially for applications in emergency response, urban mobility, and environmental monitoring. Automated GeoAI workflows that can process and analyze geospatial data in real time are essential for addressing time-sensitive challenges, such as natural disasters, disease outbreaks, and traffic management. The integration of edge computing and Internet of Things (IoT) technologies with GeoAI can further enhance the capacity for real-time data collection and analysis, enabling rapid decision-making in critical situations. Real-time GeoAI analytics will also be instrumental in enabling "smart cities," where data from various urban sensors can be continuously processed to optimize urban systems, such as energy consumption, waste management, and public safety. Future research should explore the integration of IoT networks, edge computing, and GeoAI to create responsive urban environments that can adapt dynamically to the changing needs of their residents.

References

Agustín García, P., Adegboyega, O., Edward, C., & Lukasz, P. (2020). Data acquisition and processing for GeoAI models to support sustainable agricultural practices. In *Hawaii International Conference on System Sciences*. https://doi.org/10.24251/hicss.2020.115

Bolón-Canedo, V., Morán-Fernández, L., Cancela, B., & Alonso-Betanzos, A. (2024). A review of green artificial intelligence: Towards a more sustainable future. *Neurocomputing, 599*, 128096. https://doi.org/10.1016/j.neucom.2024.128096

Campagna, M. (2005). *GIS for sustainable development*. Press. https://doi.org/10.1201/9781420037845.ch1

Casali, Y., Aydin, N. Y., & Comes, T. (2022). Machine learning for spatial analyses in urban areas: A scoping review. *Sustainable Cities and Society, 85*, 104050. https://doi.org/10.1016/j.scs.2022.104050

Christopher, D. W., Carly, S., Julie, L. D., Kirsten, J. M., William, E. C., Mariana Goodall, C., Jacob, R., Vanessa, P., Aaron, E., Thomas, M., Lauriana, C. G., Kara, S., Ann, B., David John, G., Nick, B., Andrea, S., & Christopher, T. (2024). Increasing the reproducibility and replicability of supervised AI/ML in the earth systems science by leveraging social science methods. *Earth and Space Science, 11*, e2023EA003364. https://doi.org/10.1029/2023ea003364

Dangermond, J., & Goodchild, M. F. (2019). Building geospatial infrastructure. *Geo-spatial Information Science, 23*(1), 1–9. https://doi.org/10.1080/10095020.2019.1698274

DiSalvo, C., Sengers, P., & Brynjarsdóttir, H. (2010). Mapping the landscape of sustainable HCI. In *International Conference on Human Factors in Computing Systems*. https://doi.org/10.1145/1753326.1753625

Fan, Z., Yan, Z., & Wen, S. (2023). Deep learning and artificial intelligence in sustainability: A review of SDGs, renewable energy, and environmental health. *Sustainability, 15*(18), 13493. https://doi.org/10.3390/su151813493

Fisher, D. H. (2012). Recent advances in AI for computational sustainability. *IEEE Intelligent Systems, 27*(4), 75–79. https://doi.org/10.1109/mis.2012.81

Fu, G. (2021). Geography in the Anthropocene: Transforming our world for sustainable development. *Geography and Sustainability, 3*, 1–6. https://doi.org/10.1002/sd.2162

Gao, S., Newsam, S., Zhao, L., Lunga, D., Hu, Y., Martins, B., Zhou, X., & Chen, F. (2020). GeoAI 2019 workshop report: The 3nd ACM SIGSPATIAL international workshop on GeoAI: AI for geographic knowledge discovery. *SIGSPATIAL Special, 11*(3), 23–24. https://doi.org/10.1145/3383653.3383662

Gao, S., Hu, Y., & Li, W. (2024). Introduction to geospatial artificial intelligence (GeoAI). In *Handbook of geospatial artificial intelligence* (pp. 3–16). Press. https://doi.org/10.1201/9781003308423-1

Gonzales-Inca, C., Calle, M., Croghan, D., Torabi Haghighi, A., Marttila, H., Silander, J., & Alho, P. (2022). Geospatial artificial intelligence (GeoAI) in the integrated hydrological and fluvial systems modeling: Review of current applications and trends. *Water, 14*(14), 2211. https://doi.org/10.3390/w14142211

Goodchild, M. F., Fotheringham, A. S., Kedron, P., & Li, W. (2020). Introduction: Forum on reproducibility and replicability in geography. *Annals of the American Association of Geographers, 111*(5), 1271–1274. https://doi.org/10.1080/24694452.2020.1806030

Gundersen, O. E. (2019). Standing on the feet of giants—Reproducibility in AI. *AI Magazine, 40*(4), 9–23. https://doi.org/10.1609/aimag.v40i4.5185

Haryo, L., & Pulungan, R. (2022). Smart city re-imagined: City planning and GeoAI in the age of big data. *Journal of Urban Management*. https://doi.org/10.1007/978-981-19-8069-5_12

Howe, C., & Tullis, J. A. (2022). Context for reproducibility and replicability in geospatial unmanned aircraft systems. *Remote Sensing, 14*(17), 4304. https://doi.org/10.3390/rs14174304

Hsu, C.-Y., & Li, W. (2023). Explainable GeoAI: Can saliency maps help interpret artificial intelligence's learning process? An empirical study on natural feature detection. *International Journal of Geographical Information Science, 37*(5), 963–987. https://doi.org/10.1080/13658816.2023.2191256

Hu, Y., Gao, S., Newsam, S., & Lunga, D. (2019a). GeoAI 2018 workshop report the 2nd ACM SIGSPATIAL international workshop on GeoAI. *SIGSPATIAL Special, 10*(3), 16–16. https://doi.org/10.1145/3307599.3307609

Hu, Y., Gao, S., Newsam, S., & Lunga, D. (2019b). GeoAI at ACM SIGSPATIAL. *SIGSPATIAL Special, 10*(3), 5–15. https://doi.org/10.1145/3377000.3377002

Janowicz, K. (2024). Philosophical foundations of GeoAI: Exploring sustainability, diversity, and bias in GeoAI and spatial data science. In *Handbook of geospatial artificial intelligence* (p. 17). CRC. https://doi.org/10.1201/9781003308423-3

Janowicz, K., Gao, S., McKenzie, G., Hu, Y., & Bhaduri, B. (2019). GeoAI: Spatially explicit artificial intelligence techniques for geographic knowledge discovery and beyond. *International Journal of Geographical Information Science, 34*(4), 625–636. https://doi.org/10.1080/13658816.2019.1684500

Janowicz, K., Sieber, R., & Crampton, J. (2022). GeoAI, counter-AI, and human geography: A conversation. *Dialogues in Human Geography, 12*(3), 446–458. https://doi.org/10.1177/20438206221132510

Jayant, G., Shuai, A., & Shashi, S. (2023). Towards responsible spatial data science and geo-AI. In *International Conference on Contemporary Computing*. https://doi.org/10.1145/3607947.3608060

Kedron, P., & Holler, J. (2022). Replication and the search for the laws in the geographic sciences. *Annals of GIS, 28*(1), 45–56. https://doi.org/10.1080/19475683.2022.2027011

Larsson, S. (2019). The socio-legal relevance of artificial intelligence. *Droit et Société, 103*(3), 573–593. https://doi.org/10.3917/drs1.103.0573

Li, W. (2020). GeoAI: Where machine learning and big data converge in GIScience. *Journal of Spatial Information Science, 20*, 71–77. https://doi.org/10.5311/josis.2020.20.658

Li, W., & Hsu, C.-Y. (2022). GeoAI for large-scale image analysis and machine vision: Recent Progress of artificial intelligence in geography. *ISPRS International Journal of Geo-Information, 11*(7), 385. https://doi.org/10.3390/ijgi11070385

Liu, P., & Biljecki, F. (2022). A review of spatially-explicit GeoAI applications in urban geography. *International Journal of Applied Earth Observation and Geoinformation, 112*, 102936. https://doi.org/10.1016/j.jag.2022.102936

Liu, L., & Wang, F. (2023). *Computational methods and GIS applications in social science-lab manual*. CRC.

Liu, L., Fu, X., Kötter, T., Sturm, K., Haubold, C., Guan, W. W., Bao, S., & Wang, F. (2024a). Geospatial analytics extension for KNIME. *SoftwareX, 25*, 101627.

Liu, L., Wang, F., Fu, X., Kötter, T., Sturm, K., Guan, W. W., & Bao, S. (2024b). Elevating the RRE framework for geospatial analysis with visual programming platforms: An exploration with geospatial analytics extension for KNIME. *International Journal of Applied Earth Observation and Geoinformation, 130*, 103948. https://doi.org/10.1016/j.jag.2024.103948

Marasinghe, R., Yigitcanlar, T., Mayere, S., Washington, T., & Limb, M. (2024). Towards responsible urban geospatial AI: Insights from the white and Grey literatures. *Journal of Geovisualization and Spatial Analysis, 8*(2), 24. https://doi.org/10.1007/s41651-024-00184-2

Mohammed, M. A., Ahmed, M. A., & Hacimahmud, A. V. (2023). Data-driven sustainability: Leveraging big data and machine learning to build a greener future. *Babylonian Journal of Artificial Intelligence, 2023*, 17–23. https://doi.org/10.58496/bjai/2023/005

Musgrave, M., Hallas, M., Price, R., & Pacifici, F. (2020). Accelerating sustainable development with Earth intelligence. In *IEEE International Geoscience and Remote Sensing Symposium*. https://doi.org/10.1109/igarss39084.2020.9324578

Nishant, R., Kennedy, M., & Corbett, J. (2020). Artificial intelligence for sustainability: Challenges, opportunities, and a research agenda. *International Journal of Information Management, 53*, 102104. https://doi.org/10.1016/j.ijinfomgt.2020.102104

Nüst, D., & Pebesma, E. (2020). Practical reproducibility in geography and geosciences. *Annals of the American Association of Geographers, 111*(5), 1300–1310. https://doi.org/10.1080/24694452.2020.1806028

Nüst, D., Granell, C., Hofer, B., Konkol, M., Ostermann, F. O., Sileryte, R., & Cerutti, V. (2018). Reproducible research and GIScience: an evaluation using AGILE conference papers. *PeerJournal, 6*, 2656. https://doi.org/10.7287/peerj.preprints.26561v1

Ostermann, F. O., & Granell, C. (2016). Advancing science with VGI: Reproducibility and replicability of recent studies using VGI. *Transactions in GIS, 21*(2), 224–237. https://doi.org/10.1111/tgis.12195

Persello, C., Wegner, J. D., Hansch, R., Tuia, D., Ghamisi, P., Koeva, M., & Camps-Valls, G. (2022). Deep learning and earth observation to support the sustainable development goals: Current approaches, open challenges, and future opportunities. *IEEE Geoscience and Remote Sensing Magazine, 10*(2), 172–200. https://doi.org/10.1109/mgrs.2021.3136100

Philipe, D., & Dalton, L. (2022). Embedding Ethics and Trustworthiness for Sustainable AI in Earth Sciences: Where Do We Begin? In *IEEE International Geoscience and Remote Sensing Symposium*. https://doi.org/10.1109/igarss46834.2022.9883030

Sagar, V., Siddharth, G., Hal, S., Akash, P., Shubham, G., Shweta, P., Natanael, E., Ujwal, D., Tanka Raj, J., & Nitin, B. (2022). GeoEngine: A platform for production-ready geospatial research. In *Computer vision and pattern recognition*. IEEE. https://doi.org/10.1109/cvpr52688.2022.02073

Sahana, K. S., Karthik, C. B., Sree, M. M., Chaithra, M., Abhinandithe, S. K., Snehalatha, B., Sridhar, R., Balasubramanian, S., & Madhu, B. (2023). Geospatial artificial intelligence (GeoAI): Applications in health care. *International Journal of Health and Allied Sciences, 11*(4), 4. https://doi.org/10.55691/2278-344x.1044

Shi, M., Currier, K., Liu, Z., Janowicz, K., Wiedemann, N., Verstegen, J., McKenzie, G., Graser, A., Zhu, R., & Ma, G. (2023). Thinking geographically about AI sustainability. *AGILE: GIScience Series, 4*, 1–7. https://doi.org/10.5194/agile-giss-4-42-2023

Sk, M., & Sudhir Kumar, S. (2021). Geo-intelligence role in sustainable city missions of the global south: A review. In *Geo-intelligence for sustainable development* (pp. 1–16). Springer. https://doi.org/10.1007/978-981-16-4768-0_1

Tullis, J. A., & Kar, B. (2020). Where is the provenance? Ethical replicability and reproducibility in GIScience and its critical applications. *Annals of the American Association of Geographers, 111*(5), 1318–1328. https://doi.org/10.1080/24694452.2020.1806029

VoPham, T., Hart, J. E., Laden, F., & Chiang, Y.-Y. (2018). Emerging trends in geospatial artificial intelligence (geoAI): Potential applications for environmental epidemiology. *Environmental Health, 17*(1), 40. https://doi.org/10.1186/s12940-018-0386-x

Wainwright, J. (2020). Is critical human geography research replicable? *Annals of the American Association of Geographers, 111*(5), 1284–1290. https://doi.org/10.1080/24694452.2020.1806025

Wang, S., Huang, X., Liu, P., Zhang, M., Biljecki, F., Hu, T., Fu, X., Liu, L., Liu, X., & Wang, R. (2024). Mapping the landscape and roadmap of geospatial artificial intelligence (GeoAI) in quantitative human geography: An extensive systematic review. *International Journal of Applied Earth Observation and Geoinformation, 128*, 103734. https://doi.org/10.1016/j.jag.2024.103734

Waters, N. (2020). Motivations and methods for replication in geography: Working with data streams. *Annals of the American Association of Geographers, 111*(5), 1291–1299. https://doi.org/10.1080/24694452.2020.1806027

Wilson, J. P., Butler, K., Gao, S., Hu, Y., Li, W., & Wright, D. J. (2020). A five-star guide for achieving replicability and reproducibility when working with GIS software and algorithms. *Annals of the American Association of Geographers, 111*(5), 1311–1317. https://doi.org/10.1080/24694452.2020.1806026

Wynsberghe, A. V. (2021). Sustainable AI: AI for sustainability and the sustainability of AI. *AI and Ethics, 1*(3), 213–218. https://doi.org/10.1007/s43681-021-00043-6

Yang, Y., Wu, Y., & Brian, U. (2020). Estimating the deep replicability of scientific findings using human and artificial intelligence. *Proceedings of the National Academy of Sciences of the United States of America, 117*(20), 10762–10768. https://doi.org/10.1073/pnas.1909046117

Yang, L., Driscol, J., Sarigai, S., Wu, Q., Chen, H., & Lippitt, C. D. (2022). Google earth engine and artificial intelligence (AI): A comprehensive review. *Remote Sensing, 14*(14), 3253. https://doi.org/10.3390/rs14143253

Zhang, Y., Zhang, F., & Chen, N. (2022). Migratable urban street scene sensing method based on vision language pre-trained model. *International Journal of Applied Earth Observation and Geoinformation, 113*, 102989. https://doi.org/10.1016/j.jag.2022.102989

Chapter 24
The Knowledge Politics of GeoAI

Isaac Oluoch

Abstract GeoAI is increasingly deployed in the humanitarian mapping of vulnerable communities in the Global South. Specifically in the mapping of 'slums' and informal settlements, as well as for disaster risk management (DRM). GeoAI presents several opportunities for improving the visibility of the communities, revealing the precarity they experience living in such areas. But it is also important to consider the more ethical and existential consequences that can arise from these mapping initiatives. This chapter will begin by tracing the development of cartography and its socio-political influence, drawing on insights from critical cartography and critical geography. These insights help emphasise the fact that maps and map-making are socially and politically embedded and not purely technical or objective in representing the world. This will also be a pertinent and instructive starting point for taking a more critical stance on GeoAI. This chapter will present an ethical examination of the humanitarian use of AI, addressing issues such as accountability, bias, political representation, and the potential inequalities in the creation and distribution of geospatial knowledge in the new wave of 'digital humanitarianism' (Fejerskov et al., Disasters 48(2):e12609, 2024).

Keywords GeoAI · Knowledge politics · Geoethics · Digital humanitarianism · Spatial decision-making

24.1 Introduction

> As a memetic picture of the world, the map becomes the site of debate about the world. When there are discrepancies between the map and the world, perhaps the world itself needs to be corrected.—Mathew H. Edney (2019).

I. Oluoch (✉)
Department of Behavioural Management and Social Sciences, University of Twente, Enschede, The Netherlands
e-mail: i.o.oluoch@utwente.nl

© The Author(s), under exclusive license to Springer Nature Switzerland AG 2025
X. Huang et al. (eds.), *GeoAI and Human Geography*, Springer Geography,
https://doi.org/10.1007/978-3-031-87421-5_24

Mathew Edney's (2019) assertion illustrates maps' role in informing how the world is represented. On the surface, maps appear as objective representations of the world. This conception assumes that the map is a transparent mediator between human perception and reality (Winichakul, 1997). On the other hand, referring to the map as a 'mimetic picture' as Edney does reflects the fact that what is represented on a printed paper or through visualisation software (e.g. using ArcGIS or Google Maps) is an abstraction of the world. Depending on the spatial data acquired, techniques and scale used, as well as audience, this abstraction can be very simple or very complex. For this reason, the disciplines of cartography, geography and geographic information science put heavy focus on the technical aspects of map production.

However, the maps produced by these disciplines do not only exist in a professional or technical vacuum—they play a very important role *in* the world. Wood (2010) refers to maps as 'systems of propositions, where a proposition is nothing more than a statement that affirms (or denies) the existence of something' (Wood, 2010, p. 34). But who is responsible for making these propositions? How are they given authority? Who is allowed to argue *against* these claims? Such questions highlight the necessity of thinking of maps as 'the site of debate about the world' as Edney points out. And this site of debate is complicated following the advancements and implementation of artificial intelligence (AI) in geospatial knowledge production under the nascent domain of GeoAI.

The growth of GeoAI is made possible, according to Hu et al. (2019), through three related factors. These factors are: (1) the growing volume of geospatial data for training complex AI models; (2) the development of novel AI models specifically for dealing with geographic problems, and (3) the greater availability of high-performance computer hardware to more efficiently train AI models. Scheider and Richter (2023) point out that these factors are leading to the partial replacement of more traditional approaches, as geoscientists 'are beginning to put their trust into generalised, automated learning to accomplish tasks they previously tackled with tailored methods' (Scheider & Richter, 2023, p. 17). This increased trust in AI methods is attractive to geoscientists given that 'algorithms exhibit exceptional proficiency in uncovering patterns and trends within extensive geospatial datasets' with use cases including profiling individuals and prediction of future events based on historical geospatial data (Lin, 2024).

At the same time, this trust and overreliance also carry certain negative consequences, which this chapter will explore. In the first section, the critique of maps from scholarship in critical cartography and critical geography will be presented to illustrate how map production is embedded in socio-political relations. Following this, I will focus on the growing turn to AI-assisted humanitarianism, specifically in the mapping of 'slums' and informal settlements, along with disaster risk management (DRM). Especially analysing the role of AI in the rise of 'digital humanitarianism' (Fejerskov et al., 2024). This analysis will show the importance of focusing not only on the technical promises of AI-assisted mapping initiatives but also the implicit 'knowledge politics' of these initiatives, i.e. the negotiation and contestation for how geospatial knowledge is represented (Burns, 2014). In the third section, this chapter will also assess the deontological basis for the ethical values and principles that could be used in the development of GeoAI for humanitarianism.

24.2 From Critical Geography and Cartography to Critical GeoAI

An important starting point is recognising the layered process involved in the construction of maps and in the acquisition and distribution of geospatial knowledge. Taking a historical view, scholars within critical cartography and critical geography highlight how the growth of cartography and map production was tied to state expansionism (Scott, 1998; Anderson, 2006; Biggs, 1999; Winichakul, 1997; Pickles, 2004; Wood, 2010). An expansionism that revealed the map's ability to create systems of 'totalizing classification', squaring off regions of the Earth's surface into neatly 'measured boxes' (Anderson, 2006, p. 173). Such carving out of the Earth's surface reflects what Pickles (2004) refers to as the 'cartographic gaze', which privileges 'a particular form of seeing (distanced, objective and penetrating), predicated on an epistemology and politics of mastery and control of the earth, nature and subjects' (Pickles, 2004, p. 83).

These observations paved the way for reconceptualising the map, the role of the cartographer and the very act of mapping. Looking at a map therefore involves considering the choices made in what is included on the map (and by extent what is excluded) as well as who is involved in the map's production. This is important to consider given the fact that maps have an authority—an authority that appears absolute—over the territory they represent. Krygier and Wood (2010) highlight this by describing the statements maps make as 'existential propositions', e.g. 'this is *there*' (Krygier & Wood, 2010, p. 198). Such propositions may be taken for granted in everyday life, for instance, when looking for a new restaurant or site seeing when travelling to a new city. But in more sensitive use cases, such as predictive policing or humanitarian mapping, such propositions can change how entire neighbourhoods and individuals within those neighbourhoods are seen and treated (Brayne et al., 2015; Oluoch et al., 2022). Once these propositions are repeated enough times, they solidify into facts (Wood, 2010, p. 34). Consequently, the construction and use of maps is not a purely technical concern, instead being heavily implicated within socio-political relations that necessitate ethical consideration.

Turning to the current state of map production, given the greater demand and availability of geospatial data, researchers have turned towards 'data-driven geography, which relies on AI and machine learning to enable the discovery of new geospatial knowledge' (Li et al., 2022, p. 2). Just as we need to be critical of the design, use and distribution of maps, it is also necessary to be critical of the design, use and distribution of the algorithms and datasets used in the creation of this new geospatial knowledge. Such concern is increasingly pertinent given that issues such as 'privacy, data bias, algorithmic transparency, and the responsible use of AI in decision-making processes become paramount' the more sensitive the geospatial information is (Petrochi et al., 2023, p. 285). A close look at the AI-assisted humanitarian mapping of 'slums' and informal settlements along with DRM will help illustrate the positive potential and negative consequences of the turn to AI in geographic research.

24.3 AI-Assisted Humanitarianism

24.3.1 'Slum' and Informal Settlement Mapping

Using the classification of UN-Habitat (2018), 'slum households' are characterised by five indicators: (1) lack of security of tenure, (2) lack of adequate water, (3) lack of access to sanitation, (4) inadequate structural quality of housing and (5) overcrowding within houses. I am using quotation marks around the word 'slum' because this is not a universal category—the morphological, socioeconomic and semantic description of these areas is highly heterogeneous (Mahabir et al., 2016). Thus, while there are some communities and even organisations such as Slum Dwellers International that use the word explicitly, some other communities and countries have their own terms and classification schemas (e.g. favelas in Brazil or townships in South Africa). Informal settlements are characterised by residents having the tenure status of squatters or informal rental agreements, the lack of formal basic services, and houses in proximity to hazardous areas (UN-Habitat, 2018, p. 5). In some cases, these two terms are used interchangeably, but due to the semantic sensitivity involved, I will use them separately throughout this chapter.

The traditional method of keeping account of the populations living in 'slums' and informal settlements are censuses and household surveys. But due to the rapid growth of populations living in these areas, these traditional methods are inadequate. Innovation in computational hardware and software over the past two decades has allowed researchers, non-governmental organizations (NGOs) and geographic information producing companies (e.g. Esri) to assist in improving the legibility of these cities. These initiatives are useful for several reasons, as illustrated in Fig. 24.1.

First, the mapping of these areas is performed by geo-information scientists within the fields of Earth Observation (EO) and Remote Sensing (RS), using satellite imagery, drones, GPS devices, mobile phones and geovisualization software. There are also private companies (e.g. Google and Esri), international institutions (e.g. World Bank) and NGOs (e.g. the Red Cross) performing this mapping. This research aims to improve the methods and techniques of classification and representation of 'slums' and informal settlements (Abascal et al., 2022; Ajami et al., 2019; Kohli et al., 2016; Kuffer et al., 2017, 2018; Wang, 2019). Second, from this research, a better understanding of the morphological and socioeconomic characteristics of these areas is developed (Corburn et al., 2020; Merodio Gómez et al., 2021). Thirdly, as many 'slums' and informal settlements are also found in areas prone

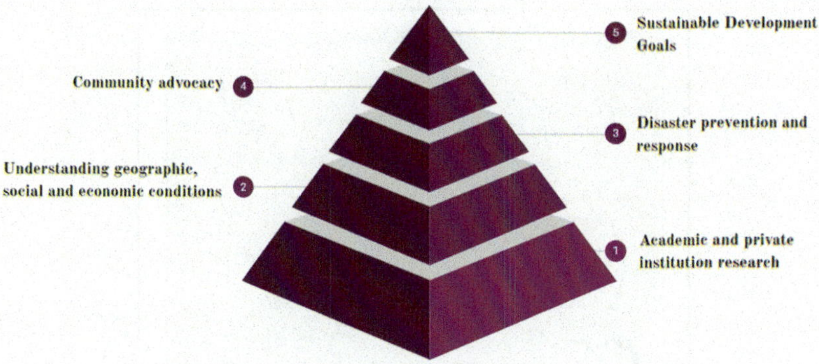

Fig. 24.1 Overview of uses of mapping 'slums' and informal settlements

to natural hazards, mapping these areas also serves for DRM initiatives. Fourthly, beyond geographic representation, the mapping of these areas can also be a means of improving political representation through community advocacy. Many of the communities living in these areas experience precarious living conditions and are often not treated as ordinary citizens, historically experiencing discrimination and stigmatisation (Brito et al., 2020; Owusu et al., 2021; Oluoch, 2022). Increasing the visibility of these communities can serve as proof of their existence (e.g., time-series imagery showing historical presence), which can help fight against the threat of forced evictions. Finally, the mapping of these communities is helps with monitoring the Sustainable Development Goals (SDGs), as these communities have inadequate standards of housing (SDG 11), inadequate access to water and sanitation (SDG 6), high rates of poverty (SDG 1), poor access to healthcare (SDG 3) as well as living in areas more exposed to adverse effects of climate change (SDG 13).

A growing area of research is the use of AI to map these areas (e.g. using Support Vector Machines as well as Deep Learning). These AI-based methods allow the transformation of geospatial data 'into different layers of abstraction which are then used for prediction and data presentation tools' (Verma et al., 2019, p. 2). An example of this can be seen in Fig. 24.2a, b from a study performed by Fallatah et al. (2022) mapping informal settlements in Jeddah, Saudi Arabia. The researchers used object-based classification from VHR imagery of the study area, using several indicators (e.g. size of dwellings, roof extent, road accessibility, building density and temporal development). The temporal analysis of the study area utilised Landsat TM and ETM+ data, enabling the researchers to study the growth of the informal settlements from 1984 to 2018.

While traditional mapping consisted of the cartographer performing much of the hard labour in the synthesis of geospatial information to craft a map, machine learning-based mapping introduces a new element. Namely, algorithms and training models perform their 'own interpretive tasks, categorizing and suggesting' predic-

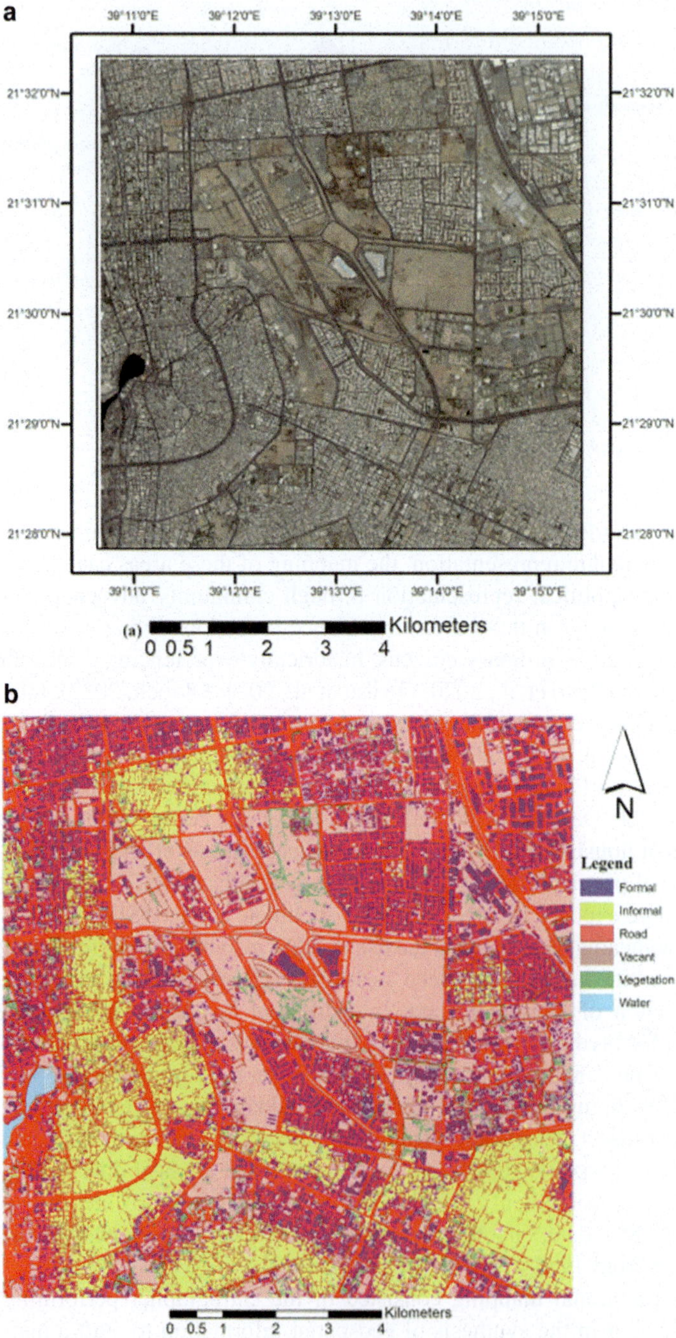

Fig. 24.2 (**a**): (Above) Case study area investigated; (**b**) (Below) The results of the machine learning workflow from object-based classification with time-series analysis. (Source: Fallatah et al., 2022)

tions (Hongladarom, 2020, p. 5). These algorithms are relied on to classify and represent the 'slums' and informal settlements being mapped. Figure 24.2b shows a very distinct delineation of where the informal settlements are in the study area, making them visible where in 2(a) they are invisible. The delineation of where these informal settlements are (and where they aren't) depends on a few social and technical factors. Factors including: (1) the quality of spatial data used (i.e. whether the data is outdated or of low resolution); (2) the validation of the spatial data (i.e. whether ground truth information such as historical maps are used for training and validation); (3) awareness of the potential biases and uncertainty within the mapping process (i.e. indicators used by researchers may lead to mapping larger settlements but may fail to capture smaller settlements due to inability to identify indicators the model is trained on). These factors reflect not only what kind of technology is used (e.g. image resolution and scale of mapping) but also issues of inclusion and accessibility.

The issue of inclusion and accessibility is important because researchers or private companies that produce maps of these areas do not always have direct contact with the communities being mapped. There are outliers to this (Gevaert et al., 2016; Thomson et al., 2020; Kuffer et al., 2021; Owusu et al., 2021). But in the instances where geospatial data is taken from these areas without any contact, this leads 'to a condition of ever-increasing legibility without better political representation' (Taylor, 2016, p. 5). Legibility is tied to mapping which communities are the most vulnerable and most reliant on aid. But if this mapping is done with the mindset of communities being nothing more than 'data subjects' and not 'data citizens' (Taylor, 2016), they may have improved geographical representation but no improvement in political representation. As a result, they will be continually left out of conversations of how the mapping process is conducted, as well as being unable to critique or refuse being represented by researchers and private companies (thus bringing issues regarding transparency, consent as well as the data rights of these communities).

24.3.2 Disaster Risk Management (DRM)

Disaster risk management (DRM) is the process of identifying and responding to the risks arising from natural as well as man-made disasters (Ghaffarian et al., 2023). These efforts help improving preparedness along with capacity building, through 'mapping populations and infrastructures exposed to disasters, identifying areas at risk, and planning emergency responses' (Gevaert et al., 2021, p. 1). Like 'slum' mapping, DRM relies on technologies including satellites, drones and GPS devices. An example of DRM mapping can be seen in Fig. 24.3 below, a snapshot of the impact of flooding that occurred in the Northwest of Nairobi County, Kenya in May

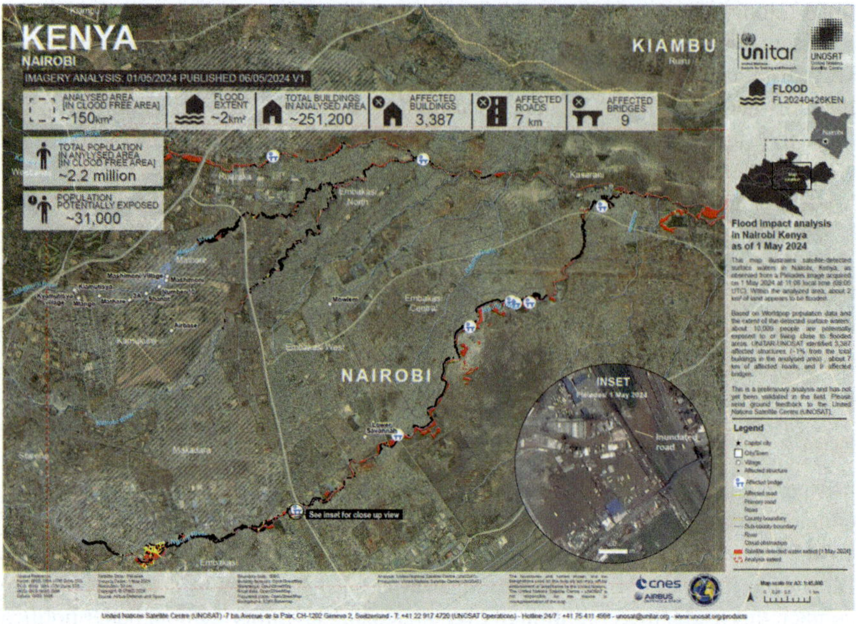

Fig. 24.3 Flood map of Nairobi County, Kenya. The red areas represent the extent of the flooding. The blue icons represent affected bridges in the study area. The inset area in the circle shows a close-up of a region in the south of the study area. (Source: UNOSAT, 2024)

2024, by the United Nations Satellite Centre (UNOSAT).[1] The snapshot utilised Pleiadas satellite data at 50 cm resolution, using waterways, road and building footprint information from OpenStreetMap. The snapshot identified 3387 affected structures, roughly 7 km of affected roads and 9 affected bridges (UNOSAT, 2024).

Researchers, governmental and non-governmental organizations are turning to AI in DRM due to the multi-modal as well as predictive capabilities that AI provides. As Velev and Zlateva (2023) point out, techniques such as image recognition and classification can help with extracting features in images such as the level of damage on buildings and the number of blocked roads. Additionally, AI for DRM is useful for long-term planning, such as helping in climate modelling to 'identify areas at risk of sea-level rise and inform coastal zone management plans' (Ghaffarian et al., 2023, p. 2). At the same time, the increased reliance on AI for DRM represents the growing trend towards 'digital humanitarianism' (Burns, 2014; Van den Homberg et al., 2020; Fejerskov et al., 2024), where mass-collaboration technologies are becoming integrated in humanitarian practices (Burns, 2014). As AI enhances the ability of humanitarian actors to collect, manage and represent larger volumes of geospatial information, 'quantification has become an essential

[1] https://unosat.org/products/3834

component of contemporary humanitarianism' (Fejerskov et al., 2024, p. 3). Quantification here refers to the use of numbers to simplify social phenomena, reducing the complexity of humanitarian emergencies (ibid). Consequently, quantification leads to the effacement of what Burns (2014) refers to as the implicit knowledge politics involved in digital humanitarianism. This knowledge politics is underscored by the struggle not simply for how geospatial knowledge is made visible but also the 'struggle for control of the *terms* of visibility' (Burns, 2014, p. 53).

Looking at the socio-political and ethical challenges posed by the turn to AI within DRM will help illustrate this implicit knowledge politics. Soden et al. (2021) identify several biases from the use of AI in humanitarian initiatives, but due to space constraints I will only focus on a few. First, representation bias can occur when models aren't truly representative of the problem that needs solving, leading to issues related to lower model transferability and lower predictive power. Second, measurement biases can occur when predictors for certain hazards are not included in the training of a model, leading to the model looking for the wrong indicators. For example, in predicting storm damages, 'the model predicts wind speed while decision makers need wind damage' (Soden et al., 2021, p. 13). Third, aggregation biases can also occur when the aggregation of large volumes of data results in useful information being lost. This presents an instance of what Fejerskov et al. (2024) refers to as non-knowledge or the ignorance and uncertainty that increases from: (i) greater reliance on digitalisation and datafication; and (ii) the expansion of actors involved in the collection and distribution of geospatial information within DRM. For example, due to the technical demands of algorithmically interpreting the data, the interpretation of this data is performed by researchers or private companies, while the people providing the data (e.g. local actors within a disaster area) have no input in how the data is classified or processed (Fejerskov et al., 2024, p. 10). This issue is also tied to questions of inclusion, transparency and accountability, as the 'associated complexity in disaster risk modelling processes may reduce the ability of government, the public, and other important stakeholders to meaningfully participate in DRM' (Gevaert et al., 2021).

The growing range of actors involved in the DRM space therefore complicates the efficacy of AI efforts quite significantly, such as the increased presence of private entrepreneurs and corporate philanthropists. If their aid is based on monetary renumeration, this can result in 'giving access to vast amounts of data [on these vulnerable groups] to actors with non-humanitarian intentions' (Van den Homberg et al., 2020, p. 459). This is especially problematic when the data gathered shows sensitive information, which may not be known by these actors as there is too much data to sort through, leading to ignorance of what data *should* be looked for (or left out). This ignorance is partially due to the aggregation bias as above mentioned. But it is also the product of the 'power bias in deciding what is to be mapped', as humanitarian actors from high-income countries (HICs) define what information should be extracted (Gevaert et al., 2021, p. 4). These actors 'may overlook important local norms and contexts [in LMICs], which are crucial for solving complex social issues' (ibid). This power bias is emblematic of a new form of data colonialism where the 'contemporary production and collection of humanitarian data [is linked] to the predatory extractive practices of historical colonialism' (Fejerskov et al., 2024, p. 10). Practices that minimise the involvement of local actors as well as the communities within the study area, and perpetuate historical inequalities.

24.4 Ethical Principles and Considerations for GeoAI

Looking at the ethical concerns within the AI-assisted humanitarian mapping of vulnerable communities is already an ongoing endeavour (Gevaert et al., 2018; Micheli et al., 2022; Gevaert, 2022; Kochupillai et al., 2022). Table 24.1 below lists 19 ethical values and principles from a review of 16 ethical guidelines on AI and 8 ethical guidelines on geographic information technologies (Oluoch, 2024). These values and principles highlight the range of issues that need to be considered in critically analysing GeoAI, especially if the geospatial knowledge produced has the potential to adversely impact certain individuals and groups.

In the lexicon of applied ethics, these values and principles follow a deontological basis, rooted in the work of Immanuel Kant. Kant's moral philosophy is based on a 'theory of human obligations' which can be used as 'the starting points for moral reasoning in actual contexts of action' (O'Neill, 2013, p. 510). This form of reasoning is best expressed by Kant's categorical imperative which states, in two parts (Kant, 2013, italics in original):

1. '*Act only in accordance with that maxim through which you can at the same time will that it become a universal law*';
2. '*Act as if the maxim of your action were to become by your will a universal law of nature.*'

In the context of AI-assisted humanitarian mapping, the underlying maxim orienting these initiatives is improving the living conditions of the vulnerable. This is reflected in the fact that most EO and RS literature begins by stressing the impoverished conditions of those living in 'slums' and informal settlements as well as disaster-prone areas, emphasising how the precarity of these communities is a global phenomenon that needs to be addressed. And one way to attend to these conditions is the development of more robust mapping tools to enable better morphological and sociodemographic understanding, which can help develop greater capacity building. And following this categorical imperative is a practical imperative, which states (Kant, 2013, p. 496, italics in original):

> *So act that you use humanity, whether in your own person or in the person of any other, always at the same time as an end, never merely as a means.*

The notion of treating humanity as 'an end, never merely as a means' refers to the fact that the autonomy of a person should always be respected (O'Neil, 2013, p. 213). This means that people should not be coerced or manipulated into any agreements or situations. As a product of this respect to autonomy, it would also

Table 24.1 Ethical values and principles for GeoAI

Values/principles	Description
Access	Ability to access spatial data, algorithms and geo-information (linked to inclusion)
Accountability	Related to responsibility, in finding who is to blame when use of geo-information technologies leads to harms
Autonomy	Respecting individuals and groups to make decisions for themselves
Beneficence	Promoting wellbeing
Bias	Related to inaccuracies in data, algorithms and geo-information technology use
Consent	Related to rights and inclusion, and reflects whether individuals and groups give their permission to providing their data or data about them being taken
Dignity	Related to autonomy, as the respect for individuals and groups
Discrimination	Individuals and groups should not have their rights or persons harmed by geo-information technologies based on their race, gender or location
Explainability/interpretability	Especially related to using AI, decisions based on geo-information technologies should be clear to those affected by these decisions
Fairness	Related to discrimination, access and justice, geo-information technologies should not increase inequalities experienced by certain individuals and groups
Justice	Upholding the rights of individuals and groups
Harm/non-maleficence	Geo-information technologies should not endanger individuals and groups
Inclusion	Related to access, accountability, consent and fairness, the design and use of geo-information technologies should be inclusionary especially of marginalised groups
Privacy	Related to consent, justice and rights, geo-information technologies should not be invasive of the lives of individuals and groups being mapped
Responsibility	Related to accountability, those using geo-information technologies need to be aware of the impacts of these technologies and potential harms
Rights	Related to justice, consent and privacy, individuals and groups have certain rights that should not be infringed upon by the use of geo-information technologies
Risks	Related to accountability, beneficence, harms and rights, the use of geo-information technologies presents many risks socially, politically and ethically
Transparency	Related to explainability/interpretability and the use of AI, decisions stemming from geo-information technologies should be made as clear as possible to those affected by these decisions
Trust	Those using geo-information technologies should aim to ensure these technologies are explainable and transparent, with minimal bias and discrimination, to increase trust

be sensible then, to consider the need for respecting the dignity, rights, consent and privacy of individuals and groups. Since without respecting the latter, it is easy to make the case that the autonomy of communities being mapped is not being respected either. Such deontological framing, though simplified here as there are several nuances to deontological ethical theory, points towards just actions being based on respect for the autonomy of others, especially the vulnerable. This notion is reflected in principles such as 'Do no harm', 'Minimise intrusion', and 'Protect the Vulnerable' from the Locus Charter (EthicalGeo & Benchmark, 2021). When such principles are not followed and the communities being mapped are not respected, this creates further potential for data colonialism, reducing vulnerable communities to being only data subjects instead of potential data citizens. Data citizens that could become more included and respected in humanitarian mapping practices.

However, one of the problems of such deontological obligations, especially if they only remain at the level of prescriptive values and principles, is that it is always possible for them to be ignored. This is because these obligations are not law-binding (Resseguier & Rodriguez, 2020; Hagendorff, 2022). It may therefore be necessary for discussions on the ethical concerns related to GeoAI to be raised to the level of legal discourse, similar to the legal discourse surrounding Big Data and AI, which produced the GDPR and AI Act in the European Union. Such discourse would help ensure that the values and principles raised by ethically minded researchers, vulnerable communities and NGOs do not fall on deaf ears. Such legal mechanisms would be particularly important in the context where humanitarian actors from the Global North (where such regulations may be present) conduct research in the Global South (where such regulations may not be present). And due to the disparate regulatory environment, these actors face no legal obligation to act in a just manner in relation to respecting the communities being mapped, as there will be negligible to no repercussions.

Bearing this in mind, it may also be necessary to consider the implementation or embedding of these values and principles in the very design of GeoAI, from the initial spatial data acquisition to the final map product. Such embedding would follow similar work in the ethical design of AI (Johnson, 2017; Floridi, 2019; Hagendorff, 2020). Thomson et al. (2020), Kuffer et al. (2020) and Owusu et al. (2021) provide potential avenues for a more ethically mindful model of humanitarian mapping with AI. Based on close contact with the communities being mapped to ascertain their requirements and recommendations. Along with paying attention to issues such as accessibility and how user-friendly the mapping tools are and observing whether certain ethical issues may arise and how to deal with them. For instance, considering the level of privacy invasion in relation to the resolution of the imagery used for mapping. Or deciding whether to use geographic masks or random perturbation to relocate points in the final map (Swanlund et al., 2020), to ensure that the communities mapped do not face harm from being easily found by malicious actors.

24.5 Conclusion

The use of GeoAI in humanitarian mapping presents a range of opportunities for improving the classification, scalability and transferability of algorithms and models. While at the same time, due to the range of actors involved in the rising wave of digital humanitarianism, there is a growing need to be more critical about GeoAI within humanitarian mapping. This chapter aimed to illustrate, through considering maps and map-making as socio-politically embedded, that the development of GeoAI is not occurring within a technical or professional vacuum. It is therefore necessary to outline what kind of obligations researchers and companies may need to comply with (whether ethical or legal) to reduce the potential harms that may be suffered, especially when the new geospatial knowledge created by GeoAI can impact the lives of vulnerable communities.

References

Abascal, A., Rothwell, N., Shonowo, A., Thomson, D. R., Elias, P., Elsey, H., Yeboah, G., & Kuffer, M. (2022). "Domains of deprivation framework" for mapping slums, informal settlements, and other deprived areas in LMICs to improve urban planning and policy: A scoping review. *Computers, Environment and Urban Systems, 93*, 101770. https://doi.org/10.1016/j.compenvurbsys.2022.101770

Ajami, A., Kuffer, M., Persello, C., & Pfeffer, K. (2019). Identifying a slums' degree of deprivation from VHR images using convolutional neural networks. *Remote Sensing, 11*(11), 1282.

Anderson, B. (2006). *Imagined communities: Reflections on the origin and spread of nationalism*. Verso Books.

Benchmark; EthicalGeo. (2021). *Locus Charter*. Retrieved November 20, 2023, from https://ethicalgeo.org/wp-content/uploads/2021/03/Locus_Charter_March21.pdf

Biggs, M. (1999). Putting the state on the map: Cartography, territory and European state formation. *Comparative Studies in Society and History: an International Quarterly, 41*, 374–405.

Brayne, S., Rosenblat, A., & Boyd, D. (2015). Predictive policing. *Data & Civil Rights: A New Era of Policing and Justice*, 2015–1027.

Brito, P. L., Kuffer, M., Koeva, M., Pedrassoli, J. C., Wang, J., Costa, F., & de Freitas, A. D. (2020). The spatial dimension of COVID-19: The potential of earth observation data in support of slum communities with evidence from Brazil. *ISPRS International Journal of Geo-Information, 9*, 557. https://doi.org/10.3390/ijgi9090557

Burns, R. (2014). Moments of closure in the knowledge politics of digital humanitarianism. *Geoforum, 53*, 51–62.

Corburn, J., Vlahov, D., Mberu, B., Riley, L., Caiaffa, W. T., Rashid, S. F., Ko, A., Patel, S., Jukur, S., Martíez-Herrera, E., Jayasinghe, S., Agarwal, S., Nguendo-Yongsi, B., Weru, J., Ouma, S., Edmundo, K., Oni, T., & Ayad, H. (2020). Slum health: Arresting COVID-19 and improving well-being in urban informal settlements. *Journal of Urban Health, 97*, 348–357. https://doi.org/10.1007/s11524-020-00438-6

Edney, M. H. (2019). *Cartography: The ideal and its history*. University of Chicago Press.

Fallatah, A., Jones, S., Wallace, L., & Mitchell, D. (2022). Combining object-based machine learning with long-term time-series analysis for informal settlement identification. *Remote Sensing, 14*(5), 1226. MDPI AG. Retrieved from http://dx.doi.org/10.3390/rs14051226

Fejerskov, A. M., Clausen, M. L., & Seddig, S. (2024). Humanitarian ignorance: Towards a new paradigm of non-knowledge in digital humanitarianism. *Disasters, 48*(2), e12609.
Floridi, L. (2019). Translating principles into practices of digital ethics: Five risks of being unethical. *Philosophy & Technology, 32*(2), 185–193.
Gevaert, C., Sliuzas, R., Persello, C., & Vosselman, G. (2016). Opportunities for UAV mapping to support unplanned settlement upgrading. *Rwanda Journal, 1.*
Gevaert, C. M. (2022). Explainable AI for earth observation: A review including societal and regulatory perspectives. *International Journal of Applied Earth Observation and Geoinformation, 112*, 102869. https://doi.org/10.1016/j.jag.2022.102869
Gevaert, C. M., Sliuzas, R., Persello, C., & Vosselman, G. (2018). Evaluating the societal impact of using drones to support urban upgrading projects. *ISPRS International Journal of Geo-Information, 7*(3), 91. https://doi.org/10.3390/ijgi7030091
Gevaert, C. M., Carman, M., Rosman, B., Georgiadou, Y., & Soden, R. (2021). Fairness and accountability of AI in disaster risk management: Opportunities and challenges. *Patterns, 2*(11), 1–9. https://doi.org/10.1016/j.patter.2021.100363
Ghaffarian, S., Taghikhah, F. R., & Maier, H. R. (2023). Explainable artificial intelligence in disaster risk management: Achievements and prospective futures. *International Journal of Disaster Risk Reduction, 98*, 104123.
Hagendorff, T. (2020). The ethics of AI ethics: An evaluation of guidelines. *Minds and Machines, 30*(1), 99–120.
Hagendorff, T. (2022). A virtue-based framework to support putting AI ethics into practice. *Philosophy & Technology, 35*(3), 55.
Hongladaron, S. (2020). Machine hermeneutics, postphenomenology, and facial recognition technology. *AI & Society, 38*(6), 2151. https://doi.org/10.1007/s00146-020-00951-x
Hu, Y., Gao, S., Lunga, D., Li, W., Newsam, S., & Bhaduri, B. (2019). GeoAI at ACM SIGSPATIAL: Progress, challenges, and future directions. *Sigspatial Special, 11*(2), 5–15.
Johnson, D. G. (2017). Can engineering ethics be taught? *The Bridge, 47*(1), 59–64.
Kant, I. (2013). Groundwork of the metaphysics of morals. In R. Shafer-Landau (Ed.), *Ethical theory: An anthology* (2nd ed.). Wiley-Blackwell.
Kochupillai, M., Kahl, M., Schmitt, M., Taubenbock, H., & Zhu, X. X. (2022). Earth observation and artificial intelligence: Understanding emerging ethical issues and opportunities. *IEEE Geoscience and Remote Sensing Magazine, 10*, 90–124. https://doi.org/10.1109/mgrs.2022.3208357
Kohli, D., Stein, A., & Sliuzas, R. (2016). Uncertainty analysis for image interpretations of urban slums. *Computers, Environment and Urban Systems, 60*, 37–49.
Krygier, J., & Wood, D. (2010). Ce n'est pas le monde (this is not the world). In M. Dodge, R. Kitchin, & C. Perkins (Eds.), *Rethinking maps* (pp. 189–220). Essay, Routledge.
Kuffer, M., Pfeffer, K., Sliuzas, R., Baud, I., & Maarseveen, M. V. (2017). Capturing the diversity of deprived areas with image-based features: The case of Mumbai. *Remote Sensing, 9*(4), 384.
Kuffer, M., Wang, J., Nagenborg, M., Pfeffer, K., Kohli, D., Sliuzas, R., & Persello, C. (2018). The scope of earth-observation to improve the consistency of the SDG slum indicator. *ISPRS International Journal of Geo-Information, 7*(11), 428.
Kuffer, M., Thomson, D. R., Boo, G., Mahabir, R., Grippa, T., Vanhuysse, S., ... & Kabaria, C. (2020). The role of earth observation in an integrated deprived area mapping "system" for low-to-middle income countries. *Remote Sensing, 12*(6), 982.
Kuffer, M., Wang, J., Thomson, D. R., Georganos, S., Abascal, A., Owusu, M., & Vanhuysse, S. (2021). Spatial information gaps on deprived urban areas (slums) in low-and-middle-income-countries: A user-centered approach. *Urban Science, 5*(4), 72.
Li, W., & Hsu, C. Y. (2022). GeoAI for large-scale image analysis and machine vision: Recent progress of artificial intelligence in geography. *ISPRS International Journal of Geo-Information, 11*(7), 385.
Lin, Y. (2024). Moving beyond anonymity: Embracing a collective approach to location privacy in data-intensive geospatial analytics. *Environment and Planning F, 3*(1–2), 45–63. https://doi.org/10.1177/26349825231224029

Mahabir, R., Crooks, A., Croitoru, A., & Agouris, P. (2016). The study of slums as social and physical constructs: Challenges and emerging research opportunities. *Regional Studies, Regional Science, 3*(1), 399–419.

Merodio Gómez, P., Juarez Carrillo, O. J., Kuffer, M., Thomson, D. R., Olarte Quiroz, J. L., Villaseñor García, E., Vanhuysse, S., Abascal, Á., Oluoch, I., Nagenborg, M., Persello, C., & Brito, P. L. (2021). Earth observations and statistics: Unlocking sociodemographic knowledge through the power of satellite images. *Sustainability, 13*(22), 12640. https://doi.org/10.3390/su132212640

Micheli, M., Gevaert, C. M., Carman, M., Craglia, M., Daemen, E., Ibrahim, R. E., Kotsev, A., Mohamed-Ghouse, Z., Schade, S., Schneider, I., et al. (2022). AI ethics and data governance in the geospatial domain of digital earth. *Big Data & Society, 9*, 20539517221138767. https://doi.org/10.1177/20539517221138767

O'Neill, O. (2013). Kantian approaches to some famine problems. In R. Shafer-Landau (Ed.), *Ethical theory: An anthology* (2nd ed.). Wiley-Blackwell.

Oluoch, I. O. (2022). Managing risk, governmentality and geoinformation: Vectors of vulnerability in the mapping of COVID-19. *Journal of Contingencies & Crisis Management, 30*, 41–49. https://doi.org/10.1111/1468-5973.12397

Oluoch, I., Nagenborg, M., & Kuffer, M. (2022). In-between the lines and pixels: Cartography's transition from tool of the state to humanitarian mapping of deprived urban areas. *Digital Society, 1*, 5. https://doi.org/10.1007/s44206-022-00008-0

Oluoch, I. (2024). Crossing boundaries: The ethics of AI and geographic information technologies. *ISPRS International Journal of Geo-Information, 13*(3), 87. https://doi.org/10.3390/ijgi13030087

Owusu, M., Kuffer, M., Belgiu, M., Grippa, T., Lennert, M., Georganos, S., & Vanhuysse, S. (2021). Towards user-driven earth observation-based slum mapping. *Computers, Environment and Urban Systems, 89*, 101681.

Petrocchi, E., Tiribelli, S., Paolanti, M., Giovanola, B., Frontoni, E., & Pierdicca, R. (2023). GeomEthics: Ethical considerations about using artificial intelligence in geomatics. In *International conference on image analysis and processing* (pp. 282–293). Springer Nature.

Pickles, J. (2004). *A history of spaces: Cartographic reason, mapping and the geo-coded world.* Routledge.

Rességuier, A., & Rodrigues, R. (2020). AI ethics should not remain toothless! A call to bring back the teeth of ethics. *Big Data & Society, 7*(2), 2053951720942541.

Scheider, S., & Richter, K. F. (2023). Pragmatic GeoAI: Geographic information as externalized practice. *KI-Künstliche Intelligenz, 37*(1), 17–31.

Scott, J. C. (1998). *Seeing like a state: How certain schemes to improve the human condition have failed.* Yale University Press.

Soden, R., Wagenaar, D., & Tijssen, A. (2021). *Responsible AI for Disaster Risk Management.* The World Bank.

Swanlund, D., Schuurman, N., Zandbergen, P., & Brussoni, M. (2020). Street masking: A network-based geographic mask for easily protecting geoprivacy. *International Journal of Health Geographics, 19*, 1–11.

Taylor, L. (2016). Data subjects or data citizens?: Addressing the global regulatory challenge of big data. In *Information, freedom and property* (pp. 81–105). Routledge.

Thomson, D. R., Kuffer, M., Boo, G., Hati, B., Grippa, T., Elsey, H., Linard, C., Mahabir, R., Kyobutungi, C., Maviti, J., Mwaniki, D., Ndugwa, R., Makau, J., Sliuzas, R., Cheruiyot, S., Nyambuga, K., Mboga, N., Kimani, N. W., de Albuquerque, J. P., & Kabaria, C. (2020). Need for an integrated deprived area "slum" mapping system (IDEAMAPS) in low- and middle-income countries (LMICs). *Social Sciences, 9*, 80. https://doi.org/10.3390/socsci9050080

UN-Habitat. (2018). *Metadata on SDGs indicator 11.1.1 indicator category: Tier 1.* UN-Habitat.

UNOSAT (2024). Flood impact analysis in Nairobi Kenya as of 1 May 2024.

Van den Homberg, M. J., Gevaert, C. M., & Georgiadou, Y. (2020). The changing face of accountability in humanitarianism: Using artificial intelligence for anticipatory action. *Politics and Governance, 8*(4), 456–467.

Velev, D., & Zlateva, P. (2023). Challenges of artificial intelligence application for disaster risk management. *The International Archives of the Photogrammetry, Remote Sensing and Spatial Information Sciences, 48*, 387–394.

Verma, D., Jana, A., & Ramamritham, K. (2019). Transfer learning approach to map urban slums using high and medium resolution satellite imagery. *Habitat International, 88*, 101981. https://doi.org/10.1016/j.habitatint.2019.04.008

Wang, J., Kuffer, M., Roy, D., & Pfeffer, K. (2019). Deprivation pockets through the lens of convolutional neural networks. *Remote Sensing of Environment, 234*, 111448. https://doi.org/10.1016/j.rse.2019.111448

Winichakul, T. (1997). *Siam mapped: A history of the geo-body of a nation*. University of Hawaii Press.

Wood, D., Fels, J., & Krygier, J. (2010). *Rethinking the power of maps*. Guilford Press.

Chapter 25
Digital Divide and GeoAI

Xiao Huang, Siqin Wang, and Yongze Song

Abstract This chapter delves into the dimensions of the digital divide specific to Geospatial Artificial Intelligence (GeoAI). GeoAI, which integrates geospatial data with artificial intelligence techniques, holds transformative potential for various sectors. In the era of GeoAI, the divide is increasingly characterized by disparities in digital literacy, technical know-how, data access, and computational resources. Yet, the uneven distribution of GeoAI benefits risks exacerbating existing social and economic inequalities. We explore how the digital divide may be further widened by the demands of GeoAI, emphasizing the need for high-quality data, advanced computational power, and specialized skills. By shedding light on the complex nature of the digital divide in the GeoAI era, this chapter aims to propose strategies that foster inclusive and equitable access to GeoAI technologies. Our goal is to ensure that the transformative potential of GeoAI can be harnessed by a broader segment of society, ultimately contributing to a more just and equitable world.

Keywords GeoAI · Digital divide · Socio-economic Inequality · Digital literacy

25.1 Introduction

Geospatial Artificial Intelligence (GeoAI), which merges geospatial technologies with artificial intelligence techniques, holds immense potential to transform various sectors including urban planning, environmental monitoring, disaster management,

X. Huang (✉)
Department of Environmental Sciences, Emory University, Atlanta, GA, USA
e-mail: xiao.huang2@emory.edu

S. Wang
Spatial Sciences Institute, University of Southern California, Los Angeles, CA, USA
e-mail: siqinwan@usc.edu

Y. Song
School of Design and the Built Environment, Curtin University, Perth, WA, Australia
e-mail: yongze.song@curtin.edu.au

© The Author(s), under exclusive license to Springer Nature Switzerland AG 2025
X. Huang et al. (eds.), *GeoAI and Human Geography*, Springer Geography,
https://doi.org/10.1007/978-3-031-87421-5_25

and public health (Wang et al., 2024a). The capabilities of GeoAI, together with the more recent Geo-Generative AI (GeoGAI), to process and analyze large volumes of spatial data can provide unprecedented insights and solutions to complex geographic problems (Wang et al., 2024b). However, the benefits of GeoAI are not uniformly distributed across society.

The digital divide in the context of GeoAI extends beyond mere access to technology. It encompasses disparities in data access and quality, computational resources, and the skills required to leverage these advanced tools. Individuals and communities that lack the necessary digital literacy, technical know-how, and access to high-quality data and computational power are at a significant disadvantage. This divide can exacerbate existing social and economic inequalities, creating a chasm between those who can utilize GeoAI effectively and those who cannot.

This chapter explores the dimensions of the digital divide specific to GeoAI and its implications for various stakeholders. We begin by defining the digital divide and identifying its potential drivers. We then examine how the GeoAI era is likely to exacerbate this issue, in terms of techniques, research, and education. Furthermore, we provide suggestions and guidance on bridging the divide in the context of GeoAI. Our objective is to shed light on the complex nature of the digital divide in the GeoAI era and propose strategies to promote more inclusive and equitable access to GeoAI technologies.

25.2 Understanding the Digital Divide

The digital divide, a term that refers to the disparity in access to and utilization of digital technologies, has emerged as a pressing global issue in the twenty-first century (Cullen, 2001). This divide not only encompasses the gap between those who have access to the Internet and those who do not, but also extends to disparities in digital literacy, skills, and the ability to realize the benefits of digital technologies. As governments and organizations worldwide increasingly implement e-government services and digital platforms, the potential impact of the digital divide on the equitable delivery of these services has become a significant concern (Bélanger & Carter, 2008).

The digital divide is a multifaceted phenomenon, with its roots in broader socio-economic and geographical inequalities. It is not simply a matter of access to hardware and infrastructure but also involves the ways in which people communicate and access information through various digital channels, including mobile phones, television, and the Internet. Thus, the complexity of the digital divide requires the consideration of social and political dimensions as well as the varied levels of access and digital literacy among people from different backgrounds and skill levels (Fuchs & Horak, 2008).

One of the primary drivers of the digital divide is socio-economic inequality. Individuals from lower-income households are less likely to afford computers, smartphones, and Internet subscriptions. The lack of access to digital technologies restricts their ability to participate in the digital economy, access information, and benefit from online education and services (Van Deursen & Van Dijk, 2011). Additionally, geographical location significantly influences digital access. Rural and remote areas often suffer from inadequate infrastructure, resulting in slower Internet speeds or a complete lack of connectivity. This geographical digital divide worsens regional inequalities, limiting economic development and access to digital services in these areas (Halford & Savage, 2010).

Educational condition is closely linked to digital literacy. Those with higher education levels are generally more proficient in using digital technologies and are better equipped to leverage the Internet for information, employment, and social engagement. Conversely, individuals with lower education levels may struggle with basic digital tasks, further deepening the divide (DiMaggio & Hargittai, 2001). Age is another critical factor in the digital divide. Older adults are less likely to use digital technologies compared to younger generations. This age-related digital divide can be attributed to a lack of familiarity and comfort with technology, as well as physical limitations that make using digital devices challenging (Hunsaker & Hargittai, 2018). Cultural and linguistic barriers also contribute to the digital divide. People from minority groups or those who speak minority languages may find fewer online resources available in their language, which can hinder their ability to fully participate in the digital world, highlighting the need for culturally and linguistically inclusive digital content (Fantin & Girardello, 2008).

The GeoAI era demands new knowledge, advanced modeling, training skills, and robust Internet facilities. As these technologies become more integral to various sectors, the digital divide risks being further exacerbated. The need for high-speed Internet and sophisticated digital skills will become even more critical, potentially widening the gap between those who have access and those who do not. Addressing the digital divide in the GeoAI era requires a multifaceted approach. Governments and organizations need to invest in digital infrastructure, particularly in underserved areas. Initiatives to provide affordable access to digital devices and Internet services are crucial. Additionally, digital literacy programs should be implemented to equip individuals with the necessary skills to navigate the digital world effectively. Policies that promote inclusive access to digital content can also help bridge cultural and linguistic gaps. Regions with poor education cannot harness the power and catch the trend of GeoAI, underscoring the importance of educational initiatives that emphasize both digital and analytical competencies. We will discuss these strategies in the following sections.

25.3 GeoAI Advancements Widen the Digital Divide

It is essential to recognize the profound implications GeoAI advancement has on the existing digital divide. GeoAI promises to revolutionize a wide range of fields. However, the benefits of GeoAI are not evenly distributed, and its implementation can exacerbate existing inequalities. This section explores the various dimensions in which GeoAI interacts with the digital divide, highlighting the challenges and disparities that arise in data access and quality, model training and computational resources, educational opportunities, policy and governance frameworks, and sociocultural contexts.

25.3.1 Data Access and Quality

In the era of GeoAI, access to high-quality data is paramount. However, disparities in data access can exacerbate the digital divide. Regions with limited Internet infrastructure and lower technological investment often lack the necessary data repositories and the capacity to collect, store, and manage geospatial data effectively. For instance, rural areas in developing countries may not have access to satellite imagery or comprehensive geospatial datasets due to inadequate Internet connectivity and financial constraints (Turner et al., 2015). In parts of sub-Saharan Africa, limited access to high-resolution satellite data impedes efforts to monitor agricultural productivity and optimize crop yields. Developed countries with robust data-sharing policies can open-source fine-grained human mobility data, facilitating various tasks such as monitoring facility usage (Wang et al., 2023; Lu et al., 2024), mitigating infectious disease challenges (Hu et al., 2021; Huang et al., 2022a, 2022b), and supporting studies on urban informatics (Kupfer et al., 2021; Huang et al., 2022a, 2022b). In contrast, many developing countries lack the resources and infrastructure to achieve similar outcomes. This lack of data access restricts the ability of underserved communities to leverage GeoAI technologies for local development, disaster management, and resource allocation. As a result, these communities may struggle to respond effectively to natural disasters, plan urban development, or manage natural resources, perpetuating cycles of poverty and underdevelopment. The inability to access and utilize advanced geospatial data can lead to inadequate disaster preparedness and recovery strategies, exacerbating the region's vulnerability and economic instability.

25.3.2 Model Training and Computational Resources

GeoAI models require significant computational power and specialized knowledge for development and training. High-performance computing resources, which are often concentrated in well-funded institutions and tech-savvy regions, are necessary to process large datasets and train sophisticated models. For example, training a GeoAI model to predict flood zones requires substantial computational resources and expertise in both hydrology and machine learning (Song et al., 2023). This process involves integrating various data sources, such as weather patterns, land elevation, and historical flood data, even with social sensing inputs, into a cohesive model that can accurately forecast flood risks (Bulti & Abebe, 2020; Huang et al., 2018a, 2018b). This level of complexity necessitates access to supercomputers or advanced cloud computing platforms, which are often beyond the reach of small municipalities or organizations in developing nations. Communities without access to these resources face substantial barriers. For instance, in many parts of Africa, universities and research institutions lack the funding and infrastructure to support high-performance computing facilities. This limitation restricts their ability to engage in cutting-edge GeoAI research and applications, thereby restricting local innovations that could address region-specific challenges. Similarly, in rural India, local governments may struggle to utilize GeoAI for agricultural optimization due to the high costs associated with computational resources and the lack of trained personnel. In Latin America, efforts to use GeoAI for monitoring deforestation in the Amazon are often stymied by insufficient computational infrastructure, limiting the effectiveness of conservation initiatives.

25.3.3 Educational Disparities

Educational disparities play a critical role in bridging the digital divide, but the GeoAI era places even greater demands on educational systems. The integration of AI with GIS requires advanced knowledge in both fields, along with strong analytical and computational skills. Regions with underfunded educational institutions may struggle to provide this level of training, leaving their populations at a disadvantage. For instance, schools in Tanzania or parts of Southeast Asia may not have the resources to offer advanced courses in GIS and AI (Sumari et al., 2018). These institutions often lack access to up-to-date software, experienced faculty, and adequate laboratory facilities, further exacerbating the educational divide. This educational gap can prevent communities from fully utilizing GeoAI technologies, limiting their participation in the digital economy. For example, in rural Latin America, the absence of specialized training programs in GeoAI means that local students and professionals are less likely to develop the skills necessary to engage in high-tech industries or research projects (Cascante-Campos, 2022). Consequently, these regions may struggle to attract investments from tech companies or participate in international collaborations, perpetuating a cycle of underdevelopment. Without adequate education, young people in these regions may miss out on opportunities in the growing field of GeoAI, further entrenching economic disparities. For instance, in Africa, where educational resources are often unevenly distributed, students in urban centers might have access to better facilities and training programs compared to their rural counterparts. This disparity can result in a significant portion of the population being excluded from the benefits of GeoAI, such as improved agricultural practices, better urban planning, and enhanced disaster management capabilities.

25.3.4 Policy and Governance Challenges

The implementation of GeoAI technologies also faces policy and governance challenges, particularly in regions with weak institutional frameworks. Effective utilization of GeoAI requires robust policies that promote data sharing, privacy protection, and equitable access to technological resources. In many developing countries, however, the absence of comprehensive geospatial policies and regulatory frameworks can lead to fragmented data governance and limited collaboration between governmental and non-governmental organizations. For instance, in several African nations, the lack of standardized policies for geospatial data collection and usage creates obstacles to the integration of GeoAI in national development plans (Osuteye et al., 2017). Additionally, bureaucratic inefficiencies and corruption can further impede the deployment of GeoAI solutions, as seen in certain Latin American countries where resource allocation for technological infrastructure is often marred by mismanagement. These governance issues not only delay the adoption of GeoAI but also perpetuate existing inequalities by failing to address the specific needs of marginalized communities. Without coherent policy frameworks, the potential of GeoAI to drive socio-economic development and environmental sustainability remains largely untapped, reinforcing the digital divide.

25.3.5 Socio-Cultural Barriers

Socio-cultural barriers also play a significant role in the unequal adoption of GeoAI technologies. Cultural attitudes toward technology, levels of digital literacy, and societal norms can influence the acceptance and utilization of GeoAI across different regions. In some traditional societies, there may be resistance to adopting new technologies due to fears of job displacement or mistrust of automated systems. For example, in certain rural communities in Asia, the introduction of GeoAI in agriculture might face opposition from farmers who rely on traditional farming practices and are skeptical of AI-driven interventions (Raj et al., 2020). Similarly, in parts of Africa, low levels of digital literacy can hinder the effective use of GeoAI tools for urban planning or disaster management. These socio-cultural challenges are often compounded by gender disparities, where women and girls have less access to education and technological resources compared to men. In many developing countries, gender norms can restrict women's participation in tech-related fields, limiting their ability to benefit from GeoAI innovations. Addressing these socio-cultural barriers requires targeted interventions that promote digital literacy, gender equality, and community engagement, ensuring that GeoAI technologies are accessible and beneficial to all segments of society.

25.4 Strategies to Bridge the Digital Divide in the GeoAI Era

To effectively address the digital divide exacerbated by the advent of GeoAI, a comprehensive and multifaceted approach is required. The following strategies highlight critical areas for intervention and policy development to ensure that the benefits of GeoAI are accessible to all, thereby promoting equitable socio-economic development and technological inclusion.

- *Investment in digital infrastructure*: Governments and organizations must prioritize extending broadband services to rural and remote areas. This includes laying fiber-optic cables to provide a stable and high-speed Internet backbone, crucial for reliable connectivity. Setting up wireless networks can complement these efforts, especially in hard-to-reach areas where physical infrastructure may be challenging to deploy. Enhancing mobile network infrastructure is also vital; upgrading existing cell towers and deploying new technologies like 5G can significantly improve Internet access in areas where fixed broadband is not feasible. Collaborations between government and private sectors can mobilize resources and expertise for large-scale infrastructure projects, ensuring sustainable and widespread digital access. Public-private partnerships can help bridge funding gaps and accelerate the deployment of advanced digital infrastructure.
- *Affordable access to digital technologies*: Providing financial aid or subsidies to low-income households can help them acquire essential digital devices such as computers and smartphones. These subsidies can be facilitated through voucher programs or direct financial assistance to ensure affordability. Establishing community centers equipped with digital technologies can offer shared access to those who cannot afford personal devices, promoting digital inclusion and providing a space for learning and collaboration. Partnering with Internet service providers to offer affordable Internet plans for underserved communities can increase digital accessibility. Additionally, creating public Wi-Fi hotspots in strategic locations such as libraries, parks, and community centers can provide free or low-cost Internet access to the public.
- *Digital literacy and education programs*: Schools should incorporate comprehensive digital literacy programs that include AI and GIS training, ensuring students acquire essential skills from an early age. These programs should be integrated into the regular curriculum and supported with up-to-date educational resources and technologies. Offering continuous learning opportunities and professional development courses for all age groups can help individuals keep up with technological advancements. Online courses, workshops, and certification programs can provide flexible learning options. Special emphasis should be placed on providing digital literacy programs to underserved and marginalized groups to bridge educational gaps. Initiatives such as mobile learning labs or community-based training sessions can make these programs more accessible to remote and disadvantaged areas.

- *Inclusive policy and governance frameworks*: Establishing clear policies for data collection, sharing, and privacy protection can facilitate collaboration and innovation while ensuring data security. Governments should develop standardized regulations that promote transparency and trust in digital transactions. Implementing anti-corruption measures and promoting transparency in resource allocation can ensure efficient use of funds and reach intended beneficiaries. Policies should specifically address the needs of marginalized groups, including women, minorities, and people with disabilities, ensuring inclusivity. Creating advisory councils or working groups that include representatives from these communities can help shape policies that are more equitable and effective.
- *Promoting cultural and linguistic inclusion*: Developing online resources in multiple languages and culturally relevant formats can help minority groups and non-native speakers access digital information. This includes translating educational content, government services, and health information into various languages and ensuring they are culturally sensitive. Awareness campaigns and community engagement initiatives can help change attitudes toward technology, fostering greater acceptance and utilization of GeoAI innovations. Workshops, seminars, and media campaigns can highlight the benefits of digital inclusion and address common concerns. Ensuring that digital platforms and tools are designed with cultural and linguistic diversity in mind can enhance their accessibility and usability. User interface designs should be intuitive and consider the diverse needs of different cultural groups.
- *Supporting local innovations and capacity building*: Providing financial support and mentorship to local startups can encourage innovation and the development of solutions tailored to regional challenges. Incubators and accelerators can offer resources, networking opportunities, and guidance to emerging tech entrepreneurs. Investing in research institutions and providing access to high-performance computing resources can foster local expertise in GeoAI. Grants, scholarships, and collaborative projects with international research bodies can enhance local research capabilities. Training programs focused on developing local skills and knowledge in GeoAI can empower communities to leverage these technologies effectively. Community-based training centers and online learning platforms can make these programs more accessible and widespread.
- *Gender equality and empowerment programs*: Offering scholarships and mentorship opportunities for women and girls in tech fields can increase their participation and representation in the digital economy. These programs can provide financial support, professional guidance, and networking opportunities to help women advance in tech careers. Creating training programs specifically for women can help bridge gender disparities in digital literacy and skills. Workshops, boot camps, and online courses designed for women can address specific challenges they face in the tech industry. Encouraging gender-inclusive policies and practices in tech-related fields ensures that the benefits of GeoAI are shared equitably across all segments of society. Organizations should promote diversity and inclusion in their hiring practices, leadership development programs, and workplace policies.

- *Community-driven approaches*: Involving local leaders, NGOs, and community organizations in the design and implementation of GeoAI projects can ensure that solutions are contextually relevant. Community consultations, participatory design workshops, and pilot projects can help tailor technologies to local needs. Engaging communities in decision-making processes fosters ownership and sustainability of digital initiatives. Regular feedback loops, community meetings, and collaborative platforms can enhance transparency and accountability. Developing GeoAI applications that address the specific needs and challenges of local communities can enhance their effectiveness and adoption. Customizing solutions for local contexts, such as using GeoAI for local agricultural practices or disaster management, can demonstrate the tangible benefits of these technologies.

25.5 Conclusion

Addressing the digital divide in the context of GeoAI is essential for fostering equitable socio-economic development and ensuring universal access to the transformative benefits of GeoAI. The fusion of geographic information systems with artificial intelligence creates unparalleled opportunities in areas such as urban planning, disaster management, and resource allocation. Nevertheless, without proactive interventions, these technological advancements could exacerbate existing social and economic disparities. The digital divide encompasses a range of inequalities including access to digital technologies, Internet connectivity, digital literacy, and the capacity to utilize digital tools for economic and social advancement. Given that GeoAI requires high-quality data, sophisticated modeling techniques, and substantial computational resources, there is a real risk that this divide could widen if underserved communities do not have the necessary infrastructure and skills.

To address this potential disparity, a holistic and multi-dimensional strategy is essential. Critical investments in digital infrastructure must be made to ensure that rural and remote communities are not overlooked. Providing affordable access to digital technologies and comprehensive digital literacy initiatives can empower these underserved populations, allowing them to fully participate in the digital economy. Furthermore, it is vital to establish inclusive policy and governance frameworks that support equitable access, ensure data security, and foster collaboration. Promoting cultural and linguistic diversity, encouraging local innovations, and enhancing regional capacities are key to driving development and adapting solutions to meet specific local challenges. Additionally, prioritizing gender equality and adopting community-driven approaches are crucial to guarantee that the benefits of GeoAI are equitably distributed among all societal segments.

References

Bélanger, F., & Carter, L. (2008). Trust and risk in e-government adoption. *The Journal of Strategic Information Systems, 17*(2), 165–176.

Bulti, D. T., & Abebe, B. G. (2020). A review of flood modeling methods for urban pluvial flood application. *Modeling Earth Systems and Environment, 6*, 1293–1302.

Cascante-Campos, A. (2022). Latin American geography education research trends in open access journals from the twenty-first century. *International Research in Geographical and Environmental Education, 31*(3), 188–204.

Cullen, R. (2001). Addressing the digital divide. *Online Information Review, 25*(5), 311–320.

DiMaggio, P., & Hargittai, E. (2001). *From the 'digital divide'to 'digital inequality': Studying internet use as penetration increases* (Vol. 4, pp. 4–2). Center for Arts and Cultural Policy Studies, Woodrow Wilson School, Princeton University.

Fantin, M., & Girardello, G. (2008). Digital literacy and cultural mediations to the digital divide. In *Digital literacy: Tools and methodologies for information society* (pp. 310–340). IGI Global.

Fuchs, C., & Horak, E. (2008). Africa and the digital divide. *Telematics and Informatics, 25*(2), 99–116.

Halford, S., & Savage, M. (2010). Reconceptualizing digital social inequality. *Information, Communication & Society, 13*(7), 937–955.

Hu, T., Wang, S., She, B., Zhang, M., Huang, X., Cui, Y., & Li, Z. (2021). Human mobility data in the COVID-19 pandemic: Characteristics, applications, and challenges. *International Journal of Digital Earth, 14*(9), 1126–1147.

Huang, X., Wang, C., & Li, Z. (2018a). Reconstructing flood inundation probability by enhancing near real-time imagery with real-time gauges and tweets. *IEEE Transactions on Geoscience and Remote Sensing, 56*(8), 4691–4701.

Huang, X., Wang, C., & Li, Z. (2018b). A near real-time flood-mapping approach by integrating social media and post-event satellite imagery. *Annals of GIS, 24*(2), 113–123.

Huang, X., Lu, J., Gao, S., Wang, S., Liu, Z., & Wei, H. (2022a). Staying at home is a privilege: Evidence from fine-grained mobile phone location data in the United States during the COVID-19 pandemic. *Annals of the American Association of Geographers, 112*(1), 286–305.

Huang, X., Zhao, Y., Wang, S., Li, X., Yang, D., Feng, Y., & Chen, B. (2022b). Unfolding community homophily in US metropolitans via human mobility. *Cities, 129*, 103929.

Hunsaker, A., & Hargittai, E. (2018). A review of internet use among older adults. *New Media & Society, 20*(10), 3937–3954.

Kupfer, J. A., Li, Z., Ning, H., & Huang, X. (2021). Using mobile device data to track the effects of the COVID-19 pandemic on spatiotemporal patterns of national park visitation. *Sustainability, 13*(16), 9366.

Lu, J., Xiao, X., Huang, X., Chuai, X., Li, Z., Wei, H., & Wang, S. (2024). Big data insights into urban park use in the pandemic: Changes in visitation patterns and exacerbated social inequalities in the US. *Cities, 152*, 105204.

Osuteye, E., Johnson, C., & Brown, D. (2017). The data gap: An analysis of data availability on disaster losses in sub-Saharan African cities. *International Journal of Disaster Risk Reduction, 26*, 24–33.

Raj, A., Dwivedi, G., Sharma, A., de Sousa Jabbour, A. B. L., & Rajak, S. (2020). Barriers to the adoption of industry 4.0 technologies in the manufacturing sector: An inter-country comparative perspective. *International Journal of Production Economics, 224*, 107546.

Song, Y., Kalacska, M., Gašparović, M., Yao, J., & Najibi, N. (2023). Advances in geocomputation and geospatial artificial intelligence (GeoAI) for map-ping. *International Journal of Applied Earth Observation and Geoinfor-mation, 120*, 103300.

Sumari, N. S., Shao, Z., & Kira, E. (2018). Challenges and opportunities for the advancement of GIS education in Tanzania. *Journal of Education and Practice, 8*, 67–75.

Turner, W., Rondinini, C., Pettorelli, N., Mora, B., Leidner, A. K., Szantoi, Z., & Woodcock, C. (2015). Free and open-access satellite data are key to biodiversity conservation. *Biological Conservation, 182*, 173–176.

Van Deursen, A., & Van Dijk, J. (2011). Internet skills and the digital divide. *New Media & Society, 13*(6), 893–911.

Wang, S., Huang, X., She, B., & Li, Z. (2023). Diverged landscape of restaurant recovery from the COVID-19 pandemic in the United States. *Iscience, 26*(6), 106811.

Wang, S., Huang, X., Liu, P., Zhang, M., Biljecki, F., Hu, T., & Bao, S. (2024a). Mapping the landscape and roadmap of geospatial artificial intelligence (GeoAI) in quantitative human geography: An extensive systematic review. *International Journal of Applied Earth Observation and Geoinformation, 128*, 103734.

Wang, S., Hu, T., Xiao, H., Li, Y., Zhang, C., Ning, H., & Ye, X. (2024b). GPT, large language models (LLMs) and generative artificial intelligence (GAI) models in geospatial science: A systematic review. *International Journal of Digital Earth, 17*(1), 2353122.

Chapter 26
Advancing Human Geography with GeoAI: Transforming Land Use and Land Cover Analysis

Elif Sertel and **Xiao Huang**

Abstract This chapter explores the pivotal role of Geospatial Artificial Intelligence (GeoAI) in advancing human geography research, particularly in the domain of Land Use and Land Cover (LULC) studies. As the Earth's environment evolves amid a growing human population, accurate LULC mapping becomes essential for understanding human–environment interactions. The chapter highlights how advancements in remote sensing technologies and the proliferation of Earth observation data have empowered geographers to monitor and analyze LULC changes with unprecedented detail. By integrating GeoAI—especially deep learning models like convolutional neural networks and transformer-based architectures—researchers can enhance the precision of LULC maps through semantic segmentation, facilitating deeper insights into urbanization, deforestation, agricultural expansion, and land abandonment. Despite these technological advancements, challenges such as the need for extensive labeled datasets and class imbalances remain. The chapter delves into the socioeconomic impacts of human-induced LULC changes on climate, hydrology, and biodiversity, emphasizing their significance in human geography. It advocates for future research to focus on developing explainable GeoAI models that incorporate socio-economic and environmental variables, aiming to enhance sustainable land management, climate resilience, and inform policy-making processes relevant to human geography.

Keywords GeoAI · Remote sensing · Land use and land cover · Earth observation

E. Sertel (✉)
Department of Geomatics Engineering, Istanbul Technical University, Sariyer, Istanbul, Turkey
e-mail: sertele@itu.edu.tr

X. Huang
Department of Environmental Sciences, Emory University, Atlanta, GA, USA
e-mail: xiao.huang2@emory.edu

© The Author(s), under exclusive license to Springer Nature Switzerland AG 2025
X. Huang et al. (eds.), *GeoAI and Human Geography*, Springer Geography,
https://doi.org/10.1007/978-3-031-87421-5_26

26.1 Introduction

Earth's resources are limited, while the human population continues to increase steadily. Therefore, sustainable management of these resources is crucial to protect the environment and provide high-quality living conditions for future generations. Human activities have dramatically altered Earth's land surface, impacting weather, climate, and the hydrological cycle; perturbing natural biogeochemical cycles; and affecting various land surface processes (Waters et al., 2016; Sertel et al., 2019). An accurate representation of land use and land cover (LULC) on Earth's surface is essential not only for estimating the impacts of land surface changes on weather, climate, hydrology, and the carbon cycle but also for understanding human–environment interactions from a human geography perspective. A comprehensive spatiotemporal understanding of LULC is crucial for analyzing the cause-and-effect dynamics of various Earth surface changes, including deforestation, urbanization, and other transformative processes that are intrinsically linked to human activities and societal development. While land cover refers to the physical characteristics of the Earth's surface, land use describes how humans utilize this land, reflecting cultural practices, economic activities, and social structures. Importantly, land cover is recognized as one of the essential climate variables within the Global Climate Observing System (GCOS), playing a critical role in monitoring and responding to climate change (Bojinski et al., 2014; Sertel et al., 2019). Additionally, LULC data play a crucial role in characterizing landscapes to analyze the spatiotemporal distribution, dynamics, and conditions of various land categories, which are fundamental aspects of human geography research. These data are not only key inputs for numerous environmental, hydrological, and climate models but also serve as valuable proxies for socio-economic variables, aiding in the study of population distribution, urban growth, and land management practices. Furthermore, LULC data hold significant potential to support several Sustainable Development Goal (SDG) indicators, helping to drive progress toward these global targets by informing policy decisions and resource allocation (Sertel et al., 2010, 2022; Kavvada et al., 2020).

The significant increase in the remote sensing (RS) platforms capable of capturing diverse types of information, combined with the recent increase in the availability of Earth observation (EO) data through globally accessible service delivery platforms, has greatly expanded access to vast amounts of RS imagery. Satellite images are among the most widely used geospatial datasets; additionally, Unmanned Aerial Vehicles (UAVs) and aircraft-based systems; conventional surveying systems; Global Navigation Satellite Systems (GNSS), such as Global Positioning System (GPS), Galileo, GLObal Navigation Satellite System (GLONASS), and BeiDou, or mobile devices or Internet of Things (IoT) sensors integrated with these technologies; crowdsourcing or volunteered geographic information (VGI). The volume of geospatial big data is in the order of Zetabytes, and platforms are also being developed for the backup and processing of such data. Geospatial artificial intelligence methods have been widely used in the processes of analyzing geospatial big data quickly and accurately and transforming it into purposeful geographic information (Lee & Kang, 2015; Ekim & Sertel, 2021; Loukili et al., 2022).

Geospatial Artificial Intelligence (GeoAI) integrates artificial intelligence (AI) and machine learning (ML) with geospatial data to analyze and interpret spatial patterns and trends. GeoAI is a powerful tool for assessing LULC changes and understanding their impacts on both the environment and human life. Considering the direct and indirect usage of multi-temporal LULC data and the complexity of RS images, the development of AI-based models to generate high-precision LULC maps of different geographies is an important challenge.

26.2 GeoAI Techniques for LULC Mapping

Machine learning-based pixel classification algorithms and geographic object-based analysis (GEOBIA) have been widely used to produce accurate LULC maps from different satellite images (Topaloğlu et al., 2021; Myint et al., 2011). Recently, image classification and segmentation methods have shifted towards deep learning-based approaches with the availability of a huge amount of EO data from a variety of platforms, such as Sentinels, Landsat, Planet, etc., and the advancement of GeoAI methods. Recent research on LULC segmentation using deep learning methods has significantly dominated the literature in remote sensing. The integration of deep learning approaches specifically convolutional neural networks (CNNs) and transformer-based architectures has resulted in highly accurate and efficient LULC map generation (Sertel et al., 2024). Deep learning (DL) models excel at extracting robust, high-level, and attributive features from images, while also implicitly approximating complex, non-linear relationships between environmental parameters through multi-layer learning (LeCun et al., 2015; Bengio et al., 2012). These models have demonstrated substantial improvements over traditional methods, especially in various domains of remote sensing, including geospatial object detection, scene understanding, and semantic segmentation (Cheng et al., 2020).

Semantic segmentation, in the case of supervision, is the task of labeling every pixel in the input image according to a predefined class legend, also known as ground truth. The CNNs aim to extract the most discriminative features out of the input image, which identifies the image characteristic. Nowadays, the common way of handling the LULC classification task is to adopt deep learning-based approaches on the nature of the recent success of deep learning in many research areas (Sertel et al., 2024; Tuia et al., 2024; Ajibola & Cabral, 2024; Sertel et al., 2023). More recently, Transformers, originally developed for natural language processing, have been adapted for LULC segmentation tasks, due to their ability to extract contextual information and long-range dependencies from images. Visual transformer (ViT), the Swin transformer, and SegFormer methods were among the first implemented Transformer-based architectures implemented in the RS domain (Dosovitskiy et al., 2010; Liu et al., 2021; Xie et al., 2021; Weng et al., 2023). However, there are still some problems exist for the remote sensing image segmentation task, such as inadequate edge structures of segments, poor segmentation accuracy, missed pixels, and false positive alarms, that should be considered while implementing different GeoAI approaches (Wang et al., 2024).

Long et al. (2015) pioneered the use of fully convolutional networks (FCNs) to extend convolutional neural networks (CNNs) for dense, pixel-level semantic segmentation. Building on this foundation, Ronneberger et al. (2015) introduced the highly influential U-Net architecture, which employs up-sampling layers and skip connections to enhance segmentation accuracy. Concurrently, Simonyan and Zisserman (2014) and Szegedy et al. (2015) demonstrated that increasing network depth and width improves performance, leading to the development of the VGG and GoogLeNet architectures, respectively. He et al. (2015) introduced the ResNet framework, where layers learn residual functions relative to their inputs, while Huang et al. (2017) developed DenseNet, utilizing shorter, direct connections between layers to promote efficient information flow.

Unlike DenseNet's emphasis on enhanced connectivity, Chaurasia and Culurciello (2017) designed LinkNet to reduce parameter complexity, making it suitable for mobile applications. Zhao et al. (2017) focused on preserving the global context by incorporating a pyramid pooling module in their Pyramid Scene Parsing Network (PSPNet). Meanwhile, Lin et al. (2016) proposed the Feature Pyramid Network (FPN), which builds feature pyramids using a combination of bottom-up and top-down approaches to capture multi-scale features effectively.

Chen et al. (2017) advanced segmentation with DeepLabv3 by integrating atrous convolutions to adjust the filter's field of view and control the detail level of feature maps. They further improved the network with DeepLabv3+ by enhancing the decoder module to better capture object boundaries (Chen et al., 2018a, 2018b). To enhance network representations, Hu et al. (2018) introduced the Squeeze-and-Excitation (SE) block, which adaptively recalibrates channel-wise feature responses. Finally, Zhou et al. (2018) redesigned U-Net with nested dense skip pathways and implemented a pruning scheme to speed up the inference, resulting in the Unet++ architecture, which offers improved segmentation performance.

Yuan et al. (2021) conducted a comprehensive literature review on deep learning methods for remote sensing image-based semantic segmentation. They highlighted Fully Convolutional Networks (FCNs), U-Net, and deep convolutional encoder–decoder architectures, such as SegNet, and DeepLab as the most commonly used segmentation architectures in the remote sensing (RS) domain (Bakirman et al., 2022; Long et al., 2015; Ronneberger et al., 2015; Badrinarayanan et al., 2017; Chen et al., 2018a, 2018b). Additionally, improved versions like Unet++ and DeepLabV3+ have gained popularity as CNN-based segmentation architectures (Sertel et al., 2022).

Beyond these, several innovative semantic segmentation models have been developed specifically for land use and land cover (LULC) mapping. These include the Dual Function Feature Aggregation Network (DFFAN) for semantic segmentation, the Multi-Level Feature Aggregation Network (MFANet), a U-Net-like Transformer for efficient segmentation of urban scene imagery (UNetFormer), and the Swin-Unet Network for aggregating contextual information and extracting spatial details in remote sensing images (CSSwin-Unet) (Ajibola & Cabral, 2024; Huang et al., 2021; Chen et al., 2021; Wang et al., 2022; Xiao et al., 2023a, 2023b).

The Segment Anything Model (SAM), developed by Meta AI, is a groundbreaking approach to image segmentation. It has demonstrated high-performance generalization capabilities on a variety of image datasets. SAM can make accurate predictions with little or no training data (Osco et al., 2023).

Multi-temporal Land Use and Land Cover (LULC) maps generated using advanced GeoAI approaches can be effectively used to accurately and precisely track and quantify land use and land cover changes within regions of interest. By leveraging these maps, spatiotemporal variations on land surface can be determined and deeper insights into the underlying patterns and drivers of land transformation can be developed. Geospatial data and artificial intelligence approaches provide a great potential for a clearer understanding of how land use evolves over time.

26.3 Socioeconomic Impacts of LULC

Human geography and the socioeconomic aspects of LULC are important fields of research to better understand the interplay between human activities and the environment. A clear understanding of human-environment relationships can facilitate the advancement of integrated approaches to land management and support the development of sustainable management strategies and policy decisions.

Global drivers, including population dynamics, economic and technological development, and the impacts of climate change, play a pivotal role in shaping global markets and migration patterns, which in turn accelerate LULC changes (Sertel et al., 2019). Population increase, urbanization, globalization, technology advancements, shifts in supply, demand, and trade patterns are considered to trigger land changes (Schirpke et al., 2023). There are natural and anthropogenic driving forces causing various types of changes in land surface. Natural disasters such as floods, wildfires, earthquakes, and hurricanes can cause significant LULC changes as well as climate change-related events such as desertification, shifting vegetation zones, and droughts can also contribute to shifts in land cover and land use. The below section focuses on human-induced LULC changes in detail.

26.4 Drivers of Human-Induced LULC Changes

One significant aspect of human geography in LULC studies is the examination of human-induced LULC changes and their impacts on the ecosystem, biodiversity, water cycle, and climate. Land use and/or land cover changes have resulted in the significant modification of natural landscapes mostly caused by anthropogenic activities due to the over-exploitation or under-utilization of natural resources. Agricultural expansion, deforestation, forest degradation, urbanization, land degradation, desertification, and agricultural land abandonment are among the most encountered changes occurring over the land areas. The abrupt increase in worldwide industrialization and population caused urbanization, deforestation, and agricultural land abandonment. (Gellrich & Zimmermann, 2007; Perpiña Castillo et al., 2020; Subedi, 2022). This section focuses on four types of land transformations, including urbanization, deforestation, agricultural expansion, and land abandonment. Multi-modal remotely sensed data and GeoAI techniques could be employed to quantify short-term or long-term LULC changes, specifically focusing on human-induced changes.

Urban expansion is a dynamic spatiotemporal process that is shaped by a diverse range of interrelated factors, including social trends, economic forces, geographic conditions, and governmental policies. The interrelationship among these factors resulted in varying patterns of urbanization in different geographical regions, which also changes over time (Feng & Wang, 2021). Demographic factors, such as population increase and migration from rural to urban areas, economic and geographic factors, and governmental policies are key drivers of urban expansion (Allan et al., 2022). Economic pressures, driven by rapid urbanization and industrial growth, significantly accelerate the landscape changes. As an example, in the Yangtze River Delta, urban expansion is closely linked to economic development and demographic trends, leading to increased competition for land resources (Yu et al., 2023). Economic growth often necessitates the conversion of agricultural land to urban uses, which can lead to a decline in cultivated land and increased land pressure (Wu et al., 2022). Topaloğlu et al. (2021) used high-resolution satellite images, remote sensing techniques, and landscape metrics to determine the human-induced changes in the Istanbul metropolitan city of Türkiye, where deforestation and agricultural land loss were determined as a result of population increase, urbanization, industrial development, and airport construction. Deep learning-based segmentation methods can be implemented for multi-temporal remote sensing data to determine the current boundaries of urban areas and quantify the spatiotemporal changes (Sertel et al., 2022). Moreover, GeoAI methods could also be tailored to derive urban-related statistics such as population. Doda et al. (2024) proposed an interpretable DL approach based on ResNet-50 architecture to generate 1 km resolution population estimations using multi-modal geospatial data, including satellite images, OpenStreetMap (OSM), Digital Elevation Models, Nighttime lights, etc.

Land abandonment is the process of gradually decreasing the intensity of land usage, followed by a period of land underutilization, and finally arriving at the point where no agronomic inputs are used for production on the land and farming is completely terminated (Subedi, 2022; Munoz-Rios et al., 2020). Osgouei et al. (2022) investigated the long-term LULC changes in Bursa, Türkiye using historical cadastral maps from 1858, aerial photographs obtained in 1955, and recent WorldView-3 satellite images. They revealed how anthropogenic activities have significantly altered the landscape over time, specifically as a result of rural depopulation, gradual afforestation, agricultural land abandonment, and deforestation. Kabadayı et al. (2022) emphasized the agricultural land abandonment problem in

Bulgaria and provided a thirty-year historical perspective using aerial photographs and satellite images. Van Vliet et al. (2015) stated that agricultural land use has considerably changed in Europe, especially in the last decades as a result of the globalization of agricultural markets, the transition from a rural to an urban society, and the shift to post-socialism in central and eastern Europe.

Deforestation has grown to be one of the most important and urgent environmental issues with substantial adverse effects on biodiversity, climate change, and human well-being (Hansen et al., 2020; Bragagnolo et al., 2021). Md Jelas et al. (2024) conducted a detailed review of the deforestation studies that implemented DL methods to satellite images and emphasized the good performance of U-Net, DeepLabV3, ResNet, SegNet, and FCN segmentation models for deforestation pattern detection. They stated that most deforestation studies benefited from satellite images as the main geospatial data source, followed by UAV images. Although GeoAI models have important strengths in forest segmentation, they present some limitations due to the challenges arising from datasets and the requirement of high computational resources. For a concrete DL model to have high transferability and generalization capability, there is a need for extensive labeled datasets (Chiang et al., 2020; Morales et al., 2018). Potential biases in training data should be carefully considered and the dataset should include balanced distribution in terms of class definitions (Lee et al., 2020). Furthermore, it could be challenging to identify deforestation in areas with varying land cover distributions or irregularly shaped forests (John & Zhang, 2022).

According to the Global Forest Resources Assessment, agricultural development is the cause of almost 90% of global deforestation, which poses serious problems for natural habitats (Ma et al., 2023). Deforestation and forest degradation due to subsistence agriculture and commodity exports, or changes caused by agricultural intensification and expansion are the main issues in the tropical agricultural-forest borders (Xiao et al. 2023). As a result of agricultural expansion, mostly natural vegetation (forests, grasslands, wetlands) is converted into agricultural lands. The main drivers for agricultural expansion are the increase in food demand, economic developments, technological advances in agriculture, government policies, and incentives. Between 2003 and 2019, cropland increased by 9% worldwide, with nearly half of this growth replacing natural habitat (Potapov et al., 2021).

26.5 Impacts of LULC Changes on Human and Environment

Anthropogenic LULC changes can significantly contribute to climate change impacts, alter the hydrological cycle, and affect air quality, which, in turn, create adverse impacts on human life and the environment.

The increase of impervious surfaces as a result of urbanization exacerbates the urban heat island causing a notable rise in local temperatures compared to surrounding rural areas; affects local climate patterns, hydrology, and biodiversity, as natural habitats are fragmented or destroyed; and intensifies urban flooding frequency (Tang et al., 2020; Dissanayake et al., 2019). Urbanization has led to increased carbon emissions primarily due to escalating energy demands. Globally, urbanization has significantly impacted carbon emissions in both developed and developing nations (Zhang et al., 2017). Albedo, surface roughness, green vegetation fraction, and stomatal resistance values vary for different land use/cover categories, leading to varying outputs from regional climate models. Urbanization due to the conversions of forest, cropland, and woodland to artificial surfaces caused warming in the Marmara Region of Türkiye due to the decreased amount of latent heat flux emitted from the surface leading to an increase of temperature values and change in surface roughness caused a decrease in 10-m wind speed (Sertel et al., 2010). Additionally, urbanization leads to higher traffic load and increased economic activities resulting in higher energy consumption, which triggers the emissions of air pollutants. Larkin et al. (2016) examined the relationships between urban characteristics and air pollution changes in 830 cities in East Asia from 2000 to 2010 and their findings emphasized that urban expansion and changes in urban characteristics are associated with increases in NO2 and PM2.5 concentrations.

Many species are threatened or become extinct due to habitat loss as a result of deforestation and agricultural expansion of forests. By 2050, global cropland is projected to increase by 26%, which will cause 17,409 species to lose their habitats (Williams et al., 2021). Deforestation considerably intensifies the climate change effects by changing carbon stocks, impacting local and regional climate patterns, and modifying hydrological cycles. Trees play a crucial role in maintaining water balance through processes such as transpiration and loss of trees as a result deforestation significantly impacts the hydrological cycle.

Kayitesi et al. (2022) reviewed several articles to emphasize the integrated impacts caused by LULC changes and climate change on the hydrological and geomorphological processes of river systems. The coupled impacts modified the components of the water balance through increased or decreased evaporation, peak flow, flooding, and river morphology.

By exchanging carbon dioxide, energy, and water vapor with the atmosphere, forests affect the climate (Schultz et al., 2017). Local and regional climate are affected as a result of deforestation since changes in surface biogeophysical characteristics will alter surface energy and water balance. Moreover, changes in surface roughness impact the sensible heat, latent heat, and horizontal advection of moisture. This reduces moisture flux convergence, leading to decreased precipitation in deforested areas, which can create a positive feedback loop that exacerbates local drying and further impacts climate. This reduction in precipitation not only affects local ecosystems but can also have far-reaching implications for agricultural productivity and water availability in surrounding regions (Pitman & Lorenz, 2016). Deforestation causes increases in both the mean and variability of temperature, resulting in daily microclimate variability which has a strong potential to impact species competition and coexistence with complex and interacting effects (Chan et al., 2023). This disruption can alter species interactions and biodiversity, which are critical components of resilient ecosystems.

The expansion of agricultural land often involves the use of fertilizers and other practices that further contribute to emissions, exacerbating the climate impacts of deforestation.

26.6 Conclusion and Future Prospects

Geospatial Artificial Intelligence (GeoAI) holds significant potential for advancing human geography research by providing more accurate, scalable, and dynamic tools for understanding and analyzing human-environment interactions, particularly through Land Use and Land Cover (LULC) patterns. By enhancing classification, monitoring, and predictive modeling, GeoAI deepens our comprehension of how LULC evolves over time, elucidating the complex interconnections between urbanization, deforestation, agricultural expansion, environmental change, and socio-economic factors. This advancement not only refines traditional methods of human geography analysis but also opens new avenues for sustainable development, disaster management, and socio-economic planning. Accurate representation of these dynamics using GeoAI methods is critical for addressing issues such as urban sprawl, resource management, and the impacts of human activities on local, regional, and global climates.

Most GeoAI research has concentrated on semantic segmentation and image classification, particularly in the context of LULC mapping, which are essential for generating accurate representations of human-induced changes on the Earth's surface. However, a significant limitation of deep learning methods is their reliance on large volumes of labeled data. These models typically operate on smaller image patches, necessitating the subsequent stitching of these patches while addressing edge artifacts to create seamless maps of the area of interest. By applying semantic segmentation to multi-temporal remote sensing data, it becomes possible to detect and analyze LULC changes over time, offering valuable insights into human geographical processes. While existing datasets and deep learning models facilitate remote sensing change detection, accurately capturing complex LULC changes in real-world scenarios often requires custom datasets tailored to specific classes and changes of interest that are significant in human geography. Notably, although GeoAI applications for urbanization and deforestation are well-represented in the literature, there is a notable gap in studies focused on cropland expansion and land abandonment—areas with profound implications for human geography that warrant further exploration.

GeoAI methods have begun to be employed in various domains relevant to human geography, including urban growth modeling, population estimations, migration pattern analysis, socioeconomic modeling, and the assessment of human impact on the environment. Recent studies have shown a growing interest in the use of Large Language Models (LLMs) in geospatial big data analysis tasks. These models can be utilized for visual recognition of remotely sensed images, modeling semantic relationships, and generating natural language descriptions of multi-temporal images and their changes over time, facilitating a deeper understanding of human geographical patterns and narratives. As GeoAI continues to evolve, its role in enhancing the accuracy and depth of human geography research is undeniable. Future research must focus on developing explainable models that can be seamlessly integrated into policy-making processes, urban planning, and resource management. Addressing data challenges and refining models to include socio-economic and environmental variables will be crucial for leveraging GeoAI in sustainable land management and urban development. By advancing these approaches, we can better support climate resilience, biodiversity preservation, and the attainment of sustainable development goals, ultimately enriching the field of human geography.

References

Ajibola, S., & Cabral, P. (2024). A systematic literature review and bibliometric analysis of semantic segmentation models in land cover mapping. *Remote Sensing, 16*, 2222. https://doi.org/10.3390/rs16122222

Allan, A., Soltani, A., Abdi, M. H., & Zarei, M. (2022). Driving forces behind land use and land cover change: A systematic and bibliometric review. *Land, 11*(8), 1222. https://doi.org/10.3390/land11081222

Badrinarayanan, V., Kendall, A., & Cipolla, R. (2017). Segnet: A deep convolutional encoder-decoder architecture for image segmentation. *IEEE Transactions on Pattern Analysis and Machine Intelligence, 39*(12), 2481–2495.

Bakirman, T., Komurcu, I., & Sertel, E. (2022). Comparative analysis of deep learning based building extraction methods with the new VHR Istanbul dataset. *Expert Systems with Applications, 202*, 117346.

Bengio, Y., Courville, A. C., & Vincent, P. (2012). Representation learning: A review and new perspectives. *IEEE Transactions on Pattern Analysis and Machine Intelligence, 35*, 1798–1828.

Bojinski, S., Verstraete, M. M., Peterson, T. C., Richter, C., Simmons, A., & Zemp, M. (2014). The concept of essential climate variables in support of climate research, applications, and policy. *Bulletin of the American Meteorological Society, 95*, 1431–1443.

Bragagnolo, L., da Silva, R. V., & Grzybowski, J. M. V. (2021). Amazon forest cover change mapping based on semantic segmentation by U-nets. *Ecological Informatics, 62*, 101279. https://doi.org/10.1016/j.ecoinf.2021.101279

Chan, S., Rubenstein, D. R., Chen, I., Fan, Y., Tsai, H., Zheng, Y., & Shen, S. (2023). Higher temperature variability in deforested mountain regions impacts the competitive advantage of nocturnal species. *Proceedings of the Royal Society B, 290*, 20230529.

Chaurasia, A., & Culurciello, E. (2017). LinkNet: Exploiting encoder representations for efficient semantic segmentation. In *2017 IEEE Visual Communications and Image Processing (VCIP), St. Petersburg, FL, USA* (pp. 1–4). https://doi.org/10.1109/VCIP.2017.8305148

Chen, L. C., Papandreou, G., Schroff, F, & Adam, H. (2017). *Rethinking atrous convolution for semantic image segmentation*. arXiv preprint arXiv:1706.05587.Retrieved December 5, 2017, from https://arxiv.org/pdf/1706.05587.pdf

Chen, K., Fu, K., Yan, M., Gao, X., Sun, X., & Wei, X. (2018a). Semantic segmentation of aerial images with shuffling convolutional neural networks. *IEEE Geoscience and Remote Sensing Letters, 15*(2), 173–177.

Chen, L. C., Zhu, Y., Papandreou, G., Schroff, F., & Adam, H. (2018b). Encoder-decoder with atrous separable convolution for semantic image segmentation. In *Computer Vision—ECCV 2018, European conference on computer vision (ECCV), Lecture Notes in Computer Science* (Vol. 11211, pp. 833–851). Springer. https://doi.org/10.1007/978-3-030-01234-2_49

Chen, B., Xia, M., & Huang, J. (2021). Mfanet: A multi-level feature aggregation network for semantic segmentation of land cover. *Remote Sensing, 13*, 731.

Cheng, G., Xie, X., Han, J., Guo, L., & Xia, G. (2020). Remote sensing image scene classification meets deep learning: Challenges, methods, benchmarks, and opportunities. *IEEE Journal of Selected Topics in Applied Earth Observations and Remote Sensing, 13*, 3735–3756. https://doi.org/10.1109/jstars.2020.3005403

Chiang, C. Y., Barnes, C., Angelov, P., & Jiang, R. (2020). Deep learning-based automated forest health diagnosis from aerial images. *IEEE Access, 8*, 144064–144076. https://doi.org/10.1109/ACCESS.2020.3012417

Dissanayake, D., Morimoto, T., Ranagalage, M., & Murayama, Y. (2019). Land-use/land-cover changes and their impact on surface urban heat islands: Case study of Kandy city, Sri Lanka. *Climate, 7*(8), 99. https://doi.org/10.3390/cli7080099

Doda, S., Kahl, M., Ouan, K., Obadic, I., Wang, Y., Taubenböck, H., & Zhu, X. X. (2024). Interpretable deep learning for consistent large-scale urban population estimation using Earth observation data. *Int. J. Appl. Earth Obs. Geoinformation, 128*, 103731.

Dosovitskiy, A., Beyer, L., Kolesnikov, A., Weissenborn, D., Zhai, X., Unterthiner, T., Dehghani, M., Minderer, M.; Heigold, G.; Gelly, S., et al. (2010). *An image is worth 16 × 16 words: Transformers for image recognition at scale*. arXiv 2020, arXiv:2010.11929.

Ekim, B., & Sertel, E. (2021). Deep neural network ensembles for remote sensing land cover and land use classification. *International Journal of Digital Earth, 14*, 1868–1881.

Feng, R., & Wang, K. (2021). Spatiotemporal effects of administrative division adjustment on urban expansion in China. *Land Use Policy, 101*, 105143.

Gellrich, M., & Zimmermann, N. E. (2007). Investigating the regional-scale pattern of agricultural land abandonment in the Swiss mountains: A spatial statistical modelling approach. *Landscape and Urban Planning, 79*(1), 65–76. https://doi.org/10.1016/j.landurbplan.2006.03.004

Hansen, J. N., Mitchard, E. T. A., & King, S. (2020). Assessing forest/non-forest separability using sentinel-1 C-band synthetic aperture radar. *Remote Sensing, 12*, 1899. https://doi.org/10.3390/rs12111899

He, K., Zhang, X., Ren, S., & Sun, J. (2015). Deep residual learning for image recognition. In *2016 IEEE Conference on Computer Vision and Pattern Recognition (CVPR)* (pp. 770–778).

Hu, J., Shen, L., & Sun, G. (2018). Squeeze-and-excitation networks. In *2018 IEEE/CVF Conference on Computer Vision and Pattern Recognition, Salt Lake City, UT, USA* (pp. 7132–7141). https://doi.org/10.1109/CVPR.2018.00745

Huang, G., Liu, Z., & Weinberger, K. Q. (2017). Densely connected convolutional networks. In *2017 IEEE Conference on Computer Vision and Pattern Recognition (CVPR)* (pp. 2261–2269).

Huang, J., Weng, L., Chen, B., & Xia, M. (2021). DFFAN: Dual function feature aggregation network for semantic segmentation of land cover. *International Journal of Geo-Information (IJGI), 10*, 125.

John, D., & Zhang, C. (2022). An attention-based U-net for detecting deforestation within satellite sensor imagery. *International Journal of Applied Earth Observation and Geoinformation, 107*, 102685. https://doi.org/10.1016/j.jag.2022.102685

Kabadayı, M. E., Osgouei, P. E., & Sertel, E. (2022). Agricultural land abandonment in Bulgaria: A Long-term remote sensing perspective, 1950–1980. *Land, 11*(10), 1855. https://doi.org/10.3390/land11101855

Kavvada, A., Metternicht, G., Kerblat, F., Mudau, N., Haldorson, M., Laldaparsad, S., Friedl, L., Held, A., & Chuvieco, E. (2020). Towards delivering on the sustainable development goals using earth observations. *Remote Sensing of Environment, 247*(2), 111930.

Kayitesi, N. M., Guzha, A. C., & Mariéthoz, G. (2022). Impacts of land use land cover change and climate change on river hydro-morphology- a review of research studies in tropical regions. *Journal of Hydrology, 615*(7), 128702.

Larkin, A., van Donkelaar, A., Geddes, J. A., Martin, R. V., & Hystad, P. (2016). Relationships between changes in urban characteristics and air quality in East Asia from 2000 to 2010. *Environmental Science & Technology, 50*(17), 9142–9149.

LeCun, Y., Bengio, Y., & Hinton, G. E. (2015). Deep learning. *Nature, 521*, 436–444. https://doi.org/10.1038/nature14539

Lee, J., & Kang, M. (2015). Geospatial big data: Challenges and opportunities. *Big Data Research, 2*, 74–81.

Lee, S. H., Han, K. J., Lee, K., Lee, K. J., Oh, K. Y., & Lee, M. J. (2020). Classification of landscape affected by deforestation using high-resolution remote sensing data and deep-learning techniques. *Remote Sensing, 12*, 3372. https://doi.org/10.3390/rs12203372

Lin, T., Dollár, P., Girshick, R. B., He, K., Hariharan, B., & Belongie, S. J. (2016). Feature pyramid networks for object detection. In *2017 IEEE Conference on Computer Vision and Pattern Recognition (CVPR)* (pp. 936–944).

Liu, Z., Lin, Y., Cao, Y., Hu, H., Wei, Y., Zhang, Z., Lin, S., & Guo, B. (2021). Swin Transformer: Hierarchical Vision Transformer using Shifted Windows. In *2021 IEEE/CVF International Conference on Computer Vision (ICCV)* (pp. 9992–10002).

Long, J., Shelhamer, E., & Darrell, T. (2015). Fully convolutional networks for semantic segmentation. In *Proceedings of the IEEE conference on computer vision and pattern recognition* (pp. 3431–3440).

Loukili, Y., Lakhrissi, Y., & Ali, S. E. B. (2022). Geospatial big data platforms: A comprehensive review. *KN Journal of Cartography and Geographic Information, 72*, 293–308.

Ma, S., Wang, L., Jiang, J., & Zhao, Y. (2023). Direct and indirect effects of agricultural expansion and landscape fragmentation processes on natural habitats. *Agriculture, Ecosystems & Environment, 353*, 108555.

Md Jelas, I., Zulkifley, M. A., Abdullah, M., & Spraggon, M. (2024). Deforestation detection using deep learning-based semantic segmentation techniques: A systematic review. *Frontiers in Forests and Global Change, 7*, 1300060. https://doi.org/10.3389/ffgc.2024.1300060

Morales, G., Kemper, G., Sevillano, G., Arteaga, D., Ortega, I., & Telles, J. (2018). Automatic segmentation of Mauritia flexuosa in unmanned aerial vehicle (UAV) imagery using deep learning. *Forests, 9*, 736. https://doi.org/10.3390/f9120736

Munoz-Rios, L. A., Vargas-Villegas, J., & Suarez, A. (2020). Local perceptions about rural abandonment drivers in the Colombian coffee region: Insights from the city of Manizales. *Land Use Policy, 91*, 104361. https://doi.org/10.1016/j.landusepol.2019.104361

Myint, S. W., Gober, P., Brazel, A., Grossman-Clarke, S., Grossman-Clarke, S., & Weng, Q. (2011). Per-pixel vs. object-based classification of urban land cover extraction using high spatial resolution imagery. *Remote Sensing of Environment, 115*, 1145–1161.

Osco, L. P., Wu, Q., Lemos, E. L., Gonçalves, W. N., Ramos, A. P. M., Li, J., & Marcato, J. (2023). The segment anything model (SAM) for remote sensing applications: From zero to one shot. *International Journal of Applied Earth Observation and Geoinformation, 124*, 103540.

Osgouei, P. E., Sertel, E., & Kabadayı, M. E. (2022). Integrated usage of historical geospatial data and modern satellite images reveal long-term land use/cover changes in bursa/Turkey, 1858–2020. *Scientific Reports, 12*(1), 9077. https://doi.org/10.1038/s41598-022-11396-1

Perpiña Castillo, C., Coll Aliaga, E., Lavalle, C., & Martínez Llario, J. C. (2020). An assessment and spatial modelling of agricultural land abandonment in Spain (2015–2030). *Sustainability, 12*(2), 560. https://doi.org/10.3390/su12020560

Pitman, A. J., & Lorenz, R. (2016). Scale dependence of the simulated impact of Amazonian deforestation on regional climate. *Environmental Research Letters, 11*(9), 094025.

Potapov, P. V., Turubanova, S., Hansen, M. C., Tyukavina, A., Zalles, V., Khan, A., Song, X., Pickens, A. H., Shen, Q., & Cortez, J. (2021). Global maps of cropland extent and change show accelerated cropland expansion in the twenty-first century. *Nature Food, 3*, 19–28.

Ronneberger, O., Fischer, P., & Brox, T. (2015). U-net: Convolutional networks for biomedical image segmentation. In *Medical image computing and computer-assisted intervention–MICCAI 2015: 18th international conference, Munich, Germany, October 5–9, 2015, proceedings, part III 18* (pp. 234–241). Springer International.

Schirpke, U., Tasser, E., Borsky, S., Braun, M., Eitzinger, J., Gaube, V., Getzner, M., Glatzel, S., Gschwantner, T., Kirchner, M., Leitinger, G., Mehdi-Schulz, B., Mitter, H., Scheifinger, H.,

Thaler, S., Thom, D., & Thaler, T. (2023). Past and future impacts of land-use changes on ecosystem services in Austria. *Journal of Environmental Management, 345*, 118728.
Schultz, N. M., Lawrence, P. J., & Lee, X. (2017). Global satellite data highlights the diurnal asymmetry of the surface temperature response to deforestation. *Journal of Geophysical Research: Biogeosciences, 122*, 903–917.
Sertel, E., Robock, A., & Ormeci, C. (2010). Impacts of land cover data quality on regional climate simulations. *International Journal of Climatology, 30*(13), 1942–1953.
Sertel, E., Imamoglu, M. Z., Cuceloglu, G., & Erturk, A. (2019). Impacts of land cover/use changes on hydrological processes in a rapidly urbanizing mid-latitude water supply catchment. *Water, 11*(5), 1075. https://doi.org/10.3390/w11051075
Sertel, E., Ekim, B., Osgouei, P. E., & Kabadayi, M. E. (2022). Land use and land cover mapping using deep learning-based segmentation approaches and VHR Worldview-3 images. *Remote Sensing, 14*(18), 4558. https://doi.org/10.3390/rs14184558
Sertel, E., Avci, C., & Kabadayı, M. E. (2023). Deep learning-based land use land cover segmentation of historical aerial images. In *IGARSS 2023–2023 IEEE International Geoscience and Remote Sensing Symposium* (pp. 2622–2625).
Sertel, E., Kabadayi, M. E., Sengul, G. S., & Tumer, I. N. (2024). HexaLCSeg: A historical benchmark dataset from hexagon satellite images for land cover segmentation. *IEEE Geoscience and Remote Sensing Magazine, 12*(3), 197–206. https://doi.org/10.1109/mgrs.2024.3394248
Simonyan, K., & Zisserman, A. (2014). *Very deep convolutional networks for large-scale image recognition*. CoRR, abs/1409.1556.
Subedi, Y. R. (2022). Paul Kristiansen, Oscar Cacho, drivers and consequences of agricultural land abandonment and its reutilisation pathways: A systematic review. *Environmental Development, 42*. https://doi.org/10.1016/j.envdev.2021.100681
Szegedy, C., et al. (2015). Going deeper with convolutions. In *2015 IEEE Conference on Computer Vision and Pattern Recognition (CVPR), Boston, MA, USA* (pp. 1–9). https://doi.org/10.1109/CVPR.2015.7298594
Tang, X., Cui, Y., Li, N., Fu, Y., Liu, X., Run, Y., & Dong, J. (2020). Human activities enhance radiation forcing through surface albedo associated with vegetation in Beijing. *Remote Sensing, 12*(5), 837. https://doi.org/10.3390/rs12050837
Topaloğlu, R. H., Aksu, G. A., Ghale, Y. A. G., & Sertel, E. (2021). High-resolution land use and land cover change analysis using GEOBIA and landscape metrics: A case of Istanbul, Turkey. *Geocarto International, 37*(25), 9071–9097. https://doi.org/10.1080/10106049.2021.2012273
Tuia, D., Schindler, K., Demir, B., Zhu, X. X., Kochupillai, M., Džeroski, S., Van Rijn, J. N., Hoos, H. H., Del Frate, F., Datcu, M., Markl, V., Saux, B. L., Schneider, R., & Camps-Valls, G. (2024). Artificial intelligence to advance earth observation: A review of models, recent trends, and pathways forward. In *IEEE Geoscience and Remote Sensing Magazine* (pp. 2–25). IEEE. https://doi.org/10.1109/mgrs.2024.3425961
Van Vliet, J., de Groot, H. L., Rietveld, P., & Verburg, P. H. (2015). Manifestations and underlying drivers of agricultural land use change in Europe. *Landscape and Urban Planning, 133*, 24–36.
Wang, L., Li, R., Zhang, C., Fang, S., Duan, C., Meng, X., & Atkinson, P. M. (2022). UNetFormer: A UNet-like transformer for efficient semantic segmentation of remote sensing urban scene imagery. *ISPRS Journal of Photogrammetry and Remote Sensing, 190*, 196–214.
Wang, R., Ma, L., He, G., Johnson, B. A., Yan, Z., Chang, M., & Liang, Y. (2024). Transformers for remote sensing: A systematic review and analysis. *Sensors, 24*, 3495. https://doi.org/10.3390/s24113495
Waters, C. N., Zalasiewicz, J., Summerhayes, C., Barnosky, A. D., Poirier, C., Gałuszka, A., Cearreta, A., Edgeworth, M., Ellis, E. C., Ellis, M. A., Jeandel, C., Leinfelder, R., McNeill, J. R., Richter, D. D., Steffen, W., Syvitski, J. P., Vidas, D., Wagreich, M., Williams, M., ... Wolfe, A. P. (2016). The Anthropocene is functionally and stratigraphically distinct from the Holocene. *Science, 351*.
Weng, L., Pang, K., Xia, M., Lin, H., Qian, M., & Zhu, C. (2023). Sgformer: A local and global features coupling network for semantic segmentation of land cover. *IEEE Journal of Selected Topics in Applied Earth Observations and Remote Sensing, 16*, 6812–6824.

Williams, D. R., Clark, M., Buchanan, G. M., Ficetola, G. F., Rondinini, C., & Tilman, D. (2021). Proactive conservation to prevent habitat losses to agricultural expansion. *Nature Sustainability, 4*(4), 314–322.

Wu, X., Wang, Y., & Hong-bo, Z. (2022). Does economic growth lead to an increase in cultivated land pressure? Evidence from China. *Land, 11*(9), 1515. https://doi.org/10.3390/land11091515

Xiao, C., Peng, L., & Zhiming, F. (2023). Agricultural expansion and forest retreat in Mainland Southeast Asia since the late 1980s. *Land Degradation & Development, 34*(17), 5606–5621.

Xiao, C., Li, P., & Feng, Z. (2023a). Agricultural expansion and forest retreat in mainland Southeast Asia since the late 1980s. *Land Degradation & Development, 34*, 5606–5621.

Xiao, D., Kang, Z., Fu, Y., Li, Z., & Ran, M. (2023b). Csswin-Unet: A Swin-Unet network for semantic segmentation of remote sensing images by aggregating contextual information and extracting spatial information. *International Journal of Remote Sensing, 44*, 7598–7625.

Xie, E., Wang, W., Yu, Z., Anandkumar, A., Alvarez, J. M., & Luo, P. (2021). SegFormer: Simple and efficient design for semantic segmentation with transformers. *Advances in Neural Information Processing Systems, 34*, 12077–12090.

Yu, Z., Chen, L., Zhang, T., Li, L., Yuan, L., Teng, G., & Chen, L. (2023). Land pressure evaluation in the yangtze river delta region: a perspective from production-living-ecology. *Land Degradation and Development, 34*, 5312. https://doi.org/10.22541/au.167699726.67209858/v1

Yuan, X., Shi, J., & Gu, L. (2021). A review of deep learning methods for semantic segmentation of remote sensing imagery. *Expert Systems with Applications, 169*, 114417. https://doi.org/10.1016/j.eswa.2020.114417

Zhang, N., Yu, K., & Chen, Z. (2017). How does urbanization affect carbon dioxide emissions? A cross-country panel data analysis. *Energy Policy, 107*, 678–687.

Zhao, H., Shi, J., Qi, X., Wang, X., & Jia, J. (2017). Pyramid Scene Parsing Network. In *2017 IEEE Conference on Computer Vision and Pattern Recognition (CVPR), Honolulu, HI, USA* (pp. 6230–6239). https://doi.org/10.1109/CVPR.2017.660

Zhou, Z., Siddiquee, M. M. R., Tajbakhsh, N., & Liang, J. (2018). UNet++: Redesigning skip connections to exploit multiscale features in image segmentation. *IEEE Transactions on Medical Imaging, 39*(6), 1856–1867. https://doi.org/10.1109/TMI.2019.2959609

Chapter 27
Cross-Disciplinary Opportunities for GeoAI and Human Geography

Song Gao

Abstract The science of convergence brings together multiple fields, to tackle global issues such as climate change, urbanization, public health, disaster resilience, and ethical implications of technologies. This chapter first reviews the reciprocal relationships between Geospatial artificial intelligence (GeoAI) and human geography and then presents a co-design framework of convergent research on GeoAI and human geography. The framework consists of four core components: (1) problem definition through stakeholder engagement, (2) knowledge and methodology integration, (3) prototype development and testing, and (4) knowledge dissemination and capacity building. This chapter highlights the importance of co-design of privacy and ethical framework for responsible and trustworthy GeoAI development, calls for the establishment of an open knowledge network, and outlines several opportunities and challenges toward cross-disciplinary developments between GeoAI and human geography.

Keywords GeoAI · Human geography · Cross-disciplinary research · Convergent science · Co-design framework

27.1 Introduction

Science has recently evolved into an era known as convergence science (Petersen et al., 2021), which involves bridging different disciplines to tackle global challenges, such as climate change, urbanization, public health crises, disaster resilience, and ethical implications of technologies. These complex issues cannot be solved by a single area of expertise. Instead, solving these challenges require interdisciplinary knowledge and innovative co-design approaches that combine the theories, technologies, methodologies, and practices of multiple disciplines. This chapter explores

---

S. Gao (✉)
Department of Geography, University of Wisconsin-Madison, Madison, WI, USA
e-mail: song.gao@wisc.edu

© The Author(s), under exclusive license to Springer Nature Switzerland AG 2025
X. Huang et al. (eds.), *GeoAI and Human Geography*, Springer Geography,
https://doi.org/10.1007/978-3-031-87421-5_27

the synergies to identify and facilitate cross-disciplinary opportunities for GeoAI and human geography.

Geospatial artificial intelligence (GeoAI) refers to the interdisciplinary field that integrates geographic studies and artificial intelligence (AI) advancements, develops spatially explicit AI methods aiming to solve scientific and engineering problems in earth systems and human–environment interactions, which is powered by theoretical and methodological breakthroughs in deep neural networks, advanced computer hardware such as graphics processing units (GPUs) and powerful high-performance computing platforms, software and tools (e.g., PyTorch, ChatGPT), and geospatial big data. The core mission of GeoAI is to advance the research and development of spatial intelligence in machines to improve spatial representation, environmental perception, intelligent reasoning and to advance scientific understanding and discovery of geographic phenomena (Gao et al., 2023).

On the other hand, human geography studies the interactions of people, communities, places, and environments at various spatial scales (Fouberg et al., 2015). Human geographers often seek to address questions like how humans interact with the physical, economic, political, and cultural environments, how humans organize space and place, and how human behaviors are shaped by, adapt to, and affect the landscapes they live (Doell et al., 2023; Jones, 2012). Such explorations have also been supported by multiple paradigms of technology advancements, such as the quantitative revolutions in 1960s and 1970s (Barnes, 2004), big data-driven paradigm in 2010s (Kitchin, 2013; Miller & Goodchild, 2015), and most recently AI-driven paradigm with deep learning and knowledge graphs (Janowicz et al., 2020).

A recent literature review of GeoAI in quantitative human geography by Wang et al. (2024) found that GeoAI models and advanced methods can help human geography studies address the complexity of human behaviors and its potential drivers. In addition, a conversation about GeoAI, counter-AI, and human geography by Janowicz et al. (2022b) discussed several opportunities and challenges in defining intelligence for machines, and the role of humans in GeoAI development and decision-making processes, emphasizing the need for ethical considerations and accountability in the development of GeoAI. In the following, I will first introduce the reciprocal relationships between GeoAI and human geography, and then propose a co-design framework and outline several opportunities and challenges toward cross-disciplinary developments.

27.2 Reciprocal Relationships Between GeoAI and Human Geography

27.2.1 GeoAI for Human Geography

The recent advancements of (geo)AI technologies (especially deep learning and generative AI) together with geospatial big data have significantly expanded the views of human geography researchers on the location, space, place, and complex human–environment interactions (Miller & Goodchild, 2015; Wang et al., 2024). Human geographers and GIScientists have explored a wide range of topics to understand how human or AI agents perceive and structure the places and environments. Some of the key research topics (but not limited to) are discussed as follows.

Place Representation and Modeling Sense of place is a fundamental concept in human geography which often refers to the human experience, perception, identity, and emotional attachment of location embedded with not only spatial coordinates but also environmental, social, and cultural characteristics (Tuan, 1979). A key interest from GIScientists is how to extract computational representations of rich place semantics and develop place-based computational models from multi-modality of geospatial data streams (Gao et al., 2017), including texts (e.g., Web blogs, social media posts, user reviews on points of interest (POIs) and neighborhoods), imagery (e.g., satellite views, street views, camera traps, crowdsourced photos), GPS trajectories, soundscapes, and georeferenced videos. A diverse array of (geo)AI methods offer new opportunities to explore and analyze the multidimensional characteristics of places, along with the associated human activities, experiences, perceptions, emotions, and movements across various contexts and geographical scales. Research efforts have been made to synthesize multi-source, multi-scale and multi-modality geospatial datasets into deep learning-friendly frameworks. For example, spatially explicit location representation techniques via deep neural networks have been utilized for automatic identification of urban functional regions using POI embeddings (i.e., multidimensional vectors). A long short-term memory (LSTM) neural network with attention mechanisms was found effective on mapping the generated regional embeddings to urban functions (Huang et al., 2022). In addition, spatial context-aware prompts via multi-modal deep learning can achieve a better performance in mixed land-use mapping than vision-based deep learning approaches (Wu et al., 2023). The development of audio-to-image stable diffusion AI models has enabled the translation of hearing to seeing experience by linking auditory and visual place perceptions (Zhuang et al., 2024). With the advent of generative AI tools (e.g., ChatGPT—a conversational AI chatbot and DALL-E—a text to image AI tool), researchers have explored the capabilities of generative AI in capturing the place identity by generating visual and textual information of different places (e.g., cities) (Jang et al., 2024). While the generated place representations from AI showed median similarity to the Wikipedia textual documents and Google Search images for the same pair of cities, the AI did generate distinguishing features between place identity in different regions (western vs. eastern). Further research is needed to explore the generative AI's potential to capture individual perspectives of place in addition to the current collective world-view of a place.

Urban Planning and Design Urban geographers and planners have been utilizing (geo)AI and machine learning techniques to extract valuable insights into urban forms, population dynamics, land-use patterns, and the interactions between social and environmental systems, for optimizing urban spatial structure, improving transportation efficiency, quality of living, and environmental sustainability (De Sabbata et al., 2023; Liu & Biljecki, 2022; Mortaheb & Jankowski, 2023). For example, machine learning on building and block morphological features (e.g., density, orientation, and compactness) and local planning expert-knowledge were combined to identify various urban block typologies (Li & Quan, 2024), which highlights the importance of integrating data-driven and knowledge-guided approaches in cross-domain collaborations. In addition, AI-powered semantic and sentiment analyses of online neighborhood review texts (Hu et al., 2019) as well as building exterior and urban streetscape images (Liang et al., 2024; Zhang et al., 2024) helped better understand the perceptions and sentiments of people toward their living neighborhoods and environments, which can be used for supporting urban planning and improving quality of life. However, the spatial heterogeneity of urban spatial structure, environmental and social characteristics in different cities often pose challenges on the transferability of AI models across cities. Interestingly, such transferred bias can be utilized to uncover the spatial mismatch between visual appearance of a place and its socioeconomic environment (e.g., house price) (Hou et al., 2024). Furthermore, the advances in generative AI models bring exciting opportunities for urban planners to facilitate the automatic rendering of urban master plans and building floorplan layouts via generative adversarial networks (GANs) (Weber et al., 2022; Ye et al., 2022), otherwise they need to rely on subjective design and labor-intensive production process. However, the generative AI also introduces massive fake data into geography (Zhao et al., 2021). Scholars need to investigate the trustworthiness of such AI-generated geospatial data and future city plans. One promising direction through cross-domain collaboration is the development of a human–AI collaborative workflow for urban planning through planning-knowledge-guided deep reinforcement learning with expert feedback and evaluation (Zheng et al., 2023).

27.2.2 Human Geography for GeoAI

Human geography rooted in understanding of human behaviors and human–environment interactions across spatial–social–technical systems can provide valuable insights into the social, cultural, economic, and ethical factors that drive the development of GeoAI in heterogeneous contexts (Li et al., 2024).

Context-Aware and Debiasing GeoAI The use of deep learning-based natural language processing and computer vision methods to understand human perception of physical and socioeconomic environments has attracted researchers' interests across disciplines such as geography, computer science, urban planning, and environmental psychology (Fan et al., 2023; Gebru et al., 2017; Hou et al., 2025; Kang et al., 2020). While human perception of places (e.g., whether a place is safe or lively) can be learned and predicted by using deep neural networks on street-view images (Zhang et al., 2018), there exist multiple types of biases in existing GeoAI-driven computational frameworks for quantifying human perceptions of places and environments: *data bias, model bias, and perception bias*. For instance, Kang et al. (2023) compared the human participant survey-based safety perception measures from local neighborhood residents with the predicted perception on street-view imagery using a GeoAI model. This research found that elder residents from local neighborhoods may underestimate the safety of their living places, which may amplify the perception bias of GeoAI models trained using their label data. In addition, the use of spatial bias (e.g., urban center vs. suburb differences) in GeoAI models may help uncover the imbalance between the developments of physical and socioeconomic environments across cities (Hou et al., 2024). Therefore, integrating human, cultural, environmental, and socioeconomic insights into pre-trained GeoAI models or fine-tuning them with local knowledge and spatiotemporal dependence may enhance AI's ability to comprehend human-place interactions and respond to the spatial heterogeneity and diverse human experiences. Such an approach not only improves the accuracy of GeoAI models in few-shot or zero-shot spatial prediction tasks but also fosters social diversity and spatial transferability of GeoAI models across socio-technical systems and geographic regions.

Human-Centered Design Designing GeoAI with a focus on humans, emphasizes the importance of considering the needs, experiences, and values of individuals involved in the process of creating GeoAI models and systems. This human-centered approach resonates with the field of human geography known for its exploration of human experiences and human–environment interactions among diverse socioeconomic groups, especially within marginalized communities. Several principles and practices such as ensuring data representativeness and data privacy, explainable and interpretable AI, interactive machine learning, keeping human or community-in-the-loop, and accountability mechanisms in GeoAI system design can be considered for addressing the social-technical challenges of AI and mitigating its potential negative impacts on human society (Janowicz et al., 2022b; Kang et al., 2024; Margetis et al., 2021). Recently, a geospatial visual analytics framework integrated with explainable AI techniques (GeoVisX) has been developed to better understanding the spatial distribution of geographical phenomena and identifying key factors impacting their distribution with visual analytics, which enhances the understanding of different spatial processes underlying spatial predictive modeling (Luo et al., 2024). Looking ahead, embracing human-centered approaches through the collaboration of human geographers and GIScientists will accelerate the design, development, and adoption of advanced GeoAI technologies in exploring human–environment interactions and place-based decision support.

27.3 Toward A Co-Design Framework of Convergent Research on GeoAI and Human Geography

Convergent research was an NSF initiative aimed to bring together researchers from diverse disciplines in natural sciences, social sciences, computational sciences, and engineering, for solving complex problems focusing on societal needs or deep scientific challenges (Ashton et al., 2024). Co-design, also known as participatory design, is a cross-domain collaborative approach to convergent research that actively involves stakeholders, which offers diverse perspectives and strategies for integrative problem-solving across multiple disciplines. In the following, I will introduce the idea of building a co-design convergent research framework of GeoAI and human geography.

Open Knowledge Network Bringing together GeoAI technological advances with human-centered viewpoints calls for an Open Knowledge Network (OKN). This OKN can host knowledge bases, people, datasets, and tools to promote cross-domain collaboration by combining knowledge and expertise from diverse fields along with quantitative tools like GIS, spatial modeling, statistical analysis, and machine learning as well as qualitative approaches such as survey, interview, discourse analysis, participant, and behavior observation (Montello & Sutton, 2006). It encourages place-based decision-making and involves stakeholders in tackling challenges related to space, place, and society. OKNs often use knowledge graphs as backbones (Janowicz et al., 2022a), which represent information as nodes (entities) and edges (relationships), to create linked machine-readable knowledge bases across targeted domains. An OKN for GeoAI and human geography will provide a shared platform and access to core knowledge, experts, datasets, case studies, and research priorities for researchers and practitioners to accelerate GeoAI research and ensure that the technical advancements are ethical, inclusive, and aligned with societal needs and human concerns.

Core Components of the Co-Design Framework The co-design convergent research framework has played an important role in addressing spatial–social issues like disaster resilience, redistricting, food insecurity, and public health responses (Ashton et al., 2024; Cai et al., 2018; Hou et al., 2021; Kruse et al., 2024). As shown in Fig. 27.1, the framework consists of the following components: (1) Problem definition through stakeholder engagement: GeoAI experts, human geographers, policymakers, industry partners, and community representatives can collaboratively define the core vision, scientific questions, and research goals that address community needs by conducting participatory-design workshops and scenario building exercises. (2) Knowledge and methodology integration: It is essential for the team to create shared vocabularies and common languages that promote understanding, along with models that merge human-centered theories with GeoAI techniques to bridge the gap between the theoretical and methodological aspects of human geography and GeoAI. (3) Prototype development and testing: After co-designing the novel methodologies, the team needs to develop and test the methods, tools, and workflows, with use cases and real-world scenarios. In this process, it is important to incorporate iterative design cycles, (geo)visual analytics, and feedback from diverse stakeholders for continuous refinement (Tu et al., 2023). (4) Knowledge dissemination and capacity building: Organizing workshops, seminars, webinars, and training programs helps the transfer of learned knowledge and skills to broader communities, which may bring new research questions and

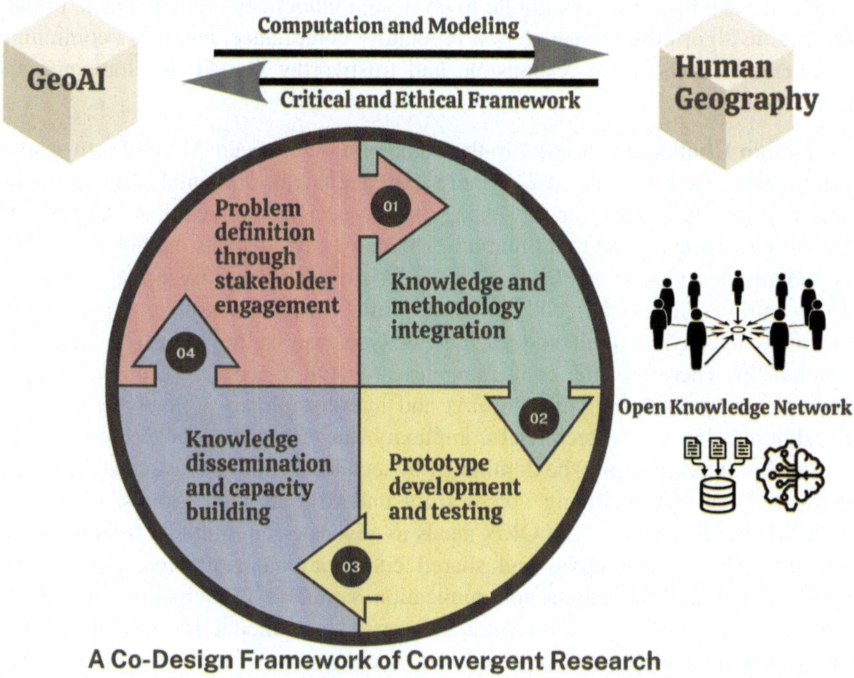

Fig. 27.1 A co-design framework of convergent research on GeoAI and human geography

community challenges. Map-based storytelling and visualization are effective tools for knowledge dissemination and interdisciplinary collaboration (Roth, 2021). A joint seminar on human geography and GeoAI at universities and community colleges might be a first step for educating next-generation students and community leaders.

Co-Design of Privacy and Ethical Framework for GeoAI Data and tools are foundations in data-driven scientific discoveries. Human geographers often provide critical concerns related to data and tool use, social constructs, power dynamics, and community trust. Their insights are invaluable for GeoAI developers who must navigate the critical and ethical implications of processing large amounts of geospatial data regarding user privacy and data security. In GeoAI development, concerns may arise due to the use of sensitive location information such as a user's home address, frequently visited places, or driving routes in the machine learning and deep learning model training process (Rao et al., 2023b). With the recent advances in (geo)AI foundation models (Mai et al., 2024), a series of data privacy and information security risks (e.g., disclosure of sensitive information, AI hallucinations, and misuse) are expected at different pre-training and fine-tuning stages, prompt-based interactions, and reinforcement learning from human or AI feedback mechanisms (Rao et al., 2023a). Moving forward, human geographers and

GIScience scholars can collaborate to co-design guidelines on data ethics, privacy protection, algorithmic transparency, regulatory compliance, and cross-community engagements that ensure responsible and trustworthy GeoAI developments and applications.

Co-Design Challenges Similar to the collaboration between AI and Earth science communities (Chen et al., 2024), there exist technical, domanial, organizational, and cultural challenges for advancing the co-design framework of GeoAI and human geography. GeoAI and human geography have distinct methodologies and epistemologies. GeoAI emphasizes on quantitative and data-driven methods which often requires large team collaboration, while human geography often focuses on qualitative, contextual, and critical analysis and values more on individual scholarship. Therefore, the convergent research OKN needs institutional support and reward for both individual efforts and interdisciplinary collaborative efforts. In addition, due to the technical complexity and computational resource limits, the co-design process may be dominated by technologically advanced stakeholders or institutions, marginalizing voices from under-represented groups, geographic regions, or communities. The OKN needs to advocate for equitable funding across domains, open-source tools, and shared computational resources. By fostering open-minded and transparent communication, equitable participation, and ethical practices, this co-design framework can serve as a model for interdisciplinary innovation that benefits both academia and society.

27.4 Conclusion

The relationship between GeoAI and human geography is reciprocal. As advancements in GeoAI continue to progress, its applications in human geography will become more diverse and impactful. At the same time, the critical insights provided by human geographers will be crucial for ensuring that GeoAI methods are developed and applied in a responsible, equitable, and ethical manner. However, this convergence requires a co-design framework that unites the theoretical and human-centric insights of human geography with the computational power and technical advances of GeoAI. Such a framework ensures that the technological advancements of GeoAI are ethical, inclusive, and aligned with societal needs.

By working together through an OKN of GeoAI and human geography, a convergent research agenda can address some of the most pressing challenges facing human–environment nexus today, including climate change and community adaption, food security, public health, disaster resilience, and equitable access to resources and services. This interdisciplinary collaboration has the potential to create a better and sustainable world for all.

References

Ashton, W. S., Sungu, A., Davis, L., Agarwalla, V., Burke, M., Benavides, E. D., et al. (2024). Whither convergence? co-designing convergent research and wrestling with its emergent tensions. *Ecology and Society, 29*(4), 1–10.

Barnes, T. J. (2004). Placing ideas: genius loci, heterotopia and geography's quantitative revolution. *Progress in Human Geography, 28*(5), 565–595.

Cai, H., Lam, N. S., Qiang, Y., Zou, L., Correll, R. M., & Mihunov, V. (2018). A synthesis of disaster resilience measurement methods and indices. *International Journal of Disaster Risk Reduction, 31*, 844–855.

Chen, M., Qian, Z., Boers, N., Creutzig, F., Camps-Valls, G., Hubacek, K., et al. (2024). Collaboration between artificial intelligence and earth science communities for mutual benefit. *Nature Geoscience, 17*(10), 949–952.

De Sabbata, S., Ballatore, A., Miller, H. J., Sieber, R., Tyukin, I., & Yeboah, G. (2023). GeoAI in urban analytics. *International Journal of Geographical Information Science, 37*(12), 2455–2463.

Doell, K. C., Berman, M. G., Bratman, G. N., Knutson, B., Kühn, S., Lamm, C., et al. (2023). Leveraging neuroscience for climate change research. *Nature Climate Change, 13*(12), 1288–1297.

Fan, Z., Zhang, F., Loo, B. P., & Ratti, C. (2023). Urban visual intelligence: Uncovering hidden city profiles with street view images. *Proceedings of the National Academy of Sciences, 120*(27), e2220417120.

Fouberg, E. H., Murphy, A. B., & De Blij, H. J. (2015). *Human geography: People, place, and culture*. Wiley.

Gao, S., Janowicz, K., Montello, D. R., Hu, Y., Yang, J.-A., McKenzie, G., & Yan, B. (2017). A data-synthesis-driven method for detecting and extracting vague cognitive regions. *International Journal of Geographical Information Science, 31*(6), 1245–1271.

Gao, S., Hu, Y., & Li, W. (2023). *Handbook of geospatial artificial intelligence*. CRC.

Gebru, T., Krause, J., Wang, Y., Chen, D., Deng, J., Aiden, E. L., & Fei-Fei, L. (2017). Using deep learning and google street view to estimate the demographic makeup of neighborhoods across the united states. *Proceedings of the National Academy of Sciences, 114*(50), 13108–13113.

Hou, X., Gao, S., Li, Q., Kang, Y., Chen, N., Chen, K., & Patz, J. A. (2021). Intra-county modeling of covid-19 infection with human mobility: Assessing spatial heterogeneity with business traffic, age, and race. *Proceedings of the National Academy of Sciences, 118*(24), e2020524118.

Hou, C., Zhang, F., Kang, Y., Gao, S., Li, Y., Duarte, F., & Li, S. (2024). Transferred bias uncovers the balance between the development of physical and socioeconomic environments of cities. *Annals of the American Association of Geographers, 115*(1), 148–166.

Hou, C., Zhang, F., Li, Y., Li, H., Mai, G., Kang, Y., et al. (2025). Urban sensing in the era of large language models. *The Innovation, 6*(1), 100749.

Hu, Y., Deng, C., & Zhou, Z. (2019). A semantic and sentiment analysis on online neighborhood reviews for understanding the perceptions of people toward their living environments. *Annals of the American Association of Geographers, 109*(4), 1052–1073.

Huang, W., Cui, L., Chen, M., Zhang, D., & Yao, Y. (2022). Estimating urban functional distributions with semantics preserved poi embedding. *International Journal of Geographical Information Science, 36*(10), 1905–1930.

Jang, K. M., Chen, J., Kang, Y., Kim, J., Lee, J., Duarte, F., & Ratti, C. (2024). Place identity: a Generative AI's perspective. *Humanities and Social Sciences Communications, 11*(1), 1–16.

Janowicz, K., Gao, S., McKenzie, G., Hu, Y., & Bhaduri, B. (2020). GeoAI: spatially explicit artificial intelligence techniques for geographic knowledge discovery and beyond. *International Journal of Geographical Information Science, 34*(4), 625–636.

Janowicz, K., Hitzler, P., Li, W., Rehberger, D., Schildhauer, M., Zhu, R., et al. (2022a). Know, know where, knowwheregraph: A densely connected, cross-domain knowledge graph and geo-enrichment service stack for applications in environmental intelligence. *AI Magazine, 43*(1), 30–39.

Janowicz, K., Sieber, R., & Crampton, J. (2022b). GeoAI, counter-AI, and human geography: A conversation. *Dialogues in Human Geography, 12*(3), 446–458.
Jones, A. (2012). *Human geography: The basics*. Routledge.
Kang, Y., Zhang, F., Gao, S., Lin, H., & Liu, Y. (2020). A review of urban physical environment sensing using street view imagery in public health studies. *Annals of GIS, 26*(3), 263–275.
Kang, Y., Abraham, J., Ceccato, V., Duarte, F., Gao, S., Ljungqvist, L., & Ratti, C. (2023). Assessing differences in safety perceptions using GeoAI and survey across neighbourhoods in stockholm, sweden. *Landscape and Urban Planning, 236*, 104768.
Kang, Y., Gao, S., & Roth, R. (2024). Artificial intelligence studies in cartography: A review and synthesis of methods, applications, and ethics. *Cartography and Geographic Information Science, 51*(1), 1–20.
Kitchin, R. (2013). Big data and human geography: Opportunities, challenges and risks. *Dialogues in Human Geography, 3*(3), 262–267.
Kruse, J., Gao, S., Ji, Y., Szabo, D. P., & Mayer, K. R. (2024). Bringing spatial interaction measures into multi-criteria assessment of redistricting plans using interactive web mapping. *Cartography and Geographic Information Science, 51*(4), 513–532.
Li, N., & Quan, S. J. (2024). Discovering urban block typologies in seoul: Combining planning knowledge and unsupervised machine learning. *Cities, 150*, 104988.
Li, W., Arundel, S., Gao, S., Goodchild, M., Hu, Y., Wang, S., & Zipf, A. (2024). GeoAI for science and the science of GeoAI. *Journal of Spatial Information Science, 29*, 1–17.
Liang, X., Chang, J. H., Gao, S., Zhao, T., & Biljecki, F. (2024). Evaluating human perception of building exteriors using street view imagery. *Building and Environment, 263*, 111875.
Liu, P., & Biljecki, F. (2022). A review of spatially-explicit GeoAI applications in Urban Geography. *International Journal of Applied Earth Observation and Geoinformation, 112*, 102936.
Luo, P., Chen, C., Gao, S., Zhang, X., Majok Chol, D., Yang, Z., & Meng, L. (2024). Understanding of the predictability and uncertainty in population distribu- tions empowered by visual analytics. *International Journal of Geographical Information Science, 39*(3), 1–31.
Mai, G., Huang, W., Sun, J., Song, S., Mishra, D., Liu, N., et al. (2024). On the opportunities and challenges of foundation models for GeoAI (vision paper). *ACM Transactions on Spatial Algorithms and Systems, 10*, 1–44.
Margetis, G., Ntoa, S., Antona, M., & Stephanidis, C. (2021). Human-centered design of artificial intelligence. In *Handbook of human factors and ergonomics* (pp. 1085–1106).
Miller, H. J., & Goodchild, M. F. (2015). Data-driven geography. *GeoJournal, 80*, 449–461.
Montello, D. R., & Sutton, P. C. (2006). *An introduction to scientific research methods in geography*. SAGE.
Mortaheb, R., & Jankowski, P. (2023). Smart city re-imagined: City planning and GeoAI in the age of big data. *Journal of Urban Management, 12*(1), 4–15.
Petersen, A. M., Ahmed, M. E., & Pavlidis, I. (2021). Grand challenges and emergent modes of convergence science. *Humanities and Social Sciences Communications, 8*(1), 1–15.
Rao, J., Gao, S., Mai, G., & Janowicz, K. (2023a). Building privacy-preserving and secure geospatial artificial intelligence foundation models. In *The 31st acm international conference on advances in geographic information systems (Sigspatial'23)* (pp. 1–4). ACM.
Rao, J., Gao, S., & Zhu, S. (2023b). Cats: Conditional adversarial trajectory synthesis for privacy-preserving trajectory data publication using deep learning approaches. *International Journal of Geographical Information Science, 37*(12), 2538–2574.
Roth, R. E. (2021). Cartographic design as visual storytelling: synthesis and review of map-based narratives, genres, and tropes. *The Cartographic Journal, 58*(1), 83–114.
Tu, Y., Wang, X., Qiu, R., Shen, H.-W., Miller, M., Rao, J., et al. (2023). An interactive knowledge and learning environment in smart foodsheds. *IEEE Computer Graphics and Applications, 43*(3), 36–47.
Tuan, Y.-F. (1979). Space and place: humanistic perspective. In *Philosophy in geography* (pp. 387–427). Springer.

Wang, S., Huang, X., Liu, P., Zhang, M., Biljecki, F., Hu, T., et al. (2024). Mapping the landscape and roadmap of geospatial artificial intelligence (GeoAI) in quantitative human geography: An extensive systematic review. *International Journal of Applied Earth Observation and Geoinformation, 128*, 103734.

Weber, R. E., Mueller, C., & Reinhart, C. (2022). Automated floorplan generation in architectural design: A review of methods and applications. *Automation in Construction, 140*, 104385.

Wu, M., Huang, Q., Gao, S., & Zhang, Z. (2023). Mixed land use measurement and mapping with street view images and spatial context-aware prompts via zero- shot multimodal learning. *International Journal of Applied Earth Observation and Geoinformation, 125*, 103591.

Ye, X., Du, J., & Ye, Y. (2022). Masterplangan: Facilitating the smart rendering of urban master plans via generative adversarial networks. *Environment and Planning B: Urban Analytics and City Science, 49*(3), 794–814.

Zhang, F., Zhou, B., Liu, L., Liu, Y., Fung, H. H., Lin, H., & Ratti, C. (2018). Mea- suring human perceptions of a large-scale urban region using machine learning. *Landscape and Urban Planning, 180*, 148–160.

Zhang, F., Salazar-Miranda, A., Duarte, F., Vale, L., Hack, G., Chen, M., & Ratti, C. (2024). Urban visual intelligence: Studying cities with artificial intelligence and street-level imagery. *Annals of the American Association of Geographers, 114*(5), 876–897.

Zhao, B., Zhang, S., Xu, C., Sun, Y., & Deng, C. (2021). Deep fake geography? when geospatial data encounter artificial intelligence. *Cartography and Geographic Information Science, 48*(4), 338–352.

Zheng, Y., Lin, Y., Zhao, L., Wu, T., Jin, D., & Li, Y. (2023). Spatial planning of urban communities via deep reinforcement learning. *Nature Computational Science, 3*(9), 748–762.

Zhuang, Y., Kang, Y., Fei, T., Bian, M., & Du, Y. (2024). From hearing to seeing: Linking auditory and visual place perceptions with soundscape-to-image gener- ative artificial intelligence. *Computers, Environment and Urban Systems, 110*, 102122.

Chapter 28
Shaping Tomorrow: The Future of GeoAI in Human Geography

Xiao Huang, Siqin Wang, John Wilson, and Peter Kedron

Abstract This chapter explores the transformative potential of GeoAI in shaping the future of human geography. It emphasizes the necessity of building inclusive systems that democratize access to GeoAI technologies, ensuring that their benefits are equitably distributed across diverse communities. Addressing ethical dilemmas arising from rapid technological advancements, the chapter advocates for balancing innovation with the protection of individual rights and societal values, highlighting data justice as a core principle. It underscores the importance of breaking down disciplinary silos through cross-disciplinary collaboration, recognizing that complex spatial phenomena require integrated approaches. To promote broader adoption within the discipline, the chapter calls for simplifying complex GeoAI tools to make them accessible to human geographers with varying technical backgrounds. By synthesizing these themes, it outlines a vision for a future where GeoAI advances human geography in an ethically sound, socially equitable, and intellectually enriching manner.

Keywords GeoAI · Human geography · Data justice · Responsibility

28.1 Introduction

The convergence of Geographic Information Systems (GIS) and Artificial Intelligence (AI) has ushered in a new era for human geography through the advent of

---

X. Huang (✉)
Department of Environmental Sciences, Emory University, Atlanta, GA, USA
e-mail: xiao.huang2@emory.edu

S. Wang · J. Wilson
Spatial Sciences Institute, University of Southern California, Los Angeles, CA, USA
e-mail: siqinwan@usc.edu; jpwilson@usc.edu

P. Kedron
Department of Geography, University of California, Santa Barbara, Santa Barbara, CA, USA
e-mail: peterkedron@ucsb.edu

© The Author(s), under exclusive license to Springer Nature Switzerland AG 2025

X. Huang et al. (eds.), *GeoAI and Human Geography*, Springer Geography,
https://doi.org/10.1007/978-3-031-87421-5_28

GeoAI. This transformative fusion advances analysis and understanding of complex spatial dynamics, offering powerful tools to tackle pressing global challenges. As we navigate this evolving landscape, it is imperative to critically assess how GeoAI can be harnessed responsibly to advance the discipline while promoting reproducibility, replicability, social equity, and ethical integrity (Kedron et al., 2021; Wilson et al., 2021).

This chapter reviews the multifaceted aspects that will define the trajectory of GeoAI within the field. We begin by emphasizing the importance of building inclusive systems that democratize access to GeoAI technologies, ensuring that their benefits are equitably distributed across diverse communities. The discussion highlights the need to overcome barriers such as the digital divide and the underrepresentation of marginalized groups in GeoAI development. We then explore the ethical dilemmas that arise from rapid technological advancement, particularly the tension between innovation and the responsibility to build reproducible and replicable workflows and protect individual rights and societal values. The concept of data justice is introduced as a core principle, underscoring the necessity of deploying the FAIR (Findable, Accessible, Interoperable, and Reusable) Guiding Principles to prevent the perpetuation of existing inequalities (Wilkinson et al., 2016). The chapter further advocates for breaking down disciplinary silos through cross-disciplinary collaboration, recognizing that the complexities of spatial phenomena require integrated approaches. By simplifying complex GeoAI tools and making them accessible to human geographers with varying technical backgrounds, we aim to foster broader adoption and integration of these technologies within the discipline.

In synthesizing these themes, we outline a vision for the future of GeoAI in human geography—one that is ethically sound, socially equitable, and intellectually enriching. Addressing the challenges and embracing the opportunities presented by GeoAI will enhance our understanding of human-environment interactions and contribute meaningfully to societal advancement.

28.2 GeoAI for all: Building Inclusive Systems

Building inclusive GeoAI systems is essential to ensure that the benefits of these technologies are equitably distributed, and they serve as tools for empowerment rather than instruments that exacerbate existing disparities. Inclusivity in GeoAI encompasses not only who has access to these technologies but also who is involved in their development and how they are applied in various social contexts (Janowicz et al., 2022).

One of the primary barriers to inclusivity in GeoAI is the uneven distribution of resources and technological infrastructure. Communities in underrepresented or marginalized settings often lack the necessary tools, Internet connectivity, and educational opportunities to engage with GeoAI effectively (Carter et al., 2020).

This digital divide can lead to a scenario where the insights and benefits derived from GeoAI are accessible only to a select few, exacerbating social and economic inequalities. Addressing this requires concerted efforts to invest in infrastructure development, provide affordable access to technology, and implement educational programs that build GeoAI literacy across different disciplinary groups (Lee, 2023). Efforts to increase the inclusivity of GeoAI are already emerging, but remain largely in their infancy. The NSF Institute for Geospatial Understanding through Integrative Discovery Environment (I-GUIDE) is developing a GeoAI-related curriculum that includes Geo-ethics as a core principle. As stressed here, identified issues include privacy, consent, equity, transparency, and assessment of the harms and benefits of this emerging technology.

Promoting diversity within GeoAI development teams is another crucial aspect of building inclusive systems. Diverse teams bring a range of perspectives, experiences, and problem-solving approaches that can help identify and mitigate biases in data and algorithms. By involving individuals from various cultural, socioeconomic, gender, and disciplinary backgrounds, GeoAI tools can be designed to be more culturally sensitive and applicable to a broader range of contexts, as witnessed in the current literature (Wang et al., 2024). Encouraging participation from underrepresented groups in GeoAI research and industry not only enriches the field but also ensures that the technologies developed are attuned to the needs of diverse communities.

Designing user-centric GeoAI systems that are accessible to individuals with varying levels of technical expertise is essential for inclusivity. Many human geographers and practitioners may lack advanced computational skills, and complex interfaces can be a deterrent to adoption (Castree et al., 2013). By developing intuitive interfaces, incorporating visual programming environments, and providing comprehensive documentation and support, GeoAI tools can be made more accessible. Governments and institutions need to promote the FAIR guiding principles to ensure equitable access to technology and data and protect against biases and discrimination. Educational programs at various levels can build capacity by teaching GeoAI skills and emphasizing the importance of ethics and inclusivity in technology development.

28.3 Ethical Dilemmas: Balancing Innovation and Responsibility

The rapid evolution of GeoAI technologies has ushered in transformative opportunities within the field of human geography, advancing our analysis and understanding of complex spatiotemporal dynamics. The adoption and use of GeoAI in urban planning, environmental conservation, disaster response, and public health strategies have the potential to significantly improve societal well-being. However, this surge in innovation also presents profound ethical dilemmas, particularly concerning the balance between technological advancement and the responsibility to protect individual rights and social equity (Shneiderman et al., 2020).

While specific ethical challenges exist around the collection and use of geospatial data with GeoAI, perhaps a more fundamental challenge is a lack of clarity concerning the normative ethical frameworks that could, or should, be used when making applied ethical decisions about design and usage of GeoAI. Consequentialist, deontological, and virtue-based are all relevant theoretical frameworks when addressing concerns about the morality of GeoAI. However, ethical discussions of the subject rarely identify which framework, or frameworks, they are operating from or how the recommendation they make may generate conflicts across frameworks. For example, a consequentialist approach rooted in utilitarianism and applied to GeoAI might well advocate for the development of algorithms that produce broad benefits and rare negative outcomes for small number of individuals. This approach would, however, conflict with a deontological approach that emphasizes a duty to not treat individuals as a means to maximizing benefits for a larger group. These types of apparently paradoxical concerns are a hallmark of ethics education in other disciplines. Teaching these normative frameworks and such paradoxes in the context of GeoAI would likely improve ethical discussions of the issue and the development of governance systems. Beyond concerns about the normative ethical foundations of GeoAI, a wide variety of applied ethical questions remain open. One of the central applied ethical challenges of GeoAI development arises from the extensive use of sophisticated data collection and analysis methods. While the aggregation of vast geospatial datasets empowers researchers to derive valuable insights, it simultaneously raises concerns about privacy, consent, and surveillance. The ability of GeoAI systems to process detailed personal and location-based information can inadvertently infringe upon individual autonomy, especially when individuals are unaware of how their data is collected, analyzed, or shared (Hu et al., 2021; Huang et al., 2021). Advanced algorithms may reconstruct personal behaviors or predict sensitive attributes without explicit consent, undermining expectations of privacy and potentially leading to unintended harm (Rocher et al., 2019). Moreover, the integration of GeoAI with other emerging technologies—such as big data analytics—amplifies the potential for unintended consequences. For instance, combining geospatial data with social media analytics and natural language processing can result in the exposure of confidential information, social profiling, or even manipulation of public discourse (Koohang et al., 2021; Huang et al., 2022). These scenarios jeopardize individual privacy and pose risks to societal trust, democratic processes, and the equitable distribution of resources.

Addressing these ethical dilemmas requires a comprehensive and proactive approach that encompasses robust data governance, ethical design principles, and inclusive policy-making. A critical starting point is the simple recognition that a GeoAI product consists of both the algorithm and the data and procedures used to train that algorithm. Identifying the products of GeoAI research as more than the algorithms they create makes apparent the wider set of ethical commitments made by those developing and using GeoAI. For example, data used for GeoAI development is often sourced from datasets whose construction may not be permitted by a university institutional review board. The obligations using this data creates for the researcher building the GeoAI application remains unclear and unresolved in the literature.

To meet these commitments, researchers and practitioners must prioritize transparency in data collection methodologies, ensuring that individuals are informed about how their data will be used and have the agency to provide or withhold consent. Having and exercising agency requires that individuals are not only informed, but understand the consequences of their choice. Informing individuals about complicated technologies is always challenging and can require substantial education. Informing individuals about emerging technologies when the developers of those technologies themselves necessarily have a limited understanding of potential consequences is even more challenging. This situation not only foregrounds a need for caution in GeoAI development but also the need for a concerted community effort to develop standards and protocols for informed consent when uncertainty about benefits and harms is high.

In addition to transparency and consent, implementing privacy-preserving techniques—such as data anonymization, encryption, and secure storage—is essential to protect personal information from unauthorized access and misuse. Emerging legal frameworks also emphasize the need to allow individuals access to their information and the right to erasure (see GDRP Articles 15 and 17). Applied broadly, these rights raise difficult ethical questions for developers and users of GeoAI. For example, if certain people are more likely to exercise their right to erasure for a particular data source, a selection bias can be created in that dataset which may affect the prediction capacity of GeoAI applications trained on that data. Identifying and addressing such bias can be particularly challenging when data is accessed from commercial sources or scraped from open sources with no systematic accounting of how data is collected or processed.

In essence, the ethical dilemmas presented by the advancement of GeoAI in human geography reflect a broader tension between the pursuit of technological progress and the imperative to uphold ethical standards. By proactively engaging with these challenges, the GeoAI community can strive to harness the transformative potential of these technologies while ensuring that innovation does not come at the expense of individual privacy, social equity, and public trust. It is through this balanced approach that GeoAI can contribute positively to society, advancing human geography in a manner that is both responsible and inclusive.

28.4 Data Justice: The Core of Ethical GeoAI

As GeoAI continues to revolutionize human geography by providing unprecedented insights into spatiotemporal patterns and processes, data justice emerges as a fundamental pillar for ensuring ethical and equitable outcomes. Data justice takes the FAIR guiding principles a step further, and focuses on the fair and equitable treatment of all individuals and communities in the collection, analysis, and application of data, ensuring that no group is disadvantaged or marginalized. For GeoAI, this principle is critical because spotty and sometimes fake data can skew results, perpetuate inequalities, and cause other adverse outcomes if not properly addressed. Bias can also be introduced unintentionally, or even by introducing ethical protections during the data collection process (e.g., the right to erasure may induce selection bias that impacts algorithm equity).

GeoAI models are inherently data-driven, relying on vast and diverse datasets to train algorithms that interpret and predict geographical phenomena. These datasets drive machine learning models, enabling them to identify patterns and make predictions that inform policy and decision-making. However, these datasets often reflect systemic biases, underrepresenting marginalized communities and less-developed regions due to historical neglect or lack of data collection infrastructure (Huang et al., 2024). These inequalities lead to models that are less accurate—or even entirely ineffective—in contexts outside well-represented areas. For example, a GeoAI model trained predominantly on urban data may fail to accurately assess rural dynamics, leading to misinformed policies that could exacerbate existing inequalities. Such disparities underscore the necessity to ensure we have sufficient data to create models that are inclusive and effective across diverse contexts.

The digital divide exacerbates this issue by limiting access to digital technologies and the Internet for significant portions of the global population, particularly in rural areas and developing countries (Oluoch, 2024). Communities without robust digital infrastructures are not only excluded from data collection efforts but also from the benefits that GeoAI technologies can offer (Mzuza et al., 2020). This state of affairs raises ethical concerns about who is included in technological advancements and who is left behind, reinforcing existing social and economic disparities. Moreover, the lack of representation in data can lead to policies and services that do not address the needs of these communities, perpetuating a cycle of neglect and marginalization that undermines social cohesion and equitable development.

Addressing data equity is therefore paramount to achieving ethical GeoAI that serves all segments of the population fairly. Ensuring that data is representative involves actively seeking out and incorporating data from underrepresented groups and regions, thereby enriching datasets and enhancing model performance across diverse contexts. Engaging directly with communities and fostering collaboration with local stakeholders can help capture nuanced geographical information that might otherwise be overlooked. Additionally, transparency about data sources, methodologies, and potential biases is essential for fostering trust and accountability. Open communication regarding the limitations of datasets and models can assist users in interpreting results appropriately and encourage efforts to improve data quality and representation. Accomplishing these tasks will likely require the development of novel data collection strategies designed to engage individuals at their present level of understanding using means of communication they are familiar with. How to create effective and inclusive consent and communication protocols that facilitate transparency and equity during and after the life of GeoAI development is a question in need of further research.

Data justice also involves acknowledging and mitigating the unintended consequences of GeoAI applications, which can sometimes cause harm despite well-meaning intentions. This includes safeguarding privacy by ensuring that personal or sensitive information is protected against misuse or unauthorized access. Obtaining informed consent from individuals and communities when collecting data respects their autonomy and legal rights. Being sensitive to the cultural and social contexts of the data subjects is crucial to avoid misinterpretation or misuse of data that could lead to discrimination or stigmatization. Ethical guidelines and regulations should be established and rigorously enforced to govern data handling practices, emphasizing the rights and dignity of individuals and communities. These regulations should include the right to not only exit a project, but to access, receive, object, restrict, and erase their data from a project at any point.

Placing data justice at the core of GeoAI allows researchers and practitioners to work toward creating systems that advance technological capabilities and promote fairness, inclusivity, and respect for all communities. This approach ensures that the transformative power of GeoAI contributes positively to society, addressing global challenges while upholding ethical standards.

28.5 Fostering Collaboration: Toward Cross-Disciplinary GeoAI

As the complexities of spatiotemporal phenomena become increasingly apparent, the convergence of diverse fields such as computer science, geography, sociology, and statistics becomes essential. GeoAI, with its advanced analytical capabilities, offers human geographers unprecedented tools to tackle multifaceted issues that transcend various disciplinary perspectives.

One of the primary drivers for cross-disciplinary engagement is the multifaceted nature of the challenges that human geography seeks to address. Issues like climate change, urbanization, migration, and resource management are inherently complex, involving intertwined social, economic, and environmental factors. GeoAI provides the computational power and sophisticated algorithms needed to analyze large and heterogeneous datasets, but unlocking this potential requires insights from various disciplines. The adoption of GeoAI in human geography opens up opportunities for novel research methodologies and applications. By collaborating with geographic information scientists, data scientists, and AI specialists, human geographers can enhance their analytical frameworks, enabling more precise modeling of spatiotemporal dynamics. These interdisciplinary approaches allow for the development of predictive models that can inform policy-making, urban planning, and disaster management with greater accuracy and efficiency. Additionally, incorporating perspectives from fields like ethics and law ensures that the deployment of GeoAI technologies aligns with societal values and legal standards.

The shift toward cross-disciplinary GeoAI research significantly promotes innovation by bringing together diverse expertise and problem-solving approaches. By integrating knowledge from fields such as computer science, geographic information science, geography, sociology, and environmental science, collaborative projects can lead to the development of new theories, and methodologies that push the boundaries of spatiotemporal analysis. This convergence allows for a more comprehensive understanding and tackling of complex spatiotemporal phenomena that single-discipline approaches might overlook.

However, embracing cross-disciplinary collaboration is not without its challenges. Differences in terminology, research paradigms, and methodological preferences can pose significant barriers to effective communication and cooperation (Michael et al., 2013). Methodological discrepancies—such as the divergence between quantitative and qualitative approaches—can create friction among collaborators who prioritize different types of data or analytical techniques (Gelo et al., 2008). The implicit adoption of potentially conflicting normative ethical frameworks might similarly lead to conflicts between collaborators. While overcoming these obstacles is undeniably difficult, it is imperative that we make concerted efforts to bridge these divides.

28.6 GeoAI for Human Geographers: Unraveling the Complex

The integration of GeoAI into human geography offers immense opportunities to advance the discipline, but it also presents notable challenges that need to be addressed thoughtfully. One of the primary obstacles is that many human geographers are not traditionally trained in the spatial sciences and often have limited coding or computational backgrounds. This skills gap can render the advanced technical aspects of GeoAI inaccessible or daunting, potentially hindering their adoption within the field. Therefore, it is imperative to develop GeoAI tools and platforms that are user-friendly and tailored to the needs of human geographers, enabling them to leverage these technologies without requiring extensive technical expertise.

To tackle this challenge, software developers and GeoAI practitioners should focus on creating intuitive interfaces, comprehensive documentation, and educational resources that lower the barrier to entry (Wang et al., 2024). Incorporating visual programming environments, drag-and-drop functionalities, and pre-built analytical modules can make GeoAI tools more approachable for those without a technical background (Li et al., 2021). Additionally, integrating GeoAI capabilities into familiar GIS platforms can facilitate smoother transitions for human geographers accustomed to traditional methods (Wilson, 2024).

Another significant challenge lies in bridging the gap between traditional human geography and the rapidly evolving field of GeoAI. There is often a lack of understanding or trust regarding the development and applicability of GeoAI methods within the broader human geography community. This skepticism may stem from concerns about the "black-box" nature of certain AI algorithms, potential over-reliance on quantitative data, or fears that technological approaches might overshadow the qualitative and theoretical foundations of the discipline (Graham & Shleton, 2013). To overcome these concerns, establishing robust communication mechanisms is essential. Interdisciplinary workshops, collaborative research projects, and open forums can facilitate dialogue between GeoAI experts and human geographers. Such engagements can demystify GeoAI methodologies, address misconceptions, and highlight the complementary strengths of combining traditional geographic theories with advanced analytical tools.

Moreover, many concepts in human geography were established decades ago and form the foundational knowledge upon which the discipline is built. Understandably, practitioners might be deeply rooted in these traditional perspectives and the methodologies that lead to their discovery. However, GeoAI can serve as a valuable tool to validate, refine, and even rejuvenate these longstanding concepts and the accompanying interpretations of spatiotemporal dynamics (Huang, 2024). GeoAI offers the ability to analyze large and complex datasets, uncover hidden patterns, and generate new insights that were previously unattainable with conventional methods. By illustrating concrete examples where GeoAI has successfully contributed to theoretical advancements in human geography—such as revealing nuanced spatial relationships or validating hypotheses with empirical data—we can encourage a more receptive attitude toward its adoption.

Embracing GeoAI does not necessitate abandoning the rich heritage of human geography. Instead, it offers an opportunity to expand upon it, using advanced tools to explore questions in greater depth and with new dimensions. Indeed, the challenges raised throughout this chapter highlight the need for statistically, computationally, theoretically, and community-focused researchers to collaborate and build many of the foundational standards and protocols necessary for the ethical development of GeoAI. As the discipline continues to evolve, the collaboration between human geographers and GeoAI practitioners will be instrumental in addressing complex societal challenges and advancing our collective knowledge of space and place in everyday life (Cf. Wilson, 2024).

28.7 Conclusion

The integration of GeoAI into human geography stands at a pivotal crossroads, heralding a new era of possibilities while posing significant challenges that demand our collective attention. As we embrace these opportunities, it becomes clear that the true potential of GeoAI lies not only in its advanced analytical capabilities but also in its capacity to deepen our understanding of complex human-environment interactions. By prioritizing inclusivity, ethical responsibility, data justice, cross-disciplinary collaboration, and accessibility, we can ensure that GeoAI serves as a catalyst for positive transformation within the discipline.

This convergence calls for a paradigm shift—a reimagining of how we conduct research, teach, and apply human geography in an increasingly data-rich world. It challenges us to move beyond traditional boundaries, fostering a culture of openness and adaptability. By integrating GeoAI thoughtfully and ethically, we can uncover new insights, validate and refine existing theories, and address pressing global issues with greater precision and empathy. The journey ahead is not without obstacles, but the likely rewards are profound. By working collectively to simplify complex technologies and bridge gaps between diverse communities of scholars and practitioners, we can lay the groundwork for a more dynamic and responsive human geography. This collaborative spirit will enable us to harness the full potential of GeoAI, ensuring it becomes a tool for empowerment rather than division.

Here, we affirm our commitment to a future where GeoAI enhances human geography in ways that are ethically sound, socially equitable, and intellectually enriching. The path forward is one of innovation grounded in responsibility—a journey that holds the promise of not only advancing our discipline but also contributing meaningfully to the broader quest for understanding and improving societal needs.

References

Carter, L., Liu, D., & Cantrell, C. (2020). Exploring the intersection of the digital divide and artificial intelligence: A hermeneutic literature review. *AIS Transactions on Human-Computer Interaction, 12*(4), 253–275.

Castree, N., Kitchin, R., & Rogers, A. (2013). *A dictionary of human geography*. Oxford University Press.

Gelo, O., Braakmann, D., & Benetka, G. (2008). Quantitative and qualitative research: Beyond the debate. *Integrative Psychological and Behavioral Science, 42*, 266–290.

Graham, M., & Shelton, T. (2013). Geography and the future of big data, big data and the future of geography. *Dialogues in Human Geography, 3*(3), 255–261.

Hu, T., Wang, S., She, B., Zhang, M., Huang, X., Cui, Y., & Li, Z. (2021). Human mobility data in the COVID-19 pandemic: Characteristics, applications, and challenges. *International Journal of Digital Earth, 14*(9), 1126–1147.

Huang, X. (2024). Computational urban science needs to go beyond computational. *Computational Urban Science, 4*(1), 1–5.

Huang, X., Li, Z., Jiang, Y., Ye, X., Deng, C., Zhang, J., & Li, X. (2021). The characteristics of multi-source mobility datasets and how they reveal the luxury nature of social distancing in the US during the COVID-19 pandemic. *International Journal of Digital Earth, 14*(4), 424–442.

Huang, X., Wang, S., Zhang, M., Hu, T., Hohl, A., She, B., & Li, Z. (2022). Social media mining under the COVID-19 context: Progress, challenges, and opportunities. *International Journal of Applied Earth Observation and Geoinformation, 113*, 102967.

Janowicz, K., Sieber, R., & Crampton, J. (2022). GeoAI, counter-AI, and human geography: A conversation. *Dialogues in Human Geography, 12*(3), 446–458.

Kedron, P., Li, W., Fotheringham, A. S., & Goodchild, M. F. (2021). Reproducibility and replicability: Opportunities and challenges for geospatial research. *International Journal of Geographical Information Science, 35*(3), 427–445.

Koohang, A., Floyd, K., Yerby, J., & Paliszkiewicz, J. (2021). Social media privacy concerns, security concerns, trust, and awareness: Empirical validation of an instrument. *Issues in Information Systems, 22*(2), 133–145.

Lee, J. (2023). Beyond geospatial inquiry—How can we integrate the latest technological advances into geography education? *Education Sciences, 13*(11), 1128.

Li, Z., Huang, X., Hu, T., Ning, H., Ye, X., Huang, B., & Li, X. (2021). ODT FLOW: Extracting, analyzing, and sharing multi-source multi-scale human mobility. *PLoS One, 16*(8), e0255259.

Michael, O., Crowley, S., Eigenbrode, S. D., & Wulfhorst, J. D. (Eds.). (2013). *Enhancing communication & collaboration in interdisciplinary research*. SAGE.

Mzuza, M. K., & Van der Westhuizen, C. (2020). Inclusion of GIS in student teacher training and its significance in higher education in southern African countries. *International Research in Geographical and Environmental Education, 29*(4), 332–346.

Oluoch, I. (2024). Crossing boundaries: The ethics of AI and geographic information technologies. *ISPRS International Journal of Geo-Information, 13*(3), 87.

Rocher, L., Hendrickx, J. M., & De Montjoye, Y. A. (2019). Estimating the success of re-identifications in incomplete datasets using generative models. *Nature Communications, 10*(1), 1–9.

Shneiderman, B. (2020). Bridging the gap between ethics and practice: Guidelines for reliable, safe, and trustworthy human-centered AI systems. *ACM Transactions on Interactive Intelligent Systems, 10*(4), 1–31.

Wang, S., Huang, X., Liu, P., Zhang, M., Biljecki, F., Hu, T., & Bao, S. (2024). Mapping the landscape and roadmap of geospatial artificial intelligence (GeoAI) in quantitative human geography: An extensive systematic review. *International Journal of Applied Earth Observation and Geoinformation, 128*, 103734.

Wilkinson, M. D., Dumontier, M., Aalbersberg, I. J., Appleton, G., Axton, M., Baak, A., & Mons, B. (2016). The FAIR guiding principles for scientific data management and stewardship. *Scientific Data, 3*, 160018.

Wilson, J. P. (2024). *Spatial data science*. Esri Press.

Wilson, J. P., Butler, K., Gao, S., Hu, Y., Li, W., & Wright, D. J. (2021). A five-star guide for achieving replicability and reproducibility when working with GIS software and algorithms. *Annals of the American Association of Geographers, 111*(5), 1311–1317.

Index

© The Editor(s) (if applicable) and The Author(s), under exclusive license to Springer Nature Switzerland AG 2025
X. Huang et al. (eds.), *GeoAI and Human Geography*, Springer Geography,
https://doi.org/10.1007/978-3-031-87421-5

The manufacturer's authorised representative in the EU is Springer Nature Customer Service Centre GmbH, Europaplatz 3, 69115 Heidelberg, Germany. If you have any concerns regarding our products, please contact ProductSafety@springernature.com

Printed and bound by CPI Group (UK) Ltd, Croydon, CR0 4YY
30/09/2026
02225571-0015